Outstanding Contributions to Logic

Volume 10

More information about this series at http://www.springer.com/series/10033

Eugenio G. Omodeo · Alberto Policriti
Editors

Martin Davis on Computability, Computational Logic, and Mathematical Foundations

Springer

Editors
Eugenio G. Omodeo
University of Trieste
Trieste
Italy

Alberto Policriti
University of Udine
Udine
Italy

ISSN 2211-2758 ISSN 2211-2766 (electronic)
Outstanding Contributions to Logic
ISBN 978-3-319-41841-4 ISBN 978-3-319-41842-1 (eBook)
DOI 10.1007/978-3-319-41842-1

Library of Congress Control Number: 2016954526

Printed on acid-free paper

This Springer imprint is published by Springer Nature
The registered company is Springer International Publishing AG
The registered company address is: Gewerbestrasse 11, 6330 Cham, Switzerland

This book is dedicated
to the memory of
Hilary Putnam.

Foreword

I am honored to write a brief foreword to this volume dedicated to the many logical issues on which Martin has shed light during his illustrious career. I first met Martin at the Cornell 1957 Summer Institute in Symbolic Logic. This meeting proved very significant to the mathematical lives of many people who met there, including both of us: for him, Hilary Putnam and for me, J. Barkley Rosser. This meeting precipitated the formation of an international research community in mathematical logic whose influence on logic and computer science is strong even after sixty years. What I learned then about Martin was the universality of his interests, his utter concentration on fundamental problems, and his insatiable urge to learn new things. These are the signal marks of his long career. Some years ago, a university in a Western state with no history in any area he represents and, without warning, offered him a prestigious post. He called me and asked why. I told him they wanted an icon. He said he did not want to be an icon and promptly turned it down. But he *is* an icon, whether he likes it or not!

Ithaca, New York
April 2016

Anil Nerode

Preface

> It is reasonable to hope that the relationship between computation and mathematical logic will be as fruitful in the next century as that between analysis and physics in the last. The development of this concern demands a concern for both applications and mathematical elegance.
>
> (John McCarthy 1963)

The dozen students or so gathered in the lovely Italian town of Perugia were amazed when Martin David Davis first showed up, wearing heavy shoes and short pants seemingly more apt for trekking than for teaching a graduate level course. His accent from the Bronx, together with a little hole, may be produced by the ember of a cigarette, over one shoulder of his red T-shirt, seized the attention of the class. His hair *à la* Queen of Sheba further increased the mismatch between Martin as a person and the stereotype unavoidably associated with his reputation as a distinguished scholar.

Admiration quickly prevailed over astonishment when Martin began his exposition of computability. For the entire one-month duration of his course, concepts remained clear and accessible. At times, when confronted with some odd question coming from his audience, Martin turned his hands upward and disarmingly said "I cannot understand"; far more often, he answered with extreme precision. He indulged in vivid images, such as "brand-new variable" or "crystal-clear proof," but his repertoire of idiomatic expressions also included "gory detail," when technicalities were inescapable.

Through mathematics, the class felt, Martin was also addressing issues of philosophical relevance and depth. On one memorable occasion, the philosophical side of his scientific inquiry showed through a lesson boldly offered in Italian. Forty years have elapsed since, and alas, the tape recording of that lesson, dealing with Turing machines and universality, has by now faded away.

Once Martin was invited to an assembly of all students participating in the Perugia summer school. Unhesitatingly, he joined the crowd, coming hand in hand with his wife Virginia; he even took the floor at some point, rational and quiet, not in the least dismayed by the excited, somehow "revolutionary," atmosphere of the event.

One of the editors of this volume dedicated to Martin was a student in his computability course in Perugia, and this explains why such an anecdotal episode has been recounted here. Many similar fascinating stories could certainly be reported by many: Over the years, Martin lectured in several countries (to cite a few: Japan, India, England, Russia, and Mexico—see Fig. 1), and they have—along

Fig. 1 Poster announcing Martin's first lecture in Mexico

with his publications—exerted a wide influence. This book will testify to this influence by focusing on scientific achievements in which Martin was involved in the first person and on further achievements, studies, and reflections in which work and vision consonant with his have played a role. Our task has been to collect testimonies of Martin's contributions to computability, computational logic, and mathematical foundations.

Three chapters are devoted to a problem that Martin said he found "irresistibly seductive" when still an undergraduate (Fig. 2) and which progressively became his "lifelong obsession": Hilbert's tenth problem—H10 for short. One of the three contains a narrative essay by the Mexican mathematician Dr. Laura Elena Morales Guerrero, telling us how a negative solution to H10 came to light through the joint effort of four protagonists (one being Martin, of course). There are two epic events in that story: One is when Julia Bowman Robinson eliminates "a serious blemish" from a proof by Martin and Hilary Putnam by showing how to avoid a hypothesis that was unproved at the time; the other is when, in 1970, notes of a talk given in Novosibirsk reach Martin in New York. Laura Elena reports to the reader the key equations in those notes based on Yuri Matiysevich's decisive work, which Martin echoed a few months later by his own use of these newly developed methods to obtain an alternative system of equations leading to the same result. See Fig. 3.

If we could also prove that the class of diophantine predicates is closed under bounded universal quantification, then it would follow from 2.7 that every recursively enumerable predicate is diophantine.

But this clearly would lead to the unsolvability of the general decision problem for diophantine equations in non-negative integers, and hence--by the remarks of §1--of Hilbert's tenth problem.

Now, evidently every diophantine predicate is recursively enumerable. If we also knew the converse of this result, then we should of course be able to answer the above question in the affirmative, with all the attendant consequences. Hence our investigation of the relation between diophantine predicates and recursively enumerable predicates.

Fig. 2 Are all recursively enumerable sets Diophantine? (From Martin's Ph.D. thesis)

Fig. 3 In all of these equations, variables range over $\mathbb{N}$. Equations (I)–(X) with parameters u, v, a have a solution for $a > 1$ if and only if $v = y_u(a)$, where $X = x_u(a)\,, Y = y_u(a)$ is the $u+1$st solution, over $\mathbb{N}$, of the Pell equation $X^2 - (a^2 - 1)Y = 1$. Equations (I)–(XV) with parameters α, β, u have a solution for $\beta \geq 1$ if and only if $\alpha = \beta^u$

(I)	$u+j = v$
($\mathrm{II_a}$)	$p+(a-1)\,q = v+r+1$
($\mathrm{II_b}$)	$g = v+t+1$
(III)	$p^2-(a^2-1)\,q^2 = 1$
($\mathrm{IV_a}$)	$h+(a+1)\,g = b\,(p+(a+1)\,q)^2$
($\mathrm{IV_b}$)	$h+(a-1)\,g = c\,(p+(a-1)\,q)^2$
(V)	$h^2-(a^2-1)\,g^2 = 1$
(VI)	$m = (h+(a+1)\,g)\,z+a$
(VII)	$m = (p+(a-1)\,q)\,f+1$
(VIII)	$x^2-(m^2-1)\,y^2 = 1$
(IX)	$y = d\,(p+(a-1)\,q)+u$
(X)	$y = e\,(h+(a+1)\,g)+v$
(XI)	$w^2-(a^2-1)\,v^2 = 1$
(XII)	$(w-(a-\beta)\,v-\alpha)^2 = \gamma^2(2a\beta-\beta^2-1)^2$
(XIII)	$\alpha+\tau+1 = 2a\beta-\beta^2-1$
(XIV)	$\eta = \beta+\zeta+1 = u+\xi+1$
(XV)	$a^2-(\eta^2-1)\,(\eta-1)^2\,(\delta+1)^2 = 1$

Another chapter on H10 is by Yuri V. Matiyasevich himself, the "clever young Russian" whose appearance Martin had predicted and who, by producing a Diophantine predicate of exponential growth, first obtained that negative solution. Yuri explains how Julia Robinson, Martin Davis, and Hilary Putnam had—through extraordinary insights—paved the way for his decisive mathematical contribution. Far from being exhausted, the field of research triggered by H10 abounds today with unanswered questions, some fairly old (e.g., does the equation of the report shown in Fig. 4 admit finitely many solutions?), other quite contemporary, and in fact, Yuri's article, after disclosing a formidable landscape of open issues in front of us, terminates with a conjecture raised by Martin in 2010.

A third related contribution is by Prof. Alexandra Shlapentokh, who enriches the landscape with extensions of H10 to recursive rings. When referring his Tenth Problem to integers, in 1900, Hilbert may have thought that he was posing the most difficult among variants of the same problem with respect to other rings. Nowadays, we know that H10 as originally posed is unsolvable, but we are in the difficult position of not being able to draw any conclusion about the analog of this problem for, say, the ring $\mathbb{Q}$ of rational numbers.

Martin has been a trailblazer of the field today known as "automated reasoning." The summer of 1954 sees him at work on a JOHNNIAC machine, implementing a logical decision procedure for integer arithmetic. In the late 1950s, a seminal report on computational methods in the propositional calculus arises from his collaboration with Hilary Putnam, the brilliant philosopher with whom Martin enjoyed

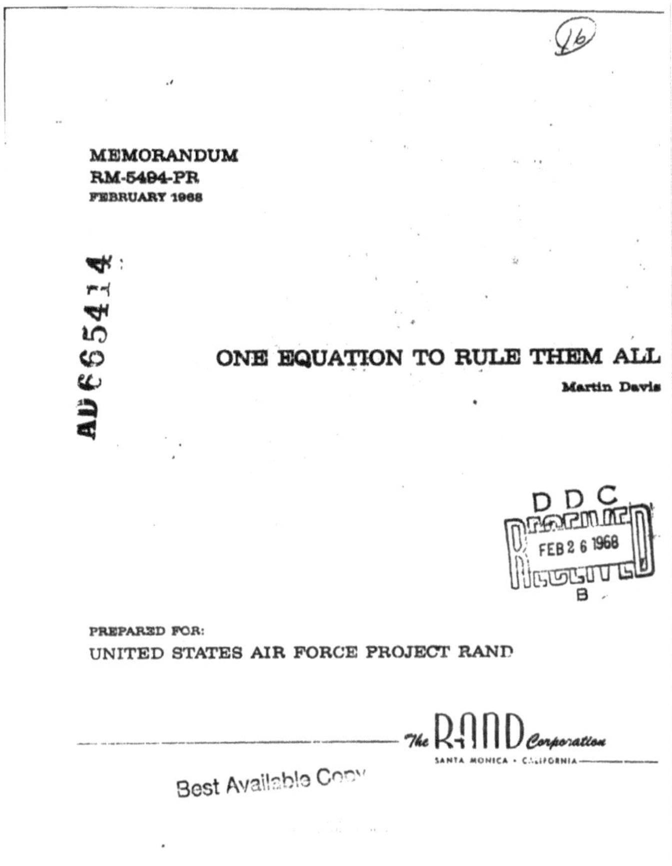

16

MEMORANDUM
RM-5494-PR
FEBRUARY 1968

AD 665414

ONE EQUATION TO RULE THEM ALL

Martin Davis

DDC
FEB 2 6 1968
B

PREPARED FOR:
UNITED STATES AIR FORCE PROJECT RAND

The RAND Corporation
SANTA MONICA • CALIFORNIA

Best Available Copy

Fig. 4 A report on an intriguing equation

discussing "all day long about everything under the sun, including Hilbert's tenth problem." The Davis–Putnam–Logemann–Loveland procedure, to date so basic in the architecture of fast Boolean satisfiability solvers, was rooted in that study, and Donald Loveland, who contributed to its pioneering implementation in the early 1960s, coauthors in this book, with Ashish Sabharwal and with Professor Bart Selman, a paper reviewing historical developments and the state of the art of propositional theorem provers.

It is slightly less known that in the early 1960s, the *most-general unification* mechanism for first-order logic was available in the working implementation of Martin's *linked conjunct* proof procedure, a forerunner of decadelong efforts to automatize reasoning in quantification theory. Unification has evolved, subsequently, into a well-established theory that proceeds hand in hand with the topic of rewriting systems. This is why dedicating a paper on a new trend in this field to

Martin seemed appropriate: Jörg Siekmann coauthors in this book, with Peter Szabó and Michael Hoche, a survey on "essential unification."

Udi Boker and Nachum Dershowitz, who dedicate an essay on "*honest computability*" to Martin, contend that "a nefarious representation can turn ... the intractable into trivial," whereas "demanding of an implementation that it also generates its internal representations of the input from an abstract term description of that input ... obviates cheating on complexity problems by giving away the answer in the representation." This attitude is akin to considerations made in the above-cited report by Davis and Putnam (1958) concerning propositional satisfiability, e.g.: "Even if the system has hundreds or thousands of formulas, it can be put into conjunctive normal form 'piece by piece', without any 'multiplying out.' This is a feasible (if laborious) task even for *hand* computation ⋯"

The centennial of Frege's *Begriffsschrift*, Martin reports, "fundamentally changed the direction of my work": Being invited to place some contemporary trends in a proper historical context, he finds "trying to trace the path from ideas and concepts developed by logicians ⋯ to their embodiment in software and hardware ⋯ endlessly fascinating."

Martin actually cultivated, since long, a keen interest in the history and philosophy of computing: The first edition of *The Undecidable*, his anthology of basic papers on unsolvable problems and computable functions, is dated 1965. One of Martin's heroes is Alan Mathison Turing; he also devoutly edited the collected works of Emil Leon Post, who had supervised his beginnings in logic at City College. In a recent paper, Martin and Wilfried Sieg have discussed a *conceptual confluence* between Post and Turing in 1936; in this book, Sieg coauthors with Máté Szabó and Dawn McLaughin, a paper addressing the question: *Did Post have Turing's Thesis?*

Yiannis N. Moschovakis unravels the history of another crucial confluence of ideas. Stephen C. Kleene, Emil Leon Post, and Andrzej Mostowski had raised questions which would influence profoundly the development of the theory of unsolvability when Martin, in the central part of his Ph.D. thesis, moved "*on into the transfinite!*", thus playing a very important role in defining natural extensions of the arithmetical hierarchy. The author skillfully alternates notes about the historical development of the subject with some carefully chosen technical details. This makes for a paper which is really a pleasure to read.

Martin once called "attention to the relevance for the foundational problems in quantum theory of some recent mathematical discoveries" arisen from logic. One of the diverse contributions dedicated to Martin, by Andreas Blass and Yuri Gurevich, aims at explaining certain sorts of *anyons*, "rather mysterious physical phenomena" which may provide a basis for quantum computing, by means of category theory.

Don Perlis's contribution speculates on the concept of infinity and distinguishes several modes of use of infinities in physics. In particular, quantum mechanics, he observes, provides intriguing examples on the subject. Nonstandard analysis—on which Martin wrote a classic—appears to shed light on some such phenomena.

"*Banishing ultrafilters from our consciousness*," the title of the paper contributed by Domenico Cantone with the editors of this book, echoes a comment by Martin in his *Applied Nonstandard Analysis* (1977). Martin then pointed out that the intricacies of the ultrapower construction of a nonstandard universe can be completely forgotten in favor of a few principles relating standard/nonstandard, internal/external objects. Bearing Martin's motto in mind, this paper recounts, and aided by a proof-checker embodying constructs for proof engineering, the authors have undertaken a verification of key results of the nonstandard approach to analysis.

The reader will also attend, inside this volume, Martin Davis and Hilary Putnam resuming some threads of their juvenile philosophical discussion. Martin has recently written about realism in mathematics (partly because Harvey Friedman had judged him "an extreme Platonist"), and Hilary cannot resist to amicably respond to his fascinating essay *Pragmatic Platonism* (also included in this book) and to discuss the relation between Martin's view and the views Hilary defends. In his turn, Martin comments on Hilary's remarks on his essay and takes the opportunity to say a little more about his view about certain topics such as mathematics and natural science, and new axioms for set theory.

The first chapter is an autobiographic essay by the eminent logician to whom the entire book is devoted. An earlier version of this essay, published in 1999, was titled "From Logic to Computer Science and Back." Martin reports that his debut as a computer programmer takes place in 1951, "without [him] realizing it," while he —a recent Ph.D. from Princeton, teaching recursive function theory at Champaign–Urbana—is designing Turing machines; he then gets recruited for a project on an automated system for navigating airplanes, with the task of writing code for an ORDVAC machine. Short afterward, Martin conceives the idea of writing his first book on computability; then, planning an extended visit to the Institute for Advanced Study in Princeton, he proposes to work on connections between logic and information theory. In the following decades, he frequently moves across the USA, teaching in various academic institutions and working on computability, on Hilbert's tenth problem, on computational logic, etc.

Multifaceted life and publications, but a substantial unity: in the new title chosen for his enriched autobiography, Martin regards himself simply as a logician.

Trieste and Udine
February 2016

Eugenio G. Omodeo
Alberto Policriti

Acknowledgements

The editors gratefully acknowledge invaluable help from Martin Davis and Yuri Matiyasevich for many aspects of the preparation of this book. Martin, in particular, supplied old manuscripts (his Ph.D. thesis, the research reports jointly written with Hilary Putnam which are included as appendices) and re-read certain parts of the draft of this book. Yuri offered great support for the preparation of Martin Davis's bibliography (Chap. 16).

Dominique Pastre (emeritus professor of computer science at UFR de mathématiques et informatique, Université Paris Descartes) kindly accepted to be an "anonymity server" to referees of her choice.

The anonymous second readers for this volume include the following:

Samson Abramsky, Department of Computer Science University of Oxford;
Johan van Benthem, University of Amsterdam and Stanford University;
Alessandro Berarducci, Dipartimento di Matematica, Università di Pisa;
Maria Paola Bonacina, Dip. di Informatica, Università degli Studi di Verona;
Patrick Cégielski, Université Paris Est Créteil;
Pietro Corvaja, Dip. di Matematica e Informatica, Università di Udine;
Liesbeth De Mol, Université de Lille 3, Villeneuve d'Ascq Cedex;
Francesco M. Donini, Università della Tuscia, Viterbo;
David Finkelstein, School of Physics, Georgia Institute of Technology, Atlanta;
Andrea Formisano, Università di Perugia;
Miriam Franchella, Dipartimento di Filosofia, Università degli Studi di Milano;
Jean-Pierre Keller, Sarl Kepler, Paris;
Gabriele Lolli, Scuola Normale Superiore di Pisa;
Stefano Mancini, School of Science and Technology, University of Camerino;
Alberto Marcone, Dip. di Matematica e Informatica, Università di Udine;
Alberto Martelli, Dipartimento di Informatica, Università degli Studi di Torino;
Daniele Mundici, Dept. of Mathematics and Computer Science "Ulisse Dini," University of Florence;

Andrea Sorbi, Dip. di Ingegneria dell'Informazione e Scienze Matematiche, Università degli Studi di Siena; and
Carlo Toffalori, Dip. Matematica e Fisica, Univ. of Camerino.

Contents

Contributors and Editors

Contributors

Andreas Blass is a professor of mathematics at the University of Michigan. He received his Ph.D. degree from Harvard University in 1970 and has been at the University of Michigan ever since except for brief leaves. He has also been a visiting researcher at Microsoft Research during one of those leaves and during each of the last 13 summers. His primary research area is set theory, but he has also worked in other branches of logic, in theoretical computer science, and in finite combinatorics. He is a fellow of the American Mathematical Society.

Udi Boker is a senior researcher in the Interdisciplinary Center (IDC), Herzliya, working in the field of logic and verification. His main research areas concern the foundations of computation, temporal logic, and automata.

Domenico Cantone is a professor of computer science since 1990. He is currently at the University of Catania, Italy, where he moved from the University of L'Aquila, Italy, in 1991. He received his Ph.D. degree from New York University in 1987, under the supervision of Prof. Jacob T. Schwartz. Since 1995, he has been a member of the Board of Directors of the journal "Le Matematiche." His main scientific interests include the following: computable set theory, automated deduction in various mathematical theories, description logic, string matching and algorithmic engineering, and, more recently, rational choice theory from a logical point of view. In the field of computable set theory, he has coauthored three monographs: *Computable Set Theory* (Clarendon Press, 1989), *Set Theory for Computing—From Decision Procedures to Declarative Programming with Sets* (Springer, 2001), and *Computational Logic and Set Theory: Applying Formalized Logic to Analysis*, (Springer, 2011).

Nachum Dershowitz is professor of computational logic at Tel Aviv University, where he has been since 1998. Prior to that, he was on the faculty of the University of Illinois at Urbana–Champaign. He coauthored the book, Calendrical Calculations (Cambridge University Press, 1997), with Edward Reingold, which won Choice's

Outstanding Academic Title Award (2002) and is going into its fourth edition. He is also the author of The Evolution of Programs (Birkhäuser, 1983), coauthor of Calendrical Tabulations (Cambridge University Press, 2002), and editor of a dozen other volumes. His research interests include foundations of computing, computational logic, computational humanities, and combinatorial enumeration. He has received the Herbrand Award for Distinguished Contributions to Automated Reasoning (2011), the Logic in Computer Science (LICS) Test-of-Time Award (2006), the Rewriting Techniques and Applications (RTA) Test-of-Time Award (2014), and the Conference on Automated Deduction (CADE) Skolem Award (2015) and was elected to Academia Europaea in 2013.

Yuri Gurevich is a principal researcher at Microsoft Research in Redmond, Washington, USA. He is also Professor Emeritus at the University of Michigan, ACM Fellow, Guggenheim Fellow, EATCS Fellow, a foreign member of Academia Europaea, and Dr. Honoris Causa of a Belgian and Russian universities.

Michael Hoche studied computer science at University of Stuttgart with visits at József Attila University, Szeged, and University Pierre and Marie Curie, Paris. He is currently working at Airbus Defense and Space, where he serves as analyst and technologist in advanced computer science and its applications. He holds a Ph.D. in computer science, which he obtained from University of Stuttgart. His current interest comprises intelligent systems applications with a focus on data integration, machine learning, and networks. Most recently, he contributes to the creation of the research agenda of Airbus Defense and Space.

Donald Loveland attended Oberlin College, MIT, and New York University (NYU) where he wrote his Ph.D. thesis under Martin Davis. He was on the faculties of NYU (Bronx Campus) where Martin was briefly a colleague, CMU, and Duke University where he was the first chairman of the Computer Science Department. The author of Automated Theorem Proving: a Logical Basis (1978) is also one of three authors of the undergraduate textbook Three Views of Logic (2014). He is author or coauthor of papers in areas that include automated deduction (primary area), complexity theory, testing procedures, logic programming, and expert systems; an ACM Fellow; and an AAAI Fellow, and he also received the Herbrand Award (2001). He is now retired.

Yuri Matiyasevich has got (Russian analog of) Ph.D. from Leningrad (now St. Petersburg) Division of Steklov Mathematical Institute in 1970 and ever since works there, today as the Head of Laboratory of mathematical logic. He is mostly known for his contribution to the (negative) solution of Hilbert's tenth problem, about which he wrote a book translated into English and French. He worked also in theoretical computer science, graph theory, and number theory. He is full member of Russian Academy of Sciences, corresponding member of Bavarian Academy of Sciences, and member of Academia Europaea and Docteur Honoris Causa de l'Université d'Auvergne et de Université Pierre et Marie Curie (Paris-6), France.

Dawn McLaughlin is a Ph.D. candidate in the philosophy department at Carnegie Mellon University and did early work on Emil Post; however, her academic focus is now on logic education. She is a collaborator on Wilfried Sieg's AProS project, in particular regarding the online logic course Logic and Proofs; she is the lead developer of the LogicLab. Her dissertation is exploring the history and practice of technology supported learning of logic.

Laura Elena Morales Guerrero was the second in a Mexican family of seven children living in Tampico, Mexico. After receiving her undergraduate degree at the University of Wisconsin, she began graduate studies in physics and mathematics at the Centro de Investigación y Estudios Avanzados (Cinvestav, IPN) in México City. Her doctorate under Jerzy Plebański involved finding exact solutions of the field equations of Einstein's general relativity. She worked in the nuclear engineering industry in Mexico and The Netherlands. While rearing her twin daughters, she started a successful retail business in Mexico City. She remains an independent researcher, and her most recent interests are in the history of mathematics, especially logic and number theory. She connected with Martin Davis through the email list HM (History of Mathematics) and arranged for him to lecture at the Universidad Nacional Autónoma de México (UNAM) where she was a researcher at the time. His talk stimulated her interest in Hilbert's tenth problem and led her to write the essay on the subject in this book. In his preface to her still unpublished book on the ancient problem of squaring the circle, he wrote, "In this charming book, Dra. Morales has traced the history of this problem, pausing along the way to explore many fascinating sidelights."

Yiannis N. Moschovakis emigrated to the USA from Greece in 1956. After four years at MIT, three years at the University of Wisconsin and one year at Harvard, he moved to UCLA in 1964 and has been there ever since, with long, annual sojourns to Greece including a halftime position at the University of Athens (UOA) from 1996 until 2005. He has written two monographs and a textbook and has supervised or cosupervised the doctoral dissertations of 22 students from the mathematics and computer science Departments of UCLA, UOA, and the Graduate Program for logic, algorithms, and computation in Athens. He retired in 2010 but has continued teaching part-time as a Distinguished Research Professor.

Moschovakis has worked in abstract and higher-type recursion; in classical and effective descriptive set theory, including the consequences of determinacy hypotheses; and in philosophical logic, especially the logic of meaning and the foundations of the theory of algorithms.

Don Perlis is a professor of computer science at the University of Maryland, College Park. He has Ph.D.s in mathematics (NYU, 1972) and computer science (Rochester, 1981). While his interests span a wide variety of areas, he recently has come to realize that most of his work relates to self-reference in one form or another, whether in logic, artificial intelligence, cognitive science, or philosophy.

Hilary Putnam is Cogan University Professor Emeritus at Harvard University. His most recent publications include philosophy in an age of science (ed. by M. De Caro and D. Macarthur, Harvard UP 2012) and naturalism, realism, and normativity (ed. by M. De Caro, Harvard UP 2016). He is the subject of The Philosophy of Hilary Putnam (ed. by R.E. Auxier, D.R. Anderson, and L.E. Hahn, Library of Living Philosophers, Open Court 2015). He holds 12 honorary degrees and is a past president of the American Philosophical Association, a fellow of the American Academy of Arts and Sciences, and a corresponding fellow of the British Academy and of the French Académie des Sciences Politiques et Morales. Recently, he has been awarded the Prometheus Prize, the Rolf Schock Prize in logic and philosophy, the Lauener Prize for analytical philosophy, and the Nicholas Rescher Prize for systematic philosophy. His interests cover most philosophical areas.

Ashish Sabharwal investigates scalable and robust methods for probabilistic and combinatorial inference, graphical models, and discrete optimization, as a research scientist at the Allen Institute for AI (AI2), especially as they apply to assessing machine intelligence through standardized examinations in science and math. Prior to joining AI2, Ashish spent over three years at IBM Watson and five years at Cornell University, after obtaining his Ph.D. from the University of Washington in 2005. Ashish has coauthored over 70 publications, been part of winning teams in international SAT competitions, and received five best paper awards and runner-up prizes at venues such as AAAI, IJCAI, and UAI.

Bart Selman is a professor of computer science at Cornell University. He was previously at AT&T Bell Laboratories. He specializes in artificial intelligence, with an emphasis on efficient reasoning procedures, planning, knowledge representation, and connections between computer science and statistical physics. He has (co) authored over 150 publications, including six best paper awards. His papers have appeared in venues spanning Nature, Science, Proc. Natl. Acad. of Sci., and a variety of conferences and journals in AI and computer science. He has received the Cornell Stephen Miles Excellence in Teaching Award, the Cornell Outstanding Educator Award, an NSF Career Award, and an Alfred P. Sloan Research Fellowship. He is a fellow of the American Association for Artificial Intelligence (AAAI) and a fellow of the American Association for the Advancement of Science (AAAS). He received the inaugural IJCAI John McCarthy award in 2015.

Alexandra Shlapentokh is a professor of mathematics in East Carolina University in Greenville, NC. She got her Ph.D. in mathematics in NYU in 1988 under Professor Harold N. Shapiro. In graduate school, Alexandra Shlapentokh also had the privilege of having Martin Davis as one of her professors. It is in his class that she was introduced for the first time to Hilbert's tenth problem which became one of her lifelong interests, eventually encompassing many questions of definability and computability in number theory. She described some of the developments in this thriving field in her book: "Hilbert's Tenth Problem: Diophantine Classes and Other Extensions to Global Fields" (Cambridge University Press).

Wilfried Sieg is a patrick suppes professor of Philosophy at Carnegie Mellon University and a fellow of the American Academy of Arts and Sciences. He joined Carnegie Mellon's faculty in 1985 as a founding member of the University's Philosophy Department and served as its head from 1994 to 2005. He is internationally known for mathematical work in proof and computation theory, historical work on modern logic and mathematics, and philosophical essays on the nature of mathematics. A collection of essays joining the three aspects of his research was published as *Hilbert's Programs and Beyond* (Oxford University Press, 2013).

Jörg Siekmann studied mathematics at Göttingen University and computer science at Essex University, England, where he received his Ph.D. in *unification theory* in 1976. He was a scientific assistant at Karlsruhe University and became professor for Artificial Intelligence at Kaiserslautern in 1983. In 1991, he was appointed as a full professor for computer science and AI at Saarbrücken University. He founded with friends the German Research Centre for Artificial Intelligence (DFKI) and became one of its directors. He is now a senior professor at the computer science department and DFKI in Saarbrücken.

Máté Szabó is a Ph.D. student in the philosophy department at Carnegie Mellon University. His main interests concern the history and philosophy of computing, in particular the interpretation of Gödel's Incompleteness Theorems, Church's and Turing's Undecidability Theorems, and of the Church-Turing Thesis. Related to these issues, he has been exploring the life and work of Emil Post and László Kalmár. He is writing his dissertation on human and machine computation with Wilfried Sieg as his advisor.

Peter Szabó studied computer science at the University of Karlsruhe where he received his diploma in 1975 and worked there afterward as a scientific assistant. He received his Ph.D. in *unification theory* from Karlsruhe University in 1982. From 1983 to 2011, he was a research engineer for software in the R&D department of the German Telecommunication company SEL, which later became part of Alcatel-Lucent. Since 2003, he is a voluntary member of our research team dealing with unification theory.

Andreas Blass Mathematics Department, University of Michigan, Ann Arbor, MI, USA

Udi Boker School of Computer Science, Interdisciplinary Center, Herzliya, Israel

Domenico Cantone DMI, Università di Catania, Catania, Italy

Martin Davis Department of Mathematics, University of California, Berkeley, CA, USA; Courant Institute of Mathematical Sciences, New York University, New York, NY, USA

Nachum Dershowitz School of Computer Science, Tel Aviv University, Ramat Aviv, Israel

Yuri Gurevich Microsoft Research, Redmond, WA, USA

Michael Hoche Airbus Defense and Space, Immenstaad, Germany

Donald Loveland Duke University, Durham, USA

Yuri Matiyasevich Laboratory of Mathematical Logic, St. Petersburg Department of V.A. Steklov Institute of Mathematics (POMI), Russian Academy of Sciences, St. petersburg, Russia

Dawn McLaughlin Carnegie Mellon University, Pittsburgh, USA

Laura Elena Morales Guerrero Centro de Investigación y de Estudios Avanzados del Instituto Politécnico Nacional in Zacatenco, Ciudad de México, Mexico

Yiannis N. Moschovakis Department of Mathematics, University of California, Los Angeles, USA

Eugenio G. Omodeo DMG/DMI, Università di Trieste, Trieste, Italy

Don Perlis University of Maryland, College Park, USA

Alberto Policriti DMIF, Università di Udine, Udine, Italy

Hilary Putnam University of Harvard, Cambridge, USA

Ashish Sabharwal Allen Institute for AI, Seattle, USA

Bart Selman Cornell University, Ithaca, USA

Alexandra Shlapentokh East Carolina University, Greenville, USA

Wilfried Sieg Carnegie Mellon University, Pittsburgh, USA

Jörg Siekmann Saarland University/DFKI, Saarbrücken, Germany

Peter Szabo Pforzheim, Germany

Máté Szabó Carnegie Mellon University, Pittsburgh, USA

About the Editors

Eugenio G. Omodeo studied mathematics at the University of Padua and then computer science at the New York University, GSAS, where he earned Ph.D. (1984) under the supervision of Martin Davis. From 1981 to 1989, he was employed by companies belonging to ENI, the National Hydrocarbon Group of Italy: There, after 1984, he coordinated R&D activities of Enidata in various projects funded by the European Commission (CEC), mainly focused on declarative programming and on quick prototyping. From 1989 to present, he has been a professor in various Italian universities (Udine, "La Sapienza" of Rome, Salerno, L'Aquila, Trieste). He has contributed to computational logic with the discovery of inference methods based on set theory, some of which have been implemented in a large-scale proof verifier developed with Jacob T. Schwartz (NYU). He coauthored three scientific monographs on computable set theory.

Alberto Policriti received his degree in mathematics from the Universtiy of Turin in 1984 and his Ph.D. in computer science under the supervision of M. Davis in 1990. From 1989, he is at the University of Udine, where he is currently professor of computer science at the Department of Mathematics, Computer Science, and Physics. His main research interests are related to computational logic and algorithms: set-theoretic and combinatorial algorithms and problems, modal and temporal logics, and algorithms and models for bioinformatics. He has coauthored two monographs and has supervised or co-supervised 15 doctoral dissertations in Logic, Algorithms, and Bioinformatics. He is one of the four founders of the "Istituto di Genomica Applicata," has been member of the scientific committee of GNCS—"Istituto di Alta Matematica," and he is currently member of the scientific committee of the EATCS.

Chapter 1
My Life as a Logician

Martin Davis

"My father and mother were honest, though poor –"
"Skip all that!" cried the Bellman in haste.
"If it once becomes dark, there's no chance of a snark–
We have hardly a minute to waste!"

"I skip forty years," said the Baker, in tears,
"And proceed without further remark ..."
–Lewis Carroll's "The Hunting of the Snark"

Abstract This brief autobiography highlights events that have had a significant effect on my professional development.

I was just over a year old when the great stock market crash occurred. My parents, Polish Jews, had immigrated to the United States after the First World War. My father's trade was machine embroidery of women's apparel and bedspreads. During the depression, embroidery was hardly in great demand, so we were dependent on home relief—what today would be called "welfare". Only with the upturn of the economy coming with the outbreak of war in 1939, was my father able to find steady work. In his spare time, he was a wonderful self-taught painter. (One of his paintings is in the collection of the Jewish Museum in New York and two others are at the Judah Magnus Museum in Berkeley.) My mother, eager to contribute to the family income, taught herself the corsetiere's craft. Until I left New York for graduate school, the room where she conducted her business by day was my bedroom at night.

This is a revision and expansion to the present of an earlier autobiographical essay [22], much of which is included verbatim. I am grateful to Springer Verlag for their permission.

M. Davis (✉)
Department of Mathematics, University of California, Berkeley, CA, USA
e-mail: martin@eipye.com

M. Davis
Courant Institute of Mathematical Sciences, New York University, New York, NY, USA

E.G. Omodeo and A. Policriti (eds.), *Martin Davis on Computability, Computational Logic, and Mathematical Foundations*,
Outstanding Contributions to Logic 10, DOI 10.1007/978-3-319-41842-1_1

In the New York City public schools, I was an adequate, but not at all exceptional, student. I've always enjoyed writing, but an interest in numbers came early as well. I remember trying to find someone who could teach me long division before I encountered it in school. My parents, whose schooling was minimal, could not help. My first "theorem" was the explanation of a card trick. I learned the trick from a friend who had no idea why it worked; I was delighted to see that I could use the algebra I was being taught in junior high school to explain it.

It was at the Bronx High School of Science that I first found myself with young people who shared the interests I had been developing in mathematics and physics. My burning ambition was to really understand Einstein's theory of relativity. There were a number of books available in the local public library as well in the school library, but I couldn't understand many of the equations. Somehow I got the idea that it was calculus I needed to learn, so I got a textbook and taught myself. When I arrived at City College as a freshman, I was able to begin with advanced calculus. During those years, the math majors at City College were an enthusiastic talented group many of whom eventually became professional mathematicians. The faculty, on the other hand, was badly overworked, and, with a few notable exceptions, had long since lost their enthusiasm. Even by the standards of the time, teaching loads were excessive, and none of the usual amenities of academic life (such as offices and secretarial help) were available. Only very few of the most determined faculty members remained active researchers. In addition to these obstacles, Emil Post struggled against physical and psychological handicaps: his left arm had been amputated in childhood and he suffered from periodically disabling manic-depressive disease. Nevertheless, Post not only continued a program of important fundamental research, but also willingly accepted students for special advanced studies on top of his regular teaching load (16 contact hours). I absorbed his belief in the overriding importance of the computability concept and especially of Turing's formulation.

At City College my academic performance was hardly outstanding. I allowed myself the luxury of working hard only on what interested me. My A grades were in mathematics, German, history, and philosophy. My worst class was a required general biology course. I hated the amount of memorization of names of plant and animal parts I had no desire to know, and found genuinely difficult the "practicums", in which we were asked to identify specimens we viewed under the microscope. I actually failed the course, and even on the second try only managed a C.

During my Freshman and Sophomore years, my passionate interest was in the foundations of real analysis. I learned various alternate approaches and proofs of the main theorems. I spent weeks working out the convergence behavior of the sequence

$$s_0 = 1; \quad s_{n+1} = x^{s_n}$$

for $x > 0$. (It converges for $(1/e)^e < x < e^{1/e}$. The case $0 < x < 1$ is tricky because, although the even-numbered terms and the odd-numbered terms each converge, when $x < (1/e)^e$ their limits are different.) I liked sequences and saw how to prove that every sequence of real numbers has a monotone subsequence as a way of obtaining the basic theorems. I even wrote quite a few chapters of a proposed textbook.

My fellow student John Stachel and I began to be interested in logic, and at his suggestion, we approached Post about a reading course in mathematical logic. Thus, in my junior year, we began studying an early version of Alonzo Church's textbook under Post's supervision. Unfortunately, it only lasted a few weeks: Post had made his discovery of the existence of incomparable degrees of unsolvability, the excitement precipitated a manic episode, and he was institutionalized. The following year Post was back and we spent a great deal of time talking about logic. He gave me a collection of his reprints and also referred me to Kleene's paper [37]. This was a paper Kleene had written in haste to get some results in publishable form before he was requisitioned for war work. For me this was a boon because it was written in a relatively informal style quite unlike Kleene's usual more opaque exposition. I spent a lot of time filling in the gaps, and in the process became enamored of the Herbrand-Gödel-Kleene equation formalism. In considerable part, my dissertation developed from that paper.

Kleene's paper showed that the sets definable in the language of arithmetic[1] formed a natural hierarchy in terms of alternating strings

$$\exists\forall\exists\ldots \quad \text{or} \quad \forall\exists\forall\ldots$$

of quantifiers applied to a computable relation: each additional quantifier makes it possible to define new sets. This result was applied to give short incisive proofs of Gödel's incompleteness theorem and the unsolvability results of Alonzo Church.

I would undoubtedly have remained at City College for my graduate studies to work with Post if that option had been available. But City College was strictly an undergraduate school, and I had to look elsewhere. I had offers of financial support from Princeton, where I could work with Church, and from the University of Wisconsin, where Kleene would have been my mentor. Post advised me to go to Princeton, and that is what I did. There was quite a culture clash between my New York Jewish working-class background and the genteel Princeton atmosphere, and at one point it seemed that my financial support would not be renewed for a second year for reasons having nothing to do with my academic performance. Although eventually I was given support for a second year, the unpleasantness made me eager to leave. Fortunately, the requirements at Princeton were sufficiently flexible that it was quite possible to obtain a Ph.D. in just 2 years, and that is what I did.

The problem that I knew would readily yield results was the extension of Kleene's arithmetic hierarchy into the constructive transfinite, what later became known as the *hyperarithmetic* sets. Post had shown that the successive layers of Kleene's hierarchy could also be generated using the jump operator,[2] and it was easy to see how to extend this method into the transfinite. But the problem that I found irresistibly seductive

---

[1]That is, the language using the symbols $\neg \supset \vee \wedge \exists \forall =$ of elementary logic together with the symbols $0 \ 1 + \times$ of arithmetic.

[2]The *jump* of a set A of natural numbers may be understood as the set of (numerical codes of) those Turing machines that will eventually halt when starting with a blank tape and able to obtain answers to any question of the form "$n \in A$?".

was Hilbert's tenth problem, the problem of the existence of integer solutions to polynomial Diophantine equations. Post had declared that the problem "begs for an unsolvability proof" and I longed to find one. Not being at all expert in number theory, I thought that it was foolish to spend my time on Diophantine equations when I had a dissertation to write and a sure thing to work on. But I couldn't keep away from Hilbert's tenth problem.

Diophantine problems often occur with parameters. In general one can consider a polynomial equation

$$p(a_1, \ldots, a_m, x_1, \ldots, x_n) = 0$$

where p is a polynomial with integer coefficients, $a_1, \ldots, a_m$ are parameters whose range is the natural numbers, and $x_1, \ldots, x_n$ are unknowns. I began to study *Diophantine sets,* that is, sets that could be defined by such an equation as the set of m-tuples of values of the parameters for which the corresponding equation has a solution in natural numbers.[3] Another way to say this is that Diophantine sets are those definable by an expression of the form

$$(\exists x_1 \ldots x_n)[p(a_1, \ldots, a_m, x_1, \ldots, x_n) = 0].$$

It was not hard to see that the class of Diophantine sets is not only a sub-class of the class of recursively enumerable (r.e.) sets,[4] but also shares a number of important properties with that class. In particular, both classes are easily seen to be closed under union and intersection, and under existential quantification of the defining expressions. A crucial property of the class of r.e. sets, a property that leads to unsolvability results, is that the class is *not* closed under taking complements. I was quite excited when I realized that the class of Diophantine sets has the same property. This was because if the Diophantine sets were closed under complementation, then the de Morgan relation

$$\forall = \neg\exists\neg$$

would lead to the false conclusion that all of the sets in Kleene's hierarchy, all arithmetically definable sets, are Diophantine. (False because there are arithmetically definable sets that are not r.e. and hence certainly not Diophantine.) Although this proof is quite non-constructive,[5] the result certainly suggested that the classes of r.e. sets and of Diophantine sets might be one and the same. If every r.e. set were indeed Diophantine, there would be a Diophantine set that is not computable which would lead at once to the unsolvability of Hilbert's tenth problem in a particularly strong form. So, I began what turned into a 20 year quest, the attempt to prove that

[3]For example, the "Pell" equation $(x + 1)^2 - d(y + 1)^2 = 1$ has natural number solutions in x, y just in case d belongs to the set consisting of 0 and all positive integers that are not perfect squares; hence that latter set is Diophantine.

[4]A set of natural numbers is *r.e.* if it is the set of inputs to some given Turing machine for which that machine eventually halts.

[5]It furnishes no example of a Diophantine set whose complement is not Diophantine.

every r.e. set is Diophantine, what Yuri Matiyasevich much later called my "daring hypothesis".

During the summer between my 2 years at Princeton I was able to prove that every r.e. set is definable by an expression of the form

$$(\exists y)(\forall k)_{\leq y}(\exists x_1 \ldots x_n)[p(k, y, a_1, \ldots, a_m, x_1, \ldots, x_n) = 0]$$

where p is a polynomial with integer coefficients. From a purely formal point of view, this result (later known as "Davis normal form") seemed tantalizingly close to my conjecture; the only difference was the presence of the bounded universal quantifier $(\forall k)_{\leq y}$. However, there was no method in sight for getting rid of this quantifier, and I couldn't help agreeing with Church's assessment when he expressed disappointment that the result was not stronger.

Meanwhile, I had a dissertation to write. I didn't think at the time that my normal form by itself would suffice, although in retrospect I think it likely would have been accepted. In any case, I worked out an extension of Kleene's hierarchy into the constructive transfinite using Kleene's system of notations for ordinals.[6] Kleene had defined a set O of natural numbers and a partial well-ordering $<_O$ on this set. Each $m \in O$ represented an ordinal $|m|$, and

$$m <_O n \Longleftrightarrow |m| < |n|.$$

With each $m \in O$ I associated a set L_m in such a way that $m <_O n$ implied that L_m is computable relative to L_n as oracle, but not the other way around. Then to extend Kleene's hierarchy, it was only necessary to consider the sets many-one reducible to the L_m. I studied their representation in terms of second order quantification and obtained the ridiculously weak result that up to ω^2 all of these sets were indeed so representable.[7] In addition I defined a constructive ordinal γ to be a *uniqueness ordinal* if whenever $|m| = |n| = \gamma$ the Turing degrees of L_m and L_n are the same. I proved that every $\gamma < \omega^2$ is indeed a uniqueness ordinal.[8]

I presented the results from my dissertation in brief talks at two professional meetings. The Diophantine result was given at a small meeting of the Association for Symbolic Logic in Worcester, Massachusetts in December 1949, which I attended with my first wife a few days after our marriage. Eight months later I attended the first post-war International Congress of Mathematicians at Harvard University, and spoke about my results on hyperarithmetic sets. This time I was alone—our marriage had proved short-lived; my wife had left me shortly before the Congress. At the Congress I met the great logician Alfred Tarski who showed considerable interest in my work, and, of particular significance, I also met Julia and Raphael Robinson. I had studied

---

[6] Actually, Kleene's system S_3.

[7] Actually without any bound on the ordinal all the sets in the hierarchy are representable with only one second order function quantifer.

[8] Clifford Spector showed that the result remains true for all constructive ordinals in his dissertation, written a few years later under Kleene's supervision.

some of their published work, and was very pleased to meet them. I was surprised to find that Julia was presenting a short contributed paper on Diophantine sets. It turned out that we had approached the subject from opposite directions. While I had been trying to find a general representation for arbitrary r.e. sets, as close as possible to a Diophantine definition, she had been seeking such definitions for various particular sets. Her result that turned out to have the most important consequences was that from the existence of a single Diophantine equation with two parameters, one of which grows exponentially as a function of the other, she could obtain a Diophantine definition of $\{\langle a, b, c\rangle \mid c = a^b\}$.

I would like to say that I expressed my pleasure at finding another Hilbert's tenth problem enthusiast. However, in Julia's sister Constance Reid's memoir, *The Autobiography of Julia Robinson*,[9] based on conversations with Julia shortly before her tragic death of leukemia, she quotes Julia as remembering me saying when we met that I couldn't see how her work "could help solve Hilbert's problem, since it was just a series of examples". I do not want to believe that I said anything so ungracious and so foolish. Julia is also quoted as remembering my "presenting a ten minute paper" at that Congress on my Diophantine results, and as that was not the case, I can comfort myself with the thought that her recollection of what I had said may also have been mistaken.

A few days after the Congress, I was on a plane from New York to Chicago, my first experience of air travel. After considerable difficulty in landing a job in a tight market, with my specialization in logic a definite disadvantage, I had had a stroke of luck. My former fellow student Richard Kadison having received a coveted National Research Fellowship, turned down the offer from the University of Illinois at Champaign-Urbana to be "Research Instructor". As their second choice, the position was offered to me, and I was delighted to accept. Research Instructors were expected to be recent Ph.D.'s and were required to teach only one course per semester at a time when the regular faculty taught three. In addition, we were given the opportunity to teach a second graduate course in our own specialty if we wished. I was very happy to take advantage of this possibility: I taught mathematical logic in the fall and recursive function theory in the spring. In this second course, I decided to begin with Turing machines. Kleene had applied Gödel's methods of arithmetic coding to develop his results for the equation calculus. I saw that the same could be done for Turing machines and that this had certain technical advantages. However, in order to develop the necessary machinery, I had to design Turing machines to carry out various specific operations; without realizing it, I was being a computer programmer!

Edward Moore (later known for his basic work on sequential machines), also a very recent Ph.D. in mathematics, was an auditor in my course. He came up to the front of the room after one of my classes and showed me how one of the Turing programs I had written on the blackboard could be improved. Then he said something very much like the following: "You should come across the street; we've got one of those machines there." In fact a superb engineering group were just finishing a computer called ORDVAC of the "johnniac class" on the University campus. I had been paying

[9]In [40] p. 61.

no attention to computers, and up to that moment had not considered that Turing's abstract machines might have some relation to real world computing machines. It would make a better story if I said that the next day I took Ed up on his invitation. But the truth is that it was the Korean War and the hot breath of the draft that led me to take that walk "across the street" some weeks later. It was clear to me that if I remained in my faculty position, I would be inducted into the army, and it was equally clear to me that that was something I wanted to avoid.

A faculty group, led by the physicist Frederick Seitz, determined to contribute to the war effort and convinced of the military significance of automated systems, started a project within the university called the Control Systems Laboratory (C.S.L.). I was recruited for the project and, with the promise of a draft exemption, accepted. My boss was the mathematician Abe Taub, an expert in relativity theory and shock waves. It was a heady time. Norbert Wiener's *Cybernetics* heralding a new age of information and control had appeared a few years earlier, von Neumann had developed the basic computer architecture still used today and was investigating the use of redundancy to obtain reliable results from unreliable components, and the transistor had just been developed at Bell labs. There was much discussion of all this at the C.S.L., and after some vacillation, a report from the battlefield on the need for better fighter plane support for the front line troops decided the direction of the first major effort.

A working model was to be produced of an automated system for navigating airplanes in real time. The "brain" of the system was to be the newly constructed ORDVAC. And the job of writing the code fell to me. My instruction in the art of computer programming was delivered by Taub in less than five minutes of "This is how it is done". I also had as textbook the basic reports by von Neumann and Goldstine with many sample programs. Of course, the project was ludicrously over-ambitious given the technology available in 1951. The ORDVAC had 5 kB of RAM; memory access required 24 ms. Addition time was 44 ms, and multiplication time a hefty millisecond. From a programmer's point of view, interpreters, compilers, or even assembly language were all non-existent. There were no index registers. Inductive loops had to be coded by incrementing the address portion of the instructions themselves. And of course all the code had to be written in absolute binary. The RAM was implemented as static charge on the surface of cathode ray tubes, which tended to decay rapidly, and was continuously being refreshed. This worked so long as the programmer was careful not to write loops so tight that the same position on the CRT's was bombarded by electrons too rapidly for the refreshing cycle to prevent spillover of charge to neighboring positions. To a contemporary programmer, these conditions seem nightmarish, but in fact it was lots of fun (especially when I let myself forget what it was all supposed to be for).

My experience as an ORDVAC programmer led me to rethink what I had been doing with Turing machines in the course I had just finished teaching. I began to see that Turing machines provided an abstract mathematical model of real-world computers. (It wasn't until many years later that I came to realize that Alan Turing himself had made that connection long before I did.) I conceived the project of writing a book that would develop recursive function theory (or as I preferred to think of it: computability theory) in such a way as to bring out this connection. I hardly imagined that 7 years would go by before I held in my hand a printed copy of *Computability & Unsolvability*. I enticed a group of my C.S.L. colleagues into providing an audience for a series of lectures on computability; the notes I provided for the lectures were a rough draft of the first part of the book.

Champaign-Urbana in the early 1950 s was not an ideal locale for a young bachelor looking for a social life. In the university community, young men outnumbered young women by something like 10 to 1. (Even among undergraduates the ratio was 4 to 1.) But I was lucky enough to attract the interest of Virginia Palmer, a graduate student. By the spring of my first year there, she had moved into my apartment, an arrangement far more unusual in those days than it would be today. In fact, the university administration took an active interest in students' intimate lives. Female graduate students (and only female students) were subject to expulsion if they were found cohabiting with a male. So our menage was somewhat dangerous, especially as Virginia's parents didn't find me a particularly desirable suitor. We planned to marry on the earliest date after the legal formalities officially dissolving my first marriage were complete. That date turned out to be the first day of autumn just about a year after my arrival in Champaign-Urbana; we were married by the local Unitarian minister in a simple ceremony with only three friends present. My second marriage has proved somewhat more durable than the first; as I write this, we have recently celebrated our 63rd anniversary.

Christmas week 1951 provided an occasion to drive East and introduce Virginia to my New York friends. It also enabled me to attend a mathematical meeting in Providence where I heard Kurt Gödel deliver his astonishing lecture in which he proposed that reflecting on his undecidability results would force one to adopt ontological assumptions characteristic of idealistic philosophy. The lecture was published only recently, after Gödel's death, but the audacious ideas he propounded have remained with me ever since I heard the lecture.

The spring of 1952 marked a major change for the ORDVAC. It had been built under contract for Army Ordnance, and it was time for its delivery to their Proving Grounds in Aberdeen, Maryland. The computer group had been busy working on a twin (not quite identical) to the ORDVAC dubbed the ILLIAC (later ILLIAC I). But here I was with my code and no computer to debug it on until the ILLIAC came on line. So I was sent to Aberdeen. Virginia came with me and we stayed in a motel in the nearby town, Havre de Grace. It was in that motel that we conceived our first child.

The ORDVAC had been installed in the the building housing the Ballistics Research Laboratory along with two older, indeed historic, computers: the EDVAC and the ENIAC. The ENIAC consisted of racks of vacuum tube circuits and plugboards such as were used by telephone switchboards, filling the four walls of a large room. The building was locked from the inside and one could only leave by first going to the ENIAC room and asking one of the people there to unlock the door. The ORDVAC was in use by Aberdeen people until 4 p.m., after which it was made available to me. Instead of the watchful crew in Urbana used to babying their creation, the computer operator was a sergeant whose main qualification was that in civilian life he had been an amateur radio operator. I was soon operating the machine myself, something I never would have been permitted to do in Urbana. One evening, I noticed that the machine seemed to be making many errors. I also noticed that I was getting very warm, but it didn't occur to me to connect these facts. Finally, when I saw a 0 change to 1 on a CRT at a time that the computer was not executing any instructions, I gave up and left. The folks back in Urbana were furious with me. The air conditioning had broken down, and there had been a very real danger that the ORDVACS could have been destroyed by the heat. It should have been powered down at once.

Back in Urbana, I found myself increasingly unhappy with what I was doing at the C.S.L. The Office of Naval Research came to my rescue with a small grant that enabled me to spend 2 years as a visiting member at the Institute for Advanced Study in Princeton. I thought that with that sponsorship, I would probably be safe from the draft. My proposal was to work on connections between logic and information theory. That was a really good idea: the great Russian mathematician Kolmogoroff and Gregory Chaitin showed what could be done with it quite a few years later. However, I found myself moving in other directions.

The Institute for Advanced Study in those years was directed by J. Robert Oppenheimer. On the faculty were Albert Einstein, Kurt Gödel, and John von Neumann. Einstein and Gödel, good friends, were often seen walking to or from the Institute buildings together. I well remember the first time we encountered them walking down the middle of Olden Lane together: Einstein dressed like a tramp accompanied by Gödel in a suit and tie carrying his briefcase. "Einstein and his lawyer" was Virginia's vivid characterization.

I had met Norman Shapiro as an undergraduate in Urbana. He had come to Princeton University as a graduate student and was writing a thesis on recursive functions. He and I organized a logic seminar. Among the regular attendees were Henry Hiz, John Shepherdson, and Hao Wang. Hilary Putnam, with whom I was later to do some of my best work, gave a philosophical talk which Norman and I mercilessly attacked. In my research, I was struck by the fact that the phenomenon of undecidability in logic could be understood abstractly in terms of the way each particular logical system provided a mapping from recursively enumerable sets[10] to subsets of their complements. I was particularly struck by the fact that Gödel's famous result about the unprovability of consistency could be expressed simply as the fact that the iteration of this map always produces the empty set. Some years later I told one of my first doctoral students, Robert Di Paola, about this, and he based his dissertation on studying that mapping. Gödel himself was uninterested when I summoned the courage to tell him about my ideas.

I occasionally thought about Hilbert's tenth problem, and I worked on my book. The chapter on applications of computability theory to logic gave me particular trouble. The problem I faced was giving a coherent exposition without writing a whole book on logic. I rewrote that chapter many times before I was satisfied. A problem of another kind was the difficulty I had in getting the Institute typists to produce a decent copy from my handwritten manuscripts. Our son was born in

[10]Actually, indices of r.e. sets.

January 1953. After he was weaned, a year later, Virginia took a job at the Princeton Public Library. I imagined I could take care of the baby and work on my book at the same time. Of course this did not work out very well.

My arrangement with the Office of Naval Research left me free to seek employment during the summer months. We certainly needed the extra money. I was able to spend the summer of 1953 working at Bell LabsBell Labs, a short commute from Princeton. My boss was Shannon, the inventor of information theory, and I was able to renew my acquaintance with Ed Moore. Shannon had recently constructed a universal Turing machine with only two states. He posed the question of giving a well defined criterion for specifying when a Turing machine could be said to be universal. I liked that question and wrote two short papers dealing with it.[11] The intellectual atmosphere at Bell Labs was stimulating and open to fundamental research. I could well understand how a fundamental breakthrough like the transistor could develop in such an environment. Shannon himself was treated like the star he was. He had a small shop with two technicians available to build any of his whimsical gadgets. His "mouse" that successfully solved mazes was already famous. Less well known was his desk calculator "Throwback I" that accepted its input in the form of Roman numerals. Shannon was also an expert unicycle rider. One day he brought his unicycle to the labs and created mass disruption by riding it down the long corridors and even into and out of elevators, bringing swarms of Bell LabsBell Labs employees streaming out of their offices to watch. Another thing I remember about that summer is the excitement of a real workers uprising in East Berlin against the Communist regime.

For the summer of 1954 I thought about applying the programming skills I had learned in Urbana to a logical decision procedure. My first choice was Tarski's quantifier elimination algorithm for the first order theory of real closed fields. But on second thought I saw that this was going to be too difficult for a first try, and instead I settled on Presburger's procedure for integer arithmetic without multiplication, since this was a much simpler quantifier elimination procedure. Had I known the Fischer-Rabin super-exponential lower bound for Presburger arithmetic (proved 20 years later), I would presumably have hesitated. But I went blithely ahead with the blessing of the Office of Ordnance Research of the U.S. Army which agreed to support the effort. I was able to do the work without leaving Princeton, using the original JOHNNIAC at the Institute for Advanced Study. To my dismay the code used all of the 5 kB of RAM available and was only able to deal with the simplest statements on the order of "The sum of two even numbers is even". My report on the program, duly delivered to the Army on its completion and included as well in the Proceedings of an important Summer Institute of Logic at Cornell in 1957 (about which, more later), ended with the understatement[12]:

> The writer's experience would indicate that with equipment presently available, it will prove impracticable to code any decision procedures considerably more complicated than Presburger's. Tarski's procedure for elementary algebra falls under this head.

---

[11] [1, 2].

[12] The report was reprinted in [43], pp. 41–48.

An anthology [43] of "classical papers on computational logic 1957–1966" published in 1983 begins its preface with the sentence:

> In 1954 a computer program produced what appears to be the first computer generated mathematical proof: Written by M. Davis at the Institute of Advanced Studies (sic), USA, it proved a number theoretic theorem in Presburger Arithmetic.

In the spring of 1954 my 2 years at the Institute were drawing to a close, and I needed to find a job. Again the market was rather tight. We had a few possibilities, but opted for the one that took us furthest west: an Assistant Professorship at the University of California at Davis. For the first time we experienced what was to be repeated over a dozen times in our lives: the trip by automobile across the United States with a stopover in Kansas City to visit Virginia's parents. As we approached our new home, the road signs seemed to be directing us: "Davis use right lane".

The liberal arts program was newly instituted at U.C. Davis which had been originally devoted to agriculture. In 1954, the population of Davis was just about 5000 people. It was not a cultural center. Amusements were in such short supply that we would drive to the local soft ice-cream drive-in just to watch the customers come and go. It was not a year in which I accomplished much scientifically. The teaching load was quite modest: just two courses per semester. When I taught calculus (to students majoring in agricultural engineering), I had to speak in a loud voice to be heard above the clatter of the turkeys in the building next door.

Virginia was pregnant again and we needed to find an obstetrician. There were none in Davis itself, but there was a hospital at the county seat, Woodland, a few miles away. Virginia found the local obstetrician there quite unacceptable. Sacramento, the state capital was perhaps 18 miles away, but we had heard too many obstetrical horror stories coming from that quarter. So we headed for progressive Berkeley 80 miles away, where Virginia found an excellent obstetrician. Today there is an excellent superhighway connecting Davis with Berkeley, but in 1954 the drive took at least 2 h. The highway ended in Richmond with the rest of the route being through city streets. Virginia's first labor had been swift and uneventful, so we knew that we had to be prepared for the possibility of not making it to Berkeley in time. We obtained a government pamphlet for farmers on delivering babies, and bought a second-hand obstetrics textbook. We were in Berkeley a few days before our Nathan made his appearance, and Virginia was assured that all was well. Back in Davis, we were awakened at 2 a.m. by a flood of amniotic fluid drenching the sheets. By 4 a.m. the crying baby had arrived. Virginia tells people that I "delivered" him, although really I just watched. Except for one detail: Nathan was born with his umbilical cord wrapped around his neck. Before I had time to think about it, I had lifted the loop away.

People we hardly knew had strong opinions about what had happened. Those who thought we had done something praiseworthy in defense of the natural seem to have been outnumbered by those who thought we had behaved in an irresponsible manner. There was even the suggestion that we should be imprisoned. We were convinced that Davis was not for us, and were determined to leave. I found a position at Ohio State University in Columbus and quickly accepted. For the summer, I got a job at the

Moore School of Electrical Engineering in Philadelphia where the ENIAC and the EDVAC had been built. So, we set out for Princeton in our 1951 Studebaker sedan with Harold and Nathan in the rear. Our plan was to spend the summer there, an easy commute to the Moore School.

The summer at the Moore School was a pretty complete disaster. They wanted me to prove a particular kind of theorem about certain numerical methods for solving ordinary differential equations. I knew very little about such things, but I saw no reason to believe that there was a theorem of the kind they wanted. I did not accomplish much for them. The best part of the summer was getting to know Hilary Putnam who was living in the same prefab housing complex for graduate student and junior faculty families where we had subleased an apartment for the summer. To my surprise, he was very interested in Hilbert's tenth problem and proposed that we collaborate. Nothing much came of this until a few years later.

A major plus of my new position at Ohio State University was that Kleene's student Clifford Spector was on the faculty. His brother was a close friend of a good friend of mine, and on his brother's advice, he had written me some years earlier about his interest in logic. Apparently, this interest had been actively discouraged at Columbia University where he had been informed that there are no interesting problems in logic. I had suggested a number of possibilities for graduate study in logic including Madison, Wisconsin with Kleene. Somewhat to my surprise, I detected something not entirely friendly in Clifford's welcome. It was several months before he became open. I learned that Kleene had been rather displeased with me. Kleene had gone to considerable trouble to get a fancy fellowship for me at the University of Wisconsin, and I had not only gone to Princeton instead, but had written a dissertation largely in areas where Kleene himself had been working. Kleene had given Spector the uniqueness ordinal problem left open in my thesis as an appropriate topic for his dissertation. Clifford reported that Kleene had whipped him on with the warning that "Davis is working on it" emphasizing the importance of reaching the goal first. In fact, I hadn't been thinking about uniqueness ordinals at all. In any case Spector was a more powerful mathematician than I. In his excellent dissertation, he not only proved that every constructive ordinal is a uniqueness ordinal (thus settling the question raised in my dissertation), but also proved a deep result in the theory of degrees of unsolvability.[13]

The 1 year we spent in Columbus was not a happy one. Among other difficulties, we were feeling financially pinched. I received my last paycheck from Davis at the beginning of June and the first from Columbus only in November. The money from the Moore School helped, but I had to return half of the money for moving expenses I had received from Davis because I had left after only 1 year, and Ohio State did not cover moving expenses. And apparently impossible to please, Virginia and I just didn't like life in Columbus very much. To save money, we moved into an apartment with just one bedroom that we gave to our two babies, while we slept on a convertible couch in the living room. The Chair of the department helped by offering me the

[13]The existence of "minimal" degrees. Only 31 years old, Clifford Spector died quite suddenly in 1961 of acute leukemia, a tragic loss.

opportunity to teach an off-campus advanced calculus course to Air Force officers at the nearby Wright-Patterson base in the summer. In the hot Ohio summer, I often taught wearing short pants. I later found that a Colonel had complained about my attire to the department Chair.

One morning that summer, at the breakfast table, Virginia pointed to an advertisement in the New York Times. An anonymous "long established university in the northeast of the United States" was seeking teachers of engineering subjects including calculus and differential equations. Salaries, the ad said would be "comparable to industry". I sent off a letter at once, and I was interviewed and hired. My academic year salary increased from $5100 to $7900 and we felt rich. The "long established university" turned out to be Rensselaer Polytechnic Institute (R.P.I.). The position was not at the main campus in Troy, New York, but at the Hartford Graduate Division in Eastern Connecticut. In 1956 the nation was experiencing an acute shortage of engineers. In the Connecticut valley, the United Aircraft Company with its Pratt-Whitney subsidiary (a major manufacturer of jet engines) had been finding it extremely difficult to hire the engineers it needed. To help to solve this problem, R.P.I. was asked to form the Hartford Graduate Division so United Aircraft engineering employees could take courses leading to a master's degree, with tuition to be paid by the company. This had helped, but not enough. So, liberal arts graduates who satisfied the minimum requirement of having completed a year of calculus and a year of physics were hired by United Aircraft and sent to the Hartford Graduate Division to study mechanical engineering. Those who completed the forty week program received a certificate and were put to work. They were also eligible to apply to R.P.I.'s master's program.

Faculty was needed to teach in this new program, and that was the reason for the ad I had answered. The Hartford Graduate Center was housed in a one-story, industrial-style building with a huge parking lot on the main highway between Hartford and Springfield, Massachusetts. Friends had predicted that moving to an environment with no research aspirations, to do elementary teaching would be the end of my research career. In fact it turned out to be an excellent move. From a personal point of view, Eastern Connecticut was beautiful and an easy drive to New York where there were friends, the amazing resources of that city, and my mother's apartment in the Bronx where she would cheerfully serve as baby sitter, and where we could spend the night. But it turned out very well professionally also. The student body were relatively mature interesting people of varied background who were fun to teach. And as the forty week program wound down, I moved into the master's level program, teaching a variety of courses far more interesting than what would have been available to a lowly assistant professor at Ohio State. Student notes for a course in functional analysis later became a short book.[14] The clerical staff turned out to be cheerful and competent and quite willing to turn my mangled and worked over manuscript for *Computability & Unsolvability* into a typescript I could send to publishers. There were mixed reviews, including one that derided the connection I was proposing with actual computers and included an invidious comparison with Kleene's recently published book (with

[14][7].

which, by the way, the overlap was not extensive). It turned out that McGraw-Hill had chosen Hartley Rogers as their reviewer, and he not only wrote the kind of laudatory review that gladdens an author's heart, but also produced an astonishingly detailed helpful critique. The book was published in McGraw-Hill's series on "Information Processing and Computers" appearing in 1958. It was eventually translated into Japanese and Italian,[15] and, reprinted by Dover in 1982,[16] it remains in print today [3].

The summer of 1957 was an exciting time for American logicians. A special "institute" on logic was held at Cornell University. For five weeks 85 logicians participated: established old-timers, those in mid-career, fresh Ph.D.'s, and graduate students. There was even Richard Friedberg, still an undergraduate, who had just created a sensation by proving the existence of two r.e. sets neither of which is computable relative to the other thus solving what had been called Post's problem. There were seminars all day. The gorges of Ithaca were beautiful, and swimming under Buttermilk Falls was a summertime pleasure. Hilary Putnam and I seized the opportunity to work together. Our two families shared a house with an unusual distribution of the quarters: there were three small separate apartments; the adult couples each got one of them, and the third went to the three children, our two boys and Hilary's Erika who was two days younger than our Nathan. Hilary and I talked all day long about everything under the sun, including Hilbert's tenth problem. Hilary tended to generate ideas non-stop, and some of them were very good. I tended to be cool and critical and could be counted on to shoot down ideas that were pretty obviously bad. Hilary's idea that turned out to be very good indeed was to begin with the normal form from my dissertation and to try to get rid of that bounded universal quantifier that blocked the path to my "daring hypothesis" by using the Chinese Remainder Theorem to code the finite sequences of integers that the quantifier generates.[17] Using little more than the fact that congruences are preserved under addition and multiplication, we obtained two relations with rather simple definitions about which we were able to show that their being Diophantine would imply that every r.e. set is likewise Diophantine.[18]

We resolved to seek other opportunities to work together. Hilary suggested we try to get funding so we could spend summers together. He proposed investigations of possible computer implementations of proof procedures for first order logic.[19] I guess

---

[15] The translator for the Italian version called it a "libro classico".

[16] A review of the Dover edition by David Harel referred to the book as one of the few "classics" in computer science.

[17] $(\forall k)_{\leq y}(\exists u)$ ... is equivalent to saying that there exists a sequence $u_0, u_1, \ldots u_y$ of numbers satisfying ... The use of the Chinese Remainder theorem to code finite sequences of integers had been used by Gödel to show that any recursively defined relation could be defined in terms of addition, multiplication and purely logical operations. I had used the same device in obtaining my normal form.

[18] [27].

[19] Abraham Robinson had proposed similar investigations in a talk at the Cornell Institute. I attended that talk, but Hilary didn't. Somehow, I didn't connect the two ideas until years later when I noticed Robinson's paper in the proceedings of the Institute.

we thought we'd have more luck being funded with that than with Hilbert's tenth problem. We agreed to work through R.P.I. By the time we got our proposal together, it was too late to be funded for the summer of 1958 by any of the usual agencies. Someone suggested that we try the National Security Agency (NSA). Although the NSA is now notorious, I'd never heard of them. Nevertheless I sent the proposal to them.

Our idea was to define a procedure that would generate a proof of a sentence by seeking a counter-example to its negation in what later became known as its Herbrand universeHerbrand!H. universe. This involved generating ever longer Herbrand expansions, and testing periodically for a truth-functional inconsistency. When I was called to NSA headquarters, it turned out that it was this test for truth-functional inconsistency that interested them. They told me that this was a very hard problem, and seemed dubious of our ability to make serious inroads in just one summer, but, finally, they did agree to sponsor our work. We were to provide a report at the end of the summer. However, unlike typical funding agencies, they specifically asked that their support *not* be acknowledged in the report. Told that I'd never heard of the NSA, the reply was that their "publicity department" was doing a good job.

I found a summer cottage on Lake Coventry for Hilary and his family. As I said elsewhere about my summers with Hilary:

> We had a wonderful time. We talked constantly about everything under the sun. Hilary gave me a quick course in classical European philosophy, and I gave him one in functional analysis. We talked about Freudian psychology, about the current political situation, about the foundations of quantum mechanics, but mainly we talked mathematics.[20]

My first copy of *Computability & Unsolvability*, smelling of printer's ink arrived that summer. Elated, I showed it to Hilary. He smilingly offered to find a typographical error on any page I'd select. Determined to show him, I turned to the reverse side of the title page containing the copyright notice, only six lines. Giving the page a quick glance, Hilary noted that the word "permission" was missing its first "i".

Our report for the NSA, entitled *Feasible Computational Methods in the Propositional Calculus* is dated October 1958. It emphasizes the use of conjunctive normal form for satisfiability testing[21] (or, equivalently, the dual disjunctive normal form for tautology testing). The specific reduction methods whose use together have been linked to the names Davis-Putnam are all present in this report.[22]

---

[20] [40] p. 93.

[21] What has become known as *the satisfiability problem*.

[22] These are:

1. The *one literal rule* also known as the *unit rule*.
2. The *affirmative-negative rule* also known as the *pure literal[3.] rule*.
4. The *rule for eliminating atomic formulas*
5. The *splitting rule*, called in the report, the *rule of case analysis*

The procedure proposed in our later published paper used rules 1, 2, and 3. The computer program written by Logemann and Loveland discussed below used 1, 2, and 4. The first of these is the "Davis-Putnam procedure" which has been the subject of theoretical analysis, nowadays referred to as DP. The second choice is the one generally implemented, is usually called DPLL to refer to

After that first summer, our research was supported by the U.S. Airforce Office of Scientific Research. It was in the summer of 1959 that Hilary and I really hit the jackpot. We decided to see how far we could get with the approach we had used at the Logic Institute in Ithaca, if, following Julia Robinson's lead, we were willing to permit variable exponents in our Diophantine equations. That is, we tried to show that every r.e. set could be defined by such an exponential Diophantine equation. After some very hard work, using Julia Robinson's techniques as well as a good deal of elementary analysis,[23] we had our result, but, alas, only by assuming as given, a fact about prime numbers that was certainly believed to be true, but which wasn't proved until many years later namely: *there exist arbitrarily long arithmetic progressions consisting entirely of prime numbers*.[24] As we wrote up our summer's work, we decided to include an account of a proof procedure for first order logic based on our work on the propositional calculus from the previous summer. Our report to the Air Force included the work on Hilbert's tenth problem, the proof procedure, and a separate paper on finite axiomatizability. Years later Julia Robinson brought a copy of this report[25] with her to Russia where the mathematicians to whom she showed it were astonished to learn that this work was supported by the U.S. Airforce. It was the proof procedure [28] that brought some notoriety to the Davis-Putnam partnership. It proposed to deal with problems in first order logic by beginning with a preprocessing step that became standard: The negation of the proposition to be proved was put into prenex normal form, followed by Skolemization to eliminate existential quantifiers, and then put into conjunctive normal form. Our crude algorithm generated a continuing Herbrand expansion periodically interrupted by tests for satisfiability along the lines mentioned above.

We submitted our work on Hilbert's tenth problem for publication and at the same time sent a copy to Julia Robinson. Julia responded soon afterwards with an exciting letter:

(Footnote 22 continued)
the four of us. It still seems to be useful. I might mention that I have received two awards based at least partly on this work and that I feel strongly that Hilary, at least, should have shared them.

[23]Among other matters, we needed to find an exponential Diophantine definition for the relation:

$$\frac{p}{q} = \sum_{k=1}^{n} \frac{1}{r+ks}.$$

We didn't go about it in the easiest way. We used the fact that

$$\sum_{k=1}^{n} \frac{1}{\alpha+k} = \frac{\Gamma'(\alpha+n+1)}{\Gamma(\alpha+n+1)} - \frac{\Gamma'(\alpha+1)}{\Gamma(\alpha+1)},$$

expanded Γ'/Γ by Taylor's theorem, and used an estimate for Γ'' to deal with the remainder.

[24]See [36].

[25]AFOSR TR59-124.

> I am very pleased, surprised, and impressed with your results on Hilbert's tenth problem. Quite frankly, I did not think your methods could be pushed further …
>
> I believe I have succeeded in eliminating the need for [the assumption about primes in arithmetic progression] by extending and modifying your proof.

She sent us her proof soon afterwards; it was a remarkable tour de force. She showed how to get all the primes we needed by using, instead of a then unproved hypothesis about primes in arithmetic progression, the prime number theorem for arithmetic progressions which provided a measure of how frequently primes occurred "on average" in such progressions. We proposed that we withdraw our paper in favor of a joint publication, and she graciously accepted. She undertook the task of writing up the work, and (another surprise), she succeeded in drastically simplifying the proof so only the simplest properties of prime numbers were used. Combined with Julia's earlier work, this new result showed that my "daring hypothesis" that all r.e. sets are Diophantine was equivalent to the existence of a single Diophantine equation whose solutions grow exponentially (in a suitable technical sense of the word).[26] The hypothesis that such an equation exists had been raised by Julia in her earlier work, and Hilary and I called it JR.

For years I thought of myself as an exile from New York. Now came an opportunity to move there. From the Institute of Mathematics (to be renamed a few years later the Courant Institute of Mathematical Sciences) at New York University came an invitation to visit for a year. Although this was just a visiting appointment, I was confident that we would not be returning to Connecticut. Cutting our bridges behind us, we sold the house we had bought just a year before. Virginia was as enthusiastic as I about our new life. We moved into an apartment overlooking the Hudson River in the Upper West Side of Manhattan. At NYU, I was asked to teach a graduate course in mathematical logic which was a great pleasure. One of the students in that course, Donald Loveland, later became one of my first doctoral students, and, still later, a colleague. One result of my new situation was access to an IBM 704 computer. I jumped at the chance to try out the proof procedure Hilary and I had proposed. Two graduate students Loveland and his friend George Logemann were assigned to me to do the programming. Donald was a particularly apt choice because he had been involved at IBM with Gelernter's "geometry machine", a program to prove theorems in high school geometry. Their programming effort was successful, but when the program was run it was found that the periodic tests for truth-functional consistency were generating large numbers of ever longer formulas that rapidly filled the available RAM. It was the *rule for eliminating atomic formulas* (later called *ground resolution*) which replaced a formula

$$(p \vee A) \wedge (\neg p \vee B) \wedge C$$

by

$$(A \vee B) \wedge C$$

[26] Cf. [33].

that was causing the problem. It was when the three of us met that we decided to try to use instead the *splitting rule*[27] which generates the pair of formulas

$$A \wedge C \qquad B \wedge C$$

The idea was that a stack for formulas to be tested could be kept in external storage (in fact a tape drive) so that formulas in RAM never became too large.

Much to my surprise this problem of testing for satisfiability a Boolean polynomial presented in form of a list of disjunctive clauses which Hilary and I had introduced as an adjunct to a theorem proving procedure for first order logic, has turned out to be of fundamental importance. The problem has been given the name SAT and it has been the subject of a huge literature both theoretical and pragmatic. The form of our algorithm that uses the splitting rule, currently referred to as DPLL, has proved to be very successful and has been incorporated in many implementations. In the case of our work, it proved successful in testing formulas consisting of thousands of clauses. Nevertheless the program was overwhelmed by the explosive nature of the Herbrand expansion in all but the simplest examples.

As I had expected and hoped, I was offered a regular faculty appointment at NYU. At that time, there were three more or less separate mathematics departments at NYU: the graduate department, the undergraduate department at the main campus in Greenwich Village, and another undergraduate department at the Bronx campus. The appointment I was offered was in the undergraduate department at the main campus. Although not what I had hoped for, I would certainly have accepted this offer, had not the first Sputnik gone aloft a few years earlier. The Soviet launching of a satellite in 1957 had provoked a furore in the United States. We were "falling behind" in science and technology. All at once, science became a growth industry. And that was why I received a very attractive offer from Yeshiva University.

Yeshiva College is housed in a building with a curiously Middle Eastern flavor in the part of Manhattan known as Washington Heights (because Washington fought a rear guard action against the British there as the revolutionary forces were retreating from New York City). It takes its name from the traditional East European yeshivas, institutions of advanced religious training based on the Talmud with instruction mostly in Yiddish, training that could lead to rabbinical ordination. Yeshiva College adjoined to this traditional curriculum, a liberal arts program in the American mode. Various schools were added to the complex, most of them secular, leading to the "university" designation in 1945. The Mathematics Department at Yeshiva College was the home of the periodical *Scripta Mathematica*, specializing in mathematical oddities and issued regularly beginning in 1932. Abe Gelbart was a mathematician at Syracuse University who had become involved with the Scripta Mathematica effort. He began to imagine the possibility of building a first-rate graduate program in mathematics and physics in this milieu. He was able to convince the Yeshiva University administration that in the post-Sputnik atmosphere, external funding would be

---

[27] See footnote 22.

readily available, and he received a go-ahead to found a new Graduate School of Science (later the Belfer Graduate School of Science) with himself as dean.

My teaching load at Yeshiva was to be two graduate courses per semester with every encouragement to develop a program in logic as opposed to the NYU offer which would have required three *undergraduate* courses per semester with the option to conduct a logic seminar on my own time. In addition the Yeshiva offer came with a salary of $500 more than I would make at NYU. For various reasons I would have preferred to remain at NYU, but they were unwilling to respond to the Yeshiva offer, and so, I phoned Gelbart and accepted. Late that spring I was informed that the NYU had reconsidered and was now willing to coming closer to the Yeshiva offer. However, I felt that I had made a commitment to Yeshiva that it would have been unethical to break. I was told to keep in touch and let them know if circumstances changed. It was 5 years before I took them up on that suggestion.

Gelbart found a home for the new school in a building not far from Yeshiva College. When I was taken to see it, I was quite startled. The building had previously been a catering palace, and I remembered it very well. It had been the scene of the celebration of my ill-fated marriage to my first wife a decade earlier. Gelbart turned out to be difficult and ill-tempered, but eager to please so long as his beneficence was duly acknowledged. The faculty, mathematicians and physicists all together on one floor, formed a very congenial group and a good deal of first-rate research was accomplished. I worked to develop a logic program and was successful in having an offer made to Raymond Smullyan which he accepted. Although Donald Loveland's degree was awarded by NYU, he was effectively part of our logic group at Yeshiva. From the beginning Robert Di Paola was at Yeshiva in order to work with me. Both Di Paola and Loveland received their doctorates in 1964.

I was able to publish several papers that were spin-offs of the work with Hilary and Julia (one of them joint with Hilary).[28] I also worked on proof procedures, a field that was beginning to be called automatic theorem proving (ATP). In fact my work with Logemann and Loveland was continuing after I had left NYU, with the IBM 704 continuing to be available to us. It was after the program was running and its weaknesses were apparent to us that an article arrived in the mail that had a major influence on my thinking. It was a paper by Prawitz [39] in which he showed how the kind of generation of spurious substitution instances that overwhelmed our procedure could be avoided. However, the procedure he proposed was subject to a combinatorial explosion from another direction. I set as my goal finding a procedure that combined the benefits of Prawitz's ideas with those of our procedure. I believed that we were on the right track in using as our basic data objects sets of disjunctive clauses (each consisting of *literals*) containing variables for which substitution instances could be sought. Prawitz had proposed to avoid spurious substitutions from the Herbrand universe by forming systems of equations the satisfaction of which would give the desired result. I came to realize that for problems expressed in our form, the required equalities always were such as to render literals complementary. That is given a pair of clauses one of which contains the literal $R(u_1, u_2, \ldots, u_n)$ while the other

[28] [5, 29].

contains $\neg R(v_1, v_2, \ldots, v_n)$, what was needed was to find substitutions to satisfy the system of equations

$$u_1 = v_1 \quad u_2 = v_2 \quad \ldots \quad u_n = v_n.$$

I also saw that for any system of substitutions that was successful in producing an inconsistent set of clauses, there necessarily had to be a subset of that set which was *linked* in the following sense:

> A set of clauses is *linked* if for each literal ℓ in one of the clauses of the set, the complementary literal $\neg\ell$ occurs in one of the remaining clauses.[29]

I had the opportunity to explain these ideas and to place them in the context of existing research at a symposium organized by the American Mathematical Society on *Experimental Arithmetic* held in Chicago in April 1962 to which I was invited to participate. The ideas developed in the paper that was published in the proceedings of the conference [4] turned out to be very influential (although I believe that many of those whose work was ultimately based on this paper were unaware of the fact).[30]

Around this time I had been invited to spend several hours weekly as a consultant at Bell Labs in Murray Hill, New Jersey. I was delighted to have the opportunity to see some of my ideas implemented. Doug McIlroy undertook to produce a working program for Bell Labs' IBM 7090, and did so in short order. The problem of finding solutions to the systems of equations needed to establish "links" was dealt with in McIlroy's program by using what was later called *unification*.[31] Peter Hinman joined the effort as a summer Bell LabsBell Labs employee and found and corrected some bugs in the McIlroy program. We wrote up our work and submitted it for publication. It was accepted with some rather minor changes. These changes were not made, for reasons now obscure and the paper never appeared.

In [4] I carelessly called the algorithm which Logemann and Loveland had implemented, the Davis-Putnam procedure as one of a number of early theorem-proving programs for first order logic. I saw the modification we had made to DP as a mere ad hoc adjustment, and was very slow to realize that from a practical point of view it was the DPLL algorithm for SAT that was the most important part of that effort and that the application to first order logic was really just a footnote. (The interest of the National Security Agency in just that part of the effort should have been a tipoff!) All of this may have had the inadvertent effect of depriving Logemann and Loveland of their proper share of the credit for developing it for many years.

---

[29]Here if $\ell = \neg R(c_1, c_2, \ldots, c_n)$ then it is understood that by $\neg\ell$, the literal $R(c_1, c_2, \ldots, c_n)$ is meant.

[30]It was J.A. Robinson's key insight that a theorem-proving procedure for first order logic could be based on not merely finding complementary literals by unification, but also then eliminating them—what he called "resolution"—that revolutionized the field [42]. Anyone interested in tracing the history might notice that whereas Robinson's [42] does not refer to my [4], his earlier [41] does and in its content clearly shows its influence.

[31]The example worked out in [4] shows unification in action.

The year 1963 brought great excitement to the world of logic. Paul Cohen invented a powerful new method he called forcing for constructing models of the axioms of set theory, and he had used this method to show that Cantor's continuum hypothesis could not be proved from the standard axioms for set theory, the Zermelo-Fraenkel axioms together with the axiom of choice. This settled a key question that had been tacitly posed by Gödel when more than two decades previously, he had shown that the continuum hypothesis couldn't be disproved from those same axioms. I was astonished to receive a letter from Paul Cohen dated November of that year reading in part:

> I really should thank you for the encouragement you gave me in Stockholm. You were directly responsible for my looking once more at set theory. ...Of course, the problem I solved had little to do with my original intent. In retrospect, though, the basic ideas I developed previously played a big role when I tried to think of throwing back a proof of the Axiom of Choice, as I had previously thought about throwing back a proof of contradiction.

In the summer of 1962 I had attended the International Congress of Mathematicians in Stockholm. These conferences are scheduled to occur every 4 years, but this was my first since the 1950 Congress at Harvard. At the Congress, I talked briefly with Paul Cohen. I knew that although he was not primarily a logician, he was a very powerful mathematician who had been attempting to find a consistency proof for the axioms of set theory. He indicated that some logicians he had talked to had been discouraging, and I urged him to pay no attention. That was really the total extent of my "encouragement". Of course, I was very pleased to receive the letter.

It was in 1963 that we realized that we were outgrowing our apartment overlooking the river. Our two sons had been sharing a bedroom. Now aged eight and ten, they had quite different temperaments. If we were to have any peace, they would have to have separate rooms. At this point Virginia found a "brownstone" town house a mile south of our apartment that we were able to buy. Although the price we paid was ridiculously low by later standards, the house and its renovation put an enormous strain on our budget. Of course, it turned out to be far and away the best investment we ever made. We lived there for 33 years. In order to make ends meet, I found myself becoming an electrician and a plumber. With the help of a friend (as much a novice as I), I even installed a new furnace.

A project that was absorbing a good deal of time and energy at this time was the preparation of my anthology of fundamental articles by Gödel, Church, Turing, Post, Kleene, and Rosser.[32] I wrote some of the commentaries for the book while attending a conference in the delightful town of Ravello south of Naples.

Meanwhile my relationship with Abe Gelbart was becoming more and more difficult. Things were brought to a head in the spring of 1965 when, interviewing a prospective faculty member, someone I had very much hoped would be a colleague, Gelbart behaved in an insulting manner. I decided that I had no choice but to resign my position. I called a friend at the Courant Institute and reminded him of the suggestion that I let them know if I were interested in a position at NYU. Soon enough an offer arrived. It was not quite what I had expected: the position was half at the Bronx

[32] [6].

campus and only half at the Institute. However, I was urged to regard the relative vacuum on the Bronx campus as an opportunity to develop a logic group there, and I was assured that I would be treated as a regular member of the graduate faculty. I accepted the position and remained with the Courant Institute until my retirement in September 1996.

I took to the notion of developing a logic group at the Bronx campus with avidity. My old friend and student Donald Loveland was already there, and he was soon joined by the Yasuharas, Ann and Mitsuru. I was able to provide support in the form of released time for research from teaching from my continued funding by the Air Force. During these years Norman Shapiro, with whom I had organized a logic seminar in Princeton many years earlier, and my former student Bob Di Paola were both at the RAND Corporation. Norman arranged for me to be able to spend summers at RAND. Our family found housing in the hills above Topanga Canyon near the Malibu coast, and I enjoyed the daily drive along the beach to the RAND facility in Santa Monica. I was required to obtain security clearance at the "Top Secret" level, not because I did any secret work there, but because classified documents in the building were not necessarily under lock and key. I was tempted once to use my clearance. The Cultural Revolution was in full swing in China, and I was thoroughly mystified by it. Di Paola urged me to seek enlightenment by looking at the intelligence reports from the various agencies easily available to me. I did so, feeling that I was losing my innocence, and was greatly disappointed. Not only did these "secret" reports contain nothing that couldn't be found in newspapers and magazines, they turned out to be anything but unbiased, clearly reflecting the party line of the agency from which they emanated.

What I did at RAND was work on Hilbert's tenth problem, specifically I tried to prove JR. I used the computing facility at RAND to print tables of Fibonacci numbers and solutions of the Pell equation looking for patterns that would do the trick. I also found one interesting equation:

$$9(x^2 + 7y^2)^2 - 7(u^2 + 7v^2)^2 = 2.$$

I proved that JR would follow if it were the case that this equation had only the trivial solution $x = u = 1$; $y = v = 0$.[33] In fact the equation turns out to have many non-trivial solutions, but the reasoning actually shows that JR would follow if there are only finitely many of them, and this question remains open.

In the academic year 1968–1969 I finally had a sabbatical leave. I would have been due for one at Yeshiva, and as part of my negotiation with NYU, I secured this leave. I spent the year in London loosely attached to Westfield College of the University of London where I taught a "postgraduate" course on Hilbert's tenth problem. I continued efforts to prove JR (and thereby settle Hilbert's tenth problem). I found myself working on sums of squares in quadratic rings, but I didn't make much progress. Meanwhile "Swinging London" was in full bloom with the mood of

[33] [8].

the "sixties" very much in evidence. Although not quite swept away by the mood, I did not entirely escape its influence.

While I was in London Jack Schwartz, a friend from my City College days and a colleague at NYU, was working to found a computer science department in the Courant Institute. I was pressed by the Courant Institute to become part of the new department. I accepted but not with alacrity. Among other issues, it meant abandoning my logic group in the Bronx. In fact, Fred Ficken, the amiable Chair at the Bronx campus was about to retire and the applied mathematician Joe Keller had agreed to take on this role with the intention of making the Bronx campus a bastion of applied mathematics. So my group didn't have much future. What neither Joe nor I knew was that the entire Bronx campus would be shut down a few years later because NYU found itself in a financial crunch.

I had been back in New York only a few months when I received an exciting phone call from Jack Schwartz. A young Russian had used the Fibonacci numbers to prove JR! Hilbert's tenth problem was finally settled! I had been half-jokingly predicting that JR would be proved by a clever young Russian, and, lo and behold, he had appeared. (I met the 22 year old Yuri Matiyasevich in person that summer at the International Congress in Nice.) After getting the news I quickly phoned Julia, and about a week later, I received from her John McCarthy's notes on a lecture he had just heard in Novosibirsk on the proof. It was great fun to work out the details of Yuri's lovely proof from the brief outline I had. I saw that the properties of the Fibonacci numbers that Yuri had used in his proof had analogues for the Pell equation solutions with which Julia had worked and I enjoyed recasting the proof in those terms. I also wrote a short paper in which I derived some consequences of the new result in connection with the number of solutions of a Diophantine equation.[34]

To make the proof of the unsolvability of Hilbert's tenth problem widely accessible, I wrote a survey article for the *American Mathematical Monthly* which was later reprinted as an appendix to the Dover edition of *Computability & Unsolvability* [11]. In addition, I collaborated with Reuben Hersh on a popular article on the subject for the *Scientific American* [26]. Suddenly awards were showered on me. For the *Monthly* article I received the Leroy P. Steele prize from the American Mathematical Society and the Lester R. Ford prize from the Mathematical Association of America, and for the Scientific American article Reuben Hersh and I shared the Chauvenet prize, also from the Mathematical Association of America. I was also invited by the Association to give the Hedrick lectures for 1976.

In May 1974, the Society sponsored a symposium on mathematical problems arising from the Hilbert problems, and of course, Yuri was invited to speak on the tenth. But he was unable to get permission from the Soviet authorities to come. So, Julia was invited instead, and she agreed on condition that I be invited as well to introduce her. When it came to writing a paper for the proceedings of the symposium, we agreed that it should be by the three of us. Yuri's contribution faced the bureaucratic obstacle that any of his draft documents had to be approved before being sent abroad.

[34] [9, 10].

But there was no such problem with letters. So he would send letters to Julia on his parts of the article generally beginning,

> Dear Julia,
> Today I would like to write about ....

One of his topics was "Famous Problems". The idea was that the same techniques that had been used to show that there is no algorithm for the tenth problem could also be used to show that various well-known problems were equivalent to the non-existence of solutions to certain Diophantine equations. One of Yuri's letters did this for the famous Riemann Hypothesis, an assertion about the complex zeros of the function $\zeta(s) = \sum_{n=1}^{\infty} \frac{1}{n^s}$ which remains unproved although it has important implications for the theory of prime numbers. The Hypothesis can be expressed in terms of the values of certain contour integrals, and Yuri's technique was to approximate these integrals by sums in a straightforward way. It was done in a very workman-like way, but it seemed very inelegant to me. I went to my colleague Harold Shapiro, an expert in analytic number theory, and he told me what to do. Julia was so pleased by the result that she sent me a note I kept on my bulletin board for years saying "I like your reduction of RH immensely".[35]

My joint appointment in mathematics and the new Computer Science Department defined my new situation at Courant when I returned from London. I had been flirting with computers and computer science for years. But now I had come out of the closet and identified myself as a computer scientist. The new Computer Science Department was developing its own culture and clashes with the mathematicians at Courant were not infrequent. It didn't help that among the mathematicians were some outstanding scientists that thoroughly outclassed our young department. To begin with our graduate students took the same exams as the mathematics students, and hiring was done by the same committee that hired mathematicians. The evolution towards autonomy was slow and painful. In the spring of 1973, two applied mathematicians and I constituted a hiring committee for computer science. The mathematicians were both heavily involved with scientific computing, but neither had any real appreciation or understanding of computer science as an autonomous discipline. I remember all too well a particular tense lunch meeting on a Friday in June of that year in which possible appointments were discussed in an atmosphere I did not find friendly. When I left the meeting I became aware of a sensation like a brick placed on my chest. I continued to experience this disagreeable sensation through the weekend and finally entered the hospital where a myocardial infarction was diagnosed. Although in retrospect I had done plenty to bring this about by poor diet and lack of exercise, I have always thought that that disagreeable meeting played a precipitating role. Before my heart attack, I had been an enthusiastic New Yorker even when in exile; but now I began to yearn to live someplace where I could have a rich professional life without the tension that I found in everyday life in New York.

---

[35] [35].

Over the years I continued to have doctoral students from both the Mathematics Department and the Computer Science Department. I certainly taught hard-core computer science courses, beginning programming and data structures, and on the graduate level, theory of computation, logic programming, and artificial intelligence. In the early years I could also count on teaching mathematical logic, set theory, and even nonstandard analysis. Unfortunately for me this flexibility gradually vanished, a fact that contributed to my decison to retire from NYU in 1996. But this is getting ahead of the story.

In the 1970 s the Italian universities were still not offering Ph.D. programs. Having received their bachelor degree, graduates were entitled to use the title "Dottore". The C.N.R., the agency of the Italian government involved with scientific research, concerned to do something about the inadequate training young Italian mathematicians received, set up summer programs in which they were exposed to graduate level courses. I was invited to give such a course on Computability in the lovely town of Perugia during the summer of 1975. This was the beginning of a connection between Italian computer science and the Courant Institute. A number of my students from that course became graduate students at NYU. Two in particular, Alfredo Ferro and Eugenio Omodeo, obtained Ph.D.'s in computer science from NYU. Ferro went back to his home town of Catania in Sicily where he started a computer science program at the university there, and sent his own students back to NYU. The relationship continues to be active.

As an undergraduate I had tried briefly to rehabilitate Leibniz's use of infinitesimal quantities as a foundation for calculus. It was easy enough to construct algebraic structures containing the real numbers as well as infinitesimals; the problem that baffled me was how to define the elementary functions such as *sine* and log on such structures. It was therefore with great excitement and pleasure that I heard Abraham Robinson's address before the Association for Symbolic Logic towards the end of 1961 in which he provided an elegant solution to this problem using techniques that he dubbed *nonstandard analysis*. Some years later together with Melvin Hausner, my roommate at Princeton and now a colleague, I started an informal seminar on the subject. We had available Robinson's treatise and some rather elegant lecture notes by Machover and Hirschfeld. Hausner was inspired to apply the technique to prove the existence of Haar measure. Reuben Hersh and I wrote a popular article on nonstandard analysis, also for the *Scientific American.*[36] Nonstandard analysis really tickled my fancy. As I wrote in the flush of enthusiasm:

> It is a great historical irony that the very methods of mathematical logic that developed (at least in part) out of the drive toward absolute rigor in analysis have provided what is necessary to justify the once disreputable method of infinitesimals. Perhaps indeed, enthusiasm for nonstandard methods is not unrelated to the well-known pleasures of the illicit. But far more, this enthusiasm is the result of the mathematical simplicity, elegance, and beauty of these methods and their far-reaching application.

---

[36]Actually, as I remember it, we worked on that article and the one on Hilbert's tenth problem for which we received the Chauvenet prize pretty much at the same time. The one on nonstandard analysis appeared in 1972 [25], a year before the prize-winning article.

I taught nonstandard analysis at Courant and benefited from class notes prepared by my student Barry Jacobs. In the summer of 1971, I taught it again at the University of British Columbia. Finally I wrote a book (the quotation above is from the preface).[37]

For the academic year 1976–77, I was able to go on sabbatical leave. I had spent two summers in Berkeley and was eager to try a whole year. John McCarthy (who had been a fellow student at Princeton) hired me to work for the month of July at his Artificial Intelligence Laboratory at Stanford University. I loved the atmosphere of play that John had fostered. The terminals that were everywhere proclaimed "Take me, I'm yours", when not in use. I was encouraged to work with the FOL proof checker recently developed by Richard Weyhrauch. Using this system, I developed a complete formal proof of the pigeon-hole principle from axioms for set theory. I found it neat to be able to sit at a keyboard and actually develop a complete formal proof, but I was irritated by the need to pass through many painstaking tiny steps to justify inferences that were quite obvious. FOL formalized a "natural deduction" version of **F**irst **O**rder **L**ogic. The standard paradigm for carrying out inferences was to strip quantifiers, apply propositional calculus, and replace quantifiers. I realized that from the viewpoint of Herbrand proofs, each of these mini-deductions could be carried out using no more than one substitution instance of each clause. I decided that this very possibility provided a reasonable characterization of what it means for an inference to be *obvious*. Using the LISP source code for the linked-conjunct theorem prover that had been developed at Bell LabsBell Labs, a Stanford undergraduate successfully implemented an "obvious" facility as an add-on to FOL. I found that having this facility available cut the length of my proof of the pigeon-hole principle by a factor of 10. This work was described at the Seventh Joint International Congress on Artificial Intelligence held in Vancouver in 1981 and reported in the Proceedings of that conference [15].

During the 1976–77 academic year, it was a great pleasure to be able to interact with the Berkeley logic group and especially with Julia Robinson. We worked on the analogue of Hilbert's tenth problem for strings under concatenation, but didn't make much progress. It had at one time been thought that proving this problem unsolvable would be the way to obtain the desired unsolvability result for the Diophantine problem. Julia guessed that the string problem was actually decidable, and she turned out to be right as we learned when we got word of Makanin's positive solution of the problem. At Berkeley that year, I taught two trimester courses, an undergraduate computability theory course for Computer Science and a graduate course in nonstandard analysis for Mathematics. For the nonstandard analysis course, I was able to use my newly published book as a text. It was a class of about thirty students, and a little intimidating. It was clear to me that among these Berkeley educated students were a number who were far better versed in model theory (the underlying basis for nonstandard analysis) than I.

Ever since my days with Hilary Putnam, I have had a continuing interest in the foundations of quantum mechanics. A preprint I received from the logician Gaisi Takeuti caught my attention as having important ramifications for quantum theory.

---

[37][12].

This paper explored Boolean-valued models of set theory using algebras of projections on a Hilbert space. Boolean-valued models (in which the "truth value" of a sentence can be any element of a given complete Boolean algebra, rather than being restricted to the usual two element algebra consisting of {`true, false`}), had been studied as an alternative way to view Paul Cohen's forcing technique for obtaining independence results in set theory. What Takeuti found was that the real numbers of his models were in effect just the self-adjoint operators on the underlying Hilbert space. Since a key element in "quantizing" a classical theory is the representation of "observables" by such operators, I felt that the connection was surely no coincidence. I wrote a short paper about the application of Takeuti's mathematics to quantum mechanics, and I was very pleased when it was published in the *International Journal of Theoretical Physics* [13].

I worked at John McCarthy's AI lab again, and this time John asked me to think about some questions involving so-called non-monotonic reasoning. I wrote a pair of short notes which John later arranged to be combined for publication in *Artificial Intelligence* [14].

I spent the academic year 1978–79 as a Visiting Professor at the Santa Barbara campus of the University of California. There was some mutual interest in a permanent appointment, but it all faded away as a consequence of wrangling over the status of the campus's two computer science programs: the one in the Mathematics Department and the one in Electrical Engineering. On my return to New York, I met a new faculty member Elaine Weyuker with whom I was to find a number of shared interests. Although trained as a theoretical computer scientist, she had moved into the turbulent field of software testing. Of course all software must be tested before it is released. Often, in practice this testing phase is ended simply because some deadline is reached or because funding runs out. From an academic point of view, the field invites attention to the problem of finding a more rational basis for the testing process. Elaine and I wrote two papers attempting to provide an explication for the notion of test data adequacy.[38]

I had been teaching theory of computation for many years, and had developed lecture notes for some of the topics covered. For a long time I had wanted to produce a book based on my course, but had never found the time or energy to complete the task. Elaine came to the rescue adding the needed critical dose of energy. In addition, she produced lots of exercises, and tested some of the material with undergraduates. The book was published and was sufficiently successful that we were asked to update the book for a second edition. Neither of us being willing to undertake this, we coaxed Ron Sigal, who had written a doctoral dissertation under my supervision, to join the team as a third author largely in charge of the revision [30].

The CADE (Conference on Automated Deduction) meetings were occurring annually devoted to theoretical and practical aspects of logical deduction by computer. The organizers of the February 1979 CADE meeting in Austin, realizing that year was the centennial of Frege's *Begriffsschrift* in which the rules of quantificational logic were first presented, thought that it would be appropriate to have a lecture that

[38] [31, 32].

would place their field in a proper historical context. Their invitation to me to give such a lecture fundamentally changed the direction of my work. I found that trying to trace the path from ideas and concepts developed by logicians, sometimes centuries ago, to their embodiment in software and hardware was endlessly fascinating. Since then I have devoted a great deal of time and energy to these questions. I've published a number of articles and a book and also given many lectures with a historical flavor.[39] For 1983–84, when I was again on sabbatical leave, I received support from the Guggenheim Foundation for this work. One key figure whose ideas I tried to promulgate was my old teacher Emil L. Post. I lectured on his work on a number of occasions including one talk at Erlangen in Germany. It was very much a labor of love to edit his collected works.[40]

For the two academic years 1988–90, I was Chair of the Computer Science Department at NYU. I had always felt that I would not be happy in an administrative position, and this experience did nothing to change my mind. I would have been hopelessly swamped without the help of the department's capable and ultra-conscientious adminstrative assistant Rosemary Amico. The NYU central administration had been increasingly unhappy with the fact that the Courant Institute as a whole was running an increasing deficit each year. At the same time, the CS department was encouraged to improve its national standing among research-oriented CS departments. The administration was said to be surprised and pleased that our department was rated among the top twenty in the nation, and we were urged to produce a plan showing how we could move up to the top ten. Assuming that the central administration understood that this would require their providing additional resources, the department prepared an ambitious plan calling for expansion in a number of directions. The central administration did not deign to reply.

After my term of office was over, it was time for another sabbatical leave. The fall 1990 semester was spent in Europe. Our first stop was Heidelberg where I lectured at the local IBM facility and at a logic meeting at the university. Next, a series of lectures on Hilbert's tenth problem at a conference in Cortona in the north of Italy. Then a month visiting Alfredo Ferro at the University of Catania in Sicily. The fall semester was completed with a stay at the University of Patras in Greece sponsored by Paul Spirakis, and we were home in time for Christmas. I had completed an important article the day before our departure from Patras, had printed it, and left a copy on a secretary's desk with a note asking her to make copies. Our departure was to be by car ferry to Italy scheduled for the following midnight. The next morning I arrived on the campus to discover that students had occupied the building where I'd been working, and were permitting no one to enter. This was dismaying. I had no copy of my article; it was stored in a VAX that I couldn't access, and the only hard copy was on the secretary's desk. At this point a faculty member, who had become a friend, appeared and, ascertaining the problem, spoke briefly to one of the students. Evidently a deal was struck. I got out my key to the massive doors locking the computer science section, and the three of us entered. There was the hard

[39] [16–21].

[40] [38].

copy of my article where I had left it and a copying machine, and I soon had several copies one of which my friend kept to send to the editor in Germany. Meanwhile the student helped himself to the copier to duplicate a handwritten document, doubtless a manifesto. We left and I was permitted to lock up.

Virginia and I took our friend and his wife to dinner that evening. Finally I asked him what he had said to the student that turned the trick. His reply was not at all what I had expected. "I reminded him that he was applying to Courant, and told him that you are the Chairman." Our ship due to sail at midnight didn't actually leave before 3 a.m. It turned out that the stabilizers were not functioning, and the voyage to Ancona took a day longer than scheduled with me being seasick most of the way. We drove to Paris in time for our flight back to New York. But our stay in New York was very short. Over the years Virginia had accompanied me on many trips. Now it was my turn to accompany her. Virginia had become an artist with an international reputation. She is particularly adept at researching and mastering traditional textile techniques and using them to make works of art. For 1991 she had been awarded a three month Indo-American Fellowship[41] to study textiles in India. Of course I came along.

Our scheduled departure date was January 15. That was also the date on which President Bush's ultimatum to Saddam Hussein demanding that his forces leave Kuwait was expiring. Friends urged us to abandon our travel plans at such an uncertain time, but we decided to go ahead. After a delay caused by a bomb scare at Kennedy airport, we arrived in New Delhi to learn that bombs were dropping on Baghdad. Given the chaos just outside airports in India with throngs insistently offering their services we were delighted to be met by representatives of the American Institute for India Studies (AIIS) who took us to their guest house. The next morning we found other American fellowship recipients in a state of panic. The U.S. State Department had issued an advisory to the effect that non-essential American personnel leave India at once. Most of the others agreed to postpone their fellowship periods and left. We decided to remain. So we were in India for the entire duration of the Gulf War. In an odd way, the situation was advantageous for us. The lack of tourists meant that it was easy to get reservations and services, and the AIIS guest house was always available. The U.S. embassy, which had been transformed into a virtual fortress, was the target of virtually daily vituperative demonstrations by militant Muslim groups, but we ourselves had no problems.

The textiles Virginia was most eager to study were in the state of Orissa, one of the poorest states in India, just south of Calcutta, and we spent most of our time there. I had a new job: I was Virginia's camera man. My job was to use the video camera to record textile processes; we accumulated 12 h of raw footage. There was a week-long tour of some of the the small villages of Orissa, where often, there were no hotels even minimally acceptable by U.S. standards. In one village, we were put up in the guest house of a cotton spinning factory.

In India the contrasts between the best and the worst is enormous. We saw people lying on the sidewalks of Calcutta waiting to die, and we had lunch with a matriarch

---

[41] These fellowships are administered by the CIES, the same office that manages Fullbright awards.

whose huge family estate is guarded by a private police force and whose foot was kissed by her servants when she permitted them to take the lunch leftovers home to their families. The best educational and scientific research institutions are first-rate by any standard. On my previous sabbatical, I had spent a month as the guest of the Tata Institute of Fundamental Research (TIFR) in Bombay, and I was able to visit them again briefly this time. In addition I lectured at the Indian Institute of Technology (IIT) in New Delhi, an outstanding school whose entrance examinations in mathematics are quite formidable. (At IIT and TIFR I was able to collect my email.) But I also lectured at colleges, allegedly institutions of higher learning, that were sadly weak.

On our way back to New York from India, we stopped in Europe. I spent a week at the University of Udine as the guest of Professor Franco Parlamento who had been a student in my Computability course in Perugia two decades earlier. Then we went to the wonderful mathematical research institute at Oberwolfach in Germany, an institute that started its successful life as an effort by German mathematicians to save their talented young people from becoming cannon fodder during the second world war. There are week-long conferences through the year on a great variety of mathematical subjects. On this occasion, it was on automatic theorem proving organized by Woody Bledsoe and Michael Richter, and a follow-up to a similar meeting 15 years earlier.

Back in New York, and back to teaching, I was approaching that sixty-fourth birthday the Beatles had sung about, and beginning to wonder how I wanted to spend the rest of my life. The things that really interested me seem to be of less and less importance to my colleagues. I had my very own course called *Arithmetic Undecidability*; in a whirlwind semester I covered the elements of first order logic through the Gödel completeness theorem, Hilbert's tenth problem, and the essential undecidability of Peano arithmetic. I taught it for the last time in the spring 1993 semester, and was rebuffed in my request to teach it again. I taught the introductory programming course for computer science majors, and indeed supervised the sections taught by others, for three successive years. I love to program, and at first, I enjoyed these courses. But after a while, I did ask myself: do I really want to be teaching Pascal to classes of 60 students not all of whom are especially receptive, at this stage of my life? A triple coronary bypass operation in January 1994 brought matters to a head. The operation was very successful, but it certainly forced me to face my mortality. In short I decided to investigate retirement possibilities. May 17, 1996 was "Martin Davis Day" at the Courant Institute. Organized by my old friends Jack Schwartz and Ricky Pollack, there were eight speakers: two from Italy (my student Eugenio Omodeo, a Perugia veteran, and Mimmo Cantone, one of Alfredo Ferro's protégés), my first two students Bob Di Paola and Don Loveland, Hilary Putnam, Elaine Weyuker, Ron Sigal and my college chum John Stachel.

My study is in a house in the Berkeley hills, and I am enjoying the dazzling reflection of the late afternoon sun in San Francisco Bay. My older son, his wife and their four children are here in Berkeley. Virginia and I still go dancing on Tuesdays. And as Virginia likes to say, retiring gave me time to work. I have been a "Visiting

Scholar" at the university here where Alfred Tarski had developed a world-class logic group.

Although I had been involved with Turing's abstract machines in my research and teaching, and considered myself a computer scientist as well as a mathematician, I was slow to appreciate Alan Turing's pioneering role in both the theoretical and practical side of computer science. As I came to understand that the circuits of a computer "embody the distilled insights of a remarkable collection of logicians, developed over centuries", I became eager to bring that point of view to the attention of the public. Publication of Turing's Ace report (in which he had presented the design of a computer proposed to be built at the British National Physics Laboratory) and his 1947 address to the London Mathematical Society (in which he explicitly stressed the connection of general purpose digital computers to his abstract machines) both of which had languished in obscurity, made it possible to see Turing's vision for the farsighted anticipation that it was. My essay [19] was an attempt to explain the importance of Turing's role as a computer pioneer as well as the extent to which his work leaned on the accomplishments of generations of logicians. With my retirement I could devote myself to the project of expanding this essay into a book for the general educated public. My "Universal Computer" [23, 24] was published in 2000. In it I was particularly eager to emphasize the importance of ideas being pursued for their own sake without necessarily expecting the immediate practical payoff that nowadays is generally sought.

In the early years of the new millennium, I found myself devoting considerable effort to debunking foolish claims by scholars who surely should have known better. With the energy and self-assurance that had been shown by designers of perpetual motion machines, a computer scientist, a philosopher, and a physicist, each claimed to have found a way to circumvent the limitations on what can be computed that emerged from the work of Church, Kleene, Post, and Turing in the 1930s. This took on the aspect of a movement called *hypercomputation*. I spoke at meetings and wrote a number of articles. Here is my abstract for a talk I gave at a special session on hypercomputation at a meeting of the American Mathematical Society in San Francisco in 2003:

> Hava Siegelmann claims that her neural nets go "beyond the Turing limit". Jack Copeland proposes to harness Turing's notion of "oracle" for a similar purpose. Tien D. Kieu proposes a quantum mechanical algorithm for Hilbert's 10th problem, known to be unsolvable. It will be shown that in each case the results depend on the presumed physical availability of infinite precision real numbers. In the first two examples, uncomputable outputs are obtained only by slipping uncomputable inputs into the formalisms. Kieu's method depends on a physically unrealizable form of adiabatic cooling.

2002 would have been Turing's 90th birth year, and I spoke at a conference in his honor in Lausanne. Jack Copeland (one of the three hypercomputationalists just mentioned) and Andrew Hodges (Turing's wonderful biographer) were also speakers. I spoke before and Andrew after Jack Copeland. Copeland did not come off unscathed.

A decade later, Turing's centenary, there were celebratory conferences devoted to Turing all over the world. I spoke at a math meeting in San Francisco in a panel with Andrew Hodges, and at meetings in Boston and Florida. In addition I spoke in England, Italy, and Peru. I was at a banquet at King's College, Cambridge where Turing's nephew spoke. I crossed the Atlantic six times, three going and three returning.

I have always thought of myself as rather lazy and very lucky. I've had a wonderful marriage—Virginia and I recently celebrated our 63rd anniversary, and we have six grandchildren. Although I had a heart attack when I was only 45, I'm still alive at 86 and well enough that Virginia and I go dancing once a week. My conjecture that anything computable has a Diophantine definition, strongly doubted by most experts, turned out to be correct. And amazingly, there is this book being edited by two of my former students with contributions by a remarkable group of scholars [34].

References

1. Davis, M. (1956). A note on universal turing machines. *Automata Studies*. In C.E. Shannon & J. McCarthy (Eds.), *Annals of Mathematics Studies* (pp. 167–175). Princeton University Press
2. Davis, M. (1957). The definition of universal turing machine. *Proceedings of the American Mathematical Society*, *8*, 1125–1126.
3. Davis, M. (1958). *Computability and unsolvability*. New York: McGraw-Hill. (Reprinted with an additional appendix, Dover 1983)
4. Davis, M. (1963). Eliminating the irrelevant from mechanical proofs. *Proceedings of Symposia in Applied Mathematics,15*, 15–30. (Reprinted from [43], pp. 315–330).
5. Davis, M. (1963). Extensions and corollaries of recent work on Hilbert's tenth problem. *Illinois Journal of Mathematics*, *7*, 246–250.
6. Davis, M. (Ed.). (1965). *The undecidable*. Raven Press. (Reprinted from Dover 2004).
7. Davis, M. (1967). *First course in functional analysis*. Gordon and Breach. (Reprinted from Dover 2013).
8. Davis, M. (1968). One equation to rule them all. *Transactions of the New York Academy of Sciences, Sec., II*(30), 766–773.
9. Davis, M. (1971). An explicit Diophantine definition of the exponential function. *Communications on Pure and Applied Mathematics*, *24*, 137–145.
10. Davis, M. (1972). On the number of solutions of Diophantine equations. *Proceedings of the American Mathematical Society*, *35*, 552–554.
11. Davis, M. (1973). Hilbert's tenth problem is unsolvable. *American Mathematical Monthly,80*, 233–269. (Reprinted as an appendix to the Dover edition of [3]).
12. Davis, M. (1977). *Applied nonstandard analysis*. Interscience-Wiley. (Reprinted from Dover 2005).
13. Davis, M. (1977). A relativity principle in quantum mechanics. *International Journal of Theoretical Physics*, *16*, 867–874.
14. Davis, M. (1980). The mathematics of non-monotonic reasoning. *Artificial Intelligence*, *13*, 73–80.
15. Davis, M. (1981). Obvious logical inferences. In *Proceedings of the Seventh Joint International Congress on Artificial Intelligence* (pp. 530–531).
16. Davis, M. (1982). Why Gödel didn't have Church's thesis. *Information and Control*, *54*, 3–24.
17. Davis, M. (1983). The prehistory and early history of automated deduction. In J. Siekmann & G. Wrightson (Eds.), *Automation of reasoning* (Vol. 1, pp. 1–28). Springer.
18. Davis, M. (1987). Mathematical logic and the origin of modern computers. *Studies in the History of Mathematics*, pp. 137–165. Mathematical Association of America. (Reprinted from

The universal turing machine—a half-century survey, pp. 149–174, by R. Herken, Ed., 1988, Hamburg, Berlin: Verlag Kemmerer & Unverzagt, Oxford University Press.)
19. Davis, M. (1988). Influences of mathematical logic on computer science. In R. Herken (Ed.), *The universal turing machine–a half-century survey* (pp. 315–326). Hamburg, Berlin: Verlag Kemmerer & Unverzagt, Oxford University Press.
20. Davis, M. (1989). Emil post's contributions to computer science. In Proceedings Fourth Annual Symposium (Ed.), *on Logic in Computer Science* (pp. 134–136). Washington, D.C.: IEEE Computer Society Press.
21. Davis, M. (1995). American logic in the 1920s. *Bulletin of Symbolic Logic, 1*, 273–278.
22. Davis, M. (1999). From logic to computer science and back. In C. S. Calude (Ed.), *People and ideas in theoretical computer science* (pp. 53–85). Springer.
23. Davis, M. (2000). *The universal computer: The road from Leibniz to Turing*. W.W. Norton. Turing Centenary Edition, CRC Press, Taylor & Francis 2012.
24. Davis, M. (2001). *Engines of logic: Mathematicians and the origin of the computer*. W.W. Norton. [Paperpack edition of *The Universal Computer*]
25. Davis, M., & Hersh, R. (1972). Nonstandard analysis. *Scientific American, 226*, 78–86.
26. Davis, M., & Hersh, R. (1973). Hilbert's tenth problem. *Scientific American,229*, 84–91 (Reprinted in J.C. Abbott (Ed.), *The Chauvenet papers, 2*, 555–571. Math. Assoc. America, 1978).
27. Davis, M., & Putnam, H. (1958). Reductions of Hilbert's tenth problem. *Journal of Symbolic Logic, 23*, 183–187.
28. Davis, M., & Putnam, H. (1960). A computing procedure for quantification theory. *Journal of the Association for Computing Machinery, 7*, 201–215. (Reprinted from [43], pp. 125–139).
29. Davis, M., & Putnam, H. (1963). Diophantine sets over polynomial rings. *Illinois Journal of Mathematics, 7*, 251–255.
30. Davis, M., & Weyuker, E. J. (1983). *Computability, complexity, and languages*. Second edition with Ron Sigal: Academic Press. 1994.
31. Davis, M., & Weyuker, E. J. (1983). A formal notion of program-based test data adequacy. *Information and Control, 56*, 52–71.
32. Davis, M., & Weyuker, E. J. (1988). Metric space based test data adequacy criteria. *The Computer Journal, 31*, 17–24.
33. Davis, M., Putnam, H., & Robinson, J. (1961). The decision problem for exponential Diophantine equations. *Annals of Mathematics, 74*, 425–436.
34. Davis, M., Logemann, G., & Loveland, D. (1962). A machine program for theorem proving. *Communications of the Association for Computing Machinery, 5*, 394–397. (Reprinted from [43], pp. 267–270).
35. Davis, M., Matijasevic, Y., & Robinson, J. (1976). Hilbert's tenth problem: Diophantine equations: positive aspects of a negative solution. *Proceedings of Symposia in Pure Mathematics, 28*, 323–378.
36. Green, B., & Tao, T. (2008). The primes contain arbitrarily long arithmetic progressions. *Annals of Mathematics, 167*, 481–547.
37. Kleene, S. C. (1943). Recursive predicates and quantifiers. *Transactions of the American Mathematical Society,53*, 41–73. (Reprinted from [11], pp. 254–287).
38. Post, E. L. (1994). *Solvability, provability, definability: The collected works of Emil L. Post* ed. by M. Davis with a biographical introduction. Boston, Basel and Berlin: Birkhäuser
39. Prawitz, D. (1960). *An improved proof procedure*. (Reprinted from [43] with an additional comment by the author, pp. 162–201).
40. Reid, C. (1996). *Julia: A life in mathematics*. Washington, D.C.: Mathematical Association of America.
41. Robinson, J. A. (1963). Theorem proving on the computer. *Journal of the Association for Computing Machinery,10*. (Reprinted from [43], pp. 372–383).
42. Robinson, J. A. (1965). A machine-oriented logic based on the resolution principle. *Journal of the Association for Computing Machinery,12*. (Reprinted from [43], pp. 397–415).
43. Siekmann, J., & Wrightson, G. (Eds.). (1983). *Automation of reasoning 1: Classical papers on computational logic 1957–1966*. Berlin: Springer.

Chapter 2
Martin Davis and Hilbert's Tenth Problem

Yuri Matiyasevich

Abstract The paper presents the history of the negative solution of Hilbert's tenth problem, the role played in it by Martin Davis, consequent modifications of the original proof of DPRM-theorem, its improvements and applications, and a new (2010) conjecture of Martin Davis.

Keywords Computability · Hilbert's Tenth Problem · DPRM-theorem

2.1 The Problem

Martin Davis will stay forever in the history of mathematics and computer science as a major contributor to the solution of Hilbert's tenth problem.

This was one among 23 problems which David Hilbert stated in his famous paper "Mathematical Problems" [18] delivered at the Second International Congress of Mathematicians. This meeting took place in Paris in 1900, on the turn of the century. These problems were, in Hilbert's opinion, among the most important problems that the passing nineteenth century was leaving open to the pending twentieth century.

The section of [18] devoted to the Tenth Problem is so short that it can be reproduced here in full:

> 10. DETERMINATION OF THE SOLVABILITY OF A DIOPHANTINE EQUATION
>
> Given a Diophantine equation with any number of unknown quantities and with rational integral numerical coefficients: *To devise a process according to which it can be determined by a finite number of operations whether the equation is solvable in rational integers.*

Equations from the statement of the problem have the form

$$P(x_1, x_2, \ldots, x_n) = 0 \tag{2.1}$$

Y. Matiyasevich (✉)
Laboratory of Mathematical Logic, St. Petersburg Department of V.A. Steklov Institute of Mathematics (POMI), Russian Academy of Sciences, 27 Fontanka,
St. petersburg, Russia 191023
e-mail: yumat@pdmi.ras.ru

E.G. Omodeo and A. Policriti (eds.), *Martin Davis on Computability, Computational Logic, and Mathematical Foundations*,
Outstanding Contributions to Logic 10, DOI 10.1007/978-3-319-41842-1_2

where P is a polynomial with integer coefficients. The equations are named after the Greek mathematician Diophantus who lived, most likely, in the 3rd century A.D.

The Tenth problem is the only one of the 23 Hilbert's problems that is (in today's terminology) a *decision problem*, i.e., a problem consisting of infinitely many *individual problems* each of which requires a definite answer: YES or NO. The heart of a decision problem is the requirement to find a single method that will give an answer to any individual subproblem.

Since Diophantus's time, number-theorists have found solutions for a large amount of Diophantine equations, and also they have established the unsolvability of a lot of other equations. Unfortunately, for different classes of equations, and often even for different individual equations, it was necessary to invent specific methods. In his tenth problem, Hilbert asks for a *universal method* for deciding the solvability of all Diophantine equations.

A decision problem can be solved in a positive or in a negative sense, that is, either by discovering a required algorithm or by showing that none exists. Hilbert foresaw the possibility of negative solutions to some mathematical problems, in [18] he wrote:

> Occasionally it happens that we seek the solution under insufficient hypotheses or in an incorrect sense, and for this reason do not succeed. The problem then arises: to show the impossibility of the solution under the given hypotheses, or in the sense contemplated. Such proofs of impossibility were effected by the ancients, for instance when they showed that the ratio of the hypotenuse to the side of an isosceles triangle is irrational. In later mathematics, the question as to the impossibility of certain solutions plays a preëminent part, and we perceive in this way that old and difficult problems, such as the proof of the axiom of parallels, the squaring of circle, or the solution of equations of the fifth degree by radicals have finally found fully satisfactory and rigorous solutions, although in another sense than that originally intended. It is probably this important fact along with other philosophical reasons that gives rise to conviction (which every mathematician shares, but which no one has as yet supported by a proof) that every definite mathematical problem must necessary be susceptible of an exact settlement, either in the form of an actual answer to the question asked, or by the proof of the impossibility of its solution and therewith the necessary failure of all attempts.

But in 1900 it was impossible even to state rigourously what would constitute a negative solution of Hilbert's tenth problem. The general mathematical notion of algorithm was developed by Alonzo Church, Kurt Gödel, Alan Turing, Emil Post, and other logicians only three decades after Hilbert's lecture [18].

The appearance of the general notion of algorithms gave the possibility to establish non-existence of algorithms for particular decision problems, and soon such *undecidable problems* were actually found. But these results didn't much impress "pure mathematicians" because the first discovered undecidable problems were from the realm of mathematical logic and the just emerging computer science.

The situation changed in 1947 when two mathematician, Andrei Andreevich Markov [28] in the USSR, and Emil Post [45] in the USA, independently proved that there is no algorithm for so called *Thue problem*. This problem was posed by Alex Thue [58] in 1914, much before the development of the general notion of an algorithm. Thue asked for a method for deciding, given a finitely presented semigroup and two elements from it, whether the defining relations imply the equality of these two elements or not. Thus Thue problem, known also as *word problem for semigroups*, became the very first decision problem, born in mathematics proper and proved to be undecidable.

After the success with Thue problem, researchers were inspired to establish the undecidability of other long standing open mathematical problems. In particular, both Markov and Post were interested in Hilbert's tenth problem. Already in 1944 Post wrote in [44] that Hilbert's tenth problem "begs for an unsolvability proof". Post had a student to whom this statement produced great impression and he decided to tackle the problem.

The name of this student was Martin Davis.

2.2 Martin Davis Conjecture

2.2.1 Statement and Corollaries

Very soon Martin Davis came to a conjecture, first announced in [3], that would imply the undecidability of Hilbert's tenth problem. To be able to state this conjecture, we need to introduce a bit more terminology.

Hilbert asked for solving Diophantine equations with *numerical* coefficients. One can also consider equation with *symbolic* coefficient, that is, equations with parameters. Such an equation has the form

$$P(a_1, \ldots, a_m, x_1, x_2, \ldots, x_n) = 0 \tag{2.2}$$

similar to (2.1) but now the variables are split into two groups: *parameters* $a_1, \ldots, a_m$, and *unknowns* $x_1, \ldots, x_n$.

As another (minor technical) deviation from Hilbert's statement of the problem, we will assume that both the parameters and the unknowns range over the natural numbers; following the tradition of mathematical logic, we will consider 0 as a natural number.

For some choice of the values of the parameters the Eq. (2.2) may have a solution in the unknowns, and for another choice may have no solution. We can consider the set $\mathscr{M}$ of all m-tuples of the values of the parameters for which the Eq. (2.2) has a solution in the unknowns:

$$\langle a_1, \ldots, a_m \rangle \in \mathscr{M} \iff \exists x_1 \ldots x_n \big[P(a_1, \ldots, a_m, x_1, x_2, \ldots, x_n) = 0 \big]. \tag{2.3}$$

Sets having such a *Diophantine representation* are also named *Diophantine*.

Traditionally, in Number Theory an equation is the primary object, and one is interested in a description of the set of the values of the parameters for which the equation has a solution. Martin Davis, in a sense, reversed the order of things taking sets as primary objects—he decided to give a general characterization of the whole class of all Diophantine sets.

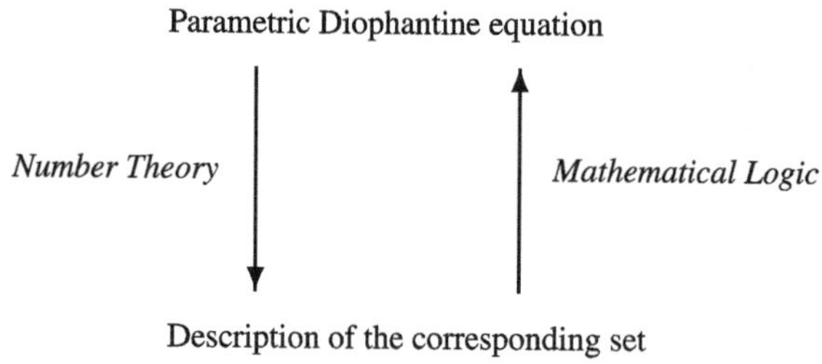

He approached this problem from a computational point of view. All Diophantine sets have one common property: they are *listable*, or, in another terminology, *effectively enumerable*. This means the following. Given a parametric Diophantine equation (2.2), we can start looking over, in some order, all possible $(m+n)$-tuples $\langle a_1, \ldots, a_m, x_1, \ldots, x_n \rangle$, for each such a tuple check if the value of the polynomial P is equal to zero, and if this happens to be the case, write down the tuple $\langle a_1, \ldots, a_m \rangle$ on some list. Sooner or later each tuple from the set (2.3) will appear on this list, maybe many times, and all tuples from the list will belong to this set.

Described above was a rather specific way of listing elements of Diophantine sets. For an arbitrary set to be listable no restriction is imposed on the method for generating its elements, the only requirement is that this should be done by an algorithm. For example, it is evident that the set of prime numbers is listable: it is easy to write a program that would print $2, 3, 5, \ldots$.

Thus, computability theory imposed an *obstacle* for a set to be Diophantine: *if a set is not listable, it cannot be Diophantine*. Martin Davis conjectured that this is the *only* obstacle.

Martin Davis conjecture. *Every listable set is Diophantine.*

I find that this was a rather daring conjecture because it has many corollaries, some of them quite striking.

For example, Martin Davis conjecture implied the existence of a one-parameter Diophantine equation

$$P(a, x_1, x_2, \ldots, x_n) = 0 \tag{2.4}$$

having a solution if and only if a is a prime number. Hilary Putnam noticed in [47] that the same would be true for the equation

$$(x_0 + 1)(1 - P^2(x_0, x_1, x_2, \ldots, x_n)) - 1 = a. \tag{2.5}$$

In other words, *the set of all prime numbers should be exactly the set of all non-negative values of the polynomial from the left-hand side of* (2.5) *assumed for all natural values of* $x_0, \ldots, x_n$. Number-theorists did not believe in such a possibility.

Some other consequences of Davis conjecture will be presented below. Of course, the undecidability of Hilbert's tenth problem is among of them. This is due to the classical fundamental result, the existence of listable sets of natural numbers for which there is no algorithm for recognizing, given a natural number a, whether it belongs to the set or not.

Davis conjecture is much stronger than what would be sufficient for proving the undecidability of Hilbert's tenth problem. Namely, it would suffice to find for any particular undecidable set $\mathcal{M}$ some representation similar to (2.2) with P being replaced by any function which becomes a polynomial in $x_1, \ldots, x_n$ after substituting numerical values for $a_1, \ldots, a_m$. For example, we could allow parameters to appear in the exponents as it was done by Anatolyi Ivanovich Mal'tsev in [27].

2.2.2 The First Step to the Proof

Martin Davis had not much informal evidence in support of his conjecture. Slight support came from the following result announced in [3] and proved in [4–6].

Theorem (Martin Davis). *For every listable set $\mathcal{M}$ there exists a polynomial Q with integer coefficients such that*

$$\langle a_1, \ldots, a_m \rangle \in \mathcal{M} \Longleftrightarrow \exists z \forall y_{\le z} \exists x_1 \ldots x_n \big[Q(a_1, \ldots, a_m, x_1, x_2, \ldots, x_n, y, z) = 0 \big]. \quad (2.6)$$

Representation of type (2.6) became known as *Davis normal form*. They can be considered as an improvement of Kurt Gödel's technique [15] of arithmetization. This technique allowed him to represent any listable set by an arithmetical formula containing, possibly, many universal quantifiers. If all of them are bounded than such an arithmetical formula defines a listable set, and this can be used as another definition of them (this is the content of Theorem 2.7 from Martin Davis dissertation [4]).

2.2.3 A Milestone

In 1959 Martin Davis and Hilary Putnam [13] managed to eliminate the single universal quantifier from Davis normal form but this was not yet a proof of Davis conjecture for two reasons.

First, they were forced to consider a broader class of *exponential Diophantine equations*. They are equations of the form

$$E_{\mathrm{L}}(a_1, \ldots, a_m, x_1, \ldots, x_n) = E_{\mathrm{R}}(a_1, \ldots, a_m, x_1, \ldots, x_n) \quad (2.7)$$

where E_L and E_R are *exponential polynomials*, that is expression constructed by traditional rules from the variables and particular positive integers by addition, multiplication and exponentiation.

Second, the proof given by Martin Davis and Hilary Putnam was conditional: they assumed that *for every k there exist an arithmetical progression of length k consisting of different prime numbers*. In 1959 this hypothesis was considered plausible but it was proved by Ben Green and Terence Green only in 2004 [16]. Thus all what Davis and Putnam needed was to wait for 45 years!

Luckily, they had not to wait for so long. Julia Robinson [49] was able to modify the construction of Davis–Putnam and get an unconditional proof. In 1961 Martin Davis, Hilary Putnam, and Julia Robinson published a joint paper [14] with the following seminal result.

DPR-theorem. *Every listable set $\mathcal{M}$ has an exponential Diophantine representation, i.e., a representation of the form*

$$\langle a_1, \ldots, a_m \rangle \in \mathcal{M} \iff$$
$$\exists x_1 \ldots x_n \big[E_L(a_1, \ldots, a_m, x_1, \ldots, x_n) = E_R(a_1, \ldots, a_m, x_1, \ldots, x_n) \big] \quad (2.8)$$

where E_L and E_R are exponential polynomials.

The elimination of the universal quantifier from Davis normal form immediately gave the undecidability of the counterpart of Hilbert's tenth problem for the broader class of exponential Diophantine equations.

2.2.4 The Last Step

The DPR-theorem was a milestone on the way to proving Davis conjecture. All what remained to do was to prove a particular case of Davis conjecture, namely, to show that exponentiation is Diophantine. Indeed, suppose that we found a particular Diophantine equation

$$A(a, b, c, x_1, \ldots, x_n) = 0 \quad (2.9)$$

which for given values of the parameters a, b, and c has a solution in $x_1, \ldots, x_n$ if and only if $a = b^c$. Using several copies of such an equation, one can easily transform an arbitrary exponential Diophantine equation into a genuine Diophantine equation (with additional unknowns) such that either both equations have solutions or none of them has.

In fact, Julia Robinson was tackling this problem from the beginning of the 1950s. It is instructive to note that her interest was originally stimulated by her teacher, Alfred Tarski, who asked one to prove that the set of all powers of 2 is *not* Diophantine. That is, the intuition of young Martin Davis was opposite to the intuition of venerable Alfred Tarski.

Julia Robinson was not able to construct the required equation (2.9) but she found [48, 50] a number of conditions sufficient for its existence. In particular she proved that in order to construct such an A, it is sufficient to have an equation

$$B(a, b, x_1, \dots, x_m) = 0 \tag{2.10}$$

which defines a relation $J(a, b)$ with the following two properties:

- for any a and b, $J(a, b)$ implies that $a < b^b$;
- for any k, there exist a and b such that $J(a, b)$ and $a > b^k$.

Julia Robinson called a relation J with these two properties *a relation of exponential growth*; Martin Davis named them *Julia Robinson predicates*.

In 1970 I [29] was able to construct the first example of a relation of exponential growth, and it was the last link in the proof of Davis conjecture. Nowadays it is often reffered to as

DPRM-theorem. *Every listable set of m-tuples of natural numbers has a Diophantine representation.*

This theorem implies, in particular, the undecidability of Hilbert's tenth problem: *There is no algorithm for deciding whether a given Diophantine equation has a solution.*

2.3 Further Modifications of Original Proofs

2.3.1 *Pell Equation*

My original construction of a Diophantine relation of exponential growth was based on the study of Fibonacci numbers defined by recurrent relations

$$\varphi_0 = 0, \qquad \varphi_1 = 1, \qquad \varphi_{n+1} = \varphi_n + \varphi_{n-1}, \tag{2.11}$$

while Julia Robinson worked with solutions of the following special kind of *Pell equation*:

$$x^2 - (a^2 - 1)y^2 = 1. \tag{2.12}$$

Solutions of this equation $\langle \chi_0, \psi_0 \rangle, \langle \chi_1, \psi_1 \rangle, \dots, \langle \chi_n, \psi_n \rangle, \dots$ listed in the order of growth, satisfy the recurrence relations

$$\chi_0 = 1, \quad \chi_1 = a, \quad \chi_{n+1} = 2a\chi_n - \chi_{n-1}, \tag{2.13}$$

$$\psi_0 = 0, \quad \psi_1 = 1, \quad \psi_{n+1} = 2a\psi_n - \psi_{n-1}. \tag{2.14}$$

Sequences $\varphi_0, \varphi_1, \ldots$ and $\psi_0, \psi_1, \ldots$ have many similar properties, for example, they grow up exponentially fast. Immediately after the acquaintance with my construction for Fibonacci numbers, Martin Davis gave in [8] a Diophantine definition of the sequence of solutions of the Pell equation (2.12). The freedom in selection of the value of the parameter a allowed Martin Davis to construct a Diophantine definition (2.9) of the exponentiation directly, that is, without using the general method proposed by Julia Robinson starting with an arbitrary Diophantine relation of exponential growth. Today the use of the Pell equation for defining the exponentiation by a Diophantine equation has become a standard.

2.3.2 Eliminating Bounded Universal Quantifier

In [13] the necessity to work with long arithmetical progressions consisting of primes only was due to the usage of a version of Gödel's technique for coding arbitrary long sequences of natural numbers via the Chinese Remainder Theorem. Julia Robinson has managed to replace such progressions by arithmetical progressions composed of pairwise relatively prime numbers having arbitrary large prime factors. Much later, in 1972, using a multiplicative version of Dirichlet principle, I [30] made further modification allowing one to work just with arithmetical progressions of arbitrary big relatively prime numbers.

In [37, Sect. 6.3] I introduced a quite different technique for eliminating bounded universal quantifier based on replacing $\forall y_{\leq z}$ by $\sum_{y=0}^{z}$ with a suitable summand allowing one to find a closed form for the corresponding sum.

2.3.3 Existential Arithmetization

The method for constructing the Davis normal form (2.6) presented in [5] starts with a representation of the set $\mathscr{M}$ by an arithmetical formula in prenex form with any number of bounded universal quantifiers constructed, for example, by Gödel's technique. Two tools are repeatedly applied to such a formula, one tool allowing us to glue two consecutive existential or universal quantifiers, and the other tool giving the possibility to change the order of consecutive universal and existential quantifiers. The footnote on page 36 in [5] tells us that the idea of this construction belongs to an unknown referee of the paper.

Nevertheless, the main theorem from [5] does belong to Martin Davis, but his original proof presented in [4] was quite different. Namely, for the initial representation of listable sets he used *normal systems* introduced by his teacher Post. Thanks to the great simplicity of the normal systems Martin Davis was able to arithmetize them in a very economical way using only one bounded universal quantifier.

While Martin Davis remarks in the same footnote that the proof presented in the paper is shorter that his original proof, the latter was very appealing to me: *now that we know that there is no need to use bounded universal quantifiers at all, could not we perform completely existential arithmetizing, thus avoiding universal quantifiers?*

Finally I was able to give such a quite different proof of the DPR-theorem based on arithmetization of the work of Turing machines [33]. In [37] I presented another way of simulating Turing machines by means of exponential Diophantine equations. Peter van Emde Boas proposed in [59] yet another, rather different way of doing it. However, James P. Jones and I found [22] that so called *register machines* are even more suitable for existential arithmetization; several versions of such "visual proof" are given in [23, 24, 35, 39].

The "advantage" of register machines over Turing machines for constructing Diophantine representations is due to the fact that the former operate directly with integers. However, register machines are not such a "classical" tool as Turing machines are. Esteemed *partial recursive function* have both properties: on the one hand they are defined on natural numbers, on the other hand, they are quite "classical". In [38] I used exponential Diophantine equations for simulating partial recursive function thus giving yet another proof of the DPR-theorem.

My paper [41] presents a unifying technique allowing one to eliminate bounded universal quantifier and simulate by means of exponential Diophantine equations Turing machines, register machines, and partial recursive functions in the same "algebraic" style.

Thus today we have quite a few very different proofs of the celebrated DPR-theorem. In contrast, it is a bit strange that no radically new techniques were found for transforming exponential Diophantine equations into genuine Diophantine ones: all known proofs in fact are minor variations of the construction presented by Martin Davis in [8] (Maxim Vsemirnov [60] made a generalization from (2.11) and (2.14) to some recurrent sequences of orders 3 and 4 but this gives no advantage for constructing Diophantine representations).

2.4 Improvements

2.4.1 Single-Fold Representations

I [32] was also able to improve the DPR-theorem in another direction, namely, to show the existence of a *single-fold* exponential Diophantine representation for every listable set, that is, a representation of the form (2.8) in which the values of $x_1, \ldots, x_n$, if they exist, are uniquely determined by the values of $a_1, \ldots, a_m$.

However, the two improvements to the DPR-theorem—to the DPRM-theorem and to single-fold representations—have not been so far combined, that is, the question about the existence of single-fold Diophantine representations for all listable sets still remains open. This is so because all today known methods of constructing Diophantine representation (2.9) are based on the study of behavior of sequences like (2.11) and (2.14) taken some modulo; clearly, this behavior is periodic and as a consequence each known Diophantine representation of exponentiation is infinite-fold—as soon as the corresponding equation (2.9) has a solution, it has infinitely many of them.

Single-fold representations have important applications (one of them is given in Sect. 2.6.2), and for this reason Martin Davis paper [7] titled *"One equation to rule them all"* remains of interest. The equation from the title is

$$9(u^2 + 7v^2)^2 - 7(r^2 + 7s^2)^2 = 2, \tag{2.15}$$

and it has a trivial solution $u = r = 1$, $v = s = 0$. Martin Davis proved that if this is the only solution, then some Diophantine relation has exponential growth. His expectations were broken by Oskar Herrman [17] who established the existence of another solution. The equation attracted interest of other researches, Daniel Shanks [53] was first in writing down two solutions explicitly and later he and Samuel S. Wagstaff, Jr. [54] found 48 more solutions.

The discovery of non-trivial solutions did not spoil Martin Davis approach completely. It fact, it can be shown that if (2.15) has only *finitely* many solutions then every listable set has a single-fold Diophantine representation.

2.4.2 Representations with a Small Number of Quantifiers

The existence of *universal listable sets* together with the DPRM-theorem implies that we can bound the number of unknowns in a Diophantine representation (2.3) of an arbitrary listable set $\mathcal{M}$; today's record $n = 9$ was obtained by me [34] (a detailed proof is presented in [19]). Accordingly, Hilbert's tenth problem remains undecidable even if we restrict ourselves to equations in 9 unknowns.

With present techniques, in order to get results for even smaller number of variables, one has to broaden the class of admissible formulas.

For example, for the DPR-theorem 3 unknowns are sufficient; originally this was proved in [36], and even for single-fold representations. Later this result was improved in [20, 21] to representations of the form

$$\begin{aligned} &\langle a_1, \dots, a_m \rangle \in \mathcal{M} \iff \\ &\exists x_1 x_2 \big[E_\mathrm{L}(a_1, \dots, a_m, x_1, x_2) \;\le\; E_\mathrm{R}(a_1, \dots, a_m, x_1, x_2) \big] \end{aligned} \tag{2.16}$$

where exponential polynomials E_L and E_R are constructed by using *unary* exponentiation 2^c only (rather than general *binary* exponentiation b^c). Harry R. Levitz proved in [26] that this result cannot be further improved to single unknown.

Soon after Martin Davis introduction of the normal form (2.6), Raphael Robinson [51] gave a rather different proof and showed that one can always take $n = 4$. In the same paper he gave another representation with 6 quantified variables, namely,

$$\langle a_1, \ldots, a_m \rangle \in \mathcal{M} \iff \exists z_1 z_2 \forall y_{\le B(a_1,\ldots,a_m,z_1,z_2)} \exists x_1 x_2 x_3 \big[Q(a_1, \ldots, a_m, x_1, x_2, x_3, y, z_1, z_2) = 0 \big]. \tag{2.17}$$

Much later, exploiting the power of the DPRM-theorem, he [52] improved the bound for Davis normal form (2.6) to $n = 3$ and showed that x_3 can be dropped from (2.17). Both of these results were further improved: in [31] to $n = 2$, in (2.6) and in [42] both x_2 and x_3 were dropped from (2.17).

More interesting is the possibility to replace the bounded universal quantifier in (2.6) and (2.17) by finite conjunction. For example, it was shown in [31] that every listable set has a representation of the form

$$\langle a_1, \ldots, a_m \rangle \in \mathcal{M} \iff \exists z_1 z_2 \&_{y=1}^{l} \exists x_1 x_2 \big[Q_y(a_1, \ldots, a_m, x_1, x_2, z_1, z_2) = 0 \big] \tag{2.18}$$

where l is a fixed number and $Q_1, \ldots, Q_l$ are polynomials with integer coefficients. Clearly, the right-hand side of (2.18) can be rewritten as a system of Diophantine equations in $2l + 2$ unknowns. While this quantity is high, each single equation has only 4 unknowns. This implies, for example, the following. Consider the class $\mathscr{D}_{2,2}$ of Diophantine sets that can be defined by formulas of the form

$$\langle x_1, x_2 \rangle \in \mathcal{M} \iff \exists z_1 z_2 \big[Q(x_1, x_2, z_1, z_2) = 0 \big]. \tag{2.19}$$

Clearly, *we cannot decide whether a given intersection of finitely many sets from class $\mathscr{D}_{2,2}$ is empty or not.* Informally, this means that among sets of pairs of natural numbers defined by Diophantine equations with just 2 unknowns there are sets with complicated structure having no "transparent" description.

In the above cited results the variables range over natural numbers; for the case of integer-valued variables corresponding results are at present somewhat weaker (in terms of the number of unknowns).

2.5 "Positive Aspects of a Negative Solution"

The title of this section reproduces part of the title of [12], the joint paper of Martin Davis, Julia Robinson, and myself written for the Proceedings of Symposium on Hilbert's problems [2]. The undecidability of Hilbert's tenth problem is just one of the corollaries of the DPRM-theorem. Actually it can serve as bridge for transferring ideas and results from Computability Theory to Number Theory; a few of such applications are given below.

2.5.1 Speeding Up Diophantine Equations

A simplest form of such transfer is as follows: *take any theorem about listable sets and replace them by Diophantine sets*. For example, one can explicitly write down a polynomial (2.5) with the set of its positive values being exactly the set of all prime numbers; the supposed impossibility of such a definition of primes was considered by many number-theorists as an informal argument against Martin Davis Conjecture.

It is quite typical that the map "listable" → "Diophantine" produces theorems not conventional for Number Theory. For example, Martin Davis published in [10] the following Diophantine counterpart of Manuel Blum's [1] *speed-up theorem*.

Theorem *For every general recursive function $\alpha(a, w)$ there are Diophantine equations*

$$B(a, x_1, \ldots, x_n) = 0, \tag{2.20}$$

$$C(a, y_1, \ldots, y_m) = 0 \tag{2.21}$$

such that:

- *for every value of a one and only one of these two equations has a solution;*
- *if equations*

$$B'(a, x'_1, \ldots, x'_{n'}) = 0, \tag{2.22}$$

$$C'(a, y'_1, \ldots, y'_{m'}) = 0 \tag{2.23}$$

are solvable exactly for the same values of the parameter a as Eqs. (2.20) *and* (2.21) *respectively, then there is third pair of equations*

$$B''(a, x''_1, \ldots, x''_{n''}) = 0, \tag{2.24}$$

$$C''(a, y''_1, \ldots, y''_{m''}) = 0 \tag{2.25}$$

such that:

- *these equations are also solvable exactly for the same values of the parameter a as Eqs.* (2.20) *and* (2.21) *respectively;*
- *for almost all a for every solution of equation* (2.22) (*Eq.* (2.23)) *there is solution of equation* (2.24) (*respectively, Eq.* (2.25)) *such that*

$$x'_1 + \cdots + x'_{m'} > \alpha(a, x''_1 + \cdots + x''_{m''}) \tag{2.26}$$

(*or*

$$y'_1 + \cdots + y'_{n'} > \alpha(a, y''_1 + \cdots + y''_{n''}) \tag{2.27}$$

respectively).

This theorem in its full generality is about an arbitrary general recursive function; replacing it by any particular (growing fast) function we obtain theorems which are purely number-theoretical but quite non-standard for Number Theory.

2.5.2 *Universal Equations*

One of the fundamental notion in Computability Theory is that of *universal Turing machine* or, equivalently, its counterpart *universal listable set.* Now the DPRM-theorem brings the idea of such kind of universality into the realm of Diophantine equations. Namely, for every fixed n, we can construct a particular Diophantine equation

$$U_n(k, a_1, \ldots, a_n, x_1, x_2, \ldots, x_m) = 0 \tag{2.28}$$

which is *universal* in the following sense: *solving an arbitrary Diophantine equation with n parameters*

$$D(a_1, \ldots, a_n, x_1, x_2, \ldots) = 0 \tag{2.29}$$

is equivalent to solving the equation

$$U_n(k_D, a_1, \ldots, a_n, x_1, x_2, \ldots, x_m) = 0 \tag{2.30}$$

resulting from the Eq. (2.28) *by choosing a particular value* k_D *for the first parameter, that is, for this fixed value of k and for any choice of the values of the parameters* $a_1, \ldots, a_m$ *either both of the Eqs.* (2.29) *and* (2.30) *have a solution or neither of them has any.*

What is remarkable in this reduction of one equation to another is the following: the degree and the number of unknowns of the Eq. (2.30) is fixed while the Eq. (2.29) can have any number of unknowns and be of arbitrarily large degree. This implies that hierarchies of Diophantine equations traditional for Number Theory (with 1, 2, 3, ... unknowns; of degree 1, 2, 3,...) collapse at some level.

Not only number-theorists never anticipated universal Diophantine equations, their possibility was incredible even for some logicians as it can be seen from the review in *Mathematical Reviews* on the celebrated paper by Martin Davis, Hilary Putnam, and Julia Robinson [14].

We can look at universal Diophantine equations as a purely number-theoretical result *inspired* by Computability Theory. But do we really need the general notion of listable sets for proving the existence of universal Diophantine equations or could we construct such equations by purely number-theoretical tools? In my book [37] I managed to prove the existence of universal Diophantine equations before proving the DPRM-theorem; in [41] I introduced another purely number-theoretical construction of universal Diophantine equations.

2.5.3 Hilbert's Eighth and Tenth Problems

The notion of listable set is very broad and can be found in a surprising variety of contexts. Here is one such example.

Hilbert included into his 8th problem an outstanding conjecture, the famous *Riemann's hypothesis*. In its original formulation it is a statement about complex zeros of Riemann's *zeta function* which is the analytical continuation of the series

$$\zeta(z) = \sum_{n=1}^{\infty} \frac{1}{n^z}. \tag{2.31}$$

Much as almost every great problem, Riemann's hypothesis has many equivalent restatements. Georg Kreisel [25] managed to reformulate it as an assertion about the emptiness of a particular listable set (each element of this set would produce a counterexample to the hypothesis). Respectively, we can construct a Diophantine representation of this set and obtain a particular Diophantine equation

$$R(x_0, \ldots, x_m) = 0 \tag{2.32}$$

which has no solution if and only if the Riemann hypothesis is true.

It was my share to write Sect. 2 of [12] devoted to reductions of Riemann's hypothesis and some other famous problems to Diophantine equations, but I failed to present the equation whose unsolvability is equivalent to Riemann's Hypothesis. Kreisel's main construction was very general, applicable to any analytical function, and some details of how to transfer it to a Diophantine equation were cumbersome. Luckily, Harold N. Shapiro, a colleague of Martin Davis, came to help and suggested a simpler construction, specific to the zeta function, based on the relationship of Riemann's hypothesis and distribution of prime numbers, and the corresponding part of Sect. 2 from [12] was written by Martin Davis.

In [37] I present a reductions of Riemann's hypothesis to Diophantine equations that is a bit simpler that the construction in [12], the simplification was due to certain new explicit constants related to distribution of primes that were obtained at that time in Number Theory.

Thus, Riemann's hypothesis can be viewed as a very particular case of Hilbert's tenth problem; such a relationship between it and Hilbert's eighth problem was not known before the DPRM-theorem was proved.

Hardly one can hope to prove or to disprove Riemann's hypothesis by examining a corresponding Diophantine equation. On the other hand, such a reduction gives an informal "explanation" of why Hilbert's tenth problem is undecidable: it would be rather surprising if such a long-standing open problem could be solved by a mechanical procedure required by Hilbert.

2.6 Other Impossibilities

DPR-theorem and DPRM-theorem turned out to be very powerful tools for establishing that many other things cannot be done algorithmically. Only a few examples will be mentioned here, surveys of many others can be found, for example, in [37, 40].

2.6.1 *The Number of Solutions*

In his tenth problem Hilbert demanded to find a method for deciding whether a given Diophantine equation has a solution or not. But one can ask many other similar questions, for example:

- is the number of solutions of a given Diophantine equation finite or infinite?
- is the number of solutions of a given Diophantine equation odd or even?
- is the number of solutions of a given Diophantine equation a prime number?

Martin Davis showed in [9] that Hilbert's tenth problem can be reduced to the above and analogous decision problems, and hence all of them are undecidable. Namely, the following theorem holds.

Theorem (Martin Davis). *Let* $\mathscr{N} = \{0, 1, 2, \dots, \infty\}$ *and let* $\mathscr{M}$ *be a proper subset of* $\mathscr{N}$*; there is no algorithm for deciding, for given Diophantine equation, whether the number of its solutions belongs to* $\mathscr{M}$ *or not.*

Clearly, the case $\mathscr{M} = \{0\}$ is the original Hilbert's tenth problem.

2.6.2 *Non-effectivizable Estimates*

Suppose that we have an equation

$$P(a, x_1, \dots, x_n) = 0, \tag{2.33}$$

which for every value of the parameter a has at most finitely many solutions in $x_1, \dots, x_n$. This fact can be expressed in two form:

- Equation (2.33) has at most $\nu(a)$ solutions;
- in every solution of (2.33) $x_1 < \sigma(a)$, …, $x_n < \sigma(a)$

for suitable functions ν and σ.

From a mathematical point of view these two statements are equivalent. However, they are rather different computationally. Having $\sigma(a)$ we can calculate $\nu(a)$ but not *vice versa*. Number-theorists have found many classes of Diophantine equations with computable $\nu(a)$ for which they fail to compute $\sigma(a)$. In such cases number-theorists say that "the estimate of the size of solutions is *non-effective*".

Now let us take some undecidable set $\mathscr{M}$ and construct an exponential Diophantine equation

$$E_{\mathrm{L}}(a, x_1, x_2, \dots, x_n) = E_{\mathrm{R}}(a, x_1, x_2, \dots, x_n) \tag{2.34}$$

giving a single-fold representation for $\mathscr{M}$. Clearly, Eq. (2.34) has the following two properties:

- for every value of the parameter a, Eq. (2.34) has at most one solution in $x_1, \dots, x_n$;
- for every effectively computable function σ there is a value of a for which the Eq. (2.34) has a solution $x_1, \dots, x_n$ such that $\max\{x_1, \dots, x_n\} > \sigma(a)$ (otherwise we would be able to determine whether a belongs to $\mathscr{M}$ or not).

In other words, the boundedness of solutions of equation (2.34) cannot be made effective in principle. This relationship between undecidability and non-effectivizability is one of the main stimuli to improve the DPRM-theorem to single-fold (or at least to finite-fold) representations and thus establish the existence of non-effectivizable estimates for genuine Diophantine equations.

2.6.3 Solutions in Other Rings

Most likely, Hilbert expected a positive solution of his tenth problem. This would allow us to recognize solvability of polynomial equations in many other rings, for example, in the ring of algebraic integers from any finite extension of the field of rational numbers, and in the ring of rational numbers. However, the obtained negative solution of Hilbert's tenth problem does not imply *directly* undecidability results for other rings. Nevertheless, different researchers were able to reduce the Tenth problem to solvability of equations in many classes of rings and thus establish the undecidability of analogs of the Tenth problem for them (for survey see book [55] or [56] in this volume).

Such reductions can be made by constructing a polynomial equation solvable in a considered ring if and only if the parameter is a rational integer, or, more generally, by constructing a *Diophantine model* of integers in that ring. Such an approach exploits the mere undecidability of the original Hilbert's tenth problem and does not require any new ideas from Computability Theory. However, number-theorists foresee some deep obstacles for the existence of such models for certain rings including, maybe the most interesting, the ring $\mathbb{Q}$ of rational numbers.

Recently Martin Davis proposed in [11] a quite different approach based on the existence of a special kind of undecidable sets constructed by Emil Post who named them *simple*. A listable set S is called simple if its complement to the set $\mathbb{N}$ of all natural numbers is infinite but contains no infinite listable set. In [43] Bjorn Poonen proved the undecidability of a counterpart of Hilbert tenth problem for a ring $\mathscr{U}$ of rational numbers denominators of which are allowed to contain "almost all" prime factors. His technique allows us to define a simple set S by a formula of the form

$$\{a \in \mathbb{N} \mid \exists x_1 \dots x_m \big[p(y_a, x_1, \dots, x_m) = 0\big] \tag{2.35}$$

where y_a is a computable function of a, p is a polynomial, and $x_1, \dots, x_m$ range over $\sim \mathscr{U}$. When these variables are allowed to range over all rational numbers, the same formula (2.35) defines some set $\widehat{S}_p$; clearly, $S \subseteq \widehat{S}_p$.

Martin Davis Conjecture [2010]. *There is a Diophantine definition of a simple set S for which $\mathbb{N} - \widehat{S}_p$ is infinite, so that $\widehat{S}_p$ is undecidable.*

Martin Davis wrote in [11]:

> This conjecture implies the unsolvability of H10 [Hilbert's tenth problem] over $\mathbb{Q}$. The conjecture seems plausible because although it is easy to construct simple sets, and there are a number of ways to do so, and if the conjecture is false, then no matter how S is constructed, and no matter what Diophantine definition of S is provided, $\widehat{S}_p$ differs from $\mathbb{N}$ by only finitely many elements. Because the additional primes permitted in denominators in the transition from $\mathscr{U}$ to $\mathbb{Q}$ form a sparse set, this seems implausible.

Let us believe in the wisdom of the celebrated guru.

References

1. Blum, M. (1967). A machine-independent theory of the complexity of recursive functions. *Journal of the ACM, 14*(2), 322–336.
2. Browder. (Ed.). (1976). Mathematical Developments arising from Hilbert Problems. *Proceedings of Symposia in Pure Mathematics* (vol. 28). American Mathematical Society.
3. Davis, M. (1950). Arithmetical problems and recursively enumerable predicates (abstract). *Journal of Symbolic Logic, 15*(1), 77–78.
4. Davis, M. (1950). *On the theory of recursive unsolvability*. PhD thesis, Princeton University.
5. Davis, M. (1953). Arithmetical problems and recursively enumerable predicates. *Journal of Symbolic Logic, 18*(1), 33–41.
6. Davis, M. (1958). *Computability and unsolvability*. New York: McGraw-Hill. Reprinted with an additional appendix, Dover 1983.
7. Davis, M. (1968). One equation to rule them all. *Transactions of the New York Academy of Sciences.Series II, 30*(6), 766–773.
8. Davis, M. (1971). An explicit Diophantine definition of the exponential function. *Communications on Pure and Applied Mathematics, 24*(2), 137–145.
9. Davis, M. (1972). On the number of solutions of Diophantine equations. *Proceedings of the American Mathematical Society, 35*(2), 552–554.
10. Davis, M. (1973). Speed-up theorems and Diophantine equations. In R. Rustin (Ed.), *Courant computer science symposium 7: Computational complexity* (pp. 87–95). New York: Algorithmics Press.
11. Davis, M. (2010). Representation theorems for r.e. sets and a conjecture related to Poonen's larges subring of $\mathbb{Q}$. *Zapiski Nauchnykh Seminarov Peterburgskogo Otdeleniya Matematicheskogo Instituta im. V. A. Steklova RAN (POMI), 377*, 50–54. Reproduced in: *Journal of Mathematical Sciences, 171*(6), 728–730 (2010).
12. Davis, M., Matijasevich, Yu., & Robinson, J. Hilbert's tenth problem. Diophantine equations: Positive aspects of a negative solution, pp. 323–378 in [2]. Reprinted in [51, pp. 269–324].

13. Davis, M., & Putnam, H. (1959). A computational proof procedure; Axioms for number theory; Research on Hilbert's Tenth Problem. O.S.R. Report AFOSR TR59-124, U.S. Air Force.
14. Davis, M., Putnam, H., & Robinson, J. (1961). The decision problem for exponential Diophantine equations. *Annals of Mathematics, 74*(2), 425–436. Reprinted in [51].
15. Gödel, K. (1931). Über formal unentscheidbare Sätze der Principia Mathematica und verwandter Systeme. I. Monatshefte für Mathematik und Physik bf 38(1), 173–198. Reprinted with English translation in: S. Feferman et al., (Eds.) (1986). *Kurt Gödel. Collected Works* (vol. I, pp. 144–195). Oxford University Press. English translation can also be found in: M. Davis, (Ed.) (1965). *The Undecidable* (pp. 4–38). Raven Press, Hewlett, New York and in: J. van Heijenoort, (Ed.) (1967). *From Frege to Gödel: A Source Book in Mathematical Logic, 1879–1931* (pp. 596–616). Harvard University Press, Cambridge, Massachusetts.
16. Green, B., & Tao, T. (2008). The primes contain arbitrarily long arithmetic progressions, *Annals of Mathematics, 167*, 481–547. doi:10.4007/annals.2008.167.481, arXiv:math.NT/0404188.
17. Herrman, O. (1971). A non-trivial solution of the Diophantine equation $9(x^2 + 7y^2)^2 - 7(u^2 + 7v^2)^2 = 2$. In A. O. L. Atkin & B. J. Birch (Eds.), *Computers in number theory* (pp. 207–212). London: Academic Press.
18. Hilbert, D. (1900) Mathematische Probleme. Vortrag, gehalten auf dem internationalen Mathematiker Kongress zu Paris 1900. *Nachrichten von der Königliche Gesellschaft der Wissenschaften zu Göttingen, Math.-Phys.Kl.* 253–297. See also Hilbert, D. (1935). *Gesammelte Abhandlungen* (vol. 3), Berlin: Springer. (Reprinted: Chelsea, New York (1965)). English translation: *Bulletin of the American Mathematical Society, 8*, 437–479 (1901–1902); reprinted in: [2, pp. 1–34].
19. Jones, J. P. (1982). Universal Diophantine equation. *Journal Symbolic Logic, 47*, 549–571.
20. Jones, J. P., & Matijasevič, Ju V. (1982). Exponential Diophantine representation of recursively enumerable sets. In J. Stern (Ed.), *Proceedings of the Herbrand Symposium: Logic Colloquium'81, Studies in Logic and the Foundations of Mathematics, 107, 159–177.* Amsterdam: North Holland.
21. Jones, J. P., & Matijasevič, Ju V. (1982). A new representation for the symmetric binomial coefficient and its applications. *Annales Sci. Mathém. du Québec, 6*(1), 81–97.
22. Jones, J. P., & Matijasevič, Ju V. (1983). Direct translation of register machines into exponential Diophantine equations. In L. Priese (Ed.), *Report on the 1st GTI-workshop* (pp. 117–130). Reihe Theoretische Informatik: Universität-Gesamthochschule Paderborn.
23. Jones, J. P., & Matijasevič, Ju V. (1984). Register machine proof of the theorem on exponential Diophantine representation of enumerable sets. *Journal Symbolic Logic, 49*(3), 818–829.
24. Jones, J. P., & Matijasevič, Ju V. (1991). Proof of recursive unsolvability of Hilbert's tenth problem. *American Mathematical Monthly, 98*(8), 689–709.
25. Kreisel, G. (1958). Mathematical significance of consistency proofs. *Journal of Symbolic Logic, 23*(2), 155–182.
26. Levitz, H. (1985). Decidability of some problem pertaining to base 2 exponential Diophantine equations. *Zeitschrift für Mathematische Logik und Grundlagen der Mathematik, 31*(2), 109–115.
27. Mal'tsev, A. I. (1968). On some boundary questions between algebra and logic (in Russian). *Trudy Mezhdunarodnogo Kongressa Matematikov (Moskva, 1966)*, pp. 217–231. Translated in: American Mathematical Society Translations. Series 2, Vol. 70: 31 invited addresses (8 in abstract) at the International Congress of Mathematicians in Moscow, 1966. American Mathematical Society, Providence, R.I. 1968, p. 266.
28. Markov, A. A. (1947). Impossibility of certain algorithms in the theory of associative systems (in Russian), *Doklady Akademii Nauk SSSR, 55*(7), 587–590 (1947). Translated in: *Compte rendus de l'Académie des Sciences de l'U.R.S.S., 55*, 583–586 (1947).
29. Matiyasevich, Yu. V. (1970). Enumerable sets are Diophantine (in Russian). *Dokl. AN SSSR, 191*(2), 278–282; Translated in: *Soviet Math. Doklady, 11*(2), 354–358. Correction ibid *11* (6) (1970), vi. Reprinted on pp. 269–273 in: *Mathematical logic in the 20th century*, G. E. Sacks, (Ed.), (2003). Singapore University Press and World Scientific Publishing Co., Singapore and River Edge, NJ.

30. Matiyasevich, Yu. V. (1972). Diophantine sets (in Russian). *Uspekhi Matematicheskikh Nauk, 27*(5), 185–222. English translation: *Russian Mathematical Surveys, 27*(5), 124–164 (1972).
31. Matiyasevich, Yu. V. (1972). Arithmetical representations of enumerable sets with a small number of quantifiers (in Russian). *Zapiski Nauchnykh Seminarov Leningradskogo Otdeleniya Matematicheskogo Instituta im. V. A. Steklova AN SSSR (LOMI), 32*, 77–84. English translation: *Journal of Soviet Mathematics, 6*(4), 410–416 (1976).
32. Matiyasevich, Yu. V. (1974). Existence of noneffectivizable estimates in the theory of exponential Diophantine equations (in Russian). *Zapiski Nauchnykh Seminarov Leningradskogo Otdeleniya Matematicheskogo Instituta im. V. A. Steklova AN SSSR (LOMI), 40*, 77–93; Translated in: *Journal of Soviet Mathematics, 8*(3), 299–311 (1977)
33. Matiyasevich, Yu. V. (1976). A new proof of the theorem on exponential Diophantine representation of enumerable sets (in Russian). *Zapiski Nauchnykh Seminarov Leningradskogo Otdeleniya Matematicheskogo Instituta im. V. A. Steklova AN SSSR (LOMI), 60*, 75–92; Translated in: *Journal of Soviet Mathematics, 14*(5), 1475–1486 (1980)
34. Matiyasevich, Yu. V. (1977). Some purely mathematical results inspired by mathematical logic. *Proceedings of Fifth International Congress on Logic, Methodology and Philosophy of science, London, Ontario, 1975* (pp. 121–127), Reidel, Dordrecht.
35. Matiyasevich, Yu. V. (1984). On investigations on some algorithmic problems in algebra and number theory. *Trudy Matematicheskogo instituta im. V. A. Steklova of Academy of Sciences of the USSR, 168*, 218–235. Translated in *Proceedings of the Steklov Institute of Mathematics, 168*, 227–252 (1986).
36. Matiyasevich, Yu. V. (1979). Algorithmic unsolvability of exponential Diophantine equations in three unknowns (in Russian). In A. A.Markov, & V. I.Homich (Eds), *Studies in the Theory of Algorithms and Mathematical Logic*, Computing Center Russian Academy Sci., Moscow, pp. 69–78; Translated in Selecta Mathematica Sovietica, *3*, 223–232 (1983/1984).
37. Matiyasevich, Yu. V. (1993). *Hilbert's Tenth Problem (in Russian)*. Fizmatlit, Moscow. English translation: MIT Press, Cambridge (Massachusetts) London (1993). French translation: *Le dixième Problème de Hilbert*, Masson, Paris Milan Barcelone (1995). http://logic.pdmi.ras.ru/~yumat/H10Pbook.
38. Matiyasevich, Yu. (1994). A direct method for simulating partial recursive functions by Diophantine equations. *Annals of Pure and Applied Logic, 67*, 325–348.
39. Matiyasevich, Yu. (2000). Hilbert's tenth problem: What was done and what is to be done. *Contemporary Mathematics, 270*, 1–47.
40. Matiyasevich, Yu. (2005). Hilbert's tenth problem and paradigms of computation. *Lecture Notes in Computer Science, 3526*, 310–321.
41. Matiyasevich, Yu. (2009). Existential arithmetization of Diophantine equations. *Annals of Pure and Applied Logic, 157*(2–3), 225–233.
42. Matiyasevich, Yu., & Robinson, J. (1974). Two universal 3-quantifier representations of recursively enumerable sets (in Russian). In B. A. Kushner & N. M. Nagornyĭ (Eds.), *Teoriya algorifmov i matematicheskaya logika*, Vychislitel'nyĭ Tsentr, Akademiya Nauk SSSR, Moscow, pp. 112–123. Reprinted in [51, pp. 223–234]. English translation in http://arxiv.org/abs/0802.1052.
43. Poonen, B. (2003). Hilbert's tenth problem and Mazur's conjecture for large subrings of $\mathbb{Q}$. *Journal of the American Mathematical Society, 16*(4), 981–990.
44. Post, E. L. (1944). Recursively enumerable sets of positive integers and their decision problems. *Bulletin of American Mathematical Society, 50*, 284–316. Reprinted in [46, pp. 461–494].
45. Post, E. L. (1947). Recursive unsolvability of a problem of Thue. *Journal of Symbolic Logic, 12*, 1–11. Reprinted in [46, pp. 503–513].
46. Post, E. L. (1994). *Solvability, provability, definability: The collected works of E. L. Post.* In M. Davis, (Ed.), Birkhäuser, Boston.
47. Putnam, H. (1960). An unsolvable problem in number theory. *Journal of Symbolic Logic, 25*(3), 220–232.
48. Robinson, J. (1952). Existential definability in arithmetic, *Transactions of the American Mathematical Society, 72*, 437–449. Reprinted in [51, pp. 47–59].

49. Robinson, J. B. (1960). The undecidability of exponential Diophantine equations. *Notices of the American Mathematical Society*, *7*(1), 75.
50. Robinson, J. Unsolvable Diophantine problems. *Proceedings of the American Mathematical Society, 22*(2), 534–538. Reprinted in [51, pp. 195–199].
51. Robinson, R. M. (1956). Arithmetical representation of recursively enumerable sets. *Journal of Symbolic Logic*, *21*(2), 162–186.
52. Robinson, R. M. (1972). Some representations of Diophantine sets. *Journal of Symbolic Logic*, *37*(3), 572–578.
53. Shanks, D. (1973). Five number-theoretic algorithms. In R. S. D. Thomas & H. C. Williams (Eds.), *Proceedings of the Second Manitoba Conference on Numerical Mathematics (University of Manitoba* (pp. 51–70). Winnipeg, Manitoba, 5–7 October 1972, volume VII of Congressus Numerantium Winnipeg, Manitoba: Utilitas Mathematica Publishing.
54. Shanks, D., & Wagstaff, S. S, Jr. (1995). 48 more solutions of Martin Davis's quaternary quartic equation. *Mathematics of Computation*, *64*(212), 1717–1731.
55. Shlapentokh, A. (2007). *Hilbert's tenth problem, New Mathematical Monographs* (vol. 7). Cambridge: Cambridge University Press.
56. Shlapentokh, A. Extensions of Hilbert's Tenth Problem: Definability and Decidability in Number Theory. This book, pp. 57–91
57. The collected works of Julia Robinson. (1996). *S. Feferman, edt*. Providence, RI: American Mathematical Society. ISBN 0-8218-0575-4. (Collected Works, vol. 6).
58. Thue, A. (1914). Probleme über Veränderungen von Zeichenreihen nach gegebenen Regeln. *Skrifter utgit av Videnskapsselskapet i Kristiana, I in Matematisk naturvidenskabelig klasse, Norske Videnskaps-Akademi, Oslo*, *10*, 493–524. Reprinted in: Nagell, T., Selberg, A., Selberg, S., & Thalberg, K. (Eds.). Selected Mathematical Papers of Axel Thue. Universitetsforlaget, Oslo, 1977.
59. van Emde Boas, P. (1997). The convenience of tiling. In A. Sorbi, (Ed.), Complexity, logic and recursion theory, Lecture Notes in Pure and Applied Mathematics (vol. 187, pp. 331–363).
60. Vsemirnov, M. A. (1995). Diophantine representations of linear recurrent relations (in Russian). *Zapiski Nauchnykh Seminarov Peterburgskogo Otdeleniya Matematicheskogo Instituta im. V. A. Steklova RAN*, *227*, 52–60. Translated in *Journal of Mathematical Sciences (New York)*, *89*(2), 1113–1118 (1998).

Chapter 3
Extensions of Hilbert's Tenth Problem: Definability and Decidability in Number Theory

Alexandra Shlapentokh

Abstract This chapter surveys some of the developments in the area of Mathematics that grew out of the solution of Hilbert's Tenth Problem by Martin Davis, Hilary Putnam, Julia Robinson and Yuri Matiyasevich.

Keywords Hilbert's tenth problem · Diophantine definability · Existential definability · Recursively enumerable sets

3.1 The End of the Beginning

When Yu. Matiyasevich completed the proof started by Martin Davis, Hilary Putnam and Julia Robinson, showing that all recursively enumerable sets of integers are Diophantine (see [4, 5, 22]), he added the last stone to the foundation of a new field which evolved from the solution of Hilbert's Tenth Problem. This new field intersecting Recursion Theory, Model Theory, Number Theory, Algebraic Geometry and lately parts of Analysis has sought to understand the expressive power of various dialects of the language of rings in different settings and to determine when the truth-values of sentences in these dialects can be decided algorithmically. Below we survey some of the recent developments in this field over finite and infinite algebraic extensions of $\mathbb{Q}$.

The question posed by Hilbert about rational integers can of course be asked about any recursive ring R (i.e. a ring where we know what the elements are and how to perform algorithmically the ring operations): is there an algorithm, which if given an arbitrary polynomial equation in several variables with coefficients in R, can determine whether this equation has solutions in R? Arguably, the most prominent

---

The author has been partially supported by the NSF grant DMS-1161456.

A. Shlapentokh (✉)
East Carolina University, Greenville, USA
e-mail: shlapentokha@ecu.edu
URL: http://www.myweb.ecu.edu/shlapentokha.com

E.G. Omodeo and A. Policriti (eds.), *Martin Davis on Computability, Computational Logic, and Mathematical Foundations*,
Outstanding Contributions to Logic 10, DOI 10.1007/978-3-319-41842-1_3

open questions in the area are the questions of decidability of an analog of Hilbert's Tenth Problem for $R = \mathbb{Q}$ and R equal to the ring of integers of an arbitrary number field. These questions proved to be quite hard and generated many other questions and mathematical developments. Almost simultaneously, the subject expanded to include infinite algebraic extensions of $\mathbb{Q}$, as well as function fields. In this paper we describe some of the more recent developments in the area dealing with algebraic extensions of $\mathbb{Q}$. We start our survey with a discussion of Hilbert's Tenth Problem for the field of rational numbers, but before we get there we need to consider a central notion occurring again and again in the discussion below: the notion of a Diophantine set.

3.1.1 Diophantine Subsets of a Ring

As mentioned above, a good place to start our survey is a definition of Diophantine subsets of a ring. This definition is at the center of much of the work that followed the solution of Hilbert's Tenth Problem ("HTP" in the future) and by itself is a knot of sorts holding together connections to several areas of Mathematics.

Definition 3.1 (*Diophantine Sets: A Number-Theoretic Definition*) Let R be a commutative ring with identity. (All the rings considered below satisfy these assumptions.) A subset $A \subset R^m$ is called Diophantine over R if there exists a polynomial $p(T_1, \dots T_m, X_1, \dots, X_k)$ with coefficients in R such that for any element $(t_1, \dots, t_m) \in R^m$ we have that

$$\exists x_1, \dots, x_k \in R : p(t_1, \dots, t_m, x_1, \dots, x_k) = 0$$

$$\Updownarrow$$

$$(t_1, \dots, t_m) \in A.$$

In this case we call $p(T_1, \dots, T_m, X_1, \dots, X_k)$ a *Diophantine definition* of A over R.

What was proved by Martin Davis, Hilary Putnam, Julia Robinson and Yuri Matiyasevich ("DPRM" in the future) is that *recursively enumerable subsets of integers (natural numbers) and Diophantine subsets of integers (natural numbers) were the same*. Now the connections to different areas of Mathematics emerge immediately because Diophantine sets can also be described as the sets *existentially definable in* R in the language of rings or as *projections of algebraic sets*. We define recursive and recursively enumerable sets below.

Definition 3.2 (*Recursive and Recursively Enumerable Subsets of* $\mathbb{Z}$) A set $A \subseteq \mathbb{Z}^m$ is called *recursive, computable or decidable* if there is an algorithm (or a computer program) to determine the membership in the set.

A set $A \subseteq \mathbb{Z}^m$ is called *recursively or computably enumerable* (r.e. and c.e. respectively in the rest of the paper) if there is an algorithm (or a computer program) to list the set.

The following theorem is a well-known result from Recursion Theory (see for example [43, Sect. 1.9]).

Theorem 3.1 *There exist recursively enumerable sets which are not recursive.*

Given the DPRM result we immediately obtain the following important corollary.

Corollary 3.1 *There are undecidable Diophantine subsets of* $\mathbb{Z}$.

It is easy to see that the existence of undecidable Diophantine sets implies that no algorithm as requested by Hilbert exists. Indeed, suppose $A \subset \mathbb{Z}$ is an undecidable Diophantine set with a Diophantine definition $P(T, X_1, \ldots, X_k)$. Assume also that we have an algorithm to determine the existence of integer solutions for polynomial equations. Now, let $a \in \mathbb{Z}$ and observe that $a \in A$ if and only if $P(a, X_1, \ldots, X_K) = 0$ has solutions in $\mathbb{Z}^k$. So if we can answer Hilbert's question effectively, we can determine the membership in A effectively.

It is also not hard to see that Diophantine sets are recursively enumerable. Given a polynomial $p(T, \bar{X})$ we can effectively list all $t \in \mathbb{Z}$ such that $p(t, \bar{X}) = 0$ has a solution $\bar{x} \in \mathbb{Z}^k$ in the following fashion. Using a recursive listing of $\mathbb{Z}^{k+1}$, we can plug each $(k+1)$-tuple into $p(T, \bar{X})$ to see if the value is 0. Each time we get a zero we add the first element of the $(k+1)$-tuple to the t-list. So the difficult part of the DPRM proof was to show that every r.e. subset of $\mathbb{Z}$ is Diophantine.

3.1.1.1 Properties of Diophantine Sets

Over many rings Diophantine sets have several simple but useful properties. First of all, over any domain a union of finitely many Diophantine sets is Diophantine. (A product of finitely many Diophantine definitions is a Diophantine definition of a union.) With a little bit of effort one can also show that as long as the fraction field of the domain in question is not algebraically closed, an intersection of finitely many Diophantine sets is Diophantine. Let $h(T) = a_0 + a_1T + \cdots + a_nT^n$ be a polynomial without roots in the fraction field, and let $f(t, \bar{x}), g(t, \bar{x})$ be Diophantine definitions of some subsets of the ring. In this case, it is not hard to see that the polynomial $\sum_{i=0}^{n} a_i f^i g^{n-i}$ has roots in the fraction field if and only if $f(t, \bar{x})$ and $g(t, \bar{x})$ have common roots in the field. The intersection of Diophantine sets being Diophantine is related to another important aspect of Diophantine equations over domains with fraction fields not algebraically closed: a finite system of equations is always equivalent to a single equation in the sense that both the system and the equation have the same solutions over the domain in question and in the sense that given a finite system of equations, the equivalent equation can be constructed effectively. We can use this property of finite systems to give more latitude to our

Diophantine definitions, i.e. we can let the Diophantine definitions over rings whose fraction fields are not algebraically closed consist of several polynomials without changing the nature of the relation.

Over $\mathbb{Z}$ (and many other rings) we have additional methods for writing Diophantine definitions. One surprisingly useful tool for writing Diophantine definitions has to do with an elementary property of GCD's (greatest common divisors).

Proposition 3.1 *If $a, b \in \mathbb{Z}_{\neq 0}$ with $(a, b) = 1$ then there exist $x, y \in \mathbb{Z}$ such that $ax + by = 1$.*

The GCD's can be used to show that the set of non-zero integers is Diophantine and thus allow us to require that values of variables are not equal, as well as to perform "division" as will be shown later. On a more theoretical level we can say that the *positive* existential theory of $\mathbb{Z}$ is the same as the existential theory of $\mathbb{Z}$.

Proposition 3.2 *The set of non-zero integers has the following Diophantine definition over $\mathbb{Z}$:*

$$\{t \in \mathbb{Z} | \exists x, u, v \in \mathbb{Z} : (2u - 1)(3v - 1) = tx\}$$

Proof If $t = 0$, then either $2u - 1 = 0$ or $3v - 1 = 0$ has a solution in $\mathbb{Z}$, which is impossible. Suppose now $t \neq 0$. Write $t = t_2t_3$, where t_2 is odd and $t_3 \not\equiv 0 \mod 3$. Since $(t_2, 2) = 1$ and $(t_3, 3) = 1$, by a property of GCD there exist $u, x_u, v, x_v \in \mathbb{Z}$ such that

$$2u + t_2x_u = 1$$

and

$$3v + t_3x_v = 1.$$

Thus $(2u - 1)(3v - 1) = t_2x_ut_3x_v = t(x_ux_v)$. □

Another important Diophantine definition allows us to convert inequalities into equations.

Lemma 3.1 (Diophantine definition of the set of non-negative integers) *From Lagrange's Theorem we get the following representation of non-negative integers:*

$$\{t \in \mathbb{Z} | \exists x_1, x_2, x_3, x_4 : t = x_1^2 + x_2^2 + x_3^2 + x_4^2\}$$

Before we proceed further with our discussion of HTP over $\mathbb{Q}$ we would like to point out that it is not hard to see that decidability of HTP over $\mathbb{Z}$ would imply decidability of HTP for $\mathbb{Q}$. Indeed, suppose we knew how to determine whether solutions exist over $\mathbb{Z}$. If $Q(x_1, \ldots, x_k)$ is a polynomial with integer coefficients, then

$$\exists x_1, \ldots, x_k \in \mathbb{Q} : Q(x_1, \ldots, x_k) = 0$$

$$\Updownarrow$$

$$\exists y_1, \ldots, y_k, z_1, \ldots, z_k \in \mathbb{Z} : Q(\frac{y_1}{z_1}, \ldots, \frac{y_k}{z_k}) = 0 \wedge z_1 \ldots z_k \neq 0,$$

where we remind the reader we can rewrite $z_1 \ldots z_k \neq 0$ as a polynomial equation and convert the resulting finite system of equations into a single one. So if we can determine whether the resulting equation had solutions over $\mathbb{Z}$, we can determine whether the original equation had solutions over $\mathbb{Q}$.

Unfortunately, the reverse implication does not work: we don't know of any easy way to derive the undecidability of HTP over $\mathbb{Q}$ from the analogous result over integers.

The fact that we can rewrite equations over $\mathbb{Q}$ as equations over $\mathbb{Z}$ is a specific instance of a more general phenomenon playing an important part in a discussion below.

Proposition 3.3 *If the set of non-zero elements of an integral domain R has a Diophantine definition over R, A is a Diophantine subset of F, the fraction field of R, and F is not algebraically closed, then $A \cap R$ has a Diophantine definition over R.*

Proof We sketch the main idea of a proof. Rewrite the variables ranging over F as ratios of variables ranging over R with a proviso that the denominator variables are not 0. Then clear all the denominators. □

3.1.1.2 Using a Diophantine Definition of $\mathbb{Z}$ to Show Undecidability

One of the earliest methods suggested for showing that HTP was undecidable over $\mathbb{Q}$ used Diophantine definitions. This idea can be summarized in the following lemma where the setting is somewhat more general. First we formally define recursive rings.

Definition 3.3 A countable ring is said to be recursive (or computable) if there exists a bijection $j : R \longrightarrow \mathbb{Z}_{\geqslant 0}$ such that the j-images of the graphs of R-addition and R-multiplication are recursive subsets of $\mathbb{Z}^3_{\geqslant 0}$.

Lemma 3.2 *Let R be a recursive ring of characteristic 0 (or in other words, a ring containing $\mathbb{Z}$ as a subring) with a fraction field not algebraically closed. If $\mathbb{Z}$ has a Diophantine definition $p(T, \bar{X})$ over R, then HTP is not decidable over R.*

Proof Let $h(T_1, \ldots, T_l)$ be a polynomial with rational integer coefficients and consider the following system of equations.

$$\begin{cases} h(T_1, \ldots, T_l) = 0 \\ p(T_1, \bar{X}_1) = 0 \\ \ldots \\ p(T_l, \bar{X}_l) = 0 \end{cases} \tag{3.1}$$

It is easy to see that $h(T_1, \ldots, T_l) = 0$ has solutions in $\mathbb{Z}$ if and only if system (3.1) has solutions in R. Thus if HTP is decidable over R, it is decidable over $\mathbb{Z}$. □

Unfortunately, the Diophantine definition plan for $\mathbb{Q}$ quickly ran into problems.

3.2 So What About $\mathbb{Q}$?

3.2.1 $\mathbb{Z}$ Is Probably Not Existentially Definable over $\mathbb{Q}$

In 1992 Barry Mazur formulated a series of conjectures which were to play an important role in the development of the subject (see [23–25]). One of the conjectures on the list was refuted in [1] but others are still open. Before we state two of the conjectures on the list, we need a definition.

Definition 3.4 (*Affine Algebraic Sets and Varieties.*) If $\{p_1(x_1, \ldots, x_m), \ldots, p_k(x_1, \ldots, x_m)\}$ is a finite set of polynomial equations over some field K, then the set of common zeros of these polynomials in $\tilde{K}^m$, where $\tilde{K}$ is an algebraically closed field containing K, is called an algebraic set. An algebraic set which is irreducible, i.e. is not a union of non-empty distinct algebraic sets, is called a variety. Given a variety V defined over K and a subfield K_0 of $\tilde{K}$ we often consider $V(K_0) = V \cap K_0$.

Mazur's conjectures on the topology of rational points are stated below:

Conjecture 3.1 (Topology of Rational Points) *Let V be any variety over $\mathbb{Q}$. Then the topological closure of $V(\mathbb{Q})$ in $V(\mathbb{R})$ possesses at most a finite number of connected components.*

This conjecture had an unpleasant consequence.

Conjecture 3.2 *There is no Diophantine definition of $\mathbb{Z}$ over $\mathbb{Q}$.*

Mazur's conjecture also refers to projective varieties, but it is the affine variety case which has the most consequences for HTP over $\mathbb{Q}$. We should also note that one can replace "variety" by "algebraic set" without changing the scope of the conjecture. (See Remark 11.1.2 of [56].) As a matter of fact, if Conjecture 3.1 is true, no infinite and discrete (in the Archimedean topology) set has a Diophantine definition over $\mathbb{Q}$.

Quite a few years later Königsmann [19] used different reasons to make a case that there is probably no Diophantine definition of $\mathbb{Z}$ over $\mathbb{Q}$.

Since the plan to construct a Diophantine definition of $\mathbb{Z}$ over $\mathbb{Q}$ ran into substantial difficulties, alternative ways were considered for showing that HTP had no solution over $\mathbb{Q}$. One of the alternative methods required construction of a Diophantine model of $\mathbb{Z}$.

Definition 3.5 (*Diophantine Model of $\mathbb{Z}$*) Let R be a recursive ring whose fraction field is not algebraically closed and let $\varphi : \mathbb{Z} \longrightarrow R^k$ be a recursive injection mapping Diophantine sets of $\mathbb{Z}$ to Diophantine sets of R^k. Then φ is called a Diophantine model of $\mathbb{Z}$ over R.

Remark 3.1 (*An Alternative Terminology from Model Theory*) Model theorist have an alternative terminology for a map described above. They would translate the statement that R has a Diophantine model of $\mathbb{Z}$ as $\mathbb{Z}$ being existentially definably interpretable in R. (See Chap. 1, Sect. 3 of [21].)

It is not hard to see that sending Diophantine sets to Diophantine sets makes the map automatically recursive. The recursiveness of the map follows from the fact that the φ-image of the graph of addition is Diophantine and therefore is recursively enumerable (by the same argument as over $\mathbb{Z}$). Thus, we have an effective listing of the set

$$D_{+} = \{(\varphi(m), \varphi(n), \varphi(m+n)), m, n \in \mathbb{Z}\}.$$

Assume we have computed $\varphi(r-1)$ for some positive integer r. Now start listing D_{+} until we come across a triple whose first two entries are $\varphi(r-1)$ and $\varphi(1)$. The third element of the triple must be $\varphi(r)$. We can simplify the requirements for the map further.

Proposition 3.4 *If R is a recursive ring and $\varphi : \mathbb{Z} \longrightarrow R^k$ is injective for some $k \in \mathbb{Z}_{>0}$, then φ is a Diophantine model if and only if the images of the graphs of $\mathbb{Z}$-addition and $\mathbb{Z}$-multiplication are Diophantine over R.*

This proposition can be proved by a straightforward induction argument which we do not reproduce here.

It quite easy to see that the following proposition holds.

Proposition 3.5 *If R is a recursive ring with a Diophantine model of $\mathbb{Z}$, then HTP has no solution over R.*

Proof If R has a Diophantine model of $\mathbb{Z}$, then R has undecidable Diophantine sets, and the existence of undecidable Diophantine sets over R leads us to the undecidability of HTP over R in the same way as it happened over $\mathbb{Z}$. To show that R has undecidable Diophantine sets, let $A \subset \mathbb{Z}$ be an undecidable Diophantine set and suppose we want to determine whether an integer $n \in A$. Instead of answering this question directly we can ask whether $\varphi(n) \in \varphi(A)$. By assumption $\varphi(n)$ is algorithmically computable. So if $\varphi(A)$ is a computable subset of R, we have a contradiction. □

3.2.2 Elliptic Curves and Diophantine Models

An old plan for building a Diophantine model of $\mathbb{Z}$ over $\mathbb{Q}$ involved using elliptic curves. Consider an equation of the form:

$$y^2 = x^3 + ax + b, \tag{3.2}$$

where $a, b \in \mathbb{Q}$ and $\Delta = -16(4a^3 + 27b^2) \neq 0$. This equation defines an elliptic curve (a non-singular plane curve of genus 1).

All the points $(x, y) \in \mathbb{Q}^2$ satisfying (3.2) (if any) together with O—the "point at infinity"—form an abelian group, i.e. there is a way to define addition on the points of an elliptic curve with O serving as the identity. The group law on an elliptic curve can be represented geometrically (see for example [61, Chap. III, Sect. 3]). However, what is important to us is the algebraic representation of the group law. Let $P = (x_P, y_P)$, $Q = (x_Q, y_Q)$, $R = (x_R, y_R)$ be the points on an elliptic curve E with rational coordinates. If $P +_E Q = R$ and $P, Q, R \neq O$, then $x_R = f(x_P, y_P, x_Q, y_Q)$, $y_R = g(x_P, y_P, x_Q, y_Q)$, where $f(z_1, z_2, z_3, z_4)$, $g(z_1, z_2, z_3, z_4)$ are fixed (somewhat unpleasant looking) *rational functions*. Further, $-P = (x_P, -y_P)$. Mordell-Weil Theorem (see [61, Chap. III]) tells us that the abelian group formed by points of an elliptic curve over $\mathbb{Q}$ is finitely generated, meaning it has a finite rank and a finite torsion subgroup. It is also not very difficult to find elliptic curves whose rank is one. So let E be such an elliptic curve defined over $\mathbb{Q}$ such that $E(\mathbb{Q}) \cong \mathbb{Z}$ as abelian groups. (In other words $E(\mathbb{Q})$ has no torsion points. In practice torsion points are not an impediment, but they do complicate the discussion.) Let P be a generator and consider a map sending an integer $n \neq 0$ to $[n]P = (x_n, y_n)$. (We should also take care of 0, but we will ignore this issue for the moment.) The group law assures us that under this map *the image of the graph of addition is Diophantine*. Unfortunately, it is not clear what happens to *the image of the graph of multiplication*. Nevertheless one might think that we have a starting point at least for our Diophantine model of $\mathbb{Z}$. Unfortunately, it turns out that situation with Diophantine models is not any better than with Diophantine definitions. Further a theorem of Cornelissen and Zahidi (see [3]) showed that multiplication of indices of elliptic curve points is probably not existentially definable.

Theorem 3.2 *If Mazur's conjecture on topology of rational points holds, then there is no Diophantine model of $\mathbb{Z}$ over $\mathbb{Q}$.*

This theorem left HTP over $\mathbb{Q}$ seemingly out of reach. It is often the case with difficult Mathematical problems that the search for solutions gives rise to a lot of new and interesting Mathematics, sometimes directly related to the original problem, sometimes only marginally so. People trying to resolve the Diophantine status of $\mathbb{Z}$ also proceeded in several directions. The two directions generating the most activity are the the big ring project and attempts to reduce the number of universal quantifiers in first-order definitions of $\mathbb{Z}$ over $\mathbb{Q}$. We review the big ring project first.

3.2.3 The Rings Between $\mathbb{Z}$ and $\mathbb{Q}$

We start with a definition of the rings in question whose first appearance on the scene in [49, 50] dates back to 1994.

Definition 3.6 (*A Ring in between*) Let $\mathscr{S}$ be a set of primes of $\mathbb{Q}$. Let $R_{\mathscr{S}}$ be the following subring of $\mathbb{Q}$.

$$\left\{\frac{m}{n} : m, n \in \mathbb{Z}, n \neq 0, n \text{ is divisible by primes of } \mathscr{S} \text{ only}\right\}$$

If $\mathscr{S} = \emptyset$, then $R_{\mathscr{S}} = \mathbb{Z}$. If $\mathscr{S}$ contains all the primes of $\mathbb{Q}$, then $R_{\mathscr{S}} = \mathbb{Q}$. If $\mathscr{S}$ is finite, we call the ring *small*. If $\mathscr{S}$ is infinite, we call the ring *big*.

Some of these rings have other (canonical) names: the small rings are also called rings of $\mathscr{S}$-integers, and when $\mathscr{S}$ contains all but finitely many primes, the rings are called semi-local subrings of $\mathbb{Q}$. To measure the "size" of big rings we use the natural density of prime sets defined below.

Definition 3.7 (*Natural Density*) If $\mathscr{A}$ is a set of primes, then the natural density of $\mathscr{A}$ is equal to the limit below (if it exists):

$$\lim_{X \to \infty} \frac{\#\{p \in A, p \leqslant X\}}{\#\{p \leqslant X\}}$$

The big and small rings are not hard to construct.

Example 3.1 (*A Small Ring not Equal* to $\mathbb{Z}$)

$$\left\{\frac{m}{3^a 5^b} : m \in \mathbb{Z}, a, b \in \mathbb{Z}_{>0}\right\}$$

Example 3.2 (*A Big Ring not Equal to* $\mathbb{Q}$)

$$\left\{\frac{m}{\prod p_i^{n_i}} : p_i \equiv 1 \bmod 4, n_i \in \mathbb{Z}_{>0}\right\}$$

Given a big or a small ring R we can now ask the following questions which were raised above with respect to $\mathbb{Q}$:

- Is HTP solvable over R?
- Do integers have a Diophantine definition over R?
- Is there a Diophantine model of integers over R?

Here one could hope that understanding what happens to HTP over a big ring can help to understand HTP over $\mathbb{Q}$.

3.2.4 Diophantine Properties of Big and Small Rings

Before trying to answer the questions above, one should observe that the big and small rings share many Diophantine properties with the integers:

Proposition 3.6 *1. The set of non-zero elements of a big or a small ring is Diophantine over the ring.*

2. *A finite system of polynomial equations over a big or a small ring can be rewritten effectively as a single polynomial equation such that the solution set for the system is the same as the solution set for the equation.*
3. *The set of non-negative elements of a big or a small ring R is Diophantine over R: a small modification of the Lagrange argument is required to accommodate possible denominators*

$$\{t \in R|\exists x_1, x_2, x_3, x_4, x_5 : x_5^2 t = x_1^2 + x_2^2 + x_3^2 + x_4^2 \wedge x_5 \neq 0\}$$

It turned out that we already knew everything we needed to know about small rings from the work of J. Robinson (see [39]). In particular from her work on the first-order definability of integers over $\mathbb{Q}$ one can deduce the following theorem and corollaries.

Theorem 3.3 (J. Robinson) *For every p, the ring $R_p = \{x \in \mathbb{Q}|x = \frac{m}{n}, m, n \in \mathbb{Z}, n > 0, p \not| n\}$ has a Diophantine definition over $\mathbb{Q}$.*

This theorem of J. Robinson will play a role in many other results, as we will see below. In particular, now using Proposition 3.3 and Parts 1 and 2 of Proposition 3.6 we get the following corollaries.

Corollary 3.2 *$\mathbb{Z}$ has a Diophantine definition over any small subring of $\mathbb{Q}$.*

Proof To see that J. Robinson's theorem implies this corollary, let R be any big or small ring and observe that, since $\mathbb{Q}$ is not algebraically closed, $R_p \cap R$ is Diophantine over R by Proposition 3.3 and Part 1 of Proposition 3.6. Let $g_{R,p}$ be the resulting Diophantine definition.

Now let R be a small ring with $p_1, \ldots, p_r$ being all the primes allowed in the denominators of its elements. Let $g_R(t, \bar{u}) = 0$ be the polynomial equation equivalent to the system

$$\begin{cases} g_{R,p_1}(t, \bar{z}_1) = 0, \\ \quad \ldots \\ g_{R,p_r}(t, \bar{z}_r) = 0, \end{cases}$$

and observe that $g_R(t, \bar{u})$ is a Diophantine definition of $R \cap R_{p_1} \cap \cdots \cap R_{p_r}$. (Existence of $g(t, \bar{u})$ is guaranteed by Part 2 of Proposition 3.6.) Suppose t is an element of this intersection. Since $t \in R$, the only primes that can divide the (reduced) denominator of t are $p_1, \ldots, p_r$. However, being an element of $R_{p_i}, i = 1, \ldots, r$ implies that p_i for all values of the index *does not* divide the denominator. Therefore, no prime can divide the denominator of t, and hence t is an integer. At the same time, trivially, $\mathbb{Z} \subseteq R \cap R_{p_1} \cap \cdots \cap R_{p_r}$. Thus, $\mathbb{Z}$ has a Diophantine definition over R. □

Given that $\mathbb{Z}$ has a Diophantine definition over R, we apply Lemma 3.2 to conclude the following.

Corollary 3.3 *HTP is unsolvable over all small subrings of $\mathbb{Q}$.*

Over big rings the questions turned out to be far more difficult.

3.2.5 Existential Models of $\mathbb{Z}$ over Big Rings

In 2003 Poonen [32] proved the first result on Diophantine undecidability (unsolvability of HTP) over a big subring of $\mathbb{Q}$.

Theorem 3.4 *There exist recursive sets of primes $\mathcal{T}_1$ and $\mathcal{T}_2$, both of natural density zero and with an empty intersection, such that for any set $\mathcal{S}$ of primes containing $\mathcal{T}_1$ and avoiding $\mathcal{T}_2$, the following hold:*

- *$\mathbb{Z}$ has a Diophantine model over $O_{\mathbb{Q},\mathcal{S}}$.*
- *Hilbert's Tenth Problem is undecidable over $O_{\mathbb{Q},\mathcal{S}}$.*

Poonen used elliptic curves to prove his result but the model he constructed was very different from the one envisioned by the old elliptic curve plan we described earlier. Poonen modeled integers *by approximation*. The construction of the model does start with an elliptic curve of rank one

$$E : y^2 = x^3 + ax + b \tag{3.3}$$

selected so that for a set of primes $\mathcal{S}$, except possibly for finitely many points, the only multiples of a generator P that have their affine coordinates in the ring $R_{\mathcal{S}}$ are in the sequence $[\pm\ell_i]P = (x_{\ell_i}, \pm y_{\ell_i})$ with $|y_{\ell_j} - j| < 10^{-j}$. We remind the reader that we know how to define positive numbers using a variation on Lagrange's theme (Proposition 3.6) and how to get rid of a finite set of undesirable values such as points of finite order (just say "$\neq$" as in Proposition 3.6 again). We claim that $\varphi : j \longrightarrow y_{\ell_j}$ is a Diophantine model of $\mathbb{Z}_{>0}$. In other words, setting $D = j(\mathbb{Z})$, we claim that φ is an injection, and D, as well as the following sets, is Diophantine over $R_{\mathcal{S}}$:

$$D_+ = \{(y_{\ell_i}, y_{\ell_j}, y_{\ell_k}) \in D^3 : k = i + j, k, i, j \in \mathbb{Z}_{>0}\}$$

and

$$D_2 = \{(y_{\ell_i}, y_{\ell_k}) \in D^2 : k = i^2, i \in \mathbb{Z}_{>0}\}.$$

(Note that if D_+ and D_2 are Diophantine, then $D_\times = \{(y_{\ell_i}, y_{\ell_j}, y_{\ell_k}) \in D^3 : k = ij, k, i, j \in \mathbb{Z}_{>0}\}$ is also Diophantine since $xy = \frac{1}{2}((x+y)^2 - x^2 - y^2)$.) It is easy to show that

$$k = i + j \Leftrightarrow |y_{\ell_i} + y_{\ell_j} - y_{\ell_k}| < 1/3.$$

and with the help of Lagrange this makes D_+ Diophantine. Similarly we have that

$$k = i^2 \Leftrightarrow |y_{\ell_i}^2 - y_{\ell_k}| < 2/5,$$

implying that D_2 is Diophantine.

To restrict the number of solutions to the elliptic curve equation, Poonen's construction relied to a large extent on the fact that the denominators of the coordinates of points on an elliptic curve which are multiples of a single point form a divisibility sequence: an integer sequence $\{a_n\}$ is called a divisibility sequence if $n|m$ implies $a_n|a_m$ (see Chaps. 4 and 10 of [14] for a discussion of such sequences and see the discussion of the formal group of an elliptic curve in Chap. 4 of [61] for an explanation of why the denominators form a divisibility sequence).

Poonen's method was further extended by Eisenträger and Everest [10], Perlega [29] and finally by Eisenträger et al. [11]. The theorem proved in [11] provides a "covering" of $\mathbb{Q}$ by big rings where HTP is undecidable.

Theorem 3.5 (Eisenträger, Everest, Shlapentokh) *For any finite set of positive computable real numbers (i.e. real numbers that are limits of computable sequences of rational numbers) $r_1, \ldots, r_k$ such that $r_1 + \cdots + r_k = 1$ we can partition the set of all (rational) primes into sets $\mathscr{P}_1, \ldots, \mathscr{P}_k$ such that the natural density of each $\mathscr{P}_i$ is r_i, each ring $R_{\mathscr{P}_i}$ has a Diophantine model of $\mathbb{Z}$ and therefore HTP is undecidable over each $R_{\mathscr{P}_i}$.*

The author also constructed a model of $\mathbb{Z}$ using Diophantine equivalence classes (a class model of $\mathbb{Z}$) over a big ring using the old idea of trying to make multiplication of indices Diophantine in [60].

Theorem 3.6 *Let E be an elliptic curve defined and of rank one over $\mathbb{Q}$. Let P be a generator of $E(\mathbb{Q})$ modulo the torsion subgroup, and fix an affine (Weierstrass) equation for E of the form $y^2 = x^3 + ax + b$, with $a, b \in \mathbb{Z}$. If (x_n, y_n) are the coordinates of $[n]P$ with $n \neq 0$ derived from this (Weierstrass) equation, then there exists a set of primes $\mathscr{W}$ of natural density one, and a positive integer m_0 such that the following set $\Pi \subset R_{\mathscr{W}}^{12}$ is Diophantine over $R_{\mathscr{W}}$.*

$$(U_1, U_2, U_3, X_1, X_2, X_3, V_1, V_2, V_3, Y_1, Y_2, Y_3) \in \Pi \Leftrightarrow$$
$$\exists \textit{ unique } k_1, k_2, k_3 \in \mathbb{Z}_{\neq 0} \textit{ such that}$$
$$\left(\frac{U_i}{V_i}, \frac{X_i}{Y_i}\right) = (x_{m_0 k_i}, y_{m_0 k_i}), \textit{ for } i = 1, 2, 3, \textit{ and } k_3 = k_1 k_2.$$

3.2.6 *The Other End of the Spectrum*

In this section we would like to describe some work (still in progress) which approached HTP over big rings from the other end of the spectrum, i.e. from the point of view of $\mathbb{Q}$. Before we can discuss these very new ideas we need to make a quick detour to review some basic notions of logic.

Definition 3.8 (*Turing Reducibility*) Let $A \subseteq \mathbb{Z}^k$, $B \subseteq \mathbb{Z}^m$ for some positive integers k, m. Assume also that there is an effective procedure to determine membership in A using an oracle for membership in B. In this case we say that A is Turing reducible to B and write $A \leqslant_T B$. If additionally we have $B \leqslant_T A$, then we say that B is Turing-equivalent to A. The equivalence classes of Turing equivalence are called Turing degrees.

We now need to review a definition of a famous set.

Definition 3.9 Let $\{f_i\}$ be an effective enumeration of all computable functions and let

$$H = \{(i, j) \in \mathbb{Z}^2_{>0} | f_i(x_j) \downarrow\}.$$

In this case we call H the *Halting Set.*

It is not hard to see that

(a) H is r.e.
and
(b) every r.e. set is Turing reducible to H (so in some sense H is the "hardest" r.e. set),

Our next step is to convert the question of decidability of HTP of a recursive ring R into a question of the Turing degree of a subset of $\mathbb{Z}_{>0}$. (Here we remind the reader that by a *recursive or computable ring* we mean a ring which is recursive as a set and with operations corresponding to total recursive functions. A ring *isomorphic to a recursive ring* will be called *computably presentable.*) To that end let $\{p_i(\bar{x})\}$ be an effective enumeration of all polynomials over R and let HTP(R) denote the set of indices corresponding to polynomials having a root in R. Now given DPRM result we have that HTP($\mathbb{Z}$) $\equiv_T H$. Further, any recursive or computably presentable ring R with a Diophantine model of $\mathbb{Z}$ has HTP(R) $\equiv_T H$.

At the same time, by results of Friedberg [16] and Muchnik [27] we know that there are Turing degrees containing undecidable r.e. sets not as hard as H, i.e. H is not Turing equivalent to these sets. What if HTP($\mathbb{Q}$) is one of these sets? If this were the case, there would be neither an algorithm to solve HTP over $\mathbb{Q}$ nor a Diophantine model of $\mathbb{Z}$ over $\mathbb{Q}$. So if HTP($\mathbb{Q}$) $\not\equiv_T$ HTP($\mathbb{Z}$) it makes sense to see if there are big subrings R of $\mathbb{Q}$, "infinitely" far away from $\mathbb{Q}$ with HTP(R) $\equiv_T$ HTP($\mathbb{Q}$).

In order to explain how we get "far away" from $\mathbb{Q}$ we need to discuss the notion of being integral at a prime and reconsider results of J. Robinson we introduced before with a new spin. If $x \in \mathbb{Q}, x \neq 0$, then we can write $x = \pm\frac{p_1^{a_1}\dots p_m^{a_m}}{q_1^{b_1}\dots q_k^{b_k}}$, where $p_1, \dots, p_m, q_1, \dots, q_k$ are distinct prime numbers and $a_1, \dots, a_m, b_1, \dots, b_k$ are positive integers. We define $\text{ord}_{p_i} x = a_i$, $\text{ord}_{q_j} x = -b_j$, and for a prime number

$$t \notin \{p_1, \dots, p_m, q_1, \dots, q_k\}$$

we define $\text{ord}_t x = 0$. If t is a prime and x is a rational number with $\text{ord}_t x \geqslant 0$, we say that x is *integral* at t.

We now go back to a result of Julia Robinson we used before: Theorem 3.3, proving that the set of all rational numbers integral at a given prime is Diophantine over $\mathbb{Q}$, and examine more closely what was involved in a construction of this Diophantine definition. The construction of the definition is in fact uniform in p, i.e. given a p there is an effective procedure taking p as its input and constructing the existential definition of the valuation ring of p—the set of all rational numbers integral at p. Since we can effectively combine a finite number of Diophantine definitions into one over any subring of $\mathbb{Q}$, we conclude that we have an algorithm for writing down Diophantine definitions of rings of rational numbers without a fixed finite set of primes dividing the denominators.

In the project still in progress (see [12]) K. Eisenträger, R. Miller, J. Park, and A. Shlapentokh have constructed families of computably presentable subrings R of $\mathbb{Q}$ with HTP(R) $\equiv_T$ HTP($\mathbb{Q}$). The constructed rings consist of rational numbers where an infinite set of primes is allowed to divide the denominator, but the complement of this set of primes, that is the set of primes that are not allowed to divide the denominator is also infinite. Priority method was used to make the set of inverted primes c.e. (and thus the rings computably presentable). Further, the set of primes which can occur as divisors of the denominators of elements in the ring can be arranged to have the lower natural density equal to 0. So we are truly looking at a ring "in the middle", "infinitely far away" from both $\mathbb{Z}$ and $\mathbb{Q}$. These rings also have the property that the set of inverted primes, i.e. primes allowed to divide the denominators is computable from HTP($\mathbb{Q}$). So if HTP($\mathbb{Q}$) is decidable, these prime sets are also decidable and the rings in question are computable subrings of $\mathbb{Q}$ (not just computably presentable).

The co-authors have also obtained an analog of Theorem 3.5, though a weaker one. More specifically, for any positive integer k, one can partition the set of all prime numbers into k sets $\mathscr{S}_1, \ldots, \mathscr{S}_K$, each of lower density 0, and construct rings $R_1, \ldots, R_k$ where the primes allowed to divide the denominators are precisely $\mathscr{S}_1, \ldots, \mathscr{S}_K$ respectively and such that HTP(R_i) $\equiv_T$ HTP($\mathbb{Q}$). Unfortunately, these rings are not necessarily computably presentable and we can only say that each $\mathscr{S}_i$ is Turing reducible to HTP($\mathbb{Q}$), so that again if HTP($\mathbb{Q}$) is decidable, these prime sets are also decidable and the rings in question are computable subrings of $\mathbb{Q}$.

Now if we combine the results above with results constructing big rings with HTP equivalent to the halting problem, then one can conclude that if HTP over $\mathbb{Z}$ is different from HTP over $\mathbb{Q}$, in particular if HTP($\mathbb{Q}$) is decidable, then we have an extremely strange picture of tightly intermingled recursive rings inside $\mathbb{Q}$ with different levels of difficulty for HTP. Such a picture seems unlikely, though of course we cannot rule it out without a proof.

3.3 All Together Now (with Universal Quantifiers)

So far we have discussed the existential theory of $\mathbb{Z}$, i.e. we made use of existential quantifiers only in making various statements in the ring language, the language of polynomial equations. It is natural to ask what happens if we also allow the use of universal quantifiers. If we do allow the universal quantifiers, the situation changes dramatically. As we have mentioned before, the result defining integers over $\mathbb{Q}$ using the "full force" of the first-order language is pretty old and belongs to J. Robinson (see [39]). Thus we have known for a while that the full first-order theory of rational numbers is undecidable. J. Robinson used quadratic forms and Hasse-Minkwoski Theorem to prove her result.

In a 2007 paper G. Cornelissen and K. Zahidi analyzed J. Robinson's formula and showed that it can be converted to a formula of the form $(\forall\exists\forall\exists)(F = 0)$ where the $\forall$-quantifiers run over a total of 8 variables, and where F is a polynomial. In 2008 Poonen [35] produced an improvement of the first-order definition of integers over $\mathbb{Q}$. He showed that $\mathbb{Z}$ is definable over $\mathbb{Q}$ using just two universal quantifiers in a $\forall\exists$-formula. B. Poonen used quadratic forms, quaternions and the Hasse Norm Principle. His definition of $\mathbb{Z}$ over $\mathbb{Q}$ is simple enough to be reproduced here: the set $\mathbb{Z}$ equals the set of $t \in \mathbb{Q}$ for which the following formula is true over $\mathbb{Q}$:

$$(\forall a, b)(\exists a_1, a_2, a_3, a_4, b_1, b_2, b_3, b_4, x_1, x_2, x_3, x_4, y_1, y_2, y_3, y_4, n)$$

$$(a + a_1^2 + a_2^2 + a_3^2 + a_4^2)(b + b_1^2 + b_2^2 + b_3^2 + b_4^2)\cdot$$

$$[(x_1^2 - ax_2^2 - bx_3^2 + abx_4^2 - 1)^2 + (y_1^2 - ay_2^2 - by_3^2 + aby_4^2 - 1)^2 +$$

$$+ n^2 + (n-1)^2 \ldots (n - 2309)^2 + (2x_1 + 2y_1 + n - t)^2] = 0$$

Starting with B. Poonen's results, J. Koenigsmann further reduced the number of quantifiers to one in [19]. As we pointed out above, this result could very well be the optimal one, since $\mathbb{Z}$ probably does not have a purely existential definition over $\mathbb{Q}$. In the same paper, Koenigsmann showed that $\mathbb{Z}$ has a purely universal definition over $\mathbb{Q}$ or alternatively, the set of non-integers has a Diophantine definition over $\mathbb{Q}$. We will return to the issue of definitions using all of the quantifiers in the sections below concerning finite and infinite algebraic extensions.

3.4 Up and Away

In this section we survey developments over number fields inspired by the solution of Hilbert's Tenth Problem. We start with a review of some terms.

- A number field K is a finite extension of $\mathbb{Q}$.
- We denote a fixed algebraic closure of $\mathbb{Q}$, i.e. the field containing the roots of all polynomials with coefficients in $\mathbb{Z}$, by $\tilde{\mathbb{Q}}$.
- A Galois closure of a number field K over $\mathbb{Q}$ is the smallest Galois number field containing K inside $\tilde{\mathbb{Q}}$.
- A totally real number field is a number field all of whose embeddings into its algebraic closure are real.
- A ring of integers O_K of a number field K is the set of all elements of the number field satisfying monic irreducible polynomials over $\mathbb{Z}$ or alternatively the integral closure of $\mathbb{Z}$ in the number field.

- A prime of a number field K is a non-zero prime ideal of O_K. If $x \neq 0$ and $x \in O_K$, then for any prime $\mathfrak{p}$ of K there exists a non-negative integer m such that $x \in \mathfrak{p}^m$ but $x \notin \mathfrak{p}^{m+1}$. We call m the order of x at $\mathfrak{p}$ and write $m = \text{ord}_{\mathfrak{p}} x$. If $y \in K$ and $y \neq 0$, we write $y = \frac{x_1}{x_2}$, where $x_1, x_2 \in O_K$ with $x_1 x_2 \neq 0$, and define $\text{ord}_{\mathfrak{p}} y = \text{ord}_{\mathfrak{p}} x_1 - \text{ord}_{\mathfrak{p}} x_2$. This definition is not dependent on the choice of x_1 and x_2 which are of course not unique. We define $\text{ord}_{\mathfrak{p}} 0 = \infty$ for any prime $\mathfrak{p}$ of K.
- Given $x \in K, x \neq 0$, for all but finitely many primes $\mathfrak{p}$ of K we have $\text{ord}_{\mathfrak{p}} x = 0$. We define a formal (finite) product

$$\mathfrak{n}(x) = \prod_{\text{ord}_{\mathfrak{p}} x > 0} \mathfrak{p}^{\text{ord}_{\mathfrak{p}} x}$$

and

$$\mathfrak{d}(x) = \mathfrak{n}\left(\frac{1}{x}\right) = \prod_{\text{ord}_{\mathfrak{p}} x < 0} \mathfrak{p}^{-\text{ord}_{\mathfrak{p}} x}.$$

If $x \in O_K$, then $\mathfrak{d}(x) = (1)$, the empty product. Of course the finite products of prime ideals of O_K also correspond to ideals of O_K. Further, finite products of prime ideals are called integral divisors and they form a semigroup under multiplication.
- Given an element $x \in \mathbb{Q}, x \neq 0$, we write $x = \frac{m}{n}, m, n \in \mathbb{Z}, n > 0, (m, n) = 1$ and define the height of x to be the $\max(|m|, |n|)$. Given $z \in K$, where K is a number field, we consider the monic irreducible polynomial $a_0 + a_1 T + \cdots + a_{n-1} T^{n-1} + T^n$ of z over $\mathbb{Q}$ and define the height of z, denoted by $h(z)$, to be the $\max(h(a_i), i = 0, \ldots, n)$.
- If K is a number field of degree n over $\mathbb{Q}$ and $\sigma_1 = \text{id}, \ldots, \sigma_n$ are all the embeddings of K into a fixed algebraic closure of $\mathbb{Q}$ and $x \in K$, then $\mathbf{N}_{K/\mathbb{Q}}(x) = \prod_{i=1}^{n} \sigma_i(x)$.

3.4.1 HTP over the Rings of Integers of Number Fields

The state of knowledge concerning the rings of integers and HTP is summarized in the theorem below.

Theorem 3.7 *$\mathbb{Z}$ is Diophantine and HTP is unsolvable over the rings of integers of the following fields:*

- *Extensions of degree 4 of $\mathbb{Q}$ (except for a totally complex extension without a degree-two subfield), totally real number fields and their extensions of degree 2. (See [6, 8].) Note that these fields include all abelian extensions.*
- *Number fields with exactly one pair of non-real embeddings (See [30, 48].)*
- *Any number field K such that there exists an elliptic curve E defined over $\mathbb{Q}$ with $E(\mathbb{Q})$ of positive rank with $[E(K) : E(\mathbb{Q})] < \infty$. (See [31, 33, 58].)*

- *Any number field K such that there exists an elliptic curve E defined over K with $E(K)$ of rank 1, and an abelian variety V defined over $\mathbb{Q}$ such that $V(\mathbb{Q})$ and $V(K)$ have the same rank. (See [2].)*

All the results above concerning rings of integers are derived by constructing a Diophantine definition of $\mathbb{Z}$ over the rings in question and they all follow what we have called elsewhere a "strong" or "weak vertical method".

Both methods rely on congruences to "force" an element t from a ring R of integers of a number field K to take values in $\mathbb{Z}$. These congruences are of the form

$$t \equiv n \bmod w \tag{3.4}$$

where $n \in \mathbb{Z}$ and $w \in R$ and of bigger height relative to t. In the strong version of the method the height of w is also large relative to n and this forces the equality $t = n$ to hold.

In the weak version of the method we don't have a bound on n, but we know that $w \in \mathbb{Z}$. In this case, again assuming the height of w is sufficiently large relative to t, we conclude that $t \in \mathbb{Q}$ but not necessarily equal to n.

One typical way to produce a congruence (3.4) is to isolate powers of a single unit ε in the ring of integers. (A unit is an invertible element of the ring.) If one succeeds in doing this, the elementary algebra produces the first ingredient of the congruence.

$$\frac{\varepsilon^{kn} - 1}{\varepsilon^k - 1} \equiv n \bmod (\varepsilon^k - 1)$$

in the ring of integers of the number field, where $k, n \in \mathbb{Z}_{>0}$. In other words we have a divisibility condition

$$\left(\varepsilon^k - 1\right) \Bigg| \left(\frac{\varepsilon^{kn} - 1}{\varepsilon^k - 1} - n\right)$$

in R. Thus, if we write

$$\frac{\varepsilon^{kn} - 1}{\varepsilon^k - 1} \equiv t \bmod (\varepsilon^k - 1),$$

then we are in fact writing $t \equiv n \bmod w$, with $w = \varepsilon^k - 1$, and we are part of the way there. Also, writing down equations affirming that the height of t is small relatively to $\varepsilon^k - 1$ is not that complicated. It can be done through a requirement that some polynomial in t divides $\varepsilon^k - 1$. It is not hard to show that given any algebraic integer u, there exists $k \in \mathbb{Z}_{>0}$ such that u divides $\varepsilon^k - 1$ in the ring of integers.

Now, if we want to use the strong vertical method, we need to make n small relative to the height of $\varepsilon^k - 1$. This, unfortunately, is rather hard and requires a pretty intimate knowledge of the equations involved. At the same time, if we want to use the weak vertical method, then we need a way to replace $\varepsilon^k - 1$ by some $w \in \mathbb{Z}$ so that w divides $\varepsilon^k - 1$ and at the same time is still large relative to t.

The weak vertical method can also be used to push t not necessarily all the way down to $\mathbb{Z}$ but maybe to a subfield M of the given field K, so that over M a different method, e.g. the strong one, can be used to complete the descent to $\mathbb{Z}$. If we are using the weak method just to get to a subfield M, we only need w to be in the ring of integers of M. This is often a lot easier, than satisfying the requirement that w is in $\mathbb{Z}$.

The strong vertical method was used by J. Denef over totally real number fields and by T. Pheidas and the author for the fields with one pair of non-real embeddings. (At the time of these results, the method did not have a name.) The construction of a Diophantine definition of $\mathbb{Z}$ over the ring of integers for all the other fields listed above used a weak vertical method. The equations used in all constructions were either norm or elliptic curve equations. The last result in Theorem 3.7 also used an abelian variety satisfying a stable rank condition. This condition is discussed in more detail in the next section. Here we would just like to explain briefly the use of norm equations and their limitations with respect to both methods.

The use of norm equations for both vertical methods depends on the interaction of ranks of unit groups in the rings of integers of number fields. First of all, the group of units inside every number field is of finite rank and we have a formula to compute the rank. If K is a number field of degree n over $\mathbb{Q}$ with r real embeddings and $2s$ non-real embeddings into the chosen algebraic closure of $\mathbb{Q}$, then the rank of the unit group is $r + s - 1$. (Non-real embeddings always come in pairs due to complex conjugation.)

Now if we have a totally real field K and its extension M of degree 2 such that it has exactly two real embeddings, we conclude that the difference in ranks of their unit groups is exactly one. Using the fact that the norm map $N_{M/K} : M \to K$ maps units to units and is a homomorphism of unit groups of M and K, from the rank calculation we conclude that the kernel of the map, i.e. the set of units whose norm is equal to 1 is a subgroup of the unit group of M of rank 1. (The rank of the kernel is the difference in the M and K unit group ranks.) A finitely generated multiplicative group of rank one is more or less a set of powers of a single element (possibly times elements of finite order, in our case roots of unity.) Writing down a polynomial equation computing the norm using the variables with values in K we can get a polynomial equation whose solutions effectively describe powers of a unit. This equation is quite well known under the name of Pell equation and has a number of convenient properties that we can leverage to bound the heights as described above. Thus we can proceed with the strong vertical method.

The first substantial applications of the weak vertical method (again long before the author of this narrative gave it a name) was due to Denef and Lipshitz [8] and it was also based on a calculation of the rank of unit groups comprised of solutions to norm equations. We show how the weak vertical method was used in this paper via the field diagram below where we assume that K is a totally real field of degree n over $\mathbb{Q}$, $[M : K] = 2$, $[F : K] = 2$, $F \cap M = K$, and G is the compositum of M and F (inside the chosen algebraic closure).

$$\begin{array}{ccc} M & \longrightarrow & G \\ \uparrow & & \uparrow \\ K & \longrightarrow & F \end{array}$$

Using the formula for computing ranks of unit groups one can choose a field F so that G has no real embeddings while the following equality holds:

$$\text{rank} \ker \mathbf{N}_{G/M} = \text{rank} \ker \mathbf{N}_{F/K}. \tag{3.5}$$

Indeed, let r_M be the number of real embeddings of M and $2s_M$ the number of non-real embeddings, so that $\frac{r_M}{2} + s_M = n$. Let r_F, s_F be the corresponding numbers for F with $\frac{r_F}{2} + s_F = n$. Using these notation and our assumptions on G we see that the left side of (3.5) is equal to $2n - r_M - s_M$ and the right side is equal to $r_F + s_F - n$. Thus we need $n - \frac{r_M}{2} = \frac{r_F}{2}$ or $2s_M = r_F$. In other words every embedding of K extended to a real embedding of M, should be extended to a non-real embedding of F and vice versa. Note that this condition on embeddings will also guarantee that all embeddings of G are non-real and $M \cap F = K$.

The final piece needed to use the weak vertical method comes from the following observation. Any unit of F with its K-norm equal to one is also a unit of G with the M-norm equal to one. This follows from the fact that $M \cap F = K$. Thus, $\ker \mathbf{N}_{F/K} \subseteq \ker \mathbf{N}_{G/M}$ and $\ker \mathbf{N}_{F/K}$ is of finite index in $\ker \mathbf{N}_{G/M}$ since $\ker \mathbf{N}_{G/M}$ is finitely generated.

Thus, if $\varepsilon \in \ker \mathbf{N}_{G/M}$, then for some fixed positive integer n independent of ε, it is the case that ε^n is actually an element of $\ker \mathbf{N}_{F/K} \subset F$. Now let $\varepsilon_1, \varepsilon_2 \in \ker \mathbf{N}_{G/M}$ and consider the equation

$$\frac{\varepsilon_2^r - 1}{\varepsilon_1^r - 1} = w \in O_G,$$

where $r \in \mathbb{Z}_{>0}$ and $r \equiv 0 \bmod n$. By the discussion above we can deduce that $w \in F$. So if we have a congruence

$$t \equiv w \bmod (\varepsilon_1^r - 1),$$

with the height of $\varepsilon_1^r - 1$ relatively large to $t \in O_M$, then $t \in O_M \cap F = O_K$. At the same time, if $\varepsilon_2 = \varepsilon_1^m$ with $m \in \mathbb{Z}_{>0}$, and $t = m$, then the congruence will hold. Thus we have a foundation for applying the weak vertical method in order to define O_K over O_M. Once we defined O_K, we can continue with the strong vertical method to get all the way down to $\mathbb{Z}$.

Even from this brief description of the way the norm equations are used in the construction of an existential definition of $\mathbb{Z}$, it is clear that this particular use of norm equations can work in special cases only, i.e. when the number fields are totally real or are not "far" from being totally real. So a different foundation for the vertical method is highly desirable. This new foundation is conjecturally provided by elliptic curves.

3.4.2 Positive Stable Rank Condition and Elliptic Curves

We now come back to the discussion of elliptic curves, curves defined by equation $y^2 = x^3 + ax + b$, but now with a, b possibly being algebraic integers while $\Delta = -16(4a^3 + 27b^2)$ is still not equal to 0. We will be looking for solutions to this equation in a specific number field K and will use them in a very different fashion to define integers compared to what we were doing over big rings to define a model of $\mathbb{Z}$. The idea to use elliptic curves with the weak vertical method, as the idea to use an elliptic curve for a model of $\mathbb{Z}$ over big rings, belongs to B. Poonen (see [32]).

The use of the weak vertical method is based on the following properties of points on elliptic curves. If we let P be a point of infinite order and let the affine coordinates of $[n]P$ corresponding to our equation be (x_n, y_n), then the following statements are true over any number field K:

1. Let $\mathfrak{A}$ be any integral divisor of K and let m be a positive integer. Then there exists $k \in \mathbb{Z}_{>0}$ such that $\mathfrak{A} \big| \mathfrak{d}(x_{km})$, where $\mathfrak{d}(x_{km})$ is the denominator of the divisor of x_{km} in the integral divisor semigroup of K.
2. There exists a positive integer m such that for any positive integers k, l,

$$\mathfrak{d}(x_{lm}) \Bigg| \mathfrak{n}\left(\frac{x_{lm}}{x_{klm}} - k^2\right)^2 \tag{3.6}$$

   in the integral divisor semigroup of K.

It is not hard to understand why the first assertion is true. The reasons are, in some sense, the same as for the assertion that a number field unit ε raised to a sufficiently high power is equivalent to 1 modulo any number field divisor. In both cases the reason is the finiteness of residue fields of primes.

Let P be a point of infinite order such that a prime $\mathfrak{p}$ of a number field K over which the curve is defined, does not occur in the denominators of the affine coordinates of P from some fixed Weierstrass equation of the elliptic curve. (We remind the reader that we can assume this equation is of the form $y^2 = x^3 + ax + b$, where a, b are integers of our number field. Thus, x and y have negative order at the same set of primes.)

We now consider our Weierstrass equation over the residue field of $\mathfrak{p}$ and for the sake of simplicity we will also assume that $\mathfrak{p}$ does not divide the discriminant of the original equation so that mod$\mathfrak{p}$ we are still looking at an elliptic curve. Given our assumption on the discriminant, P is mapped onto a non-zero element of the group of elliptic curve points. Since the field is finite, the group of points is finite and thus the image of P has a finite order r. Hence $[r]P$ is mapped to a point at infinity of the elliptic curve mod $\mathfrak{p}$. Therefore, $[r]P$ must have coordinates with negative order at $\mathfrak{p}$. Once $\mathfrak{p}$ makes it into the denominator, it will persist in all multiples of $[r]P$. Further, let p be the rational prime below $\mathfrak{p}$ (or the rational prime p such that $(p)O_K \subset \mathfrak{p}O_K$) and observe that properties of formal groups imply that $[pr]P$ will have a higher power of $\mathfrak{p}$ in the denominator of its coordinates. So any divisor of K will divide some multiple of P. This accounts for the first assertion above. The second assertion is a bit harder and also follows from properties of formal groups. A formal proof of both assertions can be found in [32].

Existence of a point of infinite order implies that the Mordell-Weil group, the group of points on the elliptic curve in question, is of positive rank. We will always have this assumption when discussing the use of elliptic curves for our definitional purposes. Unfortunately the properties above by themselves are not enough to make elliptic curves usable with the weak vertical method. We also need a *stable rank* assumption. We want the rank of Mordell-Weil group unchanged whether we look at points with coordinates in K or points with coordinates in some subfield L below. If the rank is unchanged then a fixed integer multiple of any point on the curve has its coordinates derived from our equation in L.

Assume for the purpose of simplification that every point on the curve has its coordinates in the field below and that the class number of K is 1, or in other words given an integral divisor $\mathfrak{A} = \mathfrak{p}_1 \dots \mathfrak{p}_K$ we can find an integer x such that $\mathfrak{n}(x) = \mathfrak{A}$ and therefore we can write any $y \in K$ as a ratio of two integers x_1 and x_2 with $\mathfrak{n}(x_1)$ being relatively prime to $\mathfrak{n}(x_2)$. Then we can write the x-coordinate of every point on the elliptic curve as a ratio of two algebraic integers which are relatively prime.

With these assumptions, we can now consider the following system of equations:

$$\left(\frac{u_i}{v_i}\right)^2 = \frac{a_i}{b_i}^3 + a\frac{a_i}{b_i} + b, i = 1, 2 \tag{3.7}$$

$$(a_3b_2) - t^2(b_3a_2) = b_3a_2b_2u \tag{3.8}$$

$$Ct^2(t^2+1)\dots(t^2+m)w = b_2 \tag{3.9}$$

Here $(\frac{a_i}{b_i}, \frac{u_i}{v_i})$, $i = 1, 2$ in (3.7) represent two points on our elliptic curve with coordinates written as ratios of integers. Equation (3.8) is the same as Eq. (3.6) but rewritten in terms of our variables taking integer values only. Finally (3.9) is the height bound equation, where m, C are positive integers depending on K only.

If for some element $t \in O_K$ we can find values for $u_1, v_1, u_2, v_2, a_1, b_1, a_2, b_2, u, w \in O_K$, then we can deduce from (3.7)–(3.9) that $t^2 - z \equiv 0 \bmod b_2$ in O_K, where $z \in O_L$ and b_2 is of much larger height than t. Thus, by the weak vertical method we conclude that $t \in L$. Note that we cannot conclude that $z \in \mathbb{Z}$. This would only follow if we also knew that our elliptic curve had rank equal to one and we would need additional equations. However, we do know that if t is an integer, than by arranging $(\frac{a_i}{b_i}, \frac{u_i}{v_i})$ to be multiples of the same infinite order point, as described above, we can find the values for other variables to satisfy the equations. Thus, applying the weak vertical method, we end up defining a subset of O_L containing $\mathbb{Z}$. This is actually enough to define O_L, because we can continue to define via ratios a subset of L containing $\mathbb{Q}$ and then using a basis of L over $\mathbb{Q}$ all elements of O_L.

The discussion above leaves us with two questions: how to get down to $\mathbb{Z}$ and when do we have a stable rank situation in the first place. We answer the second question first via a Theorem proved by B. Mazur and K. Rubin (see [26]).

Theorem 3.8 *Suppose K/L is a cyclic extension of prime degree of number fields. If the Shafarevich-Tate Conjecture is true for L, then there is an elliptic curve E over L with rank(E(L)) = rank(E(K)) = 1.*

Combining this theorem with the weak vertical method, we get an immediate corollary.

Corollary 3.4 *Suppose K/L is a cyclic extension of prime degree of number fields and the Shafarevich-Tate Conjecture is true for L. In this case O_L has a Diophantine definition over O_K.*

Returning to the first question we asked above about getting down to $\mathbb{Z}$, we are now in position to note that conditional on Shafarevich-Tate conjecture holding for all number fields, Corollary 3.4 implies that $\mathbb{Z}$ is existentially definable over O_L for all number fields L. Connecting the cyclic cases to an arbitrary extension L of $\mathbb{Q}$ takes several steps:

1. Let M be the Galois closure of L over $\mathbb{Q}$. In this case if $\mathbb{Z}$ has a Diophantine definition over O_M, then $\mathbb{Z}$ has a Diophantine definition over O_L. Thus without loss of generality, we can assume that L is Galois over $\mathbb{Q}$. The fact that we can always replace a given field by its finite extension, follows from the fact that any polynomial equation with variables ranging in a finite extension can be rewritten as an equivalent polynomial equation with variables ranging in a given field.
2. Let $L/\mathbb{Q}$ be a Galois extension of number fields. Let $K_1, \ldots, K_n$ be all the cyclic subextensions of L, i.e. all the subfields K_i of L such that L/K_i is cyclic. Observe that there are only finitely many such subextensions,
$$\bigcap_{i=1}^{n} K_i = \mathbb{Q},$$
$$\bigcap_{i=1}^{n} O_{K_i} = \mathbb{Z},$$
and therefore if each O_{K_i} has a Diophantine definition over O_L, then $\mathbb{Z}$ has a Diophantine definition over O_L. (Thus, it is enough to show that in every cyclic extension the ring of integers below has a Diophantine definition over the ring of integers above.)
3. If $L \subseteq H \subseteq M$ is a finite extension of number fields, O_H has a Diophantine definition over O_M, and O_L has a Diophantine definition over O_H, then O_L has a Diophantine definition over O_M. Thus, it is enough to consider cyclic extensions of prime degree only. The reason why this reduction works are pretty transparent. For example suppose $P_H(t, \bar{x})$ with $\bar{x} = (x_1, \ldots, x_r)$ is a Diophantine definition of O_L over O_H and let $P_M(t, \bar{y})$ be a Diophantine definition of O_H over O_M. Now consider the system

$$\begin{cases} P_H(t, \bar{x}) = 0 \\ P_M(t, \bar{y}_r) = 0 \\ P_M(x_1, \bar{y}_1) = 0 \\ \dots \\ P_M(x_r, \bar{y}_r) = 0 \end{cases}$$

Now it is not hard to see that this system has solutions over O_M if and only if $x_1, \dots, x_r \in O_H$ and $t \in O_L$. For a general discussion of reductions of this sort see [54] and Chap. 2 of [56].

The results above represent the state of our knowledge concerning the status of HTP over the rings of integers of number fields. We also know quite a few things about big subrings of number fields. One could say that the big ring problem is simultaneously easier and harder when considered over extensions. In the next section we start with the easier part.

3.4.3 Big Rings Inside Number Fields

The discussion of big rings requires a review of a few more definitions. As above K is a number field.

- Any prime ideal $\mathfrak{p}$ of O_K is maximal and the residue classes of O_K modulo $\mathfrak{p}$ form a field. This field is always finite and its size (a power of a rational prime number) is called the norm of $\mathfrak{p}$ denoted by $\mathbf{N}\mathfrak{p}$.
- If $\mathcal{W}$ is a set of primes of K, its natural density is defined to be the following limit if it exists:
$$\lim_{X \to \infty} \frac{\#\{\mathfrak{p} \in \mathcal{W}, \mathbf{N}\mathfrak{p} \leqslant X\}}{\#\{\mathbf{N}\mathfrak{p} \leqslant X\}}$$
- Let K be a number field and let $\mathcal{W}$ be a set of primes of K. Let $O_{K,\mathcal{W}}$ be the following subring of K.
$$\{x \in K : \text{ord}_{\mathfrak{p}} x \geqslant 0 \ \forall \mathfrak{p} \notin \mathcal{W}\}$$
If $\mathcal{W} = \emptyset$, then $O_{K,\mathcal{W}} = O_K$—the ring of integers of K. If $\mathcal{W}$ contains all the primes of K, then $O_{K,\mathcal{W}} = K$. If $\mathcal{W}$ is finite, we call the ring *small* (or the ring of $\mathcal{W}$-integers). If $\mathcal{W}$ is infinite, we call the ring *big*. These rings are the counterparts of the "in between" subrings of $\mathbb{Q}$.
- Given a field extension M/K and a prime ideal $\mathfrak{p}_K$ of K, we can talk about factorization of $\mathfrak{p}_K$ in M. In other words when we look at the ideal $\mathfrak{p}_K O_M$ of O_M, it might not be prime any more but a product of prime ideals of O_M. So, in general, in O_M we have $\mathfrak{p}_K = \prod_{i=1}^{k} \mathfrak{p}_{M,i}^{e_i}$, where $\mathfrak{p}_{M,1}, \dots, \mathfrak{p}_{M,k}$ are distinct prime ideals. We will call ideals of O_M occurring in the factorization of a prime ideal of

O_K conjugate over K and note the following property of the conjugate ideals. If $x \in O_K \subset O_M$ and $\text{ord}_{\mathfrak{p}_{M,i}} x < 0$ for some i, then $\text{ord}_{\mathfrak{p}_{M,j}} x < 0$ for all j.

Before discussing HTP and definitions of $\mathbb{Z}$ over big subrings of number fields, we should note that small subrings of number fields are also covered by the results of J. Robinson. To be more precise we have the following proposition (see [40]).

Proposition 3.7 *Let K be a number field and let $\mathfrak{p}_K$ be a prime of K. In this case the set $\{x \in K | ord_{\mathfrak{p}_K} x \geqslant 0\}$ has a Diophantine definition over K*

Now taking into account the fact that the set of non-zero elements has an existential definition over all small and big rings of any number field, we have the following corollary.

Corollary 3.5 *Let K be a number field and let $\mathscr{S}_K$ be a finite set of primes of K. In this case, O_K has a Diophantine definition over $O_{K,\mathscr{S}_K}$.*

Thus in all cases where we know HTP to be undecidable over the ring of integers of a number field, we also have that HTP is undecidable over any small subring of the field.

We now move on to big subrings. The main difference between the big subring situation over $\mathbb{Q}$ and over number fields is that we were able to construct a Diophantine definition of $\mathbb{Z}$ over some big subrings of *non-trivial* extensions of $\mathbb{Q}$. We describe these rings below.

Theorem 3.9 *Let K be a number field satisfying one of the following conditions:*

- *K is a totally real field.*
- *K is an extension of degree 2 of a totally real field.*
- *There exists an elliptic curve E defined over $\mathbb{Q}$ such that $[E(K) : E(\mathbb{Q})] < \infty$.*

Let $\varepsilon > 0$ be given. Then there exists a set $\mathscr{S}$ of non-archimedean primes of K such that

- *The natural density of $\mathscr{S}$ is greater $1 - \dfrac{1}{[K : \mathbb{Q}]} - \varepsilon$.*
- *$\mathbb{Z}$ is Diophantine over $O_{K,\mathscr{S}}$.*
- *HTP is unsolvable over $O_{K,\mathscr{S}}$.*

(See [52, 53, 55, 58, 59].)

One immediately notices that all the fields to which our theorem applies are the fields where we have definitions of $\mathbb{Z}$ over the rings of integers. This is not an accident of course. Given a number field extension M/K and an integrally closed subring R of M, call the problem of defining $R \cap K$ over R a "vertical" problem and call the problem of defining R over M a "horizontal" problem. (So a vertical problem involves an algebraic extension and a horizontal problem does not involve algebraic extensions, i.e. everything takes place inside the same field.) Using these terms, one could say that we learned how to solve a vertical problem over the fields mentioned in Theorem 3.9 and, using an observation concerning conjugate prime ideals, one can adapt these vertical solutions for horizontal purposes. In other words, let K be a number field and let $\mathscr{W}_K$ be a collection of prime ideals of K with the following property: all but finitely many ideals in $\mathscr{W}_K$ have a distinct conjugate over K such that this conjugate is not in $\mathscr{W}_K$. In this case $O_{K,\mathscr{W}_K} \cap \mathbb{Q} = O_{\mathbb{Q},\mathscr{S}_K}$, where $\mathscr{S}_K$ is either finite or empty and thus either $O_{\mathbb{Q},\mathscr{S}_K} = \mathbb{Z}$ or $\mathbb{Z}$ has a Diophantine definition of $O_{\mathbb{Q},\mathscr{S}_K}$. So to define $\mathbb{Z}$ over $O_{K,\mathscr{W}_K}$ with this type of $\mathscr{W}_K$ it is enough to define $O_{K,\mathscr{W}_K} \cap \mathbb{Q}$, that is to solve a vertical problem.

Relative to solving the corresponding vertical problem over the ring of integers, over $O_{K,\mathscr{W}_K}$ there are some additional difficulties related to bounding of heights, but the overall design of the weak vertical method is unchanged. One should note that by construction the density of the set of the inverted primes can never be one. To get results concerning big rings where the density of inverted primes is one we need Poonen's method and an elliptic curve of rank one.

There are various generalizations of Poonen's theorem to number fields. However the situation is more complicated over a number field and instead of constructing a model of $\mathbb{Z}$ by "approximation", what is constructed there is a model of a subset of the rational integers over which one can construct a model of $\mathbb{Z}$. In short, one constructs a "model of a model" (see [36].) There are also analogs of Theorems 3.5 and 3.6 (see [11, 60]). As in the case of the rings of integers, these big ring results extend to all number fields but only conjecturally depending as they are on Shafarevich-Tate conjecture. The situation is different, however, with the "other end of the spectrum" results. As we will see below, they extend seamlessly to all number fields. Before we get to those results, we need to say a few words about presentations of number fields, their primes and their big subrings.

So far most of the results we have discussed above concerning number fields are definitional in nature and do not require a discussion of the presentation of the object involved, just the language in which the definitions are made. The language of course is the language of rings, possibly with finitely many additional constants. However, when we start talking about undecidability we do need to worry about how the objects are presented. Of course number fields and rings of integers have very easy, naturally computable presentations in terms of an integral basis over $\mathbb{Q}$. (If we choose an integral basis, then the ring of integers can be generated as a $\mathbb{Z}$-module from the basis.) The situation becomes more complicated when we discuss big subrings (or even small subrings). Big subrings of $\mathbb{Q}$ are computable inside a standard presentation of $\mathbb{Q}$ precisely when the set of primes allowed in the denominator is computable. If the set is c.e. but not computable, then as we pointed out above, the ring has a computable presentation, just not as a part of the computable presentation of $\mathbb{Q}$.

The situation in the big subrings of number fields is similar since we have a computable way to describe the primes of a number field. If a number field K is given by the minimal polynomial of its generator (inside a computable presentation of $\tilde{\mathbb{Q}}$, this generator can be given explicitly), and we choose a rational prime p, then within the standard computable presentation of K, using the power basis of the field generator, we can algorithmically determine the number of distinct factors p has in K. Further for each factor we can effectively find an algebraic integer such that this integer has order one at this factor but not at any other factors of p. For a factor $\mathfrak{p}$

of p let $a(\mathfrak{p})$ be this algebraic integer. Now we can represent the prime $\mathfrak{p}$ by the pair $(p, a(\mathfrak{p}))$, where $a(\mathfrak{p})$ is given by its coordinates with respect to the fixed basis of K over $\mathbb{Q}$. Further, given an element x of K, we can effectively compute $\mathfrak{n}(x)$ and $\mathfrak{d}(x)$ in terms of our presentation of primes and assuming that $\mathscr{W}_K$ is a computable set of primes, we can determine whether $x \in O_{K,\mathscr{W}_K}$.

We can re-use the identification of HTP of a particular ring with a subset of positive integers containing the indices of all polynomials with coefficients in the ring having a root in the ring. Our next observation is that HTP(K) $\leqslant_T$ HTP($\mathbb{Q}$) since we can rewrite any polynomial equation over K with variables ranging over K as a system of polynomial equations over $\mathbb{Q}$ with variables ranging over $\mathbb{Q}$. Further, as over $\mathbb{Q}$, we also have for any $\mathscr{W}_K$ that HTP($O_{K,\mathscr{W}_K}$) $\geqslant_T$ HTP(K). Finally, we also have the following results from [12].

Theorem 3.10 *For any number field K there exist a c.e. set $\mathscr{W}_K$ of primes of K of lower density equal to zero such that $HTP(O_{K,\mathscr{W}_K}) \equiv_T HTP(K) \leqslant_T HTP(\mathbb{Q})$ and $\mathscr{W}_K \leqslant_T HTP(K)$.*

Theorem 3.11 *For any number field K and any positive integer m there exist sets $\mathscr{W}_1, \ldots, \mathscr{W}_m$ of primes of K such that $\mathscr{W}_i \leqslant_T HTP(K)$, each $\mathscr{W}_i$ has a lower density zero and $\mathscr{W}_1 \cup \cdots \cup \mathscr{W}_m$ is a partition of all primes of K.*

As over $\mathbb{Q}$ we can also also ask what do we know about definability and decidability using the full first-order theory.

3.4.4 Universal and Existential Together in Extensions

One could argue that J. Robinson solved most of the natural first-order definability and decidability questions over number fields. Before describing this aspect of her results, we should note that in addition to the old questions of decidability and definability of $\mathbb{Z}$ and the rings of integers, we also have a question of uniformity of definitions across all number fields. The question of uniformity is a new question in our discussion. It naturally does not arise when we discuss $\mathbb{Q}$ only, and as far as existential definitions over number fields are concerned, we are very far away from being able to address such questions. However, as we will see below, the situation is different when we use the full first-order language.

In [40], J. Robinson constructed a first-order definition of $\mathbb{Z}$ over the ring of integers for every number field. Amazingly she used only *one* universal quantifier to do it. These definitions were not, however, uniform across number fields. Later on in [41], J. Robinson constructed a definition which was uniform across all number fields but was using more universal quantifiers. Rumely [45] constructed a version of these definitions uniform across global fields (number fields and function fields over a finite set of constants). Finally, In the same paper where B. Poonen constructed a two-universal-quantifier definition of $\mathbb{Z}$ over $\mathbb{Q}$, he constructed a uniform two-universal-quantifier definition of the ring of integers across all number fields.

So far J. Koenigsmann's one-universal-quantifier result has not been extended to any number fields, but J. Park constructed a purely universal definition of the rings of integers over all number fields in [28].

3.4.5 Open Questions over Number Fields

We hope that from our narrative it is clear that there is no shortage of open problems. In fact one could get an impression that for every question answered at least two open ones appear. There are many ways to organize these questions. We choose to divide them into two main collections: questions of definability and questions of Turing reducibility, including questions of decidability. Now the questions of definability can also be divided into many other categories. Given two rings R_1 and R_2 with fraction fields K_1 and K_2 being number fields, one can pose a number of definability problems.

1. If $R_1 \subset R_2$ we can ask whether R_1 has a Diophantine definition over R_2. If $R_1 \not\subset R_2$ then one can ask whether $R_1 \leqslant_{Dioph} R_2$ or whether R_1 is Dioph-generated over R_2. Diophantine generation is defined as follows. Let K be a number field containing both K_1 and K_2 and let $\omega_1, \ldots, \omega_n$ be any basis of K over K_2. Now consider the question of existence of a Diophantine subset $A \subset R_2^{n+1}$ such that $(a_1, \ldots, a_n, b) \in A \Rightarrow b \neq 0$ and $R_1 = \{\sum_{i=1}^n \frac{a_i}{b}\omega_i | (a_1, \ldots, a_n, b) \in A\}$. (For more details on Diophantine generation see [56].)
2. More generally, we can ask whether R_2 has a Diophantine model of R_1 or a class Diophantine model of R_1. A class Diophantine model corresponds to what model theorists call a Diophantine interpretation and is a map which establishes a correspondence between R_1 and equivalence classes of elements of R_2 under a Diophantine equivalence relation. Further, there should be a Diophantine description of the class of the products and sums. (For an example of a class Diophantine model see [60].)

Under any of these definability relations between R_1 and R_2 we can conclude that

$$\text{HTP}(R_1) \leqslant_T \text{HTP}(R_2).$$

Of course we can ask the weaker question of Turing reducibility directly about R_1 and R_2. Further, it would be interesting to see an example where we have Turing reducibility but no definability. As discussed above, we suspect that something like this may be true with respect to $\mathbb{Q}$ and $\mathbb{Z}$ but no version of the assertion claiming Turing reducibility without definability has been proved so far. One simple example which illustrates the difficulty of these questions is described below.

Question 3.1 Let $R \subset \mathbb{Q}$ be a big ring. Let p be a rational prime number not inverted in R, and let $\hat{R} = R[\frac{1}{p}]$. In this case is $\hat{R}$ definable in any way over R (via Dioph-generation, Diophantine model, or Diophantine interpretation)? Is $\text{HTP}(\hat{R}) \leqslant_T \text{HTP}(R)$?

Note that if $\mathbb{Z}$ is Diophantine over R, than all of these questions can easily be answered in the affirmative. If we can define $\mathbb{Z}$, then we can define powers of p and thus $\hat{R}$. In some rings we can generate powers of some primes without defining $\mathbb{Z}$ but the general case remains quite vexing.

3.5 Infinite Extensions

In this section we want to discuss some of the things we know about infinite algebraic extensions of $\mathbb{Q}$ and point out the differences and similarities with the finite extension case. We start with a description of the global picture to the best of our understanding.

Let $\tilde{\mathbb{Q}}$ be a fixed algebraic closure of $\mathbb{Q}$ and consider a journey from $\mathbb{Q}$ to its algebraic closure, passing through the finite extensions of $\mathbb{Q}$ first, then through its infinite extensions fairly "far" from the algebraic closure, and finally through the infinite extensions of $\mathbb{Q}$ fairly "close" to $\tilde{\mathbb{Q}}$.

As we get closer to $\tilde{\mathbb{Q}}$, the language of rings looses more and more of its expressive power, i.e. sets which were definable before (in either full first-order language or existentially) would become undefinable and simultaneously some problems which were undecidable before would become decidable. For the author the ultimate goal of the infinite extension investigation in this setting is to describe this transition. Of course, the completion of this project is probably far away. The boundary (in terms of extensions of $\mathbb{Q}$) where previously undecidable things become decidable (e.g. the first-order theory of fields) and previously definable things become undefinable (e.g. rings of integers over their fields of fractions using the full first-order language) is likely to have a very complex description.

Further, the decidability issue is muddled by the following aspect of the problem which does not manifest itself over finite extensions. It can be shown that an algebraic extension of $\mathbb{Q}$ with a decidable existential theory (a fortiori a decidable first-order theory) must have an isomorphic computable copy inside a given algebraic closure of $\mathbb{Q}$. (See [18].) Thus, a field can have an undecidable theory (existential or elementary) simply because it has no decidable conjugate (under the action of the absolute Galois group) and not because of, should we say, "arithmetic" reasons. We are tempted to call such fields as having a "trivially" undecidable theory.

A simple example of a field with a trivially undecidable theory can be described as follows. Consider a computable sequence of prime numbers $\{p_i\}$ and choose an undecidable subset A of $\mathbb{Z}_{>0}$. Now let K be an algebraic extension of $\mathbb{Q}$ formed by adding a square root of every p_i such that $i \in A$. It is not hard to see that this field is not computable as a subfield of $\tilde{\mathbb{Q}}$, but if A is c.e. it is computably presentable. It is Galois and has no other conjugates besides itself. So by the argument above the existential theory of this field (in the language of rings) is undecidable, but surely this is not a very interesting case. A related point which should be made here is that if we consider uncountably many isomorphism classes of fields, then "most" of them will have undecidable theories simply because we have only countably many decidable theories in the language of rings.

Having moved the trivial considerations aside and concentrating on computable fields, we discover a relatively patchy state of knowledge concerning definability and decidability. If we look at the fields "close" to the algebraic closure, we see a number of decidability results. Here, of course, the problem concerning the full first-order theory is the more difficult one as compared to the problem of the existential theory. One of the more influential decidability results is due to Rumely [46], where he showed that Hilbert's Tenth Problem is decidable over the ring of all algebraic integers. This result was strengthened by L. van den Dries proving in [63] that the first-order theory of this ring was decidable. Another remarkable result is due to Fried et al. [15], where it is shown that the first-order theory of the field of all totally real algebraic numbers is decidable. This field quite possibly is a part of the decidability/undecidabilty boundary we talked about above, since J. Robinson showed in [41] that the first-order theory of the ring of all totally real integers is undecidable. Together these two results imply that the ring of integers of this field is not first-order definable (in any way) over the field.

Among other famous decidability results is the result due to A. Prestel who building on a result of A. Tarski showed that the elementary theory of the field of all real algebraic numbers is decidable (see [38, 62]). Further, due to Yu. Ershov, we know that the field of all $\mathscr{S}$-adic algebraic numbers is decidable provided $\mathscr{S}$ is a finite set of rational primes. (The field of all $\mathscr{S}$-adic algebraic numbers is the intersection of all $\tilde{\mathbb{Q}} \cap \mathbb{Q}_p$, $p \in \mathscr{S}$ with $\tilde{\mathbb{Q}}$ being some fixed algebraic closure of $\mathbb{Q}$. See [13].) The rings of integers of the fields of real and p-adic algebraic numbers are decidable too. (See [37].)

We now turn our attention to definability and undecidability results. We have already mentioned a well-known result of J. Robinson proving that the ring of integers of the field of all totally real integers is undecidable. In the same paper, J. Robinson also outlined a plan for showing undecidability of families of rings of integers. Using some of these ideas, their further elaboration by C.W. Henson (see [63, p. 199]), and R. Rumely's method for defining integrality at a prime, C. Videla produced the first-order undecidability results for a family of infinite algebraic extensions of $\mathbb{Q}$ in [64–66]. More specifically, C. Videla showed that the first-order theory of some totally real infinite quadratic extensions, any infinite cyclotomic extension with a single ramified prime, and some infinite cyclotomic extensions with finitely many ramified primes is undecidable. C. Videla also produced the first result concerning definability of the ring of integers over an infinite algebraic extension of $\mathbb{Q}$: he showed that if all finite subextensions are of degree equal to a product of powers of a fixed (for the field) finite set of primes, then the ring of integers is first-order definable over the field.

In a recent paper [17], K. Fukuzaki, also using R. Rumely's method, proved that a ring of integers is definable over an infinite Galois extension of the rationals such that every finite subextension has odd degree over the rationals and its prime ideals dividing 2 are unramified. He then used one of the results of J. Robinson to show that a large family of totally real fields contained in cyclotomics (with infinitely many ramified primes) has an undecidable first-order theory.

In another recent paper (see [47]), the author attempted to determine some general structural conditions allowing for a first-order definition of the ring of integers over its fraction field over infinite algebraic extensions of $\mathbb{Q}$. As we speculated above, a definitive description of such conditions is probably far away, but one candidate is the presence or the absence of what we called "q-boundedness" for all rational primes q. We offer an informal description of this condition below.

Given an infinite algebraic extension K_{inf} of $\mathbb{Q}$ we consider what happens to the local degrees of primes over $\mathbb{Q}$ as we move through the factor tree within K_{inf}. A rational prime p is called *q-bounded* if it lies on a path through the factor tree in K_{inf} where the local degrees of its factors over $\mathbb{Q}$ are **not** divisible by arbitrarily high powers of q. If every descendant of p in every number field contained in K_{inf} has the same property, then we say that p is hereditarily q-bounded.

For q itself we require a stronger condition: the local degrees along all the paths of the factor tree should have uniformly bounded order at q. If this condition is satisfied, we say that q (or some other prime in question) is completely *q-bounded.* If all the primes $p \neq q$ are hereditarily q-bounded and q is completely q-bounded, we say that the field K_{inf} itself is q-bounded, and we show that the ring of integers is definable in such a field. Rings of integers are also definable under some modifications of the q-boundedness assumptions, such as an assumption that all primes $p \neq q$ are hereditarily q-bounded and q is completely t-bounded for some prime $t \neq q$, etc.

It is not hard to see that the fields considered by C. Videla and K. Fukuzaki are in fact q-bounded. As mentioned above, C. Videla's results concerned infinite Galois extensions of number fields, where all the finite subextensions are of degree divisible only by primes belonging to a fixed *finite* set of primes A. Consequently, in the fields considered by C. Videla all the primes are completely q-bounded for any $q \notin A$, and thus all these fields are certainly q-bounded. K. Fukuzaki's fields are 2-bounded. However, the examples constructed by C. Videla and K. Fukuzaki do not exhaust all the q-bounded fields. One example not covered by these authors is any field where for some fixed rational prime q and some fixed $m \in \mathbb{Z}_{>0}$ we can adjoin to $\mathbb{Q}$ all ℓ^n-th roots of unity for any positive integer n and for any rational prime ℓ such that q^m does not divide $\ell - 1$.

We suspect that q-boundedness or a similar condition is necessary for definability of the ring of integers. While non-definability examples are scarce over infinite extensions, we offer the following ones: the field of all totally real numbers is not q-bounded and as we mentioned above has the ring of integers not definable over the field. Further, the field of real algebraic numbers is also not q-bounded and its ring of integers is not definable over the field by a result of Tarski [62].

3.5.1 Defining Integers Via Norm Equations

In this section we explain some ideas behind our definitions of integers in [47]. The central part of our construction is a norm equation which has no solutions if a field element in question has "forbidden" poles. (In an effort to simplify terminology we transferred some function field terms to this number field setting.) The first practitioners of this method were J. Robinson using quadratic forms and R. Rumely using more general norm equations. The author of this paper has generally employed a distinct variation of the norm method. More specifically, as explained below, the bottom field in the norm equation is not fixed, but is allowed to vary depending on the elements involved. As long as the degree of all extensions involved is bounded, such a "floating" norm equation is still (effectively) translatable into a system of polynomial equations over the given field. To set up the norm equation, let

- q be a rational prime number,
- K be a number field containing a primitive q-th root of unity,
- $\mathfrak{p}_K$ be a prime of K not dividing q,
- $b \in K$ be such that $\text{ord}_{\mathfrak{p}_K} b = -1$,
- $c \in K$ be such that c is integral at $\mathfrak{p}_K$ and is not a q-th power in the residue field of $\mathfrak{p}_K$,

and consider $bx^q + b^q$. Note that $\text{ord}_{\mathfrak{p}_K}(bx^q + b^q)$ is divisible by q if and only if $\text{ord}_{\mathfrak{p}_K} x \geqslant 0$. Further, if x is an integer, all the poles of $bx^q + b^q$ must be poles of b and are divisible by q. Assume also that all zeros of $bx^q + b^q$ and all zeros and poles of c are of orders divisible by q and $c \equiv 1 \mod q^3$. Finally, to simplify the situation further, assume that either K has no real embeddings or $q > 2$. Now consider the norm equation

$$\mathbf{N}_{K(\sqrt[q]{c})/K}(y) = bx^q + b^q. \tag{3.10}$$

Since $\mathfrak{p}_K$ does not split in this extension, if x has a pole at $\mathfrak{p}_K$, then $\text{ord}_{\mathfrak{p}_K} bx^q + b^q \not\equiv 0 \mod q$, and the norm equation has no solution y in $K(\sqrt[q]{c})$. Further, if x is an integer, given our assumptions, using the Hasse Norm Principle we can show that this norm equation does have a solution. Our conditions on c insure that the extension is unramified, and our conditions on $bx^q + b^q$ in the case x is an integer make sure that locally *at every prime not splitting in the extension* the element $bx^q + b^q$ is equal to a q-th power of some element of the local field times a unit. By the Local Class Field Theory, this makes $bx^q + b^q$ a norm locally at every prime.

For an arbitrary b and $c \equiv 1 \mod q^3$ in K, we will not necessarily have all zeros of $bx^q + b^q$ and all zeros and poles of c of orders divisible by q. For this reason, given $x, b, c \in K$ we consider our norm equation in a finite extension L of K and this extension L depends on x, b, c and q. We choose L so that all primes occurring as zeros of $bx^q + b^q$ or as zeros or poles of c are ramified with ramification degree divisible by q. We also take care to split $\mathfrak{p}_K$ completely in L, so that in L we still have that c is not a q-th power modulo any factor of $\mathfrak{p}_L$. This way, as we run through all $b, c \in K$ with $c - 1 \equiv 0 \mod q^3$, we "catch" all the primes that do not divide q and occur as poles of x.

Unfortunately, we will not catch factors of q that may occur as poles in this manner, because our assumption on c forces all the factors of q to split into distinct factors in the extension. Splitting factors of q into distinct factors protects us from a situation where such primes may ramify and cause the norm equation not to have solutions even when x is an integer. Elimination of factors of q from the denominators of the divisors of the elements of the rings we define is done separately.

The end result of this construction is essentially a uniform definition of the form $\forall\forall\exists\ldots\exists$ of the ring of $\mathcal{Q}$- integers, with $\mathcal{Q}$ containing factors of q, across all number fields containing the q-th primitive roots of unity.

Putting aside for the moment the issue of defining the set of all elements c integral at q and equivalent to 1 mod q^3, and the related issue of defining integrality at factors of q in general, we now make the transition to an infinite q-bounded extension K_{inf} by noting the following. Let $K \subset K_{\text{inf}}$, let $\mathfrak{p}_K$ be a prime of K such that $\mathfrak{p}_K$ does not divide q, let $x \in K$ and let $\text{ord}_{\mathfrak{p}_K} x < 0$. Since by assumption $\mathfrak{p}_K$ is q-bounded, it lies along a path in its factor tree within K_{inf}, where the order at q of local degrees eventually stabilizes. To simplify the situation once again, we can assume that it stabilizes immediately past K. So let N be another number field with $K \subset N \subset K_{\text{inf}}$. In this case for some prime $\mathfrak{p}_N$ above $\mathfrak{p}_K$ in N, we have that $\text{ord}_q e(\mathfrak{p}_N/\mathfrak{p}_K) = \text{ord}_q f(\mathfrak{p}_N/\mathfrak{p}_K) = 0$. (Here $e(\mathfrak{p}_N/\mathfrak{p}_K)$ is the ramification degree and $f(\mathfrak{p}_N/\mathfrak{p}_K)$ is the relative degree.) Now, let $b, c \in K$ be as above and observe that c is not a q-th power in the residue field of $\mathfrak{p}_N$ while $\text{ord}_{\mathfrak{p}_N}(bx^q + b^q) \not\equiv 0 \bmod q$. Thus the corresponding norm equation with K replaced by N and eventually by K_{inf} in (3.10) has no solution. Of course when x is an integer and we have a solution to our norm equation in K, we also have a solution in K_{inf}.

Note that for each prime $\mathfrak{p}_K$ of K, at every higher level of the tree we need just one factor with the local degree not divisible by q to make the norm equation unsolvable when $\mathfrak{p}_K$ appears in the denominator of the divisor of x. Hence having one q-bounded path per every prime of K is enough to make sure that no prime of K not dividing q occurs as a pole of any element of K in our set.

Unfortunately, if we go to an extension of K inside K_{inf}, some primes of K will split into distinct factors and can occur independently in the denominators of the divisors of elements of extensions of K. Thus, in the extensions of K inside K_{inf} we have to block each factor separately. This is where the "hereditary" part comes in. We need to require the same condition of q-boundedness for every descendant in the factor tree of every prime of K not dividing q, insuring integrality at all factors of all K-primes not dividing q.

The main reason that only one q-bounded path per prime not dividing q is enough to construct a definition of integers, is that the failure of the norm equation to have a solution locally at any one prime is enough for the equation not to have solutions globally. Conversely, in order to have solutions globally, we need to be able to solve the norm equations locally at all primes. As already mentioned above, the reason we require c to be integral at q and equivalent to 1 mod q^3 is to make sure that factors of q do not ramify when we take the q-th root of c. Just making c have order divisible by q at all primes does not in general guarantee that factors of q do not ramify in such an extension. If any factor of q does ramify, then not all local units at this factor are norms in the extension, and making sure that the right side of the norm equation has order divisible by q at all primes might not be enough to guarantee a global solution. Hence we need to control the order of $c - 1$ at *all* factors of q at every level of the factor tree simultaneously, necessitating a stronger assumption on q, than on other primes.

Depending on the field we might have a couple of options as far as integrality at q goes. If q happens to be completely p-bounded in our infinite extension for some $p \neq q$, then we can pretty much use the same method as above with p-th root replacing the q-th root. The only difference is that, assuming we have the primitive p-th root of unity in the field, by definition of a complete p-boundedness, we can fix an element c of the field such that c is not a p-th power modulo any factor of q in any finite subextension of K_{inf} containing some fixed number field. We can also fix an element b of the field such that the order of b at any factor of q is not divisible by p in any finite subextension of K_{inf} containing the same fixed number field as above. Using such elements c and b we can get an *existential* definition of a subset of the field containing all elements with the order at any factor of q bounded from below by a bound depending on b and p. If ramification degrees of factors of q are altogether bounded, then we can arrange for this set to be the set of all field elements integral at factors of q, but in a general case the bound from below will be negative. In this case to obtain the definition of integrality we will need one more step.

Before going back to infinite extensions, we would like to make a brief remark about the sets definable by our methods over number fields. First of all, over any number field all primes are completely p-bounded for every p, and the ramification degree of factors of q is altogether bounded. So we can produce an existential and uniform (with parameters) definition of integrality at all factors of q. Note also that the complement of such a set is also uniformly existentially definable with parameters using the same method. So, in summary, we now obtain a uniform definition of the form $\forall\forall\exists\ldots\exists$ of the ring of integers of any number field with a q-th primitive root of unity. This result is along the lines of B. Poonen's result in [35], though his method is slightly different from ours since it uses ramified primes rather than non-splitting primes to obtain integrality formulas and restricts the discussion to $q = 2$ and quadratic forms. As B. Poonen, we can also use $q = 2$ and thus have a two-universal quantifier formula uniformly covering all number fields, but in this case if K has real embeddings, we need to make sure that c satisfies some additional conditions in order for the norm equations to have solutions.

Returning now to the case of infinite extensions, we note that, assuming q is p-bounded we now have a uniform first-order definition with parameters of algebraic integers across all q-bounded algebraic extensions of $\mathbb{Q}$ where q is completely p-bounded. However, for the infinite case we may require more universal quantifiers. The number of these universal quantifiers will depend on the whether the ramification degree of factors of q is bounded and on whether q has a finite number of factors.

The only case left to consider now is the case where q is not completely p-bounded for any $p \neq q$ but is completely q-bounded. This case requires a somewhat more technically complicated definition than the case where we had a requisite p. In particular, we still need a cyclic extension (once again of degree q), where all the factors of q will not split. Such an extension does exist, but we might have to extend our field to be in a position to take advantage of it.

3.5.2 Defining $\mathbb{Z}$ Using Elliptic Curves with Finitely Generated Groups over the Given Field and One Completely q-Bounded Prime

We now return to some ideas we used over number fields: using elliptic curves and the weak vertical method. Below we give an informal description of a construction of a definition of a number field K over an infinite algebraic extension K_{inf} of $\mathbb{Q}$ using an elliptic curve with a Mordell-Weil group generated by points defined over K. This construction also requires one completely q-bounded prime p (which may equal to q). Observe that once we have a definition of K, a (first-order) definition of $\mathbb{Z}$ follows from a result of J. Robinson.

The use of elliptic curves in the context of definability over infinite extensions also has a long history, as long as the one for norm equations and quadratic forms. Perhaps the first mention of elliptic curves in the context of the first-order definability belongs to Robinson [42] and in the context of existential definability to Denef [7]. Following Denef [8], as has been mentioned above, the author also considered the situations where elliptic curves had finite rank in infinite extensions and showed that when this happens in a totally real field one can existentially define $\mathbb{Z}$ over the ring of integers of this field and the ring of integers of any extension of degree 2 of such a field (see [59]). C. Videla also used finitely generated elliptic curves to produce undecidability results. His approach, as discussed above, was based on an elaboration by C.W. Henson of a proposition of J. Robinson and results of D. Rohrlich (see [44]) concerning finitely generated elliptic curves in infinite algebraic extensions.

The main idea behind our construction can be described as follows. Given an element $x \in K_{\text{inf}}$, we write down a statement saying that x is integral at p and for every $n \in \mathbb{Z}_{>0}$ we have that x equivalent to some element of K mod p^n. By the weak vertical method, this is enough to "push" x into K. Our elliptic curve as above is the source of elements of K. Any solution to an affine equation $y^2 = x^3 + ax + b$ of our elliptic curve must by assumption be in K. Further if we let P be a point of infinite order and let the affine coordinates of $[n]P$ corresponding to our equation be (x_n, y_n), then we remind the reader that the following statements are true:

1. Let $\mathfrak{A}$ be any integral divisor of K and let m be a positive integer. Then there exists $k \in \mathbb{Z}_{>0}$ such that $\mathfrak{A} \big| \mathfrak{d}(x_{km})$, where $\mathfrak{d}(x_{km})$ is the denominator of the divisor of x_{km} in the integral divisor semigroup of K.
2. There exists a positive integer m such that for any positive integers k, l,

$$\mathfrak{d}(x_{lm}) \Big| \mathfrak{n}\left(\frac{x_{lm}}{x_{klm}} - k^2\right)^2$$

in the integral divisor semigroup of K. Here $\mathfrak{d}(x_{lm})$ as above refers to the denominator of the divisor of x_{lm} and $\mathfrak{n}\left(\frac{x_{lm}}{x_{klm}} - k^2\right)$ refers to the numerator of the divisor of $\frac{x_{lm}}{x_{klm}} - k^2$.

Given $u \in K_{\text{inf}}$ integral at some fixed K-prime $\mathfrak{p}_K$, we now consider a statement of the following sort: $\forall z \in K_{\text{inf}}$ there exists $x, y, \hat{x}, \hat{y} \in K_{\text{inf}}$ such that $(x, y), (\hat{x}, \hat{y})$ satisfy the chosen elliptic curve equation and both $\frac{1}{zx}$ and $x(u^2 - \frac{x}{\hat{x}})^2$ are integral at $\mathfrak{p}_K$ implying that $\frac{(u^2-\frac{x}{\hat{x}})^2}{z}$ is integral at $\mathfrak{p}_K$.

If u satisfies this formula, then since $\frac{x}{\hat{x}} \in K$, by the weak vertical method we have that $u \in K$. Further, if u is a square of an integer, this formula can be satisfied. Thus we can proceed to make sure our definition includes all integers, followed by a definition including all rational numbers as ratios of integers, and eventually all of K through a basis of K over $\mathbb{Q}$. Consequently, at the end of this process we obtain a first-order definition of K over K_{inf} and thus obtain a first-order definition of $\mathbb{Z}$ over K_{inf}. Finally, being able to define $\mathbb{Z}$ implies undecidability of the first-order theory of the field.

3.5.3 Converting Definability to Undecidability over infinite extensions

For certain totally real fields one can easily convert definability results into undecidability results. A result of J. Robinson implies that if a ring of integers has a certain invariant which C. Videla called a "Julia Robinson number", one can define a first-order model of $\mathbb{Z}$ over the ring. The Julia Robinson number s of a ring R of totally real integers is a real number s or ∞, such that $(0, s)$ is the smallest interval containing infinitely many sets of conjugates of numbers of R, i.e., infinitely many $x \in R$ with all the conjugates (over $\mathbb{Q}$) in $(0, s)$. A result of Kronecker implies that $s \geqslant 4$ (see [20]), and therefore if a totally real ring of integers in question contains the real parts of infinitely many distinct roots of unity, the Julia Robinson number for the ring is indeed 4, and we have the desired undecidability result. Thus using our definability results we can conclude that for any fixed rational prime q and positive integer m the elementary theory of the largest totally real subfield of the cyclotomic field $\mathbb{Q}(\xi_{\ell^n}, n \in \mathbb{Z}_{>0}, \ell - 1 \not\equiv \bmod q^m)$ is undecidable.

One can also obtain undecidability results for elementary theory of fields where we know the integral closure of some rings of $\mathscr{S}$-integers to be existentially undecidable. (See [51, 57]). With the help of these old existential undecidability result we obtain the following theorem.

Theorem 3.12 *Rational numbers are first-order definable in any abelian extension of $\mathbb{Q}$ with finitely many ramified primes, and therefore the first-order theory of such a field is undecidable.*

3.6 Final Remarks

This article did not discuss a great deal of progress on the analogs of HTP over different kinds of functions fields. We refer the interested reader to the following surveys and collections for more information: [9, 34, 56].

References

1. Colliot-Thélène, J.-L., Skorobogatov, A., & Swinnerton-Dyer, P. (1997). Double fibres and double covers: Paucity of rational points. *Acta Arithmetica, 79*, 113–135.
2. Cornelissen, G., Pheidas, T., & Zahidi, K. (2005). Division-ample sets and diophantine problem for rings of integers. *Journal de Théorie des Nombres Bordeaux, 17*, 727–735.
3. Cornelissen, G., & Zahidi, K. (2000). Topology of diophantine sets: Remarks on Mazur's conjectures. In J. Denef, L. Lipshitz, T. Pheidas & J. Van Geel (Eds.), *Hilbert's tenth problem: Relations with arithmetic and algebraic geometry, Contemporary mathematics* (Vol. 270, pp. 253–260). American Mathematical Society.
4. Davis, M. (1973). Hilbert's tenth problem is unsolvable. *American Mathematical Monthly, 80*, 233–269.
5. Davis, M., Matiyasevich, Y., & Robinson, J. (1976). Hilbert's tenth problem. Diophantine equations: Positive aspects of a negative solution. *Proceedings of Symposium on Pure Mathematics, 28*, 323– 378. American Mathematical Society.
6. Denef, J. (1975). Hilbert's tenth problem for quadratic rings. *Proceedings of the American Mathematical Society, 48*, 214–220.
7. Denef, J. (1980). Diophantine sets of algebraic integers II. *Transactions of American Mathematical Society, 257*(1), 227–236.
8. Denef, J., & Lipshitz, L. (1978). Diophantine sets over some rings of algebraic integers. *Journal of London Mathematical Society, 18*(2), 385–391.
9. Denef, J., Lipshitz, L., Pheidas, T., & Van Geel, J. (Eds.). (2000). *Hilbert's tenth problem: Relations with arithmetic and algebraic geometry, Contemporary mathematics* (Vol. 270). Providence, RI: American Mathematical Society. Papers from the workshop held at Ghent University, Ghent, November 2–5, 1999.
10. Eisenträger, K., & Everest, G. (2009). Descent on elliptic curves and Hilbert's tenth problem. *Proceedings of the American Mathematical Society, 137*(6), 1951–1959.
11. Eisenträger, K., Everest, G., & Shlapentokh, A. (2011). Hilbert's tenth problem and Mazur's conjectures in complementary subrings of number fields. *Mathematical Research Letters, 18*(6), 1141–1162.
12. Eisenträger, K., Miller, R., Park, J., & Shlapentokh, A. Easy as $\mathbb{Q}$. Work in progress.
13. Ershov, Y. L. (1996). Nice locally global fields. I. *Algebra i Logika, 35*(4), 411–423, 497.
14. Everest, G., van der Poorten, A., Shparlinski, I., & Ward, T. (2003). *Recurrence sequences* (Vol. 104). Mathematical Surveys and Monographs Providence, RI: American Mathematical Society.
15. Fried, M. D., Haran, D., & Völklein, H. (1994). Real Hilbertianity and the field of totally real numbers. In *Arithmetic geometry (Tempe, AZ, 1993)Contemporary Mathematics* (Vol. 174, pp. 1–34). Providence, RI: American Mathematical Society.
16. Friedberg, R. M. (1957). Two recursively enumerable sets of incomparable degrees of unsolvability (solution of Post's problem, 1944). *Proceedings of the National Academy of Sciences U.S.A., 43*, 236–238.
17. Fukuzaki, K. (2012). Definability of the ring of integers in some infinite algebraic extensions of the rationals. *MLQ Mathematical Logic Quarterly, 58*(4–5), 317–332.
18. Jarden, M., & Shlapentokh, A. On decidable fields. Work in progress.

19. Koenigsmann, J. Defining $\mathbb{Z}$ in $\mathbb{Q}$. *Annals of Mathematics*. To appear.
20. Kronecker, L. (1857). Zwei sätze über gleichungen mit ganzzahligen coefficienten. *Journal für die Reine und Angewandte Mathematik, 53*, 173–175.
21. Marker, D. (2002). *Model theory: An introduction, Graduate texts in mathematics* (Vol. 217). New York: Springer.
22. Matiyasevich, Y.V. (1993). *Hilbert's tenth problem*. Foundations of computing series. Cambridge, MA: MIT Press. Translated from the 1993 Russian original by the author, With a foreword by Martin Davis.
23. Mazur, B. (1992). The topology of rational points. *Experimental Mathematics, 1*(1), 35–45.
24. Mazur, B. (1994). Questions of decidability and undecidability in number theory. *Journal of Symbolic Logic, 59*(2), 353–371.
25. Mazur, B. (1998). Open problems regarding rational points on curves and varieties. In A. J. Scholl & R. L. Taylor (Eds.), *Galois representations in arithmetic algebraic geometry*. Cambridge University Press.
26. Mazur, B., & Rubin, K. (2010). Ranks of twists of elliptic curves and Hilbert's Tenth Problem. *Inventiones Mathematicae, 181*, 541–575.
27. Muchnik, A. A. (1956). On the separability of recursively enumerable sets. *Doklady Akademii Nauk SSSR (N.S.), 109*, 29–32.
28. Park, J. A universal first order formula defining the ring of integers in a number field. To appear in Math Research Letters.
29. Perlega, S. (2011). Additional results to a theorem of Eisenträger and Everest. *Archiv der Mathematik (Basel), 97*(2), 141–149.
30. Pheidas, T. (1988). Hilbert's tenth problem for a class of rings of algebraic integers. *Proceedings of American Mathematical Society, 104*(2), 611–620.
31. Poonen, B. Elliptic curves whose rank does not grow and Hilbert's Tenth Problem over the rings of integers. *Private Communication*.
32. Poonen, B. (2002). Using elliptic curves of rank one towards the undecidability of Hilbert's Tenth Problem over rings of algebraic integers. In C. Fieker & D. Kohel (Eds.), *7 Number theory, Lecture Notes in Computer Science* (Vol. 2369, pp. 33–42). Springer.
33. Poonen, B. (2003). Hilbert's Tenth Problem and Mazur's conjecture for large subrings of $\mathbb{Q}$. *Journal of AMS, 16*(4), 981–990.
34. Poonen, B. (2008). Undecidability in number theory. *Notices of the American Mathematical Society, 55*(3), 344–350.
35. Poonen, B. (2009). Characterizing integers among rational numbers with a universal-existential formula. *American Journal of Mathematics, 131*(3), 675–682.
36. Poonen, B., & Shlapentokh, A. (2005). Diophantine definability of infinite discrete non-archimedean sets and diophantine models for large subrings of number fields. *Journal für die Reine und Angewandte Mathematik, 27–48*, 2005.
37. Prestel, A., & Schmid, J. (1991). Decidability of the rings of real algebraic and p-adic algebraic integers. *Journal für die Reine und Angewandte Mathematik, 414*, 141–148.
38. Prestel, A. (1981). Pseudo real closed fields. In *Set theory and model theory (Bonn, 1979), Lecture notes in mathematics* (Vol. 872, pp. 127–156). Berlin-New York: Springer.
39. Robinson, J. (1949). Definability and decision problems in arithmetic. *Journal of Symbolic Logic, 14*, 98–114.
40. Robinson, J. (1959). The undecidability of algebraic fields and rings. *Proceedings of the American Mathematical Society, 10*, 950–957.
41. Robinson, J. (1962). On the decision problem for algebraic rings. In *Studies in mathematical analysis and related topics* (pp. 297–304). Stanford, Calif: Stanford University Press.
42. Robinson, R. M. (1964). The undecidability of pure transcendental extensions of real fields. *Zeitschrift für mathematische Logik und Grundlagen der Mathematik, 10*, 275–282.
43. Rogers, H. (1967). *Theory of recursive functions and effective computability*. New York: McGraw-Hill.
44. Rohrlich, D. E. (1984). On L-functions of elliptic curves and cyclotomic towers. *Inventiones Mathematicae, 75*(3), 409–423.

45. Rumely, R. (1980). Undecidability and definability for the theory of global fields. *Transactions of the American Mathematical Society*, *262*(1), 195–217.
46. Rumely, R. S. (1986). Arithmetic over the ring of all algebraic integers. *Journal für die Reine und Angewandte Mathematik*, *368*, 127–133.
47. Shlapentokh, A. First order definability and decidability in infinite algebraic extensions of rational numbers. arXiv:1307.0743 [math.NT].
48. Shlapentokh, A. (1989). Extension of Hilbert's tenth problem to some algebraic number fields. *Communications on Pure and Applied Mathematics*, *XLII*, 939–962.
49. Shlapentokh, A. (1994). Diophantine classes of holomorphy rings of global fields. *Journal of Algebra*, *169*(1), 139–175.
50. Shlapentokh, A. (1994). Diophantine equivalence and countable rings. *Journal of Symbolic Logic*, *59*, 1068–1095.
51. Shlapentokh, A. (1994). Diophantine undecidability in some rings of algebraic numbers of totally real infinite extensions of $\mathbb{Q}$. *Annals of Pure and Applied Logic*, *68*, 299–325.
52. Shlapentokh, A. (1997). Diophantine definability over some rings of algebraic numbers with infinite number of primes allowed in the denominator. *Inventiones Mathematicae*, *129*, 489–507.
53. Shlapentokh, A. (2000). Defining integrality at prime sets of high density in number fields. *Duke Mathematical Journal*, *101*(1), 117–134.
54. Shlapentokh, A. (2000). Hilbert's tenth problem over number fields, a survey. In J. Denef, L. Lipshitz, T. Pheidas & J. Van Geel (Eds.), *Hilbert's Tenth problem: Relations with arithmetic and algebraic geometry*, *Contemporary mathematics* (Vol. 270, pp. 107–137). American Mathematical Society.
55. Shlapentokh, A. (2002). On diophantine definability and decidability in large subrings of totally real number fields and their totally complex extensions of degree 2. *Journal of Number Theory*, *95*, 227–252.
56. Shlapentokh, A. (2006). *Hilbert's Tenth problem: Diophantine classes and extensions to global fields*. Cambridge University Press.
57. Shlapentokh, A. (2007). Diophantine definability and decidability in the extensions of degree 2 of totally real fields. *Journal of Algebra*, *313*(2), 846–896.
58. Shlapentokh, A. (2008). Elliptic curves retaining their rank in finite extensions and Hilbert's tenth problem for rings of algebraic numbers. *Transactions of the American Mathematical Society*, *360*(7), 3541–3555.
59. Shlapentokh, A. (2009). Rings of algebraic numbers in infinite extensions of $\mathbb{Q}$ and elliptic curves retaining their rank. *Archive for Mathematical Logic*, *48*(1), 77–114.
60. Shlapentokh, A. (2012). Elliptic curve points and Diophantine models of $\mathbb{Z}$ in large subrings of number fields. *International Journal of Number Theory*, *8*(6), 1335–1365.
61. Silverman, J. (1986). *The arithmetic of elliptic curves*. New York, New York: Springer.
62. Tarski, T. (1986). A decision method for elementary algebra and geometry. In *Collected papers*, S. R. Givant & R. N. McKenzie (Eds.), *Contemporary mathematicians* (Vol. 3, pp. xiv+682). Basel: Birkhäuser Verlag. 1945–1957.
63. van den Dries, L. (1988). Elimination theory for the ring of algebraic integers. *Journal für die Reine und Angewandte Mathematik*, *388*, 189–205.
64. Videla, C. (1999). On the constructible numbers. *Proceedings of American Mathematical Society*, *127*(3), 851–860.
65. Videla, C. (2000). Definability of the ring of integers in pro-p extensions of number fields. *Israel Journal of Mathematics*, *118*, 1–14.
66. Videla, C. R. (2000). The undecidability of cyclotomic towers. *Proceedings of American Mathematical Society*, *128*(12), 3671–3674.

Chapter 4
A Story of Hilbert's Tenth Problem

Laura Elena Morales Guerrero

Abstract I tell a story about Martin Davis's involvement with Hilbert's tenth problem, including his attitude, motivations, and what were his main contributions. With respect to Yuri Matiyasevich, I emphasize the fundamental aspects of his work in number theory that produced the needed proof. In addition I provide a glimpse of the social, educational, and cultural environment that created the quality of person and mathematician he is.

Keywords Hilbert's tenth problem · DPRM-theorem

4.1 Introduction

Hilbert's tenth problem, one of 23 presented by him in the year 1900, concerns a fundamental question, namely, whether there is an algorithmic method for determining if a given Diophantine equation has a solution. This problem was finally solved in 1970 by Yuri Matiyasevich. His solution was, however, negative: there is no such algorithm. In fact he provided the crucial missing step in the proof of a conjecture that Martin Davis had made twenty years earlier from which the non existence of an algorithm for Hilbert's problem followed at once.

Davis's conjecture involved the notion of *Diophantine set of natural numbers* defined as follows:

> A set S of natural numbers is called *Diophantine* if there is a polynomial $P(x, y_1, \ldots, y_m)$ with integer coefficients such that a given natural number x belongs to S if and only if there exist natural numbers $y_1, \ldots, y_m$ for which $P(x, y_1, \ldots, y_m) = 0$. If the variables in P are permitted to also occur in its exponents, the set S is called *exponential Diophantine*.

The conjecture was that every set of natural numbers that can be listed by an algorithm (such sets are called *recursively enumerable*, abbreviated r.e.) is Diophantine.

L.E. Morales Guerrero (✉)
Centro de Investigación y de Estudios Avanzados del Instituto Politécnico Nacional in Zacatenco, Ciudad de México, Mexico
e-mail: lem@fis.cinvestav.mx

E.G. Omodeo and A. Policriti (eds.), *Martin Davis on Computability, Computational Logic, and Mathematical Foundations*,
Outstanding Contributions to Logic 10, DOI 10.1007/978-3-319-41842-1_4

After Matiyasevich's result, the conjecture became a theorem variously known as *Matiyasevich's Theorem, the MRDP theorem, or the DPRM theorem.*[1] By now there are a number of proofs available of this theorem. However all of these proofs consist of two separate and independent parts. One part is the proof that the exponential function is Diophantine, the other, that every r.e. set is exponential Diophantine. The first part has been called a "gem of number theory", which indeed it is, and Yuri Matiyasevich's contribution was precisely proving this part. He also found a proof (in collaboration with James Jones), using register machines, in a new and simple way, of the other part, namely, the equivalence between the exponential Diophantine and r.e. sets.

A proof that every r.e. set is exponential Diophantine first appeared in a paper [1] by Martin Davis, Hilary Putnam and Julia Robinson. This result is called the Davis-Putnam-Robinson theorem in the paper by James Jones and Yuri Matiyasevich [2] in which the register machines proof is presented. Another version of it can be found in [3]. By the way, in this book there is a curious footnote that says that Hilary Putnam and Martin Davis first produced a proof of the Davis-Putnam-Robinson theorem with a serious blemish: they had to assume that there are arbitrarily long sequences of primes such that the difference between consecutive terms of the sequence is constant, that is, in arithmetic progression. I would like to point out the following with respect to that footnote: In 1959 when Davis and Putnam did that, their assumption about primes was believed to be true but seemed far beyond what number theorists could prove. They sent their work to Julia Robinson who quickly showed how to avoid their assumption. It was a real "tour de force". She used the prime number theorem for arithmetic progressions to get enough primes to push the proof through. Later (before they published) she managed to simplify the proof greatly, essentially putting it in the form presented in [4]. We will speak in more detail about this later. Why am I calling attention to this now? Because just a decade ago (April 2004, cf. [5]) two young mathematicians proved that there are indeed arbitrarily long arithmetic progressions consisting entirely of prime numbers. So the proof with a blemish that Putnam and Davis produced so many years ago turns out, in retrospect, to have been a genuine proof after all!

Davis introduced the term *Diophantine set* and began working on them at about the same time that Julia Robinson did as well. (She called them *existentially definable*.) The proof of the Davis-Putnam-Robinson theorem made use of techniques and results that had been developed by each of them.

4.2 The Beginning

It was in his 1950 doctoral dissertation that Davis stated his conjecture, published in his 1953 paper [6], which Yuri Matiyasevich called Davis's "daring hypothesis": the Diophantine sets are precisely the recursively enumerable sets, the sets that can

[1]The letters RDP (DPR) stand for Julia Robinson, Martin Davis, and Hilary Putnam.

be generated by recursive functions or, equivalently, by a Turing machine. Since by then mathematicians already knew there is an r.e. that is not decidable, if Davis's conjecture was true they would know that there is a Diophantine set that is undecidable. Specifically, there could be no algorithm to decide whether a given member of the parametric family of equations that generates this undecidable set is solvable in natural numbers—much less for the more general question covering all Diophantine equations. What Hilbert had asked for in his tenth problem would be impossible to do. Davis's idea that the Diophantine equations could define all the recursively enumerable sets, his "daring hypothesis", was generally regarded as implausible.

The first contribution to this work was by Gödel in his celebrated 1931 paper [7]. The main point of Gödel's investigations was the existence of undecidable statements in formal systems. The undecidable statements Gödel obtained involved recursive functions and in order to exhibit the simple number theoretic character of those statements, Gödel used the Chinese Remainder Theorem to reduce them to "arithmetic" form. However, without techniques for dealing with bounded universal quantifiers (developed much later in [1]), the best result yielded by Gödel's methods is that every recursive function (and indeed every r.e. set) can be defined by a Diophantine equation preceded by a finite number of existential and bounded universal quantifiers. In Davis's doctoral dissertation [6, 8], he showed that all but one of the bounded universal quantifiers could be eliminated, so that every r.e. set could be defined in the form

$$S = \{x \mid (\exists y)(\forall k)_{\leq y}(\exists y_1, \ldots, y_m)[P(k, x, y, y_1, \ldots, y_m) = 0]\}$$

where P is a polynomial with integer coefficients.

This representation became known as the Davis normal form. Matiyasevich has shown that we can take m = 2. Whether one can always get m = 1, is open. One cannot always have m = 0.

The fact that Davis had already gotten rid of all universal quantifiers necessary to define recursive functions except for one—and it was *bounded*—and the fact that each class had certain common features—as sets they were both closed under "and", "or", and existential quantification and neither was closed under *negation*—, might have led him to pose his conjecture.

In Davis's dissertation he conjectured that two fundamental concepts arising in different areas of mathematics are equivalent. Namely, that the notion of recursive enumerable or semidecidable set of natural numbers from computability theory is equivalent to the purely number theoretic notion of Diophantine sets (his "daring hypothesis"). He saw how to improve Gödel's use of the Chinese Remainder Theorem as a coding device so as to obtain a representation for recursively enumerable sets that formally speaking seemed close to the desired result. The obstacle that remained in the so-called Davis normal form was a single bounded universal quantifier, as shown above.

Independent of his work and about the same time, Julia Robinson was also studying Davis's Diophantine sets which she called *existentially definable*. She was attempting to see what kinds of sets of whole numbers she could define using Diophantine equations and existential quantifiers. Her investigations centered about the question: Is the exponential function Diophantine? In "Existential Definability in Arithmetic" [9], Robinson investigated a variety of functions that, like the powers of two, grow rapidly. She discovered that if she could define exponentiation, specifically the relation $x = y^z$, then she could also define a number of other functions including the factorial function and binomial coefficients. More surprisingly, she found she would be able to define the statement "p is a prime" in terms of a Diophantine equation. A truism of number theory had been that there was no formula for the prime numbers. Her Diophantine equation could be regarded as such a formula. She worked hard on exponentiation. Her major result was that if one could define any function that grew sufficiently rapidly, but not too rapidly, one could use this function to define exponentiation itself. She conjectured that finding such a function was possible. This would then show that exponentiation was Diophantine. This hypothesis became known as the Julia Robinson (JR) hypothesis. This was the main result of her work. Her hypothesis has played a key role in work on Hilbert's tenth problem.

JR statement is simply:
There exists a Diophantine set D of pairs of natural numbers such that

1. $\langle u, v \rangle \in D$ implies $v \leq u^u$.
2. For each k, there is $\langle u, v \rangle \in D$ such that $v > u^k$.

Her hypothesis remained an open question for about two decades. Once the Davis-Putnam-Robinson theorem was proved, attention was focused on the JR hypothesis since it was plain that it would imply that Hilbert's tenth problem was unsolvable. However, it seemed extraordinarily difficult to produce such an equation.

Davis met Julia Robinson at the 1950 International Congress of Mathematicians in Cambridge, Massachusetts, immediately after completing his doctorate. She had approached Hilbert's tenth from a direction opposite to that of Davis. Where he had tried to simplify the arithmetic representation of arbitrary recursively enumerable sets, she had been trying to produce Diophantine definitions for various specific sets and specially for the exponential function. She had introduced then what was to become her famous "hypothesis" and shown that under that assumption the exponential function is in fact Diophantine.

The luck of Julia's mathematical journey was holding when Alfred Tarski arrived in California in 1947 and she became a graduate student working under him. Julia, as well as many others ranked Tarski with Gödel as a great logician.

She had worked hard and successfully simplifying the definitions for general recursive functions during 1946–47 while in Princeton with her husband Raphael Robinson. Tarski had proposed to Julia a problem that did not interest her and she made little progress. However, one day at lunch with Raphael Robinson, Tarski talked about whether the whole numbers were definable in a formal system for the rational numbers. When Raphael came home he mentioned it to Julia. This problem she found interesting. She did not tell Tarski she would work on it but she did and she solved it. It is typical of her best work. She works with formal systems but introduces a clever idea from number theory, in this case, using the Hasse principle from algebraic number theory in connection with a particular quadratic equation she introduced. Davis described this as "an absolutely brilliant piece of work". This result was accepted by Tarski as her dissertation and she obtained her PhD in 1948. In the same year Tarski mentioned another problem to Raphael who brought it home. The problem was to show that one could not define the powers of 2

$$2, 4, 8, 16, 32, \ldots$$

using only existential quantifiers and a Diophantine equation. If this could be accomplished it would have a bearing on Hilbert's tenth problem; in fact, the eventual negative solution turned out to depend on showing the opposite.

Julia Robinson became entranced with the problem, but not from the direction Tarski had in mind. At first she did not know she was working on Hilbert's problem. She quickly decided that she did not see how to prove that the powers of two could not be defined in this way. Instead, she decided to see if she could define the powers of two. This expanded to work on defining other sets of whole numbers. She made rapid progress and on September 4, 1950, delivered a ten-minute paper at the International Congress of Mathematics at Cambridge, Massachusetts. At the same conference Davis gave a ten-minute talk on his results on the hyperarithmetic hierarchy, his dissertation work. However, he had spoken on his results on the tenth problem the previous winter at a meeting of the Association of Symbolic Logic.

Although Hilary Putnam's career had been in philosophy, he became fascinated by the tenth problem. During the summer of 1957, there was an intensive five week "Institute for Logic" at Cornell University. The families of the logicians attended as well and Putnam and his family shared a house with Davis and his family. And so they began to talk about Hilbert's tenth problem. Putnam suggested they try to use Gödel's coding to make that one little bounded universal quantifier in the Davis normal form go away. Davis was skeptical but they worked on the problem and made some progress which resulted in a joint paper, *Reductions of Hilbert's Tenth Problem* [10]. They decided to try to get funding to enable them to work together the following summer. They worked together during the summers of 1958, 1959 and 1960. It was during the summer of 1959 that they did their main work together on Hilbert's tenth problem.

In 1959 Davis and Putnam were again trying to apply Gödel coding to deal with the bounded universal quantifier in Davis normal form as they had in their work in 1957. Gödel coding uses the Chinese Remainder Theorem which in effect means working with arithmetic congruences. One writes $x \equiv y \bmod m$ to mean that m is a divisor of $x - y$. Since congruences are preserved under addition and multiplication, and polynomials are built out of a sequence of additions and multiplications, if P is a polynomial with integer coefficients, then we can conclude that

$$x_1 \equiv y_1 \bmod m, x_2 \equiv y_2 \bmod m, \ldots, x_n \equiv y_n \bmod m$$

implies

$$P(x_1, x_2, \ldots, x_n) \equiv P(y_1, y_2, \ldots, y_n) \bmod m$$

To use the Chinese Remainder Theorem to get rid of the universal quantifier $(\forall k)_{\leq y}$, one needs a sequence of moduli m_k with $k = 0, 1, \ldots, y$ each pair of which are relatively prime. Moreover to keep the polynomial form, the function m_k should be expressible by a polynomial in k. They also needed to be able to assert that if one of their moduli was a divisor of a product that it had to necessarily divide one of the factors. And this seemed to require that the moduli be not only relatively prime in pairs, but actual prime numbers. These needs could be readily supplied if one could set $m_k = a + bk$ where each m_k is prime. Thus, in the end they were forced to assume the hypothesis (nowadays proved) that there are arbitrarily long arithmetic progressions of prime numbers. Finally in order to complete their proof that every recursively enumerable set has an exponential Diophantine definition, they found themselves with the need to find exponential Diophantine definitions for the product of the terms of a finite arithmetic progression To deal with this problem they used binomial coefficients with rational numerators, for which they could find exponential Diophantine definitions extending Julia Robinson's methods, but requiring the binomial theorem with rational exponents, an infinite power series expansion. They wrote up their work in a report to their funding agency [11], and sent a copy to Julia Robinson. She responded saying: "I am very pleased, surprised and impressed with your results on Hilbert's tenth problem. Quite frankly, I did not think your methods could be pushed further …I believe I have succeeded in eliminating the need for (the assumption about primes in arithmetic progression) by extending and modifying your proof. I have this written out for my own satisfaction but it is not yet in shape for any one else."

That was the "tour de force" mentioned above. She avoided the hypothesis about primes in arithmetic progression in an elaborate and very clear argument by making use of the prime number theorem for arithmetic progressions to obtain enough primes to permit the proof to go through. She accepted Davis and Putnam's proposal that their work (which had already been submitted for publication) be withdrawn in favor of a joint publication by the three of them. Soon afterwards she succeeded in a drastic simplification of the proof: where Putnam and Davis were trying to use Gödel's coding to obtain a logical equivalence, her elegant argument made use of the fact that the primes were only needed for the implication in one direction, and that in that direction one could make do with a prime divisor of each modulus. The joint paper appeared in 1961 [1]. This paper is very efficient, eleven pages long, and the main result—that all recursive sets are exponential Diophantine—is proved in only four pages! The only thing now necessary for a solution of the tenth problem was to prove JR (not at all an easy matter) which would imply that exponentiation is Diophantine.

With the result that every recursively enumerable set has an exponential Diophantine definition combined with Robinson's earlier work on Diophantine definitions of the exponential function, it was now clear that Davis's "daring hypothesis" of the equivalence of the two notions, recursively enumerable set and Diophantine set, was now entirely equivalent to the much weaker JR hypothesis that Julia Robinson had proposed ten years earlier. What was needed was a single Diophantine equation whose solutions satisfied a simple condition.

In the summer of 1960 Putnam and Davis tried to find a third degree equation to satisfy JR. It turned out once again that they needed information the number theorists were unable to provide, this time about the units in pure cubic extensions of the rational numbers. Although Putnam continued to do important technical work in mathematical logic, he no longer worked on number theory.

During the following years Davis continued trying to prove JR. At that time Julia had became rather pessimistic about her hypothesis, and for a brief period, she actually worked towards a positive solution of Hilbert's tenth problem. A letter from her dated April 1968, responding to Davis's report on a certain equation he had found, said: "I have enjoyed studying it, but my faith in JR has not been restored. However, for the first time, I can see how it might be proved. Indeed, maybe your equation works, but it seems to need an infinite amount of good luck!"

However, in 1969, she published a paper [12] that made some progress. In those days, when Davis was asked for his opinion, he would reply in a semi-jocular vein: "I think JR is true and it will be proved by a clever young Russian." However, the hypothesis seemed implausible to many, especially because it was realized that an immediate and surprising consequence would be the existence of an absolute upper bound for the dimensions of Diophantine sets. Thus Kreisel [13] in his review of the Davis-Putnam-Robinson paper asserted:

> It is likely the present result is not closely connected with Hilbert's tenth problem. Also it is not altogether plausible that all (ordinary) Diophantine problems are uniformly reducible to those in a fixed number of variables of fixed degree.

Early in 1970 a telephone call from his friend and colleague Jack Schwartz informed Davis that the "clever young Russian" he had predicted had actually appeared. Julia Robinson sent Davis a copy of John McCarthy's notes on a talk that Grigori Tseitin had given in Novosibirsk on the proof of the Julia Robinson hypothesis by the twenty-two-year-old Yuri Matiyasevich. Although the notes were brief, everything important was there enabling Davis and Robinson to each fill in the details and convince themselves that it was correct. Later they were able to use his methods to produce their own variants of Matiyasevich's proof.

4.3 Yuri Matiyasevich

Yuri Matiyasevich started school in 1954 at the age of seven and school was important to him from the beginning. He also found it easy, except for music. During his early years in school he had to go to the hospital twice for surgery. He had learned to add large numbers and in the hospital, he was taught how to subtract large numbers. In the fifth year, mathematics was taught by a special mathematics teacher, and soon Yuri Matiyasevich was excused from normal mathematics class work as long as he did the homework and passed the tests. He read books on radio for amateurs and was perplexed about how a heterodyne receiver worked, spending hours drawing graphs of sine waves with different frequencies and then adding them to make a third graph [14].

In January 1959 one of Yuri's friends received a kit to build a superheterodyne radio receiver[2] with four vacuum tubes. They spent hours carefully assembling the kit but it never worked. Yuri received a second kit for his birthday in March and this time, a week after his birthday, he was listening to the radio. He says that this made his father very happy because his father, a construction engineer who designed railroad bridges, was purely a theoretician, and had no talent for doing anything with his hands. But this happiness did not last. A few days after listening to the radio, Yuri's father died suddenly.

Yuri's mother served during the Second World War as a typist in the army but afterwards had dedicated her life to raising Yuri and so had no job when his father died. However, the next year in school marked the beginning of mathematics competitions and Yuri's success in these became a focus for him and a ticket to new opportunities. He did very well in mathematical Olympiads. The seventh year brought the kruzhoks or extra evening classes and he was invited to join. That was not only work but also social events.

The mathematical Olympiads were dedicated to discover which young comrades were the most gifted in mathematics, and special schools were formed often at the instigation of prominent mathematicians. Yuri attended one of those elite schools in Leningrad which officially existed to provide "worker professionals" and supposedly trained operators of mainframe computers. These special schools were called *internats*. An extra year of school for everyone before work at university was added, and experienced mathematicians invested their time teaching gifted young people two days a week.

In 1962 a summer boarding school outside Moscow was organized by A.N. Kolmogorov. Also teaching were P.S. Aleksandrov and V.I. Arnold, among others. Matiyasevich attended. In keeping with these professors' love not only of mathematics but also of vigorous physical culture, the students were encouraged to swim a wide river and to hike in the woods.

In fall of 1963 a new school was opening in Moscow for able students from outside Moscow as the result of efforts by Kolmorogov and others. An uncle of Matiyasevich who lived in Moscow offered to pay for the extra costs and so Yuri moved to Moscow. Although it was not easy for him to leave Leningrad and his mother behind, he felt that he had to do it. In the summer he had a good class with Arnold, but a geometry

---

[2]That is, the broadcast frequency is first converted to an intermediate frequency before being amplified and detected.

course that Kolmogorov taught based on spatial movements rather than lines and points was too abstract for him at age sixteen. Matiyasevich was also, by this time, on something of an Olympiad treadmill. In his own words:

> In the spring of 1964 I was rather tired, participating every Sunday in some competition. One of them was the selection of the internat team for the all-union Olympiad. I easily passed the selection. During the Olympiad itself I used half of the given time and left, being sure that again I had solved all the problems. I remember that I decided, having saved time at the Olympiad, to walk from the building of the university where the competition took place to the internat situated in the suburbs of Moscow, about two hours walk. I felt that I needed to give myself a bit of rest. I was later disappointed to discover a mistake in a solution to one of the problems.

The same year the International Olympiad was held in Moscow, and Matiyasevich was chosen for the Soviet team, despite the fact that he was only a tenth-year-student. He was not happy about his performance, but still won a diploma of the first degree. Members of the team were granted admission to the university of their choice. Yuri tried to get permission to enroll at Moscow State University, the most prestigious university, but could not make his way through bureaucratic resistance. He still had to get his attestat degree from school. Fed up, he boarded a train for Leningrad. There it was worked out he would take the exams for the attestat in his first school while studying at the university. During this first year, 1964–1965, he was busy with exams and, though he attended a few seminars on logic, he and other first-year-students were forbidden to study logic.

He stayed at Leningrad and, at the beginning of his second year, the fall of 1965, Matiyasevich was introduced to Post's canonical system and his career as a mathematician began. He immediately achieved an elegant result on a difficult problem the professor proposed. This led him to meet Maslov, the local expert on Post canonical systems. The logical community in the Soviet Union had developed along different paths than in the west. The heavily philosophical tradition of Frege, Carnap, Russell, Whitehead, Gödel, and Tarski was at odds with Communist Party doctrine. They had their own logic and it was not symbolic. Therefore, after Kolmogorov's early bold work, Russian mathematicians had not been quick to pick up and pursue the work in logic of the 1930s. Sometimes mathematical terms were changed. For example, eventually the Russians had their own version of recursive functions and effective computability, which they called the theory of algorithms. In the United States and England the emergence of electronic computers interacted with symbolic logic, while in the Soviet Union this field lagged. Post's resolutely unphilosophical versions of symbolic logic (it is about rules for generating strings of symbols) was mathematical in its perspective and therefore unsubversive. Thus there were Soviet mathematicians at work in this field.

Maslov made a number of suggestions for research that Matiyasevich quickly resolved. In late 1965 Maslov suggested a more difficult question about details of the unsolvability for Thue systems. Matiyasevich solved this problem and the opportunity to publish the result was offered, but it had to be written up in an entirely rigorous manner in the style of Markov and Post. Matiyasevich was given an Underwood typewriter of manufacture predating the Revolution (by the second wife of his grandfather). He could hardly have been able to buy one of his own. He spent a considerable part of the next year typing. Only five corrections per page were allowed and the paper was 100 pages long. He missed lectures in school, particularly, he says, in complex analysis. He was invited to give a talk at the 1966 International Congress of Mathematicians in Moscow, a major honor, and was particularly impressed to meet Kleene. Toward the end of 1965, Maslov also suggested Hilbert's tenth problem. He said that "some Americans" had done some work on this problem but that their approach was probably wrong. Matiyasevich did not read their work but, like Davis and Robinson, he was enchanted by the problem and was drawn to it again and again. Once as an undergraduate he thought he had solved it and even began a seminar presenting his solution. He soon discovered his error but became known as the undergraduate who worked on Hilbert's tenth problem, with an edge of humor. As the years of his undergraduate education passed, like Davis, he too began to think he needed to discipline himself away from this trap of a problem. He did read the work of the Americans and recognized its possible importance. If he could find a Diophantine equation whose solutions grew appropriately, exponentiation would be proved to be Diophantine, and therefore by the Davis-Putnam-Robinson theorem, all recursively enumerable sets would be shown to be Diophantine, and therefore there would exist a Diophantine set that was not decidable. Hilbert's problem would be solved.

His undergraduate years were ending. He had not done anything better than the early work he had delivered at the International Congress. In his own words:

> I was spending almost all my free time trying to find a Diophantine relation of exponential growth. There was nothing wrong when a sophomore tried to tackle a famous problem but it looked ridiculous when I continued my attempts for years in vain. One professor began to laugh at me. Each time we met he would ask: "Have you proved the unsolvability of Hilbert's tenth problem? Not yet? But then you will not be able to graduate from the university!

In the fall of 1969 when a colleague told him to rush to the library to read a new paper by Robinson, a survey on what had been achieved so far in connection with Hilbert's tenth problem [12], he stayed away. However, because he was considered an expert, he was sent the paper to review and so was forced to read it. He delivered a seminar on it on December 11, 1969. Robinson's paper had a fresh flavor and a new result, namely that if any infinite set of prime numbers is Diophantine, then the exponential function is Diophantine. Matiyasevich was caught again. He spent December 1969 obsessing over the problem. On the morning of January 3, 1970, he thought he had found a solution but with an error that he was able to fix the next morning. He was now in possession of a solution in the negative of Hilbert's tenth problem. However, he was afraid there was still an error. After all, he had once gone so far as to start giving a seminar on a solution. He wrote out a full proof and asked both Maslov and Vladimir Lifschits to check it but say nothing until they talked to him again. Matiyasevich then left for a couple of weeks with his soon-to-be wife for a ski camp. There he worked also in refining his paper. He returned to Leningrad to find that the verdict was that he had solved Hilbert's tenth problem and it was no longer a secret. Both D.K. Faddeev and Markov, famous for finding mistakes, had also checked the proof and passed it. Matiyasevich gave his first public talk on the result on January 29, 1970. News of the result moved around the country. Grigori Tseitin took a copy of the manuscript and, with Matiyasevich's permission, presented it at a conference in Novosibirisk. An American mathematician John McCarthy attended this talk and it was through him that information about the result made its way to Davis and Robinson.

Whereas Robinson had worked with the sequence of solutions of so-called Pell equations of the special form $x^2 - (a^2 - 1)y^2 = 1$, Matiyasevich preferred to use the famous Fibonacci numbers which have rather similar properties. The Fibonacci numbers are defined by the recurrence $F_0 = 0$, $F_1 = 1$, $F_{n+2} = F_{n+1} + F_n$. What Yuri had proved was the set of pairs $\langle u, v\rangle$ such that v is the $2u$-th Fibonacci number is Diophantine. Because of the exponential growth of the Fibonacci numbers, it is rather obvious that this set does satisfy the conditions of JR, and for this same reason it was natural to use them in a proof of JR. What had been missing was some kind of Diophantine relation between the number F_n and the subscript n. Some time previously, Yuri had proved the following important fact:

$$\text{If } F_n^2 \mid F_m \text{ then } F_n \mid m.$$

This finally was indeed a relation between the Fibonacci numbers and their respective subscripts, and it was apparently something that the Americans working on the problem didn't know, but it required much more than this to obtain Yuri's Diophantine definition. His proof is a wonderful tapestry, delicate and beautiful. Although it would be difficult to find a clear technical connection between Julia's paper and Yuri's breakthrough, Yuri was eager to indicate at least a psychological connection. In this connection he used a Russian word whose literal translation is "wafted", as though the influence of her paper on his accomplishment was as subtle as that of the scent of a flower.[3] We must point out that at this time, it had still been far from obvious that the tenth problem was near solution or that the solution lay in the direction of Robinson's hypothesis. Matiyasevich's adviser had told him to ignore the Americans' work. In this period even Robinson had at one point despaired of proving her hypothesis. It was necessary for Yuri to recognize that it could be made to work despite all of that.

Matiyasevich had received his kandidat degree, equivalent to a PhD, in 1970 for his early work on Post's systems. He received his doctoral degree for his work on Hilbert's tenth.

The key direct result obtained in terms of which a solution of Hilbert's tenth problem was obtained is that any set that a computer can be programmed to generate, can be generated by a specific Diophantine equation. Many of the major questions in Mathematics, including Fermat's last theorem, and the four-color map problem, only settled late in the twentieth century, and Goldbach's conjecture, and the Riemann Hypothesis, both still undecided, can each be seen to be equivalent to a specific

[3]Some of this information comes from [15].

Diophantine equation having no solution [16]. This is simply astounding. In the end, the unsolvability of Hilbert's tenth problem is richer for mathematicians than a decision process for which Hilbert asked would have been.

4.4 Matiyasevich's Solution

What Yuri Matiyasevich did in 1970 in order to prove "Davis's daring hypothesis", namely, every r.e. set is Diophantine, was to use the Fibonacci numbers to construct a Diophantine equation which constituted a Diophantine definition of the set of pairs $\langle u, v\rangle$ for which v is the $2u$-th Fibonacci number. His equation is obtained by summing the squares of the left sides of the following system of equations and setting the result equal to zero.

$$\begin{aligned}
u + w - v - 2 &= 0\\
\ell - 2v - 2a - 1 &= 0\\
\ell^2 - \ell z - z^2 - 1 &= 0\\
g - b\ell^2 &= 0\\
g^2 - gh - h^2 - 1 &= 0\\
m - c(2h + g) - 3 &= 0\\
m - f\ell - 2 &= 0\\
x^2 - mxy + y^2 - 1 &= 0\\
(d - 1)\ell + u - x - 1 &= 0\\
x - v - (2h + g)(e - 1) &= 0
\end{aligned}$$

Julia Robinson's early work had shown that JR implies that the exponential function is Diophantine. Thus, Matiyasevich's proof of JR could be applied to the Davis-Putnam-Robinson theorem to conclude that every listable set, i.e., every r.e. set, is Diophantine. In particular there is a Diophantine definition of a listable set which is not computable. And so ends the story of Hilbert's tenth problem.

Later, in 1976, Yuri presented a new proof [17] of the Davis-Putnam-Robinson theorem. And in 1984, Jones and Matiyasevich proved it once again using register machines [2]. Readers may also find interesting the brief expository article [18]. Finally, Matiyasevich has written a most complete and enjoyable book on Hilbert's tenth problem and its history [19].

References

1. Davis, M., Putnam, J., & Robinson, J. (1961). The decision problem for exponential Diophante equations. *Annals of Mathematics*, *74*, 425–436.
2. Jones, J. P., & Matiyasevich, Y. V. (1984). Register machine proof of the theorem on exponential Diophantine representation of enumerable sets. *The Journal of Symbolic Logic*, *49*(3), 818–829.
3. Schöning, U., Pruim, R. J., & Pruim, R. (1998). *Gems of theoretical computer science*. Springer.
4. Davis, M. (1973). Hilbert's tenth problem is unsolvable. *American Mathematical Monthly*, *80*, 233–269. Reprinted in the Dover edition of [8].
5. Green, B., & Tao, T. (2008). The primes contain arbitrarily long arithmetic progressions. *The Annals of Mathematics*, *167*(2), 481–547.
6. Davis, M. (1953). Arithmetical problems and recursively enumerable predicates. *The Journal of Symbolic Logic*, *18*, 33–41.
7. Gödel, K. (1931). Über formal unentscheidbare Sätze der Principia Mathematica und verwandter Systemel. *Monatsch. Math. und Physik*, *38*, 173–198. English translations: (1) Gödel, K. (1962). *On formally undecidable propositions of Principia Mathematica and related systems*, Basic Books. (2) Davis, M. (Ed.). (1965). *The undecidable* (pp. 5–38). Raven Press. (3) Van Heijenoort, J. (Ed.). (1967). *From Frege to Gödel* (pp. 596–616). Harvard University press.
8. Davis, M. (1958). *Computability and unsolvability*. New York: McGraw Hill. Reprinted Dover 1982.
9. Robinson, J. (1952). Existential definability in arithmetic. *Transactions of the American Mathematical Society*, *72*, 437–449.
10. Davis, M., & Putnam, H. (1958). Reductions of Hilbert's tenth problem. *The Journal of Symbolic Logic*, *23*, 183–187.
11. Davis, M., & Putnam, H. (1959). *On Hilbert's tenth problem*, US Air Force O. S. R. Report AFOSR TR 59-124, Part III.
12. Robinson, J. (1969). Diophantine decision problems. In: LeVeque, W. J. (Ed.), *Studies in number theory*, MAA studies in mathematics (Vol. 6, pp. 76–116). Buffalo, N. Y: MAA.
13. Kreisel, G. (1962). Review of [1], *Mathematical Reviews, 24*, 573, Part A (review number A3061).
14. Yandell, B. H. (2002). *The honors class*. Natick, Massachusetts: A. K. Peters.
15. Matiyasevich, Y. V. (2006) Hilbert's tenth problem: Diophantine equations in the twentieth century. In: A. A. Bolibruch, Yu. S. Osipov, & Ya. G. Sinai (Ed.), *Mathematical events of the twentieth century* (pp. 185–213). Berlin, PHASIS, Moscow: Springer.
16. Davis, M., Matiyasevich, Y. V., & Robinson, J. (1976). Hilbert's tenth problem. Diophantine equations. Positive aspects of a negative solution. *Proceedings of Symposia in Pure Mathematics*, *28*, 323–378.
17. Matiyasevich, Y. V. (1980). A new proof of the theorem on exponential Diophantine representation of enumerable sets. English translation, *Journal of Soviet Mathematics*, *14*, 1475–1486.
18. Davis, M., & Hersh, R. (1973). Hilbert's tenth problem. *Scientific American*, *229*, 84–91.
19. Matiyasevich, Y. V. (1993). *Hilbert's tenth problem*, MIT Press. This work was originally published in Russian by Nauka Publishers, 1993.

Chapter 5
Hyperarithmetical Sets

Yiannis N. Moschovakis

Abstract The *hyperarithmetical* sets of natural numbers were introduced (independently) in the early 1950s by Martin Davis, Andrej Mostowski and Stephen Cole Kleene and their study is surely one of the most significant developments in the theory of computability: they have a rich and interesting structure and they have found applications to many areas of mathematics, including inductive definability, higher-type recursion, descriptive set theory and even classical analysis. This article surveys the development of the subject in its formative period from 1950 to 1960, starting with a discussion of its origins and with some brief pointers to later developments. There are few proofs, chosen partly because of the importance of the results but mostly because they illustrate simple, classical methods specific to this area which are not easy to find in the literature, especially in the treatment of *uniformity*; and these are given in the spirit (if not the letter) of the methods which were available at the time. This is an elementary, expository article and includes an Appendix which summarizes the few basic facts about computability theory that it assumes.

Keyword Hyperarithmetical sets

By the early 1940s, ten years after Gödel's monumental [11], the foundations of a mathematical theory of computability had been well established, primarily by the work of Alonzo Church, Alan Turing, Emil Post and Stephen Kleene. Most significant was the formulation of the *Church-Turing Thesis*, which identifies the intuitive notion of *computable function* (on the natural numbers) with the precisely defined concept of (general) *recursive function*; this was well understood and accepted (as a *law* in Emil Post's view) by all the researchers in the area, even if not yet by all logicians.[1] The Church-Turing Thesis makes it possible to give rigorous proofs of (absolute) *unsolvability* of mathematical problems whose solution asks for an "algorithm" or a "decision procedure". Several fundamental metamathematical relations had been

[1]cf. Moschovakis [32].

Y.N. Moschovakis (✉)
Department of Mathematics, University of California, Los Angeles, USA
e-mail: ynm@math.ucla.edu

E.G. Omodeo and A. Policriti (eds.), *Martin Davis on Computability, Computational Logic, and Mathematical Foundations*,
Outstanding Contributions to Logic 10, DOI 10.1007/978-3-319-41842-1_5

shown to be *undecidable*, chief among them the relation of *first-order provability* (Hilbert's *Entscheidungsproblem*, Church [3] and Turing [53]). Moreover, a general *theory of computability* had also started to develop, especially with Kleene [13].

The most obvious next steps were to

- look for unsolvability results in "ordinary mathematics", and
- study (in general) *the unsolvable.*

The first of these was (apparently) first emphasized by Post, who said in his Post [42] that "(Hilbert's 10th Problem) begs for an unsolvability proof". Post [43] and Markov [29] proved (independently) the *unsolvability of the word problem for* (finitely generated and presented) *semigroups*, the first substantial result of this type. Martin Davis' work is an important part of this line of research which is covered extensively in other parts of this volume.

My topic is the theory of *hyperarithmetical sets*, one of the most significant developments to come out of the general theory of unsolvability in which Davis also played a very important role. I will give a survey of the development of the subject in its formative period from 1950 to 1960, starting with a discussion of its origins and with a couple of brief pointers to later developments at the end. There are few proofs, chosen partly because of the importance of the results but mostly because they illustrate simple, classical methods specific to this area which are not easy to find in the literature, especially in the treatment of *uniformity*; and I have tried to give these proofs in the spirit (if not the letter) of the methods which were available at the time—with just one, notable exception, cf. Remark 5.3.1.

The Appendix collects the few basic facts from recursion and set theory that we need and fixes notation. We refer to them by App 1, App 2, etc.

5.1 Preamble: Kleene [15], Post [42] and Mostowski [38]

The two seminal articles of Kleene and Post were published within a year of each other[2] and have had a deciding influence on the development of the theory of unsolvability up until today. Mostowski wrote his [38] in ignorance of Kleene [15], he discovered independently many of Kleene's results and he asked some questions which influenced profoundly the development of the subject. We will discuss it in Sect. 5.1.4.

Kleene and Post approached "the undecidable" in markedly different ways: they chose different ways to measure the complexity of undecidable sets, they introduced different methods of proof and they employed distinct "styles of exposition". The results in them and in the research they inspired are closely related, of course, as they are ultimately about the same objects—the undecidable relations on the natural numbers; but there is no doubting the fact that they led to two different traditions in the

[2] Kleene had presented much of his [15] in a meeting of the American Mathematical Society in September 1940. I do not know when Post obtained the results in his [42].

theory of unsolvability with many of the best researchers in one of them (sometimes) knowing very little of what has happened in the other.

The first, key question was how to *measure the unsolvability* of a set of natural numbers.

5.1.1 Post's Degrees of Unsolvability

Post [42] does it by comparing the complexity of two sets $A, B \subseteq \mathbb{N}$ using several methods of *reducing effectively* the relation of membership in A to that of membership in B. The strongest of these is *one-one reducibility*,

$$A \leqslant_e^1 B \iff \varphi_e : \mathbb{N} \rightarrowtail \mathbb{N} \text{ is a total injection and } [x \in A \iff \varphi_e(x) \in B],$$
$$A \leqslant_1 B \iff (\exists e)[A \leqslant_e^1 B],$$

close to the mildly weaker *many-one reducibility* $A \leqslant_m B$ where it is not required that φ_e be an injection. The weakest and most important is *Turing reducibility*,

$$A \leqslant_e^T B \iff \chi_A = \{e\}^B, \quad A \leqslant_T B \iff (\exists e)[A \leqslant_e^T B].$$

We will also use the strict and symmetric versions of these reducibilities,

$$A <_1 B \iff A \leqslant_1 B \,\&\, B \nleqslant_1 A, \quad A \equiv_1 B \iff A \leqslant_1 B \,\&\, B \leqslant_1 A,$$

and similarly for $<_m, \equiv_m, <_T, \equiv_T$.

The symmetric relations induce natural notions of *degrees*, e.g.,

$$\text{the } 1-1 \text{ degree of A} = \mathbf{d}_1(A) = \{B : B \equiv_1 A\},$$
$$\text{the Turing degree of } A = \mathbf{d}(A) = \{B : B \equiv_T A\};$$

and the central objects of study are these sets of degrees with their natural partial orders, most significantly the *poset of Turing degrees* $(\mathscr{D}, \leqslant_T)$ where

$$\mathbf{a} \leqslant_T \mathbf{b} \iff (\exists A, B \subseteq \mathbb{N})[\mathbf{a} = \mathbf{d}(A) \,\&\, \mathbf{b} = \mathbf{d}(B) \,\&\, A \leqslant_T B].$$

Post focusses on the study of the degrees of *recursively enumerable* sets (App 7). He introduces the "self-referential" version of Turing's *Halting Problem*

$$K = \{e : \{e\}(e)\downarrow\} = \{e : (\exists t)T_1(e, e, t)\} \tag{5.1}$$

and proves that it is *r.e. complete*, i.e., it is r.e. and every r.e. set is 1-1 reducible to it. In particular K is not recursive, and then the natural question is whether there are r.e. sets intermediate in complexity between the recursive sets and K. Post proves this for all of his reducibilities except for Turing's and asks what became known as *Post's Problem*: *is there an r.e. set A such that* $\emptyset <_T A <_T K$? Friedberg and Muchnik proved that there is, some ten years later, and this initiated a research program in the theory of degrees and r.e. degrees which is still vibrant today.

5.1.2 Kleene's Arithmetical Hierarchy

Kleene [15] focusses on the *arithmetical sets*, those which are first-order definable in the standard model of arithmetic

$$\mathbf{N} = (\mathbb{N}, 0, 1, +, \cdot) \tag{5.2}$$

and measures the complexity of a set by its *simplest definition* in $\mathbf{N}$. His crucial contribution is the choice of a useful *measure of complexity of first-order definitions* in $\mathbf{N}$: a relation $P \subseteq \mathbb{N}^n$ is Σ^0_k (or *in* Σ^0_k) if it satisfies an equivalence of the form

$$P(\boldsymbol{x}) \iff (\exists t_1)(\forall t_2)(\exists t_3)\cdots(\mathsf{Q}_k t_k)R(\boldsymbol{x}, t_1, \ldots, t_k) \quad (k \geqslant 1) \tag{5.3}$$

where $R(\boldsymbol{x}, \boldsymbol{t})$ is recursive and Q_k is $\exists$ or $\forall$ accordingly as k is odd or even. A relation $P(\boldsymbol{x})$ is in $\Pi^0_k = \neg\Sigma^0_k$ if its negation is in Σ^0_k, so that

$$P(\boldsymbol{x}) \iff (\forall t_1)(\exists t_2)(\forall t_3)\cdots(\mathsf{Q}_k t_k)R(\boldsymbol{x}, t_1, \ldots, t_k) \quad (k \geqslant 1) \tag{5.4}$$

with a recursive $R(\boldsymbol{x}, \boldsymbol{t})$, and $\Delta^0_k = \Sigma^0_k \cap \Pi^0_k$. The relations which belong to one of these classes are exactly the arithmetical ones, and that was well known after Kleene [13]. The novelty here is that by allowing a recursive matrix in (5.3) and (5.4) rather than, say, a quantifier free one, Kleene can prove robust closure properties and to construct $\mathbb{N}$-*parametrizations* for these classes of relations:

Lemma 5.1.1 *(1) Closure properties:* Σ^0_k *and* Π^0_k *are closed under recursive substitutions, &,* $\vee$ *and bounded number quantification of both kinds;* Σ^0_k *is also closed under number quantification* $\exists s$; Π^0_k *is closed under* $\forall s$; *and* Δ^0_k *is closed under negation.*

(2) The $\mathbb{N}$*-Parametrization Property: there are relations* $G^n_k \subseteq \mathbb{N}^{1+n}$ *in* Σ^0_k *and recursive injections* $S^l_n : \mathbb{N}^{1+l} \to \mathbb{N}$ *such that for every n-ary* $P(\boldsymbol{x})$ *in* Σ^0_k,

$$P(\boldsymbol{x}) \iff G^n_k(e, \boldsymbol{x}) \textit{ for some } e \in \mathbb{N}, \tag{5.5}$$

and for all $\boldsymbol{y} = (y_1, \ldots, y_l)$

$$G^{l+n}_k(e, \boldsymbol{y}, \boldsymbol{x}) \iff G^n_k(S^l_n(e, \boldsymbol{y}), \boldsymbol{x}). \tag{5.6}$$

These facts are very easy by induction, starting with $k = 1$ where they are immediate by the Normal Form and Enumeration Theorem for recursive partial functions

Fig. 5.1 The arithmetical ($i = 0$) and analytical ($i = 1$) hierarchies

App 5. They imply the *Hierarchy Theorem* for the arithmetical sets pictured in Fig. 5.1 (with $i = 0$), and they can be used very effectively to measure the complexity of a set by placing it in the arithmetical hierarchy, sometimes exactly. Such were, in fact, their first applications.[3] Its main significance, however, was that it set the stage for its non-trivial extensions into the *analytical hierarchy*, also pictured in Fig. 5.1 with $i = 1$, as well as the *hyperarithmetical hierarchy* which lies between them and is our main concern.

The closure of the arithmetical classes under recursive substitutions imply that for every n-ary relation $P(\boldsymbol{x})$,

$$P \in \Sigma_k^0 \iff \{\langle \boldsymbol{x} \rangle : P(\boldsymbol{x})\} \in \Sigma_k^0,$$

i.e., these classes are determined by the sets in them; so we will sometimes abuse notation and use Σ_k^0 to denote the class of Σ_k^0 sets—and similarly for Π_k^0, Δ_k^0.

5.1.3 Kleene [15] Versus Post [42]

There is little overlap between these two papers, except that they both characterize the recursive sets as exactly those which are r.e. and have r.e. complements (Post's Theorem). Beyond that, Post limits himself to the complexity structure of r.e. sets which comprise precisely Kleene's Σ_1^0—about which Kleene says nothing non-trivial.

Both papers are brilliant examples of *concept formation*, the identification of fundamental notions which is characteristic of some of the best work in logic. Post also proves several non-trivial technical results, some by very clever constructions; there is little of this in the Kleene paper, whose technical results are proved mostly by seemingly routine computations.

Then there is the style of exposition: Post is eloquent, even colorful. He introduces suggestive, descriptive terms (*complete, creative, simple*) which give life to

---

[3] For example, Davis [4] proves that the set $\{e : (\forall x)[\{e\}(x)\downarrow]\}$ of codes of total recursive functions is in $\Pi_2^0 \setminus \Sigma_2^0$. The Hierarchy Theorem also yields a trivial proof of *Tarski's Theorem* for $\mathbf{N}$, that *arithmetical truth is not arithmetical*.

the formulation of his results and right in his first paragraph, he declares that his purpose is

> to demonstrate by example that this concept [of recursive function] admits ...of an intuitive development which can be followed, if not indeed pursued, by a mathematician, layman though he be in this formal field.[4]

His exhortation to *explain* rather than *detail* proofs resonated strongly in the work of those who followed him, sometimes with beautiful results, e.g., in the classic Rogers [45]. At the other end, Kleene is dry, formal, and more worried about whether he has a *constructive* (intuitionistic) proof than if his proof is easily comprehensible—and to some extent, these traits persisted in the writings of those who followed him.

5.1.4 Mostowski [38] and the Analogies

Mostowski's starts with the classical notions of Descriptive Set Theory. Briefly, in modern notation and (for simplicity) only for $\mathcal{N}$:

(1) A σ-algebra is any collection $\mathcal{F} \subseteq \mathcal{P}(\mathcal{N})$ which is closed under complements and countable unions;
(2) the class $\mathbf{B}$ of Borel sets is the smallest σ-algebra which contains all the open sets;
(3) a relation $P \subseteq \mathcal{N}^m$ is $\underset{\sim}{\mathbf{\Sigma}}^1_1$ if $P = \{\boldsymbol{\alpha} : (\exists\beta)F(\boldsymbol{\alpha}, \beta)\}$ with F closed;
(4) P is $\underset{\sim}{\mathbf{\Sigma}}^1_{k+1}$ if $P = \{\boldsymbol{\alpha} : (\exists\beta)\neg Q(\boldsymbol{\alpha}, \beta)\}$ with Q in $\underset{\sim}{\mathbf{\Sigma}}^1_k$;
(5) $\underset{\sim}{\mathbf{\Pi}}^1_k = \{\mathcal{N}^m \setminus P : P \in \underset{\sim}{\mathbf{\Sigma}}^1_k\}$ and $\underset{\sim}{\mathbf{\Delta}}^1_k = \underset{\sim}{\mathbf{\Sigma}}^1_k \cap \underset{\sim}{\mathbf{\Pi}}^1_k$.

The *projective classes* $\underset{\sim}{\mathbf{\Sigma}}^1_k$, $\underset{\sim}{\mathbf{\Pi}}^1_k$, $\underset{\sim}{\mathbf{\Delta}}^1_k$ were introduced by Luzin and Sierpinski in 1925 and they fall into a hierarchy that looks exactly like the arithmetical hierarchy in Fig. 5.1 with boldface letters and superscript 1. But the most fundamental result about them is older and concerns only the first level of this hierarchy:

Theorem 5.1.1 (Suslin [52]) *A set $A \subseteq \mathcal{N}$ is $\underset{\sim}{\mathbf{\Delta}}^1_1$ if and only if it is Borel.*

This was rightfully viewed as a "construction principle" which reduces a complementary pair of quantifications over the complex set $\mathcal{N}$ to a countable iteration of taking countable unions and complements, starting with the simple neighborhoods of $\mathcal{N}$. Mostowski had not read Kleene [15] but he knew Post [42] and saw a similarity between Suslin's Theorem and Post's in the form

$$\Delta^0_1 = \text{recursive},$$

---

[4]He also said that "...with a few exceptions explicitly so noted, we have obtained formal proofs of all the consequently mathematical theorems here developed informally", and it is clear that the purely intuitive approach can only go so far: we cannot hope to prove that (say) *the word problem for semigroups is unsolvable* on the basis of our intuitions about computability, without a rigorous definition of recursive functions and an appeal to the Church-Turing Thesis.

which similarly reduces Δ^0_1 definitions to "computations". He postulated the natural "analogies"

$$\begin{aligned} \text{recursive function on } \mathbb{N} &\sim \text{continuous function on } \mathcal{N}, \\ \text{recursive subsets of } \mathbb{N} &\sim \mathbf{B}, \\ \Sigma^0_1 \text{ subsets of } \mathbb{N} &\sim \utilde{\mathbf{\Sigma}}^1_1 \text{ subsets of } \mathcal{N}, \end{aligned} \tag{5.7}$$

and using these as motivation he defined the arithmetical hierarchy and established for it basically all the results in Kleene [15], so that the analogies extend to all the levels of the two hierarchies. He knew that these are not perfect: not every injective, recursive image of $\mathbb{N}$ is recursive, while by a basic, classical result, *every injective, continuous image of* $\mathcal{N}$ *is Borel.* This, however, might be just a technical wrinkle, as *every increasing, recursive image of* $\mathbb{N}$ *is recursive*. Later, after writing this paper, he thought of another fundamental property of $\utilde{\mathbf{\Sigma}}^1_1$ sets which could test the analogy, the following generalization of Suslin's Theorem due to Lusin:

Theorem 5.1.2 ($\utilde{\mathbf{\Sigma}}^1_1$ Separation) *For any two disjoint* $\utilde{\mathbf{\Sigma}}^1_1$ *sets* $A, B \subseteq \mathcal{N}$, *there is a Borel set* C *which separates them, i.e.,*

$$A \subseteq C, \quad C \cap B = \emptyset. \tag{5.8}$$

So is it true that any two disjoint r.e. sets can be separated by a recursive set? At some time between 1947 and 1950 he mentioned the problem to Kleene who (it turned out) had already answered it but not published his result:

Theorem 5.1.3 (Kleene [16]) *There exist two disjoint, r.e. sets* $A, B \subseteq \mathbb{N}$ *such that no recursive set* C *satisfies* (5.8).

So the simple minded analogies (5.7) fail, but they did not go away: they motivated a great deal of research in the twenty years that followed and ultimately, as we will see, a corrected version of them turned out to be an important part of the story of HYP.

5.2 On into the Transfinite![5]

For any $A \subseteq \mathbb{N}$, let[5]

$$A' = \{e : \{e\}^A(e)\downarrow\} = \text{ the } \textit{jump} \text{ of } A. \tag{5.9}$$

It follows that for every B,

$$B \text{ is r.e. in } A \iff B \leqslant_1 A', \tag{5.10}$$

[5]For completeness, we will repeat in this section some parts of §7–§9 of Moschovakis [37], which goes over some of the same ground in more detail and includes several proofs.

so that in particular $A <^T A'$, and we can get a sequence of sets of increasing Turing complexity by setting recursively

$$K_0 = \emptyset, \quad K_1 = K_0', \quad K_2 = K_1', \dots. \tag{5.11}$$

Now K_1 is (recursively isomorphic with) Post's complete r.e. set K and for every $k \geqslant 1$, easily, K_k is Σ_k^0-*complete*, i.e., a set is Σ_k^0 exactly when it is 1-1 reducible to K_k. It is also easy to check that the diagonal set

$$K_\omega = \{\langle m, n\rangle : m \in K_n\} \tag{5.12}$$

is recursively isomorphic with the *truth set* for arithmetic

$$\text{Truth} = \{\ulcorner \theta \urcorner : \mathbf{N} \models \theta\},$$

where $\ulcorner \theta \urcorner$ is the Gödel number of the sentence θ in the language of arithmetic, relative to some standard coding. This is not arithmetical; and then one can continue and define ever more complex non-arithmetical sets,

$$K_{\omega+1} = K_\omega', K_{\omega+2} = K_{\omega+1}', \dots, K_{\omega\cdot 2} = \{\langle m, n\rangle : m \in K_{\omega+n}\} \dots \tag{5.13}$$

indexed by the ordinals $\xi < \omega^2$. The sequence $\{K_\xi : \xi < \omega^2\}$ was defined by Davis [4] who also showed that

$$\eta \leqslant \xi < \omega^2 \implies K_\eta \leqslant_m K_\xi \text{ and } \xi < \eta \implies K_\xi <_T K_\eta. \tag{5.14}$$

These facts are all fairly simple to verify today. They were not so easy[6] before 1955, when the theory of relative recursion had not been worked out in detail: Kleene [15, 18, 19], Davis [4, 5] and Mostowski [38, 39] all prove various versions of them, not always the cleanest or strongest, sometimes awkwardly and (in the case of Davis and Mostowski) mostly without knowing all of each other's or Kleene's work. Nevertheless, the later papers Davis [4], Mostowski [39] and Kleene [19] all take the crucial step of defining natural extensions of the arithmetical hierarchy beyond its first ω classes $\Sigma_1^0, \Sigma_2^0, \dots$, "*on into the transfinite*" in Davis' exhortation with which we headed this section.

[6]For example, to prove that K_k is Σ_k^0-complete, you need the first of the following strengthenings of (5.10): *there are recursive injections* $u(e, t), v(e)$ *such that for all* A, B and all e, t,

$$(1)\ \{e\}^A(t)\downarrow \iff u(e, t) \in A' \text{ and } (2)\ A \leqslant_e^T B \implies A' \leqslant_{v(e)}^1 B'. \tag{5.15}$$

Proof: For (1), choose $\overline{m}$ so that for any A, $\{\overline{m}\}^A(e, t, y) = \{e\}^A(t)$ and set $u(e, t) = S_1^2(\overline{m}, e, t)$. For (2) you start with a recursive $v_1(e)$ such that $A \leqslant_e^T B \implies \{e\}^A(t) = \{v_1(e)\}^B(t)$ and do a similar construction. That $u(e, t)$ and $v(e)$ are (absolutely) *recursive injections*—which has applications—depends on the fact that the functions $S_n^{l,m}$ in App 5 are independent of any function parameters and injective, which I cannot find in any of the early texts (including Kleene [17]) even for $m = 0$.

The definitions (5.11)–(5.13) of $\{K_\xi : \xi < \omega^2\}$ depend on choosing for each limit ordinal $\xi = \omega \cdot s < \omega^2$ the specific, increasing sequence $n \mapsto \omega \cdot (s \dot{-} 1) + n$ converging to ξ. This is natural enough, but not the only choice, and it is not obvious how to make a "natural" or "best" choice[7] for ordinals above ω^2. This leads us to the next, crucial bit:

5.2.1 *Notations for Ordinals, S_1 and O*

Following Kleene [14], let first

$$0_O = 1, \quad (t+1)_O = 2^{t_O}, \quad e_t = \{e\}(t_O),$$

and (by App 10), let $|\ | : \mathbb{N} \rightharpoonup$ Ordinals be the least partial function on $\mathbb{N}$ to ordinals which satisfies the following[8]:

(1) $|1| = 0$.
(2) For every t, $|2^t| = |t| + 1$.
(3) For every e, if for every t, $|e_t|\downarrow$ and $|e_t| < |e_{t+1}|$, then $|3 \cdot 5^e| = \lim_{t\to\infty} |e_t|$.

With $S_1 = \{z : |z|\downarrow\}$, the pair $\mathrm{S}_1 = (S_1, |\ |)$ is *the first* Church-Kleene *notation system* for ordinals and the only one we will use. Kleene [14] also introduced a smaller notation system $\mathrm{S}_3 = (O, |\ |_3)$ and a partial ordering $\leqslant_O$ of O such that

$$O \subsetneq S_1, \quad a \in O \implies |a|_3 = |a|, \text{ and so } a <_O b \implies |a| < |b|, \tag{5.16}$$

and then used that in all his work on the topic—as did Spector and most researchers in the field. We will occasionally refer to O and $\leqslant_O$ when we want to quote early results exactly as they were stated, but we will not use them in any essential way and so we skip their precise definition.[9]

A countable ordinal ξ is *constructive* if $\xi = |z|$ for some $z \in S_1$. Note that directly from the definition, *the constructive ordinals form an initial segment of the set of*

---

[7]Spector [48] eliminates dramatically the most obvious approach at limit ordinals: *No increasing sequence* $\mathbf{d}_0 < \mathbf{d}_1 < \cdots$ of Turing degrees has a least upper bound. Of course, this was not known to Davis, Kleene and Mostowski when they wrote these early papers.

[8]Kleene's obtuse coding (the 3 and 5 in the definition) is motivated by the plans he and Church had to develop a general "constructive theory of ordinals" beyond Cantor's first and second number classes. They never got into this, but some (non-trivial and highly technical) results were proved by others, cf. Kreider and Rogers [25], Putnam [44], Enderton-Putnam [7]. We will not cover this topic here.

[9]O and $\leqslant_O$ are defined by a (simultaneous) inductive definition as in App 10 which (in Kleene's words) "is regarded from the finitary point of view as a correction, in that it eliminates the presupposition of the classical (non-constructive) second number class". There are problems with this view, partly because many results about constructive ordinals cannot be proved (or even stated) without referring to ordinals. In any case, we will use S_1 here.

countable ordinals. Their supremum

$$\omega_1 = \sup\{|a| : a \in S_1\} \text{ (the Church-Kleene omega-1)} \tag{5.17}$$

is a "constructive analog" of the first uncountable ordinal Ω_1; it is a fundamental constant of definability theory and it can be characterized in many natural ways, including the following early result:

Theorem 5.2.1 (Markwald [30], Spector [47]) *An ordinal ξ is constructive if and only if it is finite or the order type of a recursive wellordering of* $\mathbb{N}$.[10]

5.2.2 *The H_a-sets*

By recursion on the ordinal $|a|$, we associate with each $a \in S_1$ a set $H_a \subseteq \mathbb{N}$ so that:

(H1) $H_1 = \mathbb{N}$,
(H2) $H_{2^b} = H_b'$, and
(H3) if $a = 3 \cdot 5^e$, then $x \in H_a \iff (x)_0 \in H_{e(x)_1}$.

This is exactly the definition in Kleene [19], except that he gave it for $a \in O \subsetneq S_1$. The earlier Davis [4] gave an almost identical definition (for $a \in S_1$) which differs only in the details of the coding, and Mostowski [39] gave a somewhat different and abbreviated version which seems to avoid ordinal codes, cf. Sect. 5.3.3.

Davis [4] proves that for $a, b \in S_1$, $|a|, |b| < \omega^2$,

$$|a| \leqslant |b| \implies H_a \leqslant_m H_b \text{ and } |b| < |a| \implies H_b <_T H_a, \tag{5.18}$$

so that, in particular,

$$|a| = |b| < \omega^2 \implies H_a \equiv_T H_b \quad (a, b \in S_1)$$

and asks if every constructive ordinal has this *uniqueness property*. This turned out to be a difficult problem and led some five years later to one of the first spectacular results in the area[11]:

Theorem 5.2.2 (Spector [47]) *For all $a, b \in S_1$,*

$$|a| \leqslant |b| \implies H_a \leqslant_T H_b \text{ and } |b| < |a| \implies H_b <_T H_a.$$

In particular, $|a| = |b| \implies \mathbf{d}(H_a) = \mathbf{d}(H_b)$ and if we set $\mathbf{d}_{|a|} = \mathbf{d}(H_a)$, then $\{\mathbf{d}_\xi : \xi < \omega_1\}$ is an increasing sequence of Turing degrees of length ω_1.

[10] A proof of this basic fact is included in §8 of Moschovakis [37].

[11] For a discussion of the Spector Uniqueness Theorem and an outline of its proof for S_1 see §9 of Moschovakis [37].

Much more was done with constructive ordinals and the H_a-sets in the fifties and sixties, especially by Kleene who used them as his main tool for studying the hyperarithmetical sets. We will not go much into this here, for good reasons that we will explain in due course; but before we dig into our main topic, we need to discuss briefly some important, early work that we will not cover in detail.

5.2.3 Myhill [40]

Two sets A, B are *recursively isomorphic* if one is carried onto the other by a recursive permutation of $\mathbb{N}$,

$$A \equiv B \iff A \leqslant_e B \text{ where } \varphi_e : \mathbb{N} \rightarrowtail\!\!\!\twoheadrightarrow \mathbb{N} \text{ is a bijection.}$$

Myhill [40] introduces this notion and shows (among other things) that

$$\textit{for all } A, B \subseteq \mathbb{N},\ \textit{if } A \equiv_1 B, \textit{then } A \equiv B, \tag{5.19}$$

and so *any two r.e. complete sets are recursively isomorphic.* His methods also combine easily and to significant advantage with some of the results above: for example, Davis' proof of (5.18) naturally gives the much neater[12]

$$|a| = |b| < \omega^2 \implies H_a \equiv H_b. \tag{5.20}$$

However, none of Davis, Kleene or Mostowski knew of this article of Myhill when they wrote the papers we have been discussing.

5.2.4 Effective Grounded Recursion

More significantly, neither Davis nor Mostowski refer or appeal explicitly to the following basic fact:

Theorem 5.2.3 (Kleene's 2nd Recursion Theorem) *For every recursive partial function $f(e, x_1, \ldots, x_n, \alpha_1, \ldots, \alpha_m)$, there is a number e such that*

$$\{e\}(x_1, \ldots, x_n, \alpha_1, \ldots, \alpha_m) = f(e, x_1, \ldots, x_n, \alpha_1, \ldots, \alpha_m).$$

For recursion on $\mathbb{N}$, this was stated unbilled and proved[13] in the last two lines of §2 of Kleene [14] and it is the main technical tool that Kleene used for all his work

---

[12]This strong uniqueness property cannot be extended to ω^2, cf. Moschovakis [31], Nelson [41].
[13]Choose $\overline{k}$ such that $\{\overline{k}\}(t, \boldsymbol{x}, \boldsymbol{\alpha}) = f(S^{1,m}_n(t, t), \boldsymbol{x}, \boldsymbol{\alpha})$ and take $e = S^{1,m}_n(\overline{k}, \overline{k})$.

on constructive ordinals, hyperarithmetical sets—and much more. Myhill [40] also used it, crucially, as did Spector [47] in his proof of the Uniqueness Theorem 5.2.2. Kleene and Spector use the 2nd Recursion Theorem to justify *effective grounded recursion*, which we can illustrate here with a relevant example.

Consider Davis' definition of the sets $\{L_a : a \in S_1\}$ which are his versions of the H_a-sets:

(L1) $L_1 = \emptyset$,
(L2) $L_{2^b} = L'_b$, and
(L3) if $a = 3 \cdot 5^e$, then $x \in L_a \iff (x)_1 \in H_{e(x)_2}$.

Well, L_1 is the complement of H_1 and in the limit case Davis uses $(x)_1$ and $(x)_2$ rather than Kleene's $(x)_0$ and $(x)_1$ which, together, don't amount to much of a difference. The two definitions should be equivalent up to Turing equivalence, and they are[14]:

Lemma 5.2.1 *For every $a \in S_1$, $H_a \equiv_T L_a$. In fact, there are recursive partial functions $u(a)$, $v(a)$ which converge on S_1 and satisfy*

$$H_a \leqslant^T_{u(a)} L_a, \quad L_a \leqslant^T_{v(a)} H_a \quad (a \in S_1). \tag{$*$}$$

The partial functions $u(a)$, $v(a)$ are *uniformities* which witness respectively the reducibilities $H_a \leqslant_T L_a$, $L_a \leqslant_T H_a$.

Proof The Turing equivalence $H_a \equiv_T L_a$ should be more-or-less trivial by induction on the ordinal $|a|$ and it is, when $|a|$ is 0 or a successor ordinal (granting it for its predecessor). At a limit stage $a = 3 \cdot 5^e$, however, there is no obvious way to put together the equivalences $H_{e_t} \equiv_T L_{e_t}$ supplied by the induction hypothesis to prove that $H_a \equiv_T L_a$, and it is clear that we need to formulate a stronger, "uniform" proposition which will supply a usable induction hypothesis at limit stages. For the first reducibility in ($*$), one "recursion loading device" that works is the following:

Sublemma. There is a recursive partial function $f(i, a, x)$ which converges for all i, x when $i \leqslant 1$ and $a \in S_1$ and satisfies the following:

$$x \in H_a \iff f(0, a, x) = 0 \vee [f(0, a, x) \neq 0 \;\&\; f(1, a, x) \in L_a]. \tag{$**$}$$

Proof of the Sublemma. We set $f(0, 1, x) = 0$ and $f(1, 1, x) = 1$. If $a = 2^b$ for some b, then $f(0, a, x) = 1$ and it is not hard to define $f(1, a, x)$ from $f(i, b, x)$ so that ($**$) holds using (5.15) in Footnote 6. Suppose now $a = 3 \cdot 5^e$ and ($**$) holds for all ordinals less than $|a|$. We compute the conditions that $f(i, a, x)$ must satisfy by examining the equivalences which hold if it does:

[14] In the terminology of Post [42], the proof shows that H_a and L_a are *equivalent by bounded truth tables*. Had Davis chosen to set $L_1 = \mathbb{N}$ at the basis, then these modified L_as are recursively isomorphic with Kleene's H_a sets, and by a simpler argument than the proof of this Lemma.

$$\begin{aligned} x \in H_a &\iff (x)_0 \in H_{e_{(x)_1}} \\ &\Leftrightarrow f(0, e_{(x)_1}, (x)_0) = 0 \vee [f(0, e_{(x)_1}, (x)_0) \neq 0 \,\&\, f(1, e_{(x)_1}, (x)_0) \in L_{e_{(x)_1}}] \\ &\Leftrightarrow f(0, e_{(x)_1}, (x)_0) = 0 \vee [f(0, e_{(x)_1}, (x)_0) \neq 0 \,\&\, \langle 0, f(1, e_{(x)_1}, (x)_0), (x)_1\rangle \in L_a] \\ &\iff f(0, a, x) = 0 \vee [f(0, a, x) \neq 0 \,\&\, f(1, a, x) \in L_a] \end{aligned}$$

where we have used the induction hypothesis in the second line and the definition of L_a in the third (with an irrelevant 0 put into the first position so that $f(1, a, b)$ codes a triple). So when $a = 3 \cdot 5^e$ we need to have

$$f(0, a, x) = f(0, e_{(x)_1}, (x)_0), \quad f(1, a, x) = \langle 0, f(1, e_{(x)_1}, (x)_0), (x)_1\rangle. \quad (***)$$

Now, the 2nd Recursion Theorem easily supplies us with a recursive partial function $f(i, a, x)$ which satisfies the relevant conditions for $a = 1, a = 2^b$ and $(***)$, and then the proof is completed by a routine transfinite induction on $|a|$. □ (Proof of the Sublemma)

The corresponding Sublemma for the second reducibility in $(*)$ is proved by a similar construction, and then the two Sublemmas together imply $(*)$. □

Briefly (and vaguely), to "compute" a function $f : D \to \mathbb{N}$ which is defined on $D \subseteq \mathbb{N}^n$ by the recursion

$$f(\boldsymbol{x}) = G(f \restriction \{\boldsymbol{x}' : \boldsymbol{x}' \prec \boldsymbol{x}\}, \boldsymbol{x}) \quad (\boldsymbol{x} \in D) \tag{5.21}$$

along some wellfounded relation $\prec \subset (\mathbb{N}^n \times \mathbb{N}^n)$, we use the 2nd Recursion Theorem to find a recursive partial f which converges on D and satisfies (5.21) *on the assumption that one such* f *exists*; and then we prove by induction along $\prec$ that f indeed satisfies (5.21). It is very important for the applications that *no definability assumptions are needed for* D *or* $\prec$, except as they might be used to define f; for the proof of Lemma 5.2.1, for example,

$$D = \{(i, a, x) : a \in S_1\}, \quad (i, a, x) \prec (j, b, y) \iff |a| < |b|,$$

and we have no estimate of the complexity of this D and this $\prec$, certainly not now.

The method is very general and we cannot do it justice here, but it has played a very important role in the study of hyperarithmetical sets and so I thought it important to give in full at least one proof which uses it. Another, similar but more difficult example is the *uniform version* of Spector's Uniqueness Theorem 5.2.2:

$$|a| \leqslant |b| \implies H_a \leqslant_T H_b \textit{ uniformly for all } a, b \in S_1.$$

Its precise meaning is that *there is a recursive partial function* $u(a, b)$, a *uniformity*, such that

$$a, b \in S_1 \,\&\, |a| \leqslant |b| \implies [u(a, b)\downarrow \;\&\; H_a \leqslant^T_{u(a,b)} H_b]. \tag{5.22}$$

This formulation not only gives useful, additional information, but is necessary for the proof of the Uniqueness Theorem (by effective grounded recursion).

In the sequel I will often refer to effective grounded recursion and uniformity, but with little detail and less explanation.[15]

5.3 The Basic Facts About HYP (1950–1960)

A set $A \subseteq \mathbb{N}$, relation $R \subseteq \mathbb{N}^n$ or (total) function $f : \mathbb{N}^n \to \mathbb{N}$ is *hyperarithmetical* if it is recursive in some H_a; HYP is the set of all hyperarithmetical sets, and

$$\text{if } A \leqslant_e^T H_a, \text{ then } \langle a, e\rangle \text{ is a HYP} - \mathit{code\,of}\ A. \tag{5.23}$$

To express succinctly (and prove) the basic properties of HYP-sets, it is useful to think of them as "bundled" with their codes by the following general notion:

5.3.1 *Codings and Uniformities*

A (surjective) *coding* of a set X is a pair (C, π), where $\pi : C \twoheadrightarrow X$ is a surjection of the *codeset* C onto X, and we call any $c \in C$ a *code* (or name) of the object $\pi(c) \in X$. If $C \subseteq \mathbb{N}$, we say that *the coding is in* $\mathbb{N}$. These are the only codings we will need for a while.

So $(S_1, |\ |)$ is a coding of the constructive ordinals; $(S_1, a \mapsto H_a)$ is a coding of the H_a-sets; $(S_1, a \mapsto L_a)$ is a coding of Davis' L_a-sets; and for a very elementary example, $(\mathbb{N}, e \mapsto \varphi_e)$ is a coding of the set of unary recursive partial functions. The coding of HYP we introduced by (5.23) is formally the pair

$$C = \{\langle a, e\rangle : a \in S_1 \ \&\ \{e\}^{H_a} \text{ is total}\},$$
$$\pi(\langle a, e\rangle) = \{x \in \mathbb{N} : \{e\}^{H_a}(x) = 1\}. \tag{5.24}$$

In practice we will never be so formal, in fact we will sometimes use codings which are "specified by the context" without a formal definition of C and π.

Codings are useful for expressing succinctly *uniform properties* of *coded sets*. Their general theory is technically messy, not very interesting mathematically and certainly not worth putting here.[16] We will confine ourselves to these remarks and "detail" sufficiently many claims to make the ideas clear. For example:

---

[15] Cf. Moschovakis [36, 37] for a discussion (and many examples), and 7A.4 of Moschovakis [35] for a specific result which codifies many of the applications of effective grounded recursion in Descriptive Set Theory.

[16] The interested reader may want to look at Moschovakis [36] where it was necessary to develop this generalized abstract nonsense in some detail.

Lemma 5.3.1 HYP *is uniformly closed under complements and relative recursion.*
In detail, there are recursive partial functions $u(c)$ and $v(c, e)$ such that:

(1) *If A is* HYP *with code c, then $u(c)\downarrow$ and is a* HYP*-code of* $(\mathbb{N} \setminus A)$.
(2) *If c is a* HYP*-code of a set B and $A \leqslant_e^T B$, then $v(c, e)\downarrow$ and is a* HYP*-code of A.*

This is a simple lemma, as are the similar claims of uniform closure of the hyperarithmetical relations (with their natural coding) under all first-order operations on $\mathbb{N}$. There is no use of effective effective grounded recursion in these proofs, we only need appeal to uniform properties of the jump operation like (5.15). The next result is also quite easy, but its proof requires effective grounded recursion and some auxiliary definitions on the constructive ordinals:

Lemma 5.3.2 HYP *is uniformly closed under recursive unions.*
In detail, there is a recursive partial function $u(e)$ such that if φ_e is total and for each t, $\varphi_e(t)$ is a HYP*-code of a set $A_t \subseteq \mathbb{N}$, then $u(e)\downarrow$ and is a* HYP*-code of $\bigcup_t A_t$.*

Coding invariance
Two codings $(C_1, \pi_1), (C_2, \pi_2)$ in $\mathbb{N}$ of the same set $X = \pi_1[C_1] = \pi_2[C_2]$ are *equivalent* if there are recursive partial functions $u_1(a), u_2(b)$ such that

$$a \in C_1 \implies [u_1(a)\downarrow \ \& \ u_1(a) \in C_2 \ \& \ \pi_2(u_1(a)) = \pi_1(a)]$$

and similarly with 1 and 2 interchanged. It is clear that propositions like Lemmas 5.3.1 and 5.3.2 which hold uniformly for a certain coding also hold uniformly for every equivalent coding—and for some of them the proof might be easier.[17] We exploit this idea by establishing an elegant characterization of HYP which produces a coding for it equivalent to the classical one in (5.23) but much simpler.

5.3.2 HYP *as Effective Borel*

An *effective σ-algebra on* $\mathbb{N}$ is any collection $X \subset \mathscr{P}(\mathbb{N})$ of sets of natural numbers which admits a coding (C, π) in $\mathbb{N}$ so that the following hold:

(1) Every singleton $\{\{t\}\}$ belongs to X uniformly, i.e., for some total, recursive $u_1(t)$ and every t, $u_1(t)$ is a code of $\{\{t\}\}$ in X.

[17] For a classical example, consider the coding of recursive partial functions specified by the Normal Form Theorem in App 5. Its precise definition depends on the choice of computation model that we use, Turing machines, systems of recursive equations or whatever, but all these codings are equivalent and so uniform propositions about them are *coding invariant*. §4.3–§4.5 of Rogers [45] considers this situation in some detail and formulates stronger notions of equivalence than the one we use.

(2) X is uniformly closed under complements, i.e., there is a recursive partial function $u_2(c)$ such that

$$c \in C \implies [u_2(c)\downarrow \ \& \ \pi(u_2(c)) = \mathbb{N} \setminus \pi(c)].$$

(3) X is uniformly closed under recursive unions, i.e., for some recursive partial function $u_3(e)$,

$$(\forall t)[\varphi_e(t)\downarrow \ \& \ \varphi_e(t) \in C] \\ \implies [u_3(e)\downarrow \ \& \ u_3(e) \in C \ \& \ \pi(u_3(e)) = \textstyle\bigcup_t \pi(\Phi_e(t))].$$

As in the definition of $(S_1, |\ |)$, let $\mathbf{b} : \mathbb{N} \rightharpoonup \mathscr{P}(\mathbb{N})$ be the least partial function on $\mathbb{N}$ to $\mathscr{P}(\mathbb{N})$, such that

(1) $\mathbf{b}(\langle 1, t\rangle) = \{\{t\}\}$,
(2) $\mathbf{b}(\langle 2, y\rangle) = \mathbb{N} \setminus \mathbf{b}(y)$, and
(3) if φ_e is total and for every i, $\mathbf{b}(\varphi_e(i))\downarrow$, then $\mathbf{b}(\langle 3, e\rangle) = \bigcup_i \mathbf{b}(\varphi_e(i))$

and set

$$B = \{i : \mathbf{b}(i)\downarrow\}, \quad B_i = \mathbf{b}(i) \ \ (\text{if } i \in B), \quad \mathsf{B} = \{B_i : i \in B\}, \tag{5.25}$$

the collection of *effective Borel* subsets of $\mathbb{N}$.

Lemma 5.3.3 B *is the least effective σ-algebra on $\mathbb{N}$, uniformly.*

Proof The coding $(B, i \mapsto B_i)$ witnesses that B is an effective σ-algebra on $\mathbb{N}$. To see that it is uniformly the least one, suppose (C, π) is a coding witnessing that some X is an effective σ-algebra on $\mathbb{N}$ and define by a natural effective grounded recursion a recursive partial function u such that

$$i \in B \implies [u(i)\downarrow \ \& \ u(i) \in C \ \& \ B_i = \pi(u(i))].$$

□

Theorem 5.3.1 $\mathsf{HYP} = \mathsf{B}$ *uniformly, i.e., (C, π) in (5.24) and $(B, i \mapsto B_i)$ in (5.25) are equivalent codings of* HYP.

Proof HYP is an effective σ-algebra on $\mathbb{N}$ by Lemmas 5.3.1 and 5.3.2 and a simple construction which puts into it every singleton, uniformly. By Lemma 5.3.3 then, $\mathsf{B} \subseteq \mathsf{HYP}$, uniformly. To prove $\mathsf{HYP} \subseteq \mathsf{B}$, we need to verify that *every effective σ-algebra on $\mathbb{N}$ is uniformly closed under the jump operation, relative recursion and diagonalization*, which is not difficult as these operations can be effectively reduced to complementation and the taking of recursive unions; we then use effective grounded recursion to define a uniform embedding of HYP into B. □

Remark 5.3.1 The theorem gives us a different view of hyperarithmetical sets and a simpler way to prove important properties of them which do not explicitly refer to the H_a-sets, and these include most of the important properties of HYP. I am not certain who should be credited for it: it was "in the air" in the mid-sixties and I think that it was probably first formulated by Shoenfield, but I cannot find now a specific citation. In any case, it was certainly not known in the 50s, and our use of it here is the most substantial anachronism in this exposition of what was proved then.

5.3.3 Lebesgue [28] and Mostowski [39]

The situation is actually quite similar to one that came up in classical analysis at the turn of the last century. Recall the definition of Borel subsets of $\mathcal{N}$ in (2) of Sect. 5.1.4. In modern notation, the *Borel hierarchy* $\{ \underset{\sim}{\boldsymbol{\Sigma}}^0_\xi : \xi < \Omega_1\}$ (on $\mathcal{N}$) is defined by setting

$$\underset{\sim}{\boldsymbol{\Sigma}}^0_1 = \text{the collection of all open subsets of } \mathcal{N} \tag{5.26}$$

and then by recursion on the countable ordinals,

$$A \in \underset{\sim}{\boldsymbol{\Sigma}}^0_\xi \iff A = \bigcup_i (\mathcal{N} \setminus A_i) \text{ with each } A_i \in \bigcup_{\eta<\xi} \underset{\sim}{\boldsymbol{\Sigma}}^0_\eta \quad (\xi > 1). \tag{5.27}$$

These definitions were first given (for the reals) by Lebesgue [28] who proved (among many other fascinating and much deeper things) that

$$\mathbf{B} = \bigcup_{\xi<\Omega_1} \underset{\sim}{\boldsymbol{\Sigma}}^0_\xi. \tag{5.28}$$

As it happens, most of the important applications of the Borel sets to analysis (including measure theory and integration) use only the definitions and (5.28), which is easy and handy for proving properties of Borel sets by ordinal induction. The fine structure of the Borel hierarchy is a very interesting and much-studied topic but not as fundamental as **B**.

The definition of hyperarithmetical sets in Mostowski [39] is inspired by the classical theory of Borel sets, although he does not cite Lebesgue [28] or any other "classical" work. It is a difficult paper to read, basically an outline: he appears to define his hierarchy directly on ordinals rather than notations (which is not possible with the tools he uses) and he refers cryptically to (what must be) effective grounded recursion as "*a rather developed technique which we do not wish to presuppose here*". Section 9 of Kleene [19] supplies the details which are needed to make Mostowski's construction rigorous and comes up with a precise characterization of the intended hierarchy: in modern notation

$$\Sigma_a = \{A \subseteq \mathbb{N} : A \text{ is r.e. in } H_a\} = \{A \subseteq \mathbb{N} : A \leqslant_1 H_{2^a}\} \quad (a \in S_1). \tag{5.29}$$

It is immediate from the definition that

$$\text{if } 1 \leqslant |a| = k < \omega, \text{ then } \Sigma_a = \Sigma^0_k.$$

Moreover, Σ_a depends only on the ordinal $|a|$ by the Spector Uniqueness Theorem 5.2.2 and

$$\mathsf{HYP} = \bigcup_{a \in S_1} \Sigma_a. \tag{5.30}$$

The hierarchy $\{\Sigma_a : a \in S_1\}$ has been studied even less that the Borel hierarchy $\{\boldsymbol{\Sigma}_\xi : \xi < \Omega_1\}$, partly because the topic is not easy. It is obvious that it is a hierarchy, since every Σ_a has a complete set (H_{2^a}); but to prove (for example) that *every* Σ_a *is closed under conjunction* you must use effective grounded recursion, and for more difficult questions these proofs become very complex. In any case, we will not work with it here: for what we will do, the identification $\mathsf{HYP} = \mathsf{B}$ suffices and yields simpler proofs.

5.3.4 *The Analytical Hierarchy;* $\mathsf{HYP} \subseteq \Delta^1_1$

Useful and natural as the characterization $\mathsf{HYP} = \mathsf{B}$ may be, it does not provide explicit definitions for the hyperarithmetical sets and relations. These require quantification over sets of natural numbers or, equivalently, the Baire space $\mathcal{N} = \mathbb{N}^{\mathbb{N}}$.

A relation $P(x, \beta)$ with arguments in $\mathbb{N}$ and (possibly) $\mathcal{N}$ is *analytical* if it is first-order definable in the two-sorted structure of *analysis*

$$\mathbf{N}^2 = (\mathbb{N}, \mathcal{N}, 0, 1, +, \cdot, \text{ ap}) \tag{5.31}$$

where ap $(\alpha, t) = \alpha(t)$ is the *application* operation. Kleene [19] classifies the arithmetical and analytical relations with arguments in $\mathbb{N}$ and $\mathcal{N}$ in hierarchies which look so much like the arithmetical hierarchy over $\mathbb{N}$ that we pictured them together in Fig. 5.1. We are mostly interested here in the "first level" of the analytical hierarchy, the *pointclasses*[18] Π^1_1, Σ^1_1, Δ^1_1, but it is almost as easy to define them all. Briefly, and using the notions and notation in the Appendix:

(1) $P(\boldsymbol{x}, \boldsymbol{\beta})$ with $\boldsymbol{x} = (x_1, \ldots, x_n) \in \mathbb{N}^n$ and $\boldsymbol{\beta} = (\beta_1, \ldots, \beta_m) \in \mathcal{N}^m$ is Σ^0_1 if it is the domain of convergence of a recursive partial function,

$$P(\boldsymbol{x}, \boldsymbol{\beta}) \iff f(\boldsymbol{x}, \boldsymbol{\beta})\downarrow;$$

P is Π^0_k if it is the negation of a Σ^0_k relation; P is Σ^0_{k+1} if

---

[18]A *pointclass* in this paper is any collection Γ of relations $P(\boldsymbol{x}, \boldsymbol{\alpha})$ with arguments in $\mathbb{N}$ and $\mathcal{N}$. It is an awkward term but useful, and is has been well established since the 70 s for collections of relations in various spaces typically specified by the context.

$$P(\boldsymbol{x}, \boldsymbol{\beta}) \iff (\exists t)Q(\boldsymbol{x}, t, \boldsymbol{\beta}) \quad \text{with } Q \text{ in } \Pi^0_k;$$

and $\Delta^0_k = \Sigma^0_k \cap \Pi^0_k$.

These are the arithmetical relations with arguments in $\mathbb{N}$ and $\mathcal{N}$, those which can be defined in $\mathbf{N}^2$ without using quantification over $\mathcal{N}$.

(2) $P(\boldsymbol{x}, \boldsymbol{\beta})$ is Π^1_1 if

$$P(\boldsymbol{x}, \boldsymbol{\beta}) \iff (\forall\alpha)Q(\boldsymbol{x}, \boldsymbol{\beta}, \alpha), \tag{5.32}$$

with arithmetical $Q(\boldsymbol{x}, \boldsymbol{\beta}, \alpha)$; it is Π^1_k if it is the negation of a Σ^1_k relation; and it is Σ^1_{k+1} if

$$P(\boldsymbol{x}, \boldsymbol{\beta}) \iff (\exists\alpha)Q(\boldsymbol{x}, \boldsymbol{\beta}, \alpha) \quad \text{with } Q \text{ in } \Pi^1_k;$$

and $\Delta^1_k = \Sigma^1_k \cap \Pi^1_k$.

The analytical pointclasses Π^1_k, Σ^1_k, Δ^1_k have all the closure properties of their analogs Π^0_k, Σ^0_k, Δ^0_k in the arithmetical hierarchy over $\mathbb{N}$, and they are also closed under number quantification of both kinds and under substitution of total recursive functions into $\mathbb{N}$ or $\mathcal{N}$, App 8. In addition, Π^1_k is closed under $\forall\alpha$ and Σ^1_k is closed under $\exists\alpha$. These closure properties are very easy to prove, but not without consequence[19]:

Lemma 5.3.4 *The codeset B of* $\mathsf{B} = \mathsf{HYP}$ *defined in* (5.25) *is* Π^1_1.

Proof By its definition, B is the least fixed point $\overline{\Phi}$ of the monotone operator $\Phi : \mathcal{P}(\mathbb{N}) \to \mathcal{P}(\mathbb{N})$ defined by

$$\begin{aligned} x \in \Phi(A) \iff & (\exists t)[x = \langle 1, t\rangle] \vee (\exists y)[x = \langle 2, y\rangle \;\&\; y \in A] \\ & \vee (\exists e)[x = \langle 3, e\rangle \;\&\; (\forall i)(\exists w)[\varphi_e(i) = w \;\&\; w \in A]] \end{aligned} \tag{5.33}$$

so that by (5.59),

$$i \in B \iff (\forall A)[[(\forall x)[x \in \Phi(A) \implies x \in A]] \implies i \in A].$$

If we code each set A by the 0-*set* $Z_\alpha = \{x : \alpha(x) = 0\}$ of some $\alpha \in \mathcal{N}$ and set

$$\Phi(x, \alpha) \iff x \in \Phi(Z_\alpha), \tag{5.34}$$

then $\Phi(x, \alpha)$ is arithmetical (just replace $u \in A$ by $\alpha(u) = 0$ in (5.33)); and

$$i \in B \iff (\forall\alpha)[[(\forall x)[\Phi(x, \alpha) \implies \alpha(x) = 0]] \implies \alpha(i) = 0],$$

so that B is Π^1_1. □

[19] They also suffice to prove that the notation system S_1 *is* Π^1_1, cf. Lemma 1 in the proof of Theorem 9.2 in Moschovakis [37].

This is a very general method of proof: it can be used to show that *if* Φ *is monotone on* $\mathcal{P}(\mathbb{N})$ *and the relation* $\Phi(x,\alpha)$ *associated with* Φ *by* (5.34) *is* Π^1_1, *then* $\overline{\Phi}$ *is* Π^1_1 and, of course, it can be generalized in many ways.

Much of the theory of Π^1_1 depends on the following refinement of its definition (5.32):

Theorem 5.3.2 (Normal Form for Π^1_1) *Every* Π^1_1 *relation* $P(\boldsymbol{x},\boldsymbol{\beta})$ *satisfies an equivalence*

$$P(\boldsymbol{x},\boldsymbol{\beta}) \iff (\forall\alpha)(\exists t)R(\boldsymbol{x},\boldsymbol{\beta},\overline{\alpha}(t)) \tag{5.35}$$

where $R(\boldsymbol{x},\boldsymbol{\beta},u)$ *is recursive and monotone upward on its last (sequence code) argument, i.e.,*

$$[R(\boldsymbol{x},\boldsymbol{\beta},u)\ \&\ u \sqsubseteq v] \implies R(\boldsymbol{x},\boldsymbol{\beta},v). \tag{5.36}$$

It is easy to prove, using the closure properties of Π^1_1, the somewhat unusual "dual" of the Axiom of Choice, that for every relation $R(t,s)$,

$$(\exists t)(\forall s)R(t,s) \iff (\forall\alpha)(\exists t)R(t,\alpha(t))$$

and the Normal Form Theorem for recursive partial functions, App 5. By App 5 again, it implies the analog of (2) in Lemma 5.1.1:

Lemma 5.3.5 ($\mathbb{N}$-Parametrization for Π^1_1) *For all* $n,m \geqslant 0$, *there is a* Π^1_1 *relation*

$$G(e,\boldsymbol{x},\boldsymbol{\beta}) \iff G^{n,m}(e,x_1,\dots,x_n,\beta_1,\dots,\beta_m)$$

such that for every Π^1_1 *relation* $P(\boldsymbol{x},\boldsymbol{\beta})$,

$$P(\boldsymbol{x},\boldsymbol{\beta}) \iff G(e,\boldsymbol{x},\boldsymbol{\beta})\ \textit{for some}\ e\in\mathbb{N}; \tag{5.37}$$

moreover, there are recursive injections $S^l_n : \mathbb{N}^{1+l} \to \mathbb{N}$ *such that for all tuples* $\boldsymbol{y} = y_1,\dots,y_l \in \mathbb{N}$,

$$G^{l+n,m}(e,\boldsymbol{y},\boldsymbol{x},\boldsymbol{\beta}) \iff G^{n,m}(S^l_n(e,\boldsymbol{y}),\boldsymbol{x},\boldsymbol{\beta}). \tag{5.38}$$

When (5.37) holds, we call e a Π^1_1-*code* of $P(\boldsymbol{x},\boldsymbol{\beta})$ and a Σ^1_1-*code* of its negation $\neg P(\boldsymbol{x},\boldsymbol{\beta})$; and if e is a Π^1_1-code and m a Σ^1_1-code of $P(\boldsymbol{x},\boldsymbol{\beta})$, then $\langle e,m\rangle$ is a Δ^1_1-*code* of it.

To see how the Parametrization Property is used, suppose $R(\boldsymbol{x},t)$ is a Π^1_1 relation on $\mathbb{N}$ (for simplicity) with code e and

$$P(x) \iff (\exists t)R(\boldsymbol{x},t).$$

Let $Q(m,\boldsymbol{x}) \iff (\exists t)G(m,\boldsymbol{x},t)$ (with the appropriate superscripts) and let $\overline{s}$ be a Π^1_1-code of Q; then

$$P(\boldsymbol{x}) \iff (\exists t)R(\boldsymbol{x}, t) \iff (\exists t)G(e, \boldsymbol{x}, t)$$
$$\iff Q(e, \boldsymbol{x}) \iff G(\overline{s}, e, \boldsymbol{x}) \iff G(S^1_1(\overline{s}, e), \boldsymbol{x}),$$

so $S^1_1(\overline{s}, e)$ is a code of $P(\boldsymbol{x})$. The upshot is that Π^1_1 is *uniformly closed under* $\exists s$, and by similar, trivial computations, Π^1_1, Σ^1_1 and Δ^1_1 are uniformly closed under all (reasonable) operations under which they are closed, including those listed above. This implies that *the collection of* Δ^1_1 *subsets of* $\mathbb{N}$ *is an effective* σ*-algebra on* $\mathbb{N}$, which with Lemma 5.3.3 then yields

Theorem 5.3.3 (Kleene [19]) $\mathsf{HYP} \subseteq \Delta^1_1$, *uniformly. In detail, there are relations* $H_\Sigma(i, x)$ *and* $H_\Pi(i, x)$ *in* Σ^1_1 *and* Π^1_1 *respectively, such that*

$$i \in B \implies \Big(x \in B_i \iff H_\Sigma(i, x) \iff H_\Pi(i, x)\Big).$$

Davis [4] and Mostowski [39] had already shown that every HYP-relation is analytical, but Kleene's result is a considerable improvement and begs for the converse.

5.3.5 *Kleene's Theorem,* $\mathsf{HYP} = \Delta^1_1$

This was the most important, early result about HYP and it is still the most fundamental.

Theorem 5.3.4 (Kleene [21]) $\Delta^1_1 \subseteq \mathsf{HYP}$, *uniformly, so* $\mathsf{HYP} = \Delta^1_1$.

The foundational import of Kleene's Theorem is that it reduces existential quantification $(\exists \alpha)$ over the continuum $\mathcal{N}$ to regimented iteration of first-order quantification over $\mathbb{N}$—in the very special circumstances where a set A and its complement can both be defined by just one such quantification on arithmetical relations.

There are many proofs of Kleene's Theorem, all of them ultimately based on the Normal Form Theorem 5.3.2 for Π^1_1 and using effective grounded recursion. The proof in Kleene [21] is quite complex and depends on several technical results about constructive ordinals and the H_a-sets. To outline briefly the much simpler argument in Spector [47] , put first

$$x \leqslant_f y \iff \varphi_f(x, y) = 0, \quad L = \{f : \varphi_f \text{ is total and } \leqslant_f \text{ is a linear order}\},$$
$$W = \{f \in L : \ \leqslant_f \text{ is a wellordering}\},$$
$$||f|| = \text{the order type of } \leqslant_f \quad (f \in W).$$

By Markwald's Theorem 5.2.1, $\{||f|| : f \in W\}$ is exactly the set constructive ordinals, and we set

$$W_\xi = \{f \in W : ||f|| \leqslant \xi\} \quad (\xi < \omega_1).$$

The first move is to check that the initial segments $\{f : ||f|| \leqslant ||s||\}$ of W are uniformly Δ^1_1 for $s \in W$:

Lemma 5.3.6 *There are binary relations $\leqslant_\Sigma$ and $\leqslant_\Pi$ in Σ^1_1 and Π^1_1 respectively, such that*

$$s \in W \implies \Big([f \in W \,\&\, ||f|| \leqslant ||s||] \iff f \leqslant_\Sigma s \iff f \leqslant_\Pi s\Big).$$

Proof Set

$$\begin{aligned} f \leqslant_\Sigma s &\iff f, s \in L \,\&\, \text{there is an order-preserving embedding of } \leqslant_f \text{ into } \leqslant_s,\\ f \leqslant_\Pi s &\iff f, s \in L \,\&\, \text{there is no order preserving embedding of } \leqslant_s\\ &\qquad\qquad \text{into a proper initial segment of } \leqslant_f . \end{aligned}$$

To verify that these relations do it, we code embeddings using elements of Baire space and use the closure properties of Σ^1_1 and Π^1_1. □

The second move introduces what is now called the *Kleene-Brouwer* or *Luzin-Sierpinski* ordering on finite sequences. It is used in Kleene [21] and in many proofs of Kleene's Theorem:

Lemma 5.3.7 (Spector [47]) *W is Π^1_1-complete, uniformly.*

In detail: W is Π^1_1 and there is a recursive function $u_1(a)$ such that if a is a Π^1_1-code of a set $A \subseteq \mathbb{N}$, then $\varphi_{u_1(a)}$ is injective and

$$x \in A \iff \{u_1(a)\}(x) \in W.$$

Proof W is Π^1_1 directly from its definition. To show that it is Π^1_1-complete, suppose that A is Π^1_1 with code a, so that by Theorem 5.3.2 and Lemma 5.3.5,

$$x \in A \iff G(a, x) \iff (\forall\alpha)(\exists t)R(a, x, \overline{\alpha}(t))$$

with a fixed recursive $R(a, x, v)$ (not depending on A) which is monotone upward in its last argument. Define the transitive relation

$$u \preceq^{a,x} v \iff v \sqsubseteq u \,\&\, \neg R(a, x, u)$$

and prove that

$$x \in A \iff (\forall\alpha)(\exists t)R(a, x, \overline{\alpha}(t)) \iff \preceq^{a,x} \text{ is wellfounded,} \tag{5.39}$$

most easily by checking its contrapositive

$$x \notin A \iff (\exists\alpha)(\forall t)\neg R(a, x, \overline{\alpha}(t)) \iff \preceq^{a,x} \text{ is not wellfounded.}$$

We then *linearize* $\preceq^{a,x}$ by setting

$$u \leqslant^{a,x} v \iff \neg R(a,x,u) \,\&\, \neg R(a,x,v)$$
$$\& \Big(v \sqsubseteq u \vee [u \mid v \,\&\, \min\{(u)_i : i < \mathrm{lh}(u)\} < \min\{(v)_i : i < \mathrm{lh}(v)\}]\Big);$$

verify that this is a linear ordering such that

$$x \in A \iff \leqslant^{a,x} \text{ is a wellordering},$$

in fact $\leqslant^{a,x} = \leqslant_{g(a,x)}$ with a recursive g such that for any $a, x, g(a,x) \in L$; and infer that

$$x \in A \iff \leqslant^{a,x} \text{ is a wellordering} \iff g(a,x) \in W. \tag{5.40}$$

To finish the proof we need to define a recursive $u_1(a)$ such that $\{g(a,x)\}(s) = \{\{u_1(a)\}(x)\}(s)$ and $\{u_1(a)\}$ is injective for every a, and this is done by manipulating the S^l_n- functions as usual. □

The third move is Spector's. It is what makes his proof simpler than Kleene's who worked with O rather than W.

Lemma 5.3.8 (Boundedness, Spector [47]) *Every Σ^1_1 subset of W is a subset of W_ξ for some $\xi < \omega_1$, uniformly.*

In detail: there is a recursive partial function $u_2(b)$ such that if b is a Σ^1_1-code of a set $A \subseteq \mathbb{N}$, then

$$A \subseteq W \implies [u_2(b)\downarrow, u_2(b) \in W, \text{ and } A \subseteq W_{||u_2(b)||}].$$

Proof Let $G(b,x)$ be a parametrization of the unary Π^1_1 relations by Lemma 5.3.5, so that a set $A \subseteq \mathbb{N}$ is Σ^1_1 with code b if

$$A = G^c_b = \{s : \neg G(b,s)\}.$$

Fix also by the Π^1_1-completeness of W a recursive injection $g : \mathbb{N} \to \mathbb{N}$ such that

$$G(x,x) \iff g(x) \in W. \tag{$*$}$$

The relation

$$P(b,f) \iff (\exists s)[\neg G(b,s) \,\&\, g(f) \leqslant_\Sigma s]$$

is Σ^1_1, and so by Lemma 5.3.5 again, there is a recursive injection $v(b) = S^2_1(\overline{k}, b)$ (with some $\overline{k}$) such that

$$(\exists s)[\neg G(b,s) \,\&\, g(f) \leqslant_\Sigma s] \iff \neg G(v(b), f). \tag{$**$}$$

The key observation is that

$$\textit{if } A = G_b^c \subseteq W, \textit{ then } G(v(b), v(b)) :$$

because if $A \subseteq W$ and $\neg G(v(b), v(b))$, then there is some $s \in W$ such that $g(v(b)) \leqslant_\Sigma s$; which gives $g(v(b)) \in W$ by Lemma 5.3.6; which in turn gives $G(v(b), v(b))$ by $(*)$, contradicting the hypothesis. From $G(v(b), v(b))$ we get $g(v(b)) \in W$, by $(*)$ again, and so by taking negations in $(**)$,

$$A = G_b^c \subseteq W \implies (\forall s)[s \in A \implies ||s|| < ||g(v(b))||],$$

which is what we needed to show with $u_2(b) = g(v(b.b))$. □

Outline of proof of Theorem 5.3.4. By the two lemmas, if A is Δ_1^1 with code $\langle a, b\rangle$, then

$$x \in A \iff \{u_1(a)\}(x) \in W_\xi \text{ with } \xi = ||u_2(b)||. \tag{5.41}$$

To complete the proof we need to show that $W_{||f||}$ is in B *uniformly for* $f \in W$, and this is done by a fairly straightforward effective grounded recursion along $\{(f, g) : f, g \in W \;\&\; ||f|| \leqslant ||g||\}$. □

Spector's write-up of his proof is not quite this simple because he works with the H_a-codes rather than the B-codes of HYP and (in effect) proves the uniform inclusion B $\subseteq$ HYP on the fly.

Moreover, neither Kleene nor Spector claimed explicitly the full, uniform version of Kleene's Theorem 5.3.4, although all the "mathematical facts" needed for it are in their papers.[20] Most likely they did not even think of it: in the spirit of the time, a result was formulated uniformly only when this was necessary, typically in order to prove it by effective grounded recursion. Uniform claims did not become important in themselves until the 70s, when the applications of these ideas to Descriptive Set Theory made them necessary. We will discuss this briefly in Sect. 5.4.2.

Spector's Lemmas 5.3.6–5.3.8 are important results with many applications besides their use in proving Kleene's Theorem. We state one of them here and then one more, not quite so simple in the next section.[21]

Theorem 5.3.5 (Spector [47]) *If $\preceq$ is a Δ_1^1 wellordering with field $F \subseteq \mathbb{N}$, then rank $(\preceq) < \omega_1$, uniformly.*

This is usually abbreviated by the equation

$$\delta_1^1 = \omega_1,$$

[20]What's missing in their papers is the second part in the proof of the Boundedness Lemma 5.3.8 which looks tricky at first sight but is a standard, elementary tool in this area. It computes "witnesses to counterexamples" using diagonalization in very general circumstances, and we have already used it to establish the uniform properties of the jump in Footnote 6.

[21]Cf. App 9 for the notation we use about wellorderings and ranks.

δ^1_1 being the least ordinal which is not the order type of a HYP wellordering.

Proof Suppose, towards a contradiction that $\preceq$ is a Δ^1_1 wellordering with rank $(\preceq) \geqslant \omega_1$ and set

$$f \in A \iff f \in L \ \& \text{ there is an order preserving map of } \leqslant_f \text{ into } \preceq .$$

This is a Σ^1_1 set and the hypotheses imply that $A = W$, which contradicts Lemma 5.3.8. The uniform version is proved similarly, using the uniform version of the same Lemma. □

5.3.6 Addison [1] and the Revised Analogies

Kleene's Theorem 5.3.4 is an immediate consequence of the following more general

Theorem 5.3.6 (Strong Separation for Σ^1_1, Addison [1]) *For any two disjoint, Σ^1_1 subsets of $\mathbb{N}$, there is a* HYP *set C which separates them, i.e.,*

$$A \subseteq C, \quad C \cap B = \emptyset.$$

In fact, Addison [1] claims more and less than this result: he states it for subsets A, B of any product space $\mathbb{N}^n \times \mathcal{N}^m$ rather than just $\mathbb{N}$ and his (abbreviated) proof is formulated quite abstractly and also gives the classical Separation Theorem 5.1.2 for $\utilde{\boldsymbol{\Sigma}}^1_1$; but he does not note that the result holds uniformly (in given Σ^1_1-codes of A and B), which it does, and he only says of the separating set C that "it is Δ^1_1" skipping the punchline "and hence HYP" which he certainly knows for subsets of $\mathbb{N}$. This may be partly because there was no generally accepted definition of HYP *subset of* $\mathbb{N}^n \times \mathcal{N}^m$ at the time, or because Addison's paper is about separation and not construction principles. He also does not discuss the obvious revision of the analogies (5.7)

$$\begin{gathered} \text{recursive function on } \mathbb{N} \sim \text{continuous function on } \mathcal{N}, \\ \mathsf{HYP} \sim \mathbf{B}, \\ \Pi^1_1 \text{ sets of integers } \sim \utilde{\boldsymbol{\Pi}}^1_1 \text{ subsets of } \mathcal{N}, \end{gathered} \tag{5.42}$$

which are the working hypotheses of Mostowski [39]. They are bolstered by the following result which is not hard to prove using Spector-type ordinal assignments and the method of proof of Kleene's Theorem 5.1.3:

Theorem 5.3.7 *There exists disjoint Π^1_1-sets A, B which are not* HYP*-separable, i.e., no* HYP *set C satisfies*

$$A \subseteq C, \quad C \cap B = \emptyset.$$

On the other hand, to my knowledge, Addison [1] was first to refer to *Effective Descriptive Set Theory*, which suggests that more than "analogies" are in play; and he introduced the modern *lightface* $\Sigma^1_k, \ldots$ and *boldface* $\underset{\sim}{\Sigma}{}^1_k \ldots$ notation which has been universally accepted.

5.3.7 Relativization and the Kreisel Uniformization Theorem

We mention in App 6 the method of *proof by relativization*, which works because (roughly) recursion in some fixed parameters $\boldsymbol{\beta}$ has all the properties of "absolute" recursion. It is not simple to formulate a general metatheorem which captures all its applications—especially when uniformities are involved which should be "absolutely" recursive. It is, however, a very powerful method, heavily used by the early researchers in hyp theory, especially Kleene and Spector. We illustrate it here by proving two important and useful results.

The relative forms $\Sigma^{i,\boldsymbol{\beta}}_k$, $\Pi^{i,\boldsymbol{\beta}}_k$, $\Sigma^{i,\boldsymbol{\beta}}_k$, $\Delta^{i,\boldsymbol{\beta}}_k$ of the arithmetical and analytical hierarchies are defined simply by replacing "recursive" by "recursive in $\boldsymbol{\beta}$" in their definitions, and they have all the properties of their absolute forms, including Lemma 5.1.1 (with absolutely recursive $S^{l,m}_n$ functions).

The same is true for the relativized system $S^{\boldsymbol{\beta}}_1$ of ordinal notations: we simply replace e_t in Sect. 5.2.1 by $e^{\boldsymbol{\beta}}_t = \{e\}^{\boldsymbol{\beta}}(t_O)$ and write $|a|^{\boldsymbol{\beta}}$ for the ordinal with code $a \in S^{\boldsymbol{\beta}}_1$. Markwald's Theorem 5.2.1 remains true: an ordinal ξ is less than

$$\omega^{\boldsymbol{\beta}}_1 = \sup\{|a|^{\boldsymbol{\beta}} : a \in S^{\boldsymbol{\beta}}_1\}$$

exactly when it is the order type of a wellordering (of part of $\mathbb{N}$) which is recursive in $\boldsymbol{\beta}$. We use these ordinals to define the relativized $H^{\boldsymbol{\beta}}_a$ sets by replacing (H2) in Sect. 5.2.2 by

$$(\mathrm{H2}^{\boldsymbol{\beta}})\ \ H^{\boldsymbol{\beta}}_{2^b} = \text{jump}\,(H^{\boldsymbol{\beta}}_b; \boldsymbol{\beta}) = \{e : \{e\}(e, H^{\boldsymbol{\beta}}_b, \boldsymbol{\beta})\downarrow\}$$

and we set

$$A \in \mathsf{HYP}^{\boldsymbol{\beta}} \iff (\exists a \in S^{\boldsymbol{\beta}}_1)[A \leqslant_T H^{\boldsymbol{\beta}}_a].$$

With these definitions, all the basic facts about HYP relativize, including Spector's Uniqueness Theorem 5.2.2, the characterization of $\mathsf{HYP}^{\boldsymbol{\beta}}$ as the least σ-algebra on $\mathbb{N}$ which is *effective in* $\boldsymbol{\beta}$, Theorem 5.3.1 and the uniform inclusion $\mathsf{HYP}^{\boldsymbol{\beta}} \subseteq \Delta^{1,\boldsymbol{\beta}}_1$, Theorem 5.3.3. For the converse inclusion (Kleene's Theorem), we need to relativize the basic notions of Spector [47] : we set

$$x \leqslant^{\boldsymbol{\beta}}_f y \iff \varphi_f(x, y, \boldsymbol{\beta}) = 0,$$
$$L^{\boldsymbol{\beta}} = \{f : (\forall x, y)[\varphi_f(x, y, \boldsymbol{\beta})\downarrow] \text{ and } \leqslant^{\boldsymbol{\beta}}_f \text{ is a linear order}\},$$
$$W^{\boldsymbol{\beta}} = \{f \in L^{\boldsymbol{\beta}} : \ \leqslant^{\boldsymbol{\beta}}_f \text{ is a wellordering}\},$$
$$||f||^{\boldsymbol{\beta}} = \text{the order type of } \leqslant^{\boldsymbol{\beta}}_f \quad (f \in W^{\boldsymbol{\beta}}).$$

Using these we get immediately the relativized versions of Lemma 5.3.6 and (what we need of) the relativized version of Lemma 5.3.7, basically (5.40):

(1) *There are relations* $\leqslant^{\boldsymbol{\beta}}_{\Sigma}$ *and* $\leqslant^{\boldsymbol{\beta}}_{\Pi}$ *in* Σ^1_1 *and* Π^1_1 *respectively, such that for all* $\boldsymbol{\beta}$,

$$s \in W^{\boldsymbol{\beta}} \implies \Big([f \in W^{\boldsymbol{\beta}} \;\&\; \|f\|^{\boldsymbol{\beta}} \leqslant \|s\|^{\boldsymbol{\beta}}] \iff f \leqslant^{\boldsymbol{\beta}}_{\Sigma} s \iff f \leqslant^{\boldsymbol{\beta}}_{\Pi} s\Big).$$

(2) *If* $P(\boldsymbol{x}, \boldsymbol{\beta})$ *is* Π^1_1, *then there is a total recursive function* $f(\boldsymbol{x})$ *such that*

$$P(\boldsymbol{x}, \boldsymbol{\beta}) \iff f(\boldsymbol{x}) \in W^{\boldsymbol{\beta}}.$$

These suffice to relativize Spector's proof of the non-uniform version of Kleene's Theorem 5.3.4

$$\text{for every } \boldsymbol{\beta},\ \ \mathsf{HYP}^{\boldsymbol{\beta}} = \Delta^{1,\boldsymbol{\beta}}_1,$$

and a little more detailed version of (2) gives also the uniform version.

With single sets rather than tuples of functions $\boldsymbol{\beta}$, for simplicity, we set

$$A \leqslant_h B \iff A \in \mathsf{HYP}^B \iff A \text{ is hyperarithmetical in } B.$$

The *hyperdegrees* that are induced by this reducibility have been studied extensively, cf. Sacks [46]. We will not go into this topic here, except for the following, early and important result. To appreciate what it says, notice that because W is Π^1_1-complete,

$$W \leqslant_h A \iff \text{every } \Pi^1_1 \text{ set is hyperarithmetical in } A.$$

Theorem 5.3.8 (Spector [47]) *For every set* $A \subseteq \mathbb{N}$,

$$W \leqslant_h A \iff \omega_1 < \omega^A_1,$$

and in relativized form, for all $A, B \subseteq \mathbb{N}$,

$$W^A \leqslant_h B \iff \omega^A_1 < \omega^B_1.$$

Proof Suppose first that $W \leqslant_h A$ and set

$$x \in D \iff \Big(x \in W \;\&\; (\forall y)[(y \in W \;\&\; \|y\| = \|x\|) \implies x \leqslant y\Big),$$
$$x \preceq y \iff x, y \in D \;\&\; \|x\| \leqslant \|y\|;$$

now $\preceq$ is a wellordering of rank ω_1 and it is $\Delta^{1,A}_1$, so its rank is below $\delta^{1,A}_1 = \omega^A_1$ by the relativized version of Spector's Theorem 5.3.5.

The converse is a bit easier. □

Our second example illustrates a somewhat more subtle application of the relativization technique: roughly, it proves a universal property $(\forall \boldsymbol{\beta})Q(\boldsymbol{\beta})$ by treating an arbitrary tuple $\boldsymbol{\beta}$ as a parameter, relativizing to it the proof of a simple (absolute) proposition, and then exploiting the uniform nature of the proof to infer $Q(\boldsymbol{\beta})$ with variable $\boldsymbol{\beta}$.

Theorem 5.3.9 (Π^1_1-Uniformization on $\mathbb{N}$, Kreisel [27]) *For every Π^1_1 relation $P(\boldsymbol{x}, y, \boldsymbol{\beta})$, there is a Π^1_1 relation $P^*(\boldsymbol{x}, y, \boldsymbol{\beta})$ such that*

$$P^*(\boldsymbol{x}, y, \boldsymbol{\beta}) \implies P(\boldsymbol{x}, y, \boldsymbol{\beta}) \text{ and } (\exists y)P(\boldsymbol{x}, y, \boldsymbol{\beta}) \implies (\exists! y)P^*(x, y, \boldsymbol{\beta}). \tag{5.43}$$

It follows that if $P(\boldsymbol{x}, y)$ is Π^1_1, then

$$(\forall \boldsymbol{x})(\exists y)P(\boldsymbol{x}, y) \implies (\exists f : \mathbb{N}^n \to \mathbb{N})[f \text{ is } \mathsf{HYP} \ \& \ (\forall \boldsymbol{x})P(\boldsymbol{x}, f(\boldsymbol{x}))].$$

Proof In the simple case where the list $\boldsymbol{\beta}$ of variables over $\mathcal{N}$ is empty, we choose a recursive $g : \mathbb{N}^n \to \mathbb{N}$ such that $P(\boldsymbol{x}, y) \iff g(\boldsymbol{x}, y) \in W$ and set

$$\begin{aligned} P^*(\boldsymbol{x}, y) \iff P(\boldsymbol{x}, y) \ \& \ (\forall u)[g(\boldsymbol{x}, y) \leqslant_\Pi g(\boldsymbol{x}, u)] \\ \& \ (\forall u)[g(\boldsymbol{x}, u) \leqslant_\Sigma g(\boldsymbol{x}, y) \implies y \leqslant u]. \end{aligned}$$

This also gives the second claim: check that if $(\forall \boldsymbol{x})(\exists y)P(\boldsymbol{x}, y)$, then $P^*(\boldsymbol{x}, y)$ is the graph of a function f and it is Δ^1_1, since

$$\neg P^*(\boldsymbol{x}, y) \iff (\exists z)[P^*(\boldsymbol{x}, z) \ \& \ z \neq y)].$$

To get the more useful claim with parameters, we relativize this argument using (1) and (2) above. Given a Π^1_1 relation $P(\boldsymbol{x}, y, \boldsymbol{\beta})$, choose a recursive $g(\boldsymbol{x}, y)$ such that

$$P(\boldsymbol{x}, y, \boldsymbol{\beta}) \iff g(\boldsymbol{x}, y) \in W^{\boldsymbol{\beta}}$$

and set

$$\begin{aligned} P^*(\boldsymbol{x}, y, \boldsymbol{\beta}) \iff P(\boldsymbol{x}, y, \boldsymbol{\beta}) \ \& \ (\forall u)[g(\boldsymbol{x}, y) \leqslant^{\boldsymbol{\beta}}_\Pi g(\boldsymbol{x}, u)] \\ \& \ (\forall u)[g(\boldsymbol{x}, u) \leqslant^{\boldsymbol{\beta}}_\Sigma g(\boldsymbol{x}, y) \implies y \leqslant u]. \end{aligned}$$

The check that this works is exactly as before. □

The *Kondo-Addison Uniformization Theorem* for Π^1_1 relations $P(\boldsymbol{x}, \alpha, \boldsymbol{\beta})$ (Kondo [24], Addison) is much deeper, but this simple result is also interesting and very useful.

5.3.8 HYP-*Quantification and the Spector-Gandy Theorem*

The (coded) *graph* of a function $\alpha : \mathbb{N} \to \mathbb{N}$ is the set

$$\text{Graph}\,(\alpha) = \{\langle s, t\rangle : \alpha(s) = t\} \subset \mathbb{N},$$

and we often write "$\alpha \in \mathsf{HYP}$" when we really mean "Graph$(\alpha) \in \mathsf{HYP}$", i.e., that α is hyperarithmetical. We collect here some interesting, easy (now) facts about the quantifier $(\exists\alpha \in \mathsf{HYP})$ and we also formulate the basic *Spector-Gandy Theorem* about it—which has never been easy.

It is natural to code the HYP-functions using a subset of the coding of HYP-sets as effectively Borel in (5.25):

$$\begin{aligned} B^1 = &\{i \in B : B_i = \text{ Graph}\,(\alpha) \text{ for some } \alpha\},\\ &\text{and if } i \in B^1, \text{ then } \beta_i(s) = t \iff \langle s, t\rangle \in B_i. \end{aligned} \tag{5.44}$$

The key (easy) facts about this coding is that B^1 *is* Π^1_1 by Lemma 5.3.4 and Theorem 5.3.3, and that *for each* $i \in B^1$, the relation

$$\alpha = \beta_i \iff (\forall s, t)[\alpha(s) = t \iff \langle s, t\rangle \in B_i] \tag{5.45}$$

is Δ^1_1 uniformly, by Theorem 5.3.3 again.

Theorem 5.3.10 *(1)* HYP-*Quantification Theorem, Kleene ([21], [22]). If*

$$P(\boldsymbol{x}) \iff (\exists\alpha \in \mathsf{HYP})Q(\boldsymbol{x}, \alpha) \tag{5.46}$$

and $Q(\boldsymbol{x}, \alpha)$ *is* Π^1_1, *then* $P(\boldsymbol{x})$ *is also* Π^1_1.

(2) HYP *is not a basis for* Π^0_1, *Kleene [21]. There is a non-empty,* Π^0_1 *set* $A \subseteq \mathcal{N}$ *which has no* HYP *members.*

(3) Upper classification of HYP. *As a subset of* $\mathcal{N}$, HYP *is* Π^1_1.

(4) Lower classification of HYP. *As a subset of* $\mathcal{N}$, HYP *is not* Σ^1_1.

Proof (1) Compute:

$$(\exists\alpha \in \mathsf{HYP})Q(\boldsymbol{x}, \alpha) \iff (\exists i)[i \in B^1 \;\&\; (\forall\alpha)[\alpha = \beta_i \implies Q(\boldsymbol{x}, \alpha)]].$$

(2) Towards a contradiction, assume that every non-empty, Π^0_1 set $A \subseteq \mathcal{N}$ has a HYP member and let $P \subseteq \mathbb{N}$ be an arbitrary Σ^1_1 set. By the Normal Form for Π^1_1 Theorem 5.3.2 (applied to $\neg P$),

$$P(x) \iff (\exists\alpha)(\forall t)R(x, t, \alpha) \iff A_x = \{\alpha : (\forall t)R(x, t, \alpha)\} \neq \emptyset$$

with a recursive R. Since every A_x is Π^0_1, our assumption implies that

$$P(x) \iff (\exists\alpha \in \mathsf{HYP})(\forall t)R(x,t,\alpha);$$

which by (1) means that every Σ^1_1 subset of $\mathbb{N}$ is Π^1_1, which it is not.

(3) $\alpha \in \mathsf{HYP} \iff (\exists i)[i \in B^1 \;\&\; \alpha = \beta_i]$.

(4) The relation $P(i,\alpha) \iff i \in B^1 \;\&\; \alpha = \beta_i$ is Π^1_1, so by the Kreisel Uniformization Theorem 5.3.9, there is a Π^1_1 relation $P^*(i,\alpha)$ such that

$$P^*(i,\alpha) \implies i \in B^1 \;\&\; \alpha = \beta_i, \quad \alpha \in \mathsf{HYP} \implies (\exists! i)P^*(i,\alpha).$$

Let $D(i) \iff (\exists\alpha \in \mathsf{HYP})P^*(i,\alpha)$. This is Π^1_1 by (1), but if HYP is Σ^1_1, then it is also Σ^1_1, since

$$D(i) \iff (\exists\alpha)[\alpha \in \mathsf{HYP} \;\&\; (\forall j)[P^*(j,\alpha) \implies i = j]].$$

It follows that the function

$$\beta(i) = \begin{cases} 1 \dot{-} \beta_i(i) & \text{if } D(i), \\ 0 & \text{otherwise} \end{cases}$$

is Δ^1_1 and has no code in B^1, which is absurd. □

Kleene [23] proved Part (1) of this theorem with a Π^0_1 relation $Q(\boldsymbol{x},\alpha)$ and asked whether this version of (5.46) gives a normal form for Π^1_1. Spector [49] proved that it does, and gandy [10] gave an independent proof of this basic fact after hearing of Spector's result.

Theorem 5.3.11 (Spector [49], Gandy [10]) *Every Π^1_1 relation P on $\mathbb{N}$ satisfies an equivalence*

$$P(\boldsymbol{x}) \iff (\exists\alpha \in \mathsf{HYP})(\forall t)R(\boldsymbol{x},\overline{\alpha}(t)) \tag{5.47}$$

with a recursive $R(\boldsymbol{x},u)$. In fact, $R(\boldsymbol{x},u)$ can be chosen so that

$$P(\boldsymbol{x}) \iff (\exists\alpha \in \mathsf{HYP})(\forall t)R(\boldsymbol{x},\overline{\alpha}(t)) \iff (\exists!\alpha \in \mathsf{HYP})(\forall t)R(\boldsymbol{x},\overline{\alpha}(t)).$$

Spector's proof is difficult, as is Gandy's, both of them depending on a detailed, combinatorial analysis of Π^1_1 definitions and properties of the constructive ordinals coded by O. Easier proofs and generalizations of the first claim (without the uniqueness) were found later, cf. Moschovakis [33–35].

Taken together, Kleene's HYP-Quantification and the Spector-Gandy Theorem have important foundational import, perhaps best expressed by the following

Corollary 5.3.1 (Kleene, Spector) *A relation $P(\boldsymbol{x})$ on $\mathbb{N}$ satisfies*

$$P(\boldsymbol{x}) \iff (\forall\alpha)Q_1(\boldsymbol{x}, \alpha)$$

with an arithmetical $Q_1(\boldsymbol{x}, \alpha)$ if and only if it satisfies

$$P(\boldsymbol{x}) \iff (\exists\alpha \in \mathsf{HYP})Q_2(\boldsymbol{x}, \alpha)$$

with an arithmetical $Q_2(\boldsymbol{x}, \alpha)$.

Moreover, the equivalence holds *uniformly*, i.e., Q_2 can be constructed from Q_1 and vice versa.

The Corollary reduces *one* quantification over the continuum $\mathscr{N}$ on arithmetical relations to one quantification (of the opposite kind) over the countable set $\mathsf{HYP} \subsetneq \mathscr{N}$ whose members are constructed by regimented iteration of quantification over $\mathbb{N}$.

5.3.9 The Kleene [22] HYP *hierarchy*

This is perhaps the deepest and certainly the most difficult technical work of Kleene on hyperarithmetical sets.[22]

Theorem 5.3.12 (Kleene [22]) *If the monotone operator Δ on $\mathscr{P}(\mathscr{N})$ is defined by* (5.52) *below, then*

$$\eta < \xi < \omega_1 \implies \overline{\Delta}^{\eta} \subsetneq \overline{\Delta}^{\xi} \text{ and } \mathsf{HYP} = \bigcup\nolimits_{\xi<\omega_1} \overline{\Delta}^{\xi}. \tag{5.48}$$

Even without the definition of Δ, a hierarchy of the form $\{\Delta^{\xi} : \xi < \omega_1\}$ on HYP is more satisfactory than hierarchies like (5.29), because it is constructed without reference to any codings: there is no need for results like Spector's Uniqueness Theorem to establish *coding invariance*. The specific operator Δ that we define next also gives a novel understanding of HYP and yields many interesting applications.

Definitions with Range and Basis $\mathscr{F}$

A relation $P(\boldsymbol{x})$ is Σ^1_1 *with range* $\mathscr{F} \subseteq \mathscr{P}(\mathscr{N})$ if

$$P(\boldsymbol{x}) \iff (\exists\alpha \in \mathscr{F})Q(\boldsymbol{x}, \boldsymbol{\beta}, \alpha) \tag{5.49}$$

with $\boldsymbol{\beta} = \beta_1, \ldots, \beta_m \in \mathscr{F}$ and an arithmetical Q; it is Σ^1_1 *with basis* $\mathscr{F}$ if

[22] It is also his last paper on the subject.

$$P(\boldsymbol{x}) \iff (\exists\alpha)Q(\boldsymbol{x},\boldsymbol{\beta},\alpha) \iff (\exists\alpha\in\mathscr{F})Q(\boldsymbol{x},\boldsymbol{\beta},\alpha) \tag{5.50}$$

with $\boldsymbol{\beta}\in\mathscr{F}$ and an arithmetical Q; and it is Δ^1_1 with range or basis $\mathscr{F}$ if both P and its negation $\neg P$ are Σ^1_1 with range or basis $\mathscr{F}$ respectively.

If $P(\boldsymbol{x})$ is Σ^1_1 with basis $\mathscr{F}$, then it is also Σ^1_1 with range $\mathscr{F}$, clearly. The converse is not true: because every Π^1_1 relation is Σ^1_1 with range HYP by the Spector-Gandy Theorem 5.3.11, while

$$\textit{if } P(\boldsymbol{x}) \textit{ is } \Sigma^1_1 \textit{ with basis } \mathsf{HYP},\ \textit{then } P(\boldsymbol{x}) \textit{ is } \mathsf{HYP} \tag{5.51}$$

by Kleene's HYP-Quantification Theorem 5.3.10 (1)—and Theorem 5.3.4, of course, the inclusion $\Delta^1_1\subseteq\mathsf{HYP}$ being basic to all this work. We let[23]

$$\Delta(\mathscr{F}) = \{A\subseteq\mathbb{N} : A \text{ is arithmetical or } \Delta^1_1 \text{ with basis } \mathscr{F}\}. \tag{5.52}$$

It is clear from (5.51) that $\Delta(\mathsf{HYP})=\mathsf{HYP}$, and so the least fixed point $\overline{\Delta}$ of Δ is included in HYP. For the rest of (5.48), Kleene needs to show that

(1) $\mathsf{HYP}\subseteq\overline{\Delta}$, and
(2) if $\eta<\xi<\omega_1$, then $\overline{\Delta}^{\eta}\subsetneq\overline{\Delta}^{\xi}$.

For (1), he proves (in effect) that

$$a\in O \implies H_a \textit{ is } \Delta^1_1 \textit{ with basis } \bigcup\nolimits_{|b|<|a|}\Sigma_b$$

with Σ_a defined in (5.29). The key idea for (2) is to use the *ramified analytical hierarchy* comprising the iterates of the monotone operator

$$\mathrm{An}\,(\mathscr{F}) = \{A\subseteq\mathbb{N} : \text{for some } n,\ A \text{ is } \Sigma^1_n \text{ with range } \mathscr{F}\}$$

on $\mathscr{P}(\mathscr{N})$. Kleene shows that if $\xi<\omega_1$, then $\overline{\mathrm{An}}^{\xi}\subseteq\mathsf{HYP}$; and so if $\kappa(\Delta)<\omega_1$, then $\mathsf{HYP}=\overline{\Delta}$ would be a fixed point of An which contradicts the Spector-Gandy Theorem. Both proofs are by effective grounded recursion and require more detailed, delicate formulations of (1) and (2) to go through.

To formulate one of the simplest and most elegant characterizations of HYP that comes out of Theorem 5.3.12, recall the two-sorted structure of analysis $\mathbf{N}^2=(\mathbb{N},\mathscr{N},0,1,+,\cdot,\mathrm{ap})$ we used in Sect. 5.3.4. Its formal language A^2 has variables $x,y,\ldots,s,t,\ldots$ over $\mathbb{N}$ and $\alpha,\beta,\ldots$ over $\mathscr{N}$ and symbols 0, 1, +, ·, ap. Its *standard interpretation* is $\mathbf{N}^2$. We are interested in general, *ω-models* of A^2-theories in which the number variables range over $\mathbb{N}$ and the function variables over some $\mathscr{F}\subseteq\mathscr{N}$, and for any formula φ we will write

$$\mathscr{F}\models\varphi \iff (\mathbb{N},\mathscr{F},0,1,+,\cdot,\mathrm{ap})\models\forall\!\!\!\forall\varphi$$

[23]We need to include all arithmetical sets in $\Delta(\mathscr{F})$, ow. $\Delta(\emptyset)=\emptyset$ and Δ would close at 0 and build up the empty set.

where $\forall\!\!\!\forall\, \varphi$ is the universal closure of φ. As usual, we identify sets with their representing functions in such models,

$$A \in \mathscr{F} \iff \chi_A \in \mathscr{F} \quad (A \subseteq \mathbb{N}).$$

An A^2-formula φ is *arithmetical* if no function quantifiers occur in it. As usual, by $\varphi(x, y, \beta, \gamma)$ we will denote any formula in which the variables x, y, β, γ may occur free *but do not necessarily include all the variables which occur free in* φ.

We consider three axiom schemes in A^2:

Arithmetical comprehension. With arithmetical $\varphi(s)$ (in which α does not occur free),

$$(\exists\alpha)(\forall s)[\alpha(s) = 1 \iff \varphi(s)]. \qquad (\Delta^0_\infty - \text{Comp})$$

Δ^1_1*-comprehension.* With arithmetical $\varphi(s, \gamma), \psi(s, \gamma)$ (in which α does not occur free),

$$\begin{aligned}(\forall s)[(\exists\gamma)\varphi(s,\gamma) &\iff (\forall\gamma)\psi(s,\gamma)] \\ &\implies (\exists\alpha)(\forall s)[\alpha(s) = 1 \iff (\exists\gamma)\varphi(s,\gamma)]. \quad (\Delta^1_1 - \text{Comp})\end{aligned}$$

Σ^1_1*-Choice.* With arithmetical $\varphi(s, \alpha, \gamma)$,[24]

$$(\forall s)(\exists\alpha)(\exists\gamma)\varphi(s,\alpha,\gamma) \implies (\exists\alpha)(\forall s)(\exists\gamma)\varphi(s,(\alpha)_s,\gamma). \qquad (\Sigma^1_1 - \text{Choice})$$

Clearly, (Δ^1_1-Comp) $\implies$ (Δ^0_∞-Comp), and Kreisel [26] verified that[25]

$$(\Delta^0_\infty\text{-Comp}) + (\Sigma^1_1\text{-Choice}) \implies (\Delta^1_1\text{-Comp}). \tag{5.53}$$

Theorem 5.3.13 *(1) (Kleene [22], Kreisel [26])* HYP *is the least model of (Δ^1_1-Comp).*

(2) (Kreisel [26]) HYP *satisfies (Σ^1_1-Choice).*

Proof (1) If A is Σ^1_1 with range HYP, then it is Π^1_1 by the HYP-Quantification Theorem 5.3.10 (1); and if A is also Π^1_1 with range HYP, then it is Δ^1_1 and hence HYP. This proves that HYP satisfies (Δ^1_1-Comp), if we apply it to the set $A = \{s : (\exists\gamma)\varphi(s,\gamma)\}$ and then take $\alpha = \chi_A$. To see that it is the least model of (Δ^1_1-Comp), assume that $\mathscr{F}$ satisfies (Δ^1_1-Comp) and prove by induction on ξ that $\overline{\Delta}^\xi \subseteq \mathscr{F}$ using Theorem 5.3.12.

---

[24] We assume some formal treatment of recursive substitutions into A^2 formulas. In this case, the relevant recursive function is $(\alpha, s) \mapsto (\alpha)_s$, and we use the equivalences

$$\varphi(s,(\alpha)_s,\gamma) \iff (\exists\delta)[\delta = (\alpha)_s \;\&\; \varphi(s,\delta,\gamma)] \iff (\forall\delta)[\delta = (\alpha)_s \to \varphi(s,\delta,\gamma)].$$

These are satisfied by every model $\mathscr{F}$ of (Δ^0_∞-Comp).

[25] The converse fails, cf. Steel [51].

(2) Suppose that $(\forall s)(\exists \alpha \in \mathsf{HYP})(\exists \gamma \in \mathsf{HYP})\varphi(s, \gamma, \alpha)$ with an arithmetical φ and set

$$P(s, i) \iff i \in B^1 \;\&\; (\forall \alpha)[\alpha = \beta_i \implies (\exists \gamma \in \mathsf{HYP})\varphi(s, \gamma, \alpha)].$$

This is in Π^1_1, so by the Kreisel Uniformization Theorem 5.3.9, it is uniformized by a Π^1_1 relation $P^*(s, i)$; we check easily that some $\alpha \in \mathsf{HYP}$ satisfies

$$(\alpha)_s = \beta_i \text{ for the unique } i \text{ which satisfies } P^*(s, i),$$

and then this α also satisfies the right-hand-side of $(\Sigma^1_1 - \text{Choice})$. □

Another relevant and important result that we will not discuss here in detail is the following:

Theorem 5.3.14 (Gandy et al. [9]) *A set $A \subseteq \mathbb{N}$ is* HYP *if and only if its characteristic function χ_A belongs to every $\mathcal{F} \subseteq \mathcal{P}(\mathcal{N})$ which satisfies the axiom scheme of full comprehension, i.e., for every formula $\varphi(s)$ in which α does not occur free,*

$$(\exists \alpha)(\forall s)[\alpha(s) = 1 \iff \varphi(s)]. \qquad (\Delta^1_\infty - \text{Comp})$$

Beyond these (and many other) applications, however, the importance of Theorem 5.3.12 is primarily foundational. To quote Kleene [22],

> the definition [with basis $\mathcal{F}$] means the same to persons with various universes of functions, so long as each person's universe includes at least $\mathcal{F}$ (of which he may have no exact conception).

One can argue that it presents HYP as a *potential totality* which can be comprehended by mathematicians with varying views of "the continuum", much like $\mathbb{N}$ can be understood as xza potential totality within classical and constructive mathematics alike.

5.3.10 Inductive Definability on $\mathbb{N}$

It should be clear by now that inductive definitions permeate our subject, but is was not until Spector [50] that a neat, precise result was formulated expressing the connection.

Suppose $\Phi : \mathscr{P}(\mathbb{N}^l) \to \mathscr{P}(\mathbb{N}^l)$ is a monotone operator as in App 10 and (generalizing mildly (5.34)), define the *representing relation* of Φ by

$$\Phi(\boldsymbol{y}, \alpha) \iff \boldsymbol{y} \in \Phi(Z_\alpha) \text{ where } Z_\alpha = \{\boldsymbol{y}' : \alpha(\langle \boldsymbol{y}' \rangle) = 0\}. \tag{5.54}$$

The operator Φ is *in* a pointclass Γ (such as Π^0_1 or Π^1_1) if $\Phi(\boldsymbol{x}, \alpha)$ is in Γ; and a relation $P(\boldsymbol{x})$ is Γ-*inductive on* $\mathbb{N}$ if it is many-one reducible to the least fixed point $\overline{\Phi}$ of a monotone operator in Γ.

Lemma 5.3.9 (Spector [47]) [26] *If $\Phi(A)$ is a monotone, Π^1_1 operator on $\mathscr{P}(\mathbb{N})$ and $P \subseteq \mathbb{N}$ is Π^1_1, then*

$$x \in \Phi(P) \implies (\exists H \subseteq P)[H \in \mathsf{HYP} \;\&\; x \in \Phi(H)].$$

This is not really difficult, but its simplest proof requires identifying the monotone Π^1_1 operators with those which are Π^1_1-*positive*, suitably defined, and it is a bit too lengthy to include here.

Theorem 5.3.15 *(1) (Kleene [20]).*[27] *Every Π^1_1 relation $P(\boldsymbol{x})$ is Π^0_1-inductive on $\mathbb{N}$, in fact there is a Π^0_1 monotone operator Φ on $\mathbb{N}^{1+n}$ such that*[28]

$$P(\boldsymbol{x}) \iff (1, \boldsymbol{x}) \in \overline{\Phi}. \tag{5.55}$$

(2) (Spector [50]) If $\Phi : \mathscr{P}(\mathbb{N}^l) \to \mathscr{P}(\mathbb{N}^l)$ is Π^1_1, then its least fixed point $\overline{\Phi}$ is Π^1_1 and its closure ordinal $\kappa(\Phi) \leqslant \omega_1$.

Proof (1) is basically immediate from (5.39), which expresses Kleene's key understanding of Π^1_1 definitions: for a given $P(\boldsymbol{x})$ in Π^1_1 (and adjusting the notation in (5.39)), we set

$$(u, \boldsymbol{x}) \in \Phi(A) \iff \mathrm{Seq}(u) \;\&\; \Big(R(\boldsymbol{x}, u) \vee (\forall s)(u * \langle s \rangle, \boldsymbol{x}) \in A\Big), \tag{5.56}$$

prove first by induction on ξ that

$$(u, \boldsymbol{x}) \in \overline{\Phi}^\xi \implies \mathrm{Seq}(u) \;\&\; (\forall \alpha \sqsupseteq u)(\exists t) R(\boldsymbol{x}, \overline{\alpha}(t))$$

---

[26] This is not quite explicit in Spector [47], but Sacks [46] (8.5) credits it to Spector and I think this is right.

[27] This is seriously implicit in §24 of Kleene [20], but the idea of the proof is there and Spector correctly credits Kleene for it.

[28] The "1" is necessary here, in fact *it is not the case that every Π^1_1 set is the least fixed point $\overline{\Phi}$ of an arithmetical monotone operator Φ on* $\mathbb{N}$, cf. Feferman [8] and Moschovakis [34] (8.13, falsely claimed in the 1974 edition for all "countable acceptable structures"). Feferman's result was the first applications of Cohen's forcing to arithmetic.

and then check easily that (with $1 = \langle\,\rangle$, the code of the empty sequence),

$$(1, x) \notin \overline{\Phi} \implies (\exists\alpha)(\forall t)\neg R(x, \overline{\alpha}(t)) \implies \neg P(x).$$

This gives (5.55).

(2) That $\overline{\Phi}$ is Π^1_1 if Φ is Π^1_1, we have already proved in Lemma 5.3.4. For the more difficult bound on the closure ordinal, we first check that $P = \bigcup_{\xi<\omega_1} \overline{\Phi}^{\xi}$ is Π^1_1 by effective grounded recursion and then apply the Lemma. □

Like Kreisel [26], Spector [50] was presented at the famed *Symposium on Foundations of Mathematics* held in Warsaw in 1959. It has many more (and more difficult) results, but its main significance lies in this simple characterization of Π^1_1 (and hence HYP) in terms of inductive definability.

5.3.11 HYP *as Recursive in* ^{2}E

Starting with his [23], Kleene developed a theory of absolute and relative recursion for functions with arguments in the objects of *finite type over* $\mathbb{N}$, i.e., members of the sets T_i where

$$T_0 = \mathbb{N}, \quad T_{i+1} = (T_i \to \mathbb{N}) = \text{the set of functions on } T_i \text{ to } \mathbb{N}.$$

This is a technically difficult but fascinating topic, with some important applications to Descriptive Set Theory but especially to the foundations of the theory of recursion: it was the first example where there is no natural notion of *machine computable function* that can be defined independently of "recursiveness", and so it forces an examination of the meaning of *recursive definitions* in and of themselves. We cannot go into it here, but it is worth stating one of Kleene's basic results which relate it to HYP[29]:

In Kleene's words, the following type-2 object "embodies" the operation of quantification over $\mathbb{N}$:

$$^2\mathsf{E}(\alpha) = \begin{cases} 0, & \text{if } (\exists t)[\alpha(t) = 0], \\ 1, & \text{otherwise} \end{cases} \qquad (\alpha \in \mathcal{N}).$$

Theorem 5.3.16 (Kleene [23]) *A set $A \subseteq \mathbb{N}$ is hyperarithmetical if and only if it is recursive in* ^{2}E.

[29] Cf. Kechris and Moschovakis [12] for a relatively simple introduction to recursion in higher types and Sacks [46] for a full development.

In fact, for $A \subseteq \mathbb{N}$,

$$A \in \Pi^1_1 \iff A \text{ is recursively enumerable in } {}^2\mathsf{E}$$
$$\iff A = \{x : f(x)\downarrow\} \text{ for some } f : \mathbb{N} \rightharpoonup \mathbb{N} \text{ recursive in } {}^2\mathsf{E},$$

which bolsters the understanding of Π^1_1 as an analog of Σ^0_1 in *recursion in* ${}^2\mathsf{E}$.[30]

5.4 Concluding Remarks

The main results from the period 1950–1960 that we have surveyed established HYP as a robust class of sets, those subsets of $\mathbb{N}$ which can be defined (and can be guaranteed to exist) if we accept the structure **N** of arithmetic, quantification over $\mathbb{N}$ and recursion. The main new method introduced in this work is undoubtedly effective grounded recursion, but there are also many interesting tricks, especially in computing "witnesses to counterexamples" as in the proof of Lemma 5.3.8.

There were primarily three developments which followed this work and are still extensively pursued today: *recursion in higher types* which we have already discussed and the following two.

5.4.1 IND *and* HYP *on Abstract Structures*

Of the many characterizations of HYP, the easiest to formulate for an arbitrary structure $\mathbf{A} = (A, R_1, \ldots, R_k)$ is Spector's *inductive definability* in Sect. 5.3.10, cf. Moschovakis [34]. Briefly, a relation $P \subseteq A^n$ is *inductive in* **A** if it is one of the mutual least fixed points of a finite system of positive, elementary (first-order) relations with arguments in A, and it is *hyperelementary in* **A** if both P and its negation $A^n \setminus P$ are inductive.

Part of the theory of HYP and Π^1_1 can be developed for HYP(**A**) and IND(**A**) for arbitrary **A**; some of the results require an assumption that **A** is (almost) *acceptable*, roughly meaning that **A** admits a hyperelementary coding scheme for tuples; and suitable formulations of virtually all the results in the body of this paper can be established for all *countable, acceptable structures*, including Kleene's centerpiece that $\mathsf{IND}(\mathbf{A}) = \Pi^1_1(\mathbf{A})$ and so $\mathsf{HYP}(\mathbf{A}) = \Delta^1_1(\mathbf{A})$.

Kleene's Theorem 5.3.12 holds for all acceptable structures almost exactly in the form that it is stated in Sect. 5.3.9, with ω_1 replaced by the *closure ordinal* $\kappa(\mathbf{A})$ of **A**, an important invariant. It is proved, however, by an entirely different argument

---

30 Kleene [23] does not mention this and I recall him saying (much later) that he was not certain that the notion of a recursive partial function in higher type recursion was natural, but I cannot point to a reference for this.

which is different from (and perhaps simpler) than Kleene's even for the classical structure **N** of arithmetic.

The proofs, in fact, are the most interesting aspect of this generalization of HYP theory: there is little coding and no use of effective grounded recursion. These are replaced by constructs which were first used in higher type recursion (*Stage Comparison Theorems*) and ideas from the theory of *infinite open games*.

The most interesting application of inductive definability is to the structure $\mathbf{N}^2$ of analysis in (5.31) which is intimately related to our last topic.

5.4.2 *Effective Descriptive Set Theory*

The term was coined by Addison [1] who formulated his results about the spaces $\mathbb{N}^n \times \mathcal{N}^m$ and might have still be thinking of "analogies" between the classical and the effective results; but in the 50+ years since then, effective descriptive set theory has evolved into a unified study of *definability on recursive Polish spaces* which include $\mathbb{N}$, $\mathcal{N}$ and the real numbers and has deep applications to parts of topology and analysis in addition to classical descriptive set theory and logic. A good part of it is covered in Moschovakis [35], which, however, is concerned with many other things and is not sufficiently comprehensive on this topic.

5.5 Appendix: Some Basic Facts and Notation

We list here some elementary definitions and results, primarily to establish notation.

App 1 $\mathbb{N} = \{0, 1, \ldots\}$ is the set of natural numbers and $\mathcal{N} = \mathbb{N}^{\mathbb{N}}$ is the *Baire space* of all unary functions on $\mathbb{N}$. This carries the natural product topology with $\mathbb{N}$ discrete, generated by the basic neighborhoods

$$\mathcal{N}_{k_0,\ldots,k_t} = \{\alpha \in \mathcal{N} : \alpha(0) = k_0, \ldots, \alpha(t) = k_t\}$$

and the product spaces $\mathbb{N}^n \times \mathcal{N}^m$ carry the corresponding product topologies.

In general, lower case Latin letters vary over $\mathbb{N}$ and Greek letters $\alpha, \beta, \ldots$ vary over $\mathcal{N}$.

App 2 A *partial function* $f : X \rightharpoonup Y$ is a function $f : D_f \to Y$, where $D_f \subseteq X$ is the *domain of convergence* of f. We write

$$f(x)\downarrow \iff x \in D_f, \quad f(x)\uparrow \iff x \notin D_f \quad (x \in X),$$
$$f(x) = g(x) \iff [f(x)\uparrow \;\&\; g(x)\uparrow] \vee [f(x)\downarrow \;\&\; g(x)\downarrow \;\&\; f(x) = g(x)].$$

Partial functions *compose strictly*, e.g.,

$$f(g(x), h(x)) = w \iff (\exists u, v)[g(x) = u \ \&\ h(x) = v \ \&\ f(u, v) = w].$$

It is sometimes convenient to identify $f : X \rightharpoonup Y$ with its *graph*

$$\text{Graph}\,(f) = \{(x, y) \in X \times Y : f(x) = y\}.$$

App 3 $\chi_A : X \to \mathbb{N}$ is the *characteristic function* of $A \subseteq X$ (= 1 on A and 0 on $A^c = X \setminus A$).

App 4 *Sequence coding* in $\mathbb{N}$. The following functions and relations are recursive, with p_i the $(i+1)$'th prime number:

$\langle u_0, \ldots, u_{n-1}\rangle = p_0^{u_0+1} \cdots p_{n-1}^{u_{n-1}+1}$ = the code of $(u_0, \ldots, u_{n \dot{-} 1})$;

$\text{Seq}(u) \Leftrightarrow$ *u is the code of some sequence*, and if it is, then $\text{lh}(u)$ is its length and for $i < \text{lh}(u)$, $(u)_i = u_i$;

$u \sqsubseteq v \Leftrightarrow$ *u codes an initial segment of the sequence coded by v*;

$u \sqsubsetneq v \Leftrightarrow u \sqsubseteq v \ \&\ u \neq v$;

$u \mid v \Leftrightarrow$ u and v are codes of incompatible sequences $\Leftrightarrow \neg(u \sqsubseteq v \vee v \sqsubseteq u)$;

$u * v$ = the code of the concatenation of the sequences coded by u and v;

$\overline{\alpha}(t) = \langle \alpha(0), \ldots, \alpha(t \dot{-} 1)\rangle$ (= 1 if $t = 0$).

App 5 *Kleene's Normal Form and Enumeration Theorem*: Every recursive partial function(al) $f : \mathbb{N}^n \times \mathcal{N}^m \rightharpoonup \mathbb{N}$ is $\varphi_e^{n,m}$ for some e, where

$$\begin{aligned}\varphi_e^{n,m}(x_1, \ldots, x_n, \alpha_1, \ldots, \alpha_m) &= \{e\}^{n,m}(x_1, \ldots, x_n, \alpha_1, \ldots, \alpha_m)\\ &= U(\mu t T_n^m(e, x_1, \ldots, x_n, t, \overline{\alpha}_1(t), \ldots, \overline{\alpha}_m(t)))\end{aligned} \tag{5.57}$$

with (primitive) recursive T_n^m and U, and we will skip some or all of the superscripts n, m when they are clear from the context or irrelevant. Moreover, there are (primitive) recursive *injections* $S_n^{l,m}(e, y_1, \ldots, y_l)$ such that

$$\{e\}(y_1, \ldots, y_l, x_1, \ldots, x_n, \boldsymbol{\alpha}) = \{S_n^{l,m}(e, y_1, \ldots, y_l)\}(x_1, \ldots, x_n, \boldsymbol{\alpha}). \tag{5.58}$$

We call e a *code* of $\varphi_e^{n,m}$ and we use both φ_e and $\{e\}$ interchangeably, as the desire for neat typography dictates.

To avoid (implausible) confusion, we use $\{\{t\}\}$ for the singleton set whose only member is e.

App 6 *Relativization*. It is sometimes useful to fix some of the function arguments in the Normal Form Theorem and treat them as *parameters*. We write

$$\{e\}^{\boldsymbol{\beta}}(\boldsymbol{x}, \boldsymbol{\alpha}) = \{e\}(\boldsymbol{x}, \boldsymbol{\beta}, \boldsymbol{\alpha})$$

and we say that the partial function $(\boldsymbol{x}, \boldsymbol{\alpha}) \mapsto \{e\}^{\beta}(\boldsymbol{x}, \boldsymbol{\alpha})$ is *recursive in* $\boldsymbol{\beta}$ or *relative to* $\boldsymbol{\beta}$. For recursion *relative to a set* B, we write

$$\{e\}^{B}(\boldsymbol{x}) = \{e\}(\boldsymbol{x}, \chi_{B}) \quad (B \subseteq \mathbb{N}).$$

It is often—almost always—the case that a result about (absolutely) recursive partial functions can be easily seen to be true about partial functions recursive in some $\boldsymbol{\beta}$, by *relativization*, i.e., basically adding the superscript $^{\beta}$ to all functions in the proof; it is a simple but very useful method of proof.

App 7 *Recursively enumerable sets.* A set $A \subseteq \mathbb{N}$ is *r.e. in* $B \subseteq \mathbb{N}$ if

$$x \in A \iff \{e\}^{B}(x)\downarrow \text{ for some } e,$$

and (absolutely) *r.e.* if it is r.e. in the empty set.

App 8 *Total recursive functions into* $\mathcal{N}$. A (total) function $f : \mathbb{N}^n \times \mathcal{N}^m \to \mathcal{N}$ is recursive if

$$f(\boldsymbol{x}, \boldsymbol{\alpha}) = \lambda t f_*(t, \boldsymbol{x}, \boldsymbol{\alpha})$$

for some recursive partial $f_* : \mathbb{N}^{1+n} \times \mathcal{N}^m \rightharpoonup \mathbb{N}$. Useful examples include the *tupling* and *projection functions*:

$$\langle \alpha_0, \ldots, \alpha_{k-1} \rangle = \lambda t \begin{cases} \alpha_i(s), & \text{if } t = \langle i, s \rangle \text{ for some } i < k \text{ and some } s, \\ 0, & \text{otherwise,} \end{cases}$$
$$(\beta)_i = \lambda t \beta(\langle i, t \rangle),$$

so that for $i < k$, $(\langle \alpha_0, \ldots, \alpha_{k-1} \rangle)_i = \alpha_i$.

The class of total recursive functions into $\mathbb{N}$ or $\mathcal{N}$ is closed under composition—which is not true for recursive partial functions with values in $\mathcal{N}$.

App 9 *The rank of a strict, well founded relation.* A binary relation $\prec$ on a set F is *well founded* if there is no infinite descending chain $x_0 \succ x_1 \succ \cdots$ or, equivalently, if there is a function $\rho : X \to$ Ordinals such that

$$x \prec y \implies \rho(x) < \rho(y) \quad (x, y \in F);$$

the (pointwise) least such function $\rho_{\prec}$ is the *rank function* of $\prec$ and

$$\text{rank}\,(\prec) = \sup\{\rho_{\prec}(x) + 1 : x \in F\}.$$

When we apply this to the strict part $\prec$ of a wellordering $\preceq$, we get a (unique) similarity

$$\rho_{\preceq} : \{x : x \preceq x\} = F \rightarrowtail\!\!\!\!\rightarrow \text{rank}\,(\preceq)$$

of $\preceq$ with an ordinal.

App 10 *Monotone inductive definitions.* An operator $\Phi : \mathscr{P}(X) \to \mathscr{P}(X)$ on the subsets of a space X is monotone if

$$A \subseteq B \implies \Phi(A) \subseteq \Phi(B) \quad (A, B \subseteq X).$$

The set

$$\overline{\Phi} = \bigcap\{A : \Phi(A) \subseteq A\} \tag{5.59}$$

defined inductively (or *built up*) by Φ is the *least fixed point* of Φ, and

$$\overline{\Phi} = \bigcup \overline{\Phi}^{\xi}, \text{ where for each ordinal } \xi, \overline{\Phi}^{\xi} = \Phi(\bigcup_{\eta<\xi} \overline{\Phi}^{\eta}) \tag{5.60}$$

(with the usual convention that $\bigcup \emptyset = \emptyset$). Moreover,

$$\eta < \xi \implies \overline{\Phi}^{\eta} \subseteq \overline{\Phi}^{\xi} \subseteq X,$$

and since these *iterates* cannot increase forever, there is a least ordinal $\kappa = \kappa(\Phi)$, the *closure ordinal* of Φ such that

$$\eta < \xi < \kappa(\Phi) \implies \overline{\Phi}^{\eta} \subsetneq \overline{\Phi}^{\xi} \text{ and } \overline{\Phi} = \bigcup_{\xi<\kappa(\Phi)} \overline{\Phi}^{\xi}. \tag{5.61}$$

An operator Φ is *operative on X to W* if its domain is $\mathscr{P}(X \times W)$ and

$$f : X \rightharpoonup W \implies \Phi(\text{ Graph } (f)) = \text{ Graph } (g) \text{ for some } g : X \rightharpoonup W.$$

When this holds, then $\overline{\Phi} : X \rightharpoonup W$ is (the graph of) the least partial function fixed by the operator Φ.

References

1. Addison, J. W. (1959). Separation principles in the hierarchies of classical and effective descriptive set theory. *Fundamenta Methematicae*, 46:123–135.
2. Church, A. (1935). An unsolvable problem in elementary number theory. *Bulletin of the American Mathematical Society*, 41:332–333. This is an abstract of [3].
3. Church, A. (1936). An unsolvable problem in elementary number theory. *American Journal of Mathematics*, pages 345–363. An abstract of this paper was published in [52].
4. Davis, M. (1950a). *On the theory of recursive unsolvability*. PhD thesis, Princeton University.
5. Davis, M. (1950b). Relatively recursive functions and the extended Kleene hierarchy. page 723. Proceedings of the International Congress of Mathematicians, Cambridge, Mass, 1950.
6. Davis, M. (1965). *The undecidable*. Raven Press.
7. Enderton, H. B., and Putnam, H. (1970). A note on the hyperarithmetical hierarchy. *The Journal of Symbolic Logic*, 35:429–430.
8. Feferman, S. (1965). Some applications of forcing and generic sets. *Fundamenta Methematicae*, 56:325–345.

9. Gandy, R., Kreisel, G., and Tait, W. (1960). Set existence. *Bulletin of the Polish Academy of Sciences, Series in Science, Mathematics and Astronomy*, 8:577–582.
10. Gandy, R. O. (1960). Proof of Mostowski's conjecture. *Bulletin de l'Academie Polonaise des Sciences, Série des sciences mathématiques, astronomiques et physiques*, 8:571–575.
11. Gödel, K. (1931). Über formal unentscheidbare Sätze der Principia Mathematica and verwandter Systeme, I. *Monatshefte für Mathematik und Physik*, pages 173–198. English translations in [53] and [54].
12. Kechris, A. S. and Moschovakis, Y. N. (1977). Recursion in higher types. In Barwise, J., editor, *Handbook of Mathematical Logic*, Studies in Logic, No. 90, pages 681–737. North Holland, Amsterdam.
13. Kleene, S. C. (1936). General recursive functions of natural numbers. *Mathematische Annalen*, 112:727–742.
14. Kleene, S. C. (1938). On notation for ordinal numbers. *Journal of Symbolic Logic*, 3:150–155.
15. Kleene, S. C. (1943). Recursive predicates and quantifiers. *Transactions of the American Mathematical Society*, 53:41–73.
16. Kleene, S. C. (1950). A symmetric form of Gödel's theorem. *Konin. Neder. Akad. van Weten. te Amst. Proc. of the Section of Sciences*, 53:800–802.
17. Kleene, S. C. (1952). *Introduction to metamathematics*. North Holland Co: D. Van Nostrand Co.
18. Kleene, S. C. (1953). Arithmetical predicates and function quantifiers. *Journal of Symbolic Logic*, 18:190. abstract of a talk presented at a meeting of the Association for Symbolic Logic on December 29, 1952.
19. Kleene, S. C. (1955a). åArithmetical predicates and function quantifiers. *Transactions of the American Mathematical Society*, 79:312–340.
20. Kleene, S. C. (1955b). On the form of predicates in the theory of constructive ordinals (second paper). *American Journal of Mathematics*, 77:405–428.
21. Kleene, S. C. (1955c). Hierarchies of number theoretic predicates. *Bulletin of the American Mathematical Society*, 61:193–213.
22. Kleene, S. C. (1959a). Quantification of number-theoretic functions. *Compositio Mathematica*, 14:23–40.
23. Kleene, S. C. (1959b). Recursive functionals and quantifiers of finite types I. *Transactions of the American Mathematical Society*, 91:1–52.
24. Kondo, M. (1938). Sur l'uniformization des complementaires analytiques et les ensembles projectifs de la second classe. *Japanese Journal of Mathematics*, 15:197–230.
25. Kreider, D. L., and Rogers, H, Jr. (1961). Constructive versions of ordinal number classes. *Transactions of the American Mathematical Society*, 100:325–369.
26. Kreisel, G. (1961). Set theoretic problems suggested by the notion of potential totality. *Infinitistic methods* (pp. 103–140). Pergamon, New York.
27. Kreisel, G. (1962). The axiom of choice and the class of hyperarithmetic functions. *Indagationes Mathematicae*, 24:307–319.
28. Lebesgue, H. (1905). Sur les fonctions represéntables analytiquement. *Journal de Mathématiques 6ᵉ serie*, 1:139–216.
29. Markov, A. A. (1947). On the impossibility of certain algorithms in the theory of associative systems. *Coptes rendus (Doklady) de l'Academie des Sciences de USSR*, 55:583–586.
30. Markwald, W. (1954). Zur Theorie der konstruktiven Wohlordnungen. *Mathematischen Annalen*, 127:135–149.
31. Moschovakis, Y. N. (1966). Many-one degrees of the predicates $H_a(x)$. *Pacific Journal of Mathematics*, 18:329–342.
32. Moschovakis, Y. N. (1968). Review of four papers on Church's Thesis. *The Journal of Symbolic Logic*, 33:471–472.
33. Moschovakis, Y. N. (1969). Abstract first order computability II. *Transactions of the American Mathematical Society*, 138:465–504.
34. Moschovakis, Y. N. (1974). *Elementary Induction on Abstract Structures*. North Holland, Amsterdam. Studies in Logic, No. 77. Republished by Dover Publications, Mineola, NY, 2008, with a correction to 8.3.

35. Moschovakis, Y. N. (2009). *Descriptive set theory, Second edition*, volume 155 of *Mathematical Surveys and Monographs*. American Mathematical Society.
36. Moschovakis, Y. N. (2010a). Classical descriptive set theory as a refinement of effective descriptive set theory. *Annals of Pure and Applied Logic*, 162:243–255.
37. Moschovakis, Y. N. (2010b). Kleene's amazing second recursion theorem. *The Bulletin of Symbolic Logic*, 16:189–239.
38. Mostowski, A. (1947). On definable sets of positive integers. *Fundamenta Methematicae*, 34:81–112.
39. Mostowski, A. (1951). A classification of logical systems. *Studia Philosophica*, 4:237–274.
40. Myhill, J. (1955). Creative sets. *Zeitschrifft für Mathematische Logik und Grundlagen der Mathematik*, 1:97–108.
41. Nelson, G. C. (1974). Many-one reducibility within the Turing degrees of the hyperarithmetic sets $Ha(x)$. *Transactions of the American Mathematical Society*, 191:1–44.
42. Post, E. L. (1944). Recursively enumerable sets of positive integers and their decision problems. *Bulletin of the American Mathematical Society*, 50:284–316.
43. Post, E. L. (1947). Recursive unsolvability of a problem of Thue. *The Journal of Symbolic Logic*, 12:1–11.
44. Putnam, H. (1961). Uniqueness ordinals in higher constructive number classes. *Essays on the foundations of mathematics*, pages 190–206. Magnes Press, Hebrew University, Jerusalem.
45. Rogers, Jr., H. (1967). *Theory of recursive functions and effective computability*. McGraw-Hill.
46. Sacks, G. E. (1990). *Higher Recursion Theory*. Perspectives in Mathematical Logic: Springer.
47. Spector, C. (1955). Recursive wellorderings. *Journal of Symbolic Logic*, 20:151–163.
48. Spector, C. (1956). On degress of recursive unsolvability. *Annals of Mathematics*, 64:581–582.
49. Spector, C. (1960). *Hyperarithmetical quantifiers. Fundamenta Methematicae*, 48:313–320.
50. Spector, C. (1961). Inductively defined sets of natural numbers. *Infinitistic methods*, pages 97–102. Pergamon, New York.
51. Steel, J. R. (1978). Forcing with tagged trees. *Annals of Mathematical Logic*, 15:55–74.
52. Suslin, M. (1917). Sur une definition des ensembles measurables B sans nombres transfinis. *Comptes Rendus Acad. Sci. Paris*, 164:88–91.
53. Turing, A. M. (1936). On computable numbers with an application to the Entscheidungsproblem. *Proceedings of the London Mathematical Society*, 42:230–265. *A correction*, ibid. volume 43 (1937), pp. 544–546.
54. Van Heijenoort, J., editor (1967). *From Frege to Gödel, a source book in mathematical logic, 1879–1931*. Harvard University Press, Cambridge, Massachusetts, London, England.

Chapter 6
Honest Computability and Complexity

Udi Boker and Nachum Dershowitz

> ***Goldstein:*** And what causes you to say that?
> ***Davis:*** *Honesty.*
> —Martin Davis: An Interview Conducted by Andrew Goldstein (IEEE History Center, July 18, 1991)
>
> For Martin—scientist, scholar, and thinker.

Abstract We raise some issues posed by the use of representations for data structures. A nefarious representation can turn the incomputable into computable, the non-recursively-enumerable into regular, and the intractable into trivial. To overcome such problems, we propose criteria for "honesty" of implementation. In particular, we demand that inputs to functions and queries to decision procedures be specified as constructor terms.

Keywords Computability · Representation · Encoding · Simulation · Computational models · Computational power · Computational complexity · Effectiveness · Church-Turing Thesis · Universality

---

This author's research benefited from a fellowship at the Institut d'Études Avancées de Paris (France), with the financial support of the French state, managed by the French National Research Agency's "Investissements d'avenir" program (ANR-11-LABX-0027-01 Labex RFIEA+).

U. Boker (✉)
School of Computer Science, Interdisciplinary Center, Herzliya, Israel
e-mail: udiboker@idc.ac.il

N. Dershowitz
School of Computer Science, Tel Aviv University, Ramat Aviv, Israel
e-mail: nachum@cs.tau.ac.il

E.G. Omodeo and A. Policriti (eds.), *Martin Davis on Computability, Computational Logic, and Mathematical Foundations*,
Outstanding Contributions to Logic 10, DOI 10.1007/978-3-319-41842-1_6

6.1 Honesty Is Needed

Computations have no choice but to manipulate representations of objects rather than the objects themselves. Most often, strings of symbols taken from some finite alphabet are used for the purpose. Numbers, for example, are usually denoted by sequences of decimal symbols, or binary bits, or unary strokes (like the tally numbers of paleolithic times). In logic, one therefore distinguishes between numbers n, which reside in an ideal Platonic world, and numerals $\underline{n}$, their symbolic representation as (first-order) terms. Similarly, graphs, which are set-theoretic objects, are typically either represented as lists of edges (pairs of nodes) or as binary adjacency matrices.

Given that representation is an inescapable necessity, some natural questions arise immediately:

- How much of a difference can the choice of representation make to computability or complexity measurements?

  Answer: It can make all the difference between computable and incomputable, or between tractable and intractable.

- Who gets to choose the representation: Abe who formulates the queries, or Cay who designs the program to answer them?

  Our answer: Cay may reinterpret Abe's formulation any way she sees fit, but the reinterpretation is part and parcel of the process of answering.

- What is wrong with a representation of graphs that lists nodes in the order of a Hamiltonian path, if there is such—in which case deciding the question takes linear time?

  Answer: Cay will only be able to quickly answer the specific question whether there is a Hamiltonian path, whereas she would have a much harder time performing basic graph operations, such as adding an edge.

- Is it legitimate to say that the parity of an integer (that is, whether it is even or odd) can be determined in constant time, when that is the case only for very specific representations of numbers (namely, least-significant-first binary, as opposed to ternary, say)?

  Short answer: No.

Garey and Johnson [15, pp. 9–10] address the questions of representation and computational models as they impact the measurement of computational complexity. They assert upfront that it matters little, as long as one sticks to what is considered "reasonable":

> The intractability of a problem turns out to be essentially independent of the particular encoding scheme and computer model used for determining time complexity.

They go on to explain why at length:

> Let us first consider encoding schemes. Suppose for example that we are dealing with a problem in which each instance is a graph.... Such an instance might be described by simply listing all the vertices and edges, or by listing the rows of the adjacency matrix for the graph, or by listing for each vertex all the other vertices sharing a common edge with it (a "neighbor" list). Each of these encodings can give a different input length for the same graph. However, it is easy to verify that the input lengths they determine differ at most polynomially from one another, so that any algorithm having polynomial time complexity under one of these encoding schemes also will have polynomial time complexity under all the others. In fact, the standard encoding schemes used in practice for any particular problem always seem to differ at most polynomially from one another. It would be difficult to imagine a "reasonable" encoding scheme for a problem that differs more than polynomially from the standard ones.

This discussion is followed by a caveat:

> Although what we mean here by "reasonable" cannot be formalized, the following two conditions capture much of the notion:
>
> (1) the encoding of an instance I should be concise and not "padded" with unnecessary information or symbols, and
>
> (2) numbers occurring in I should be represented in binary (or decimal, or octal, or in any fixed base other than 1).
>
> If we restrict ourselves to encoding schemes satisfying these conditions, then the particular encoding scheme used should not affect the determination of whether a given problem is intractable.

The main concern expressed in the above is that the input size should faithfully reflect the complexity of the input object. The choice of size can make a big difference, of course [6]:

> The computational complexity of a problem should not be obscured by a particular representation scheme.... Many problems are "fast" under the unary representation, as many computationally (probably) intractable problems in number theory are also "fast" under unary representation, such as factoring, discrete logarithm. But that is not *honest complexity theory*. The time is really exponential, compared to a more "reasonable" representation scheme of the information, such as in binary. [*Italics ours.*]

There are other ways in which a choice of representation may be unreasonable, besides being unnecessarily large. It could give away the answer—even if the sizes differ only polynomially, or it may harbor hints that make the task easier than it really is. That is the problem with a representation of graphs that lists nodes in Hamiltonian order, for example; it puts the solution—when there is one—right in front of one's nose. Our proposal for measuring complexity honestly will solve this problem by taking into consideration both the cost of computing a given function as well as the cost of generating the function's inputs (the nodes and edges of a graph in the Hamiltonian case).

This chapter looks at questions of "honesty" of representation in various contexts. We begin with what we feel is the underlying problem posed by representations, namely, the camouflaging of extra information (Sect. 6.2). After proposing a solution (Sects. 6.3 and 6.4), we consider how it resolves the problem of honest computability (Sects. 6.5 and 6.6) and relate honesty to Martin Davis's definition of universality (Sect. 6.7). Then we turn to see how this proposal also solves the problem of honest complexity (Sect. 6.8). With a solution in place, we analyze why considering formal languages, rather than functions, does not work (Sect. 6.9).

6.2 Dishonest Representations

The complexity attributed to the computation of a function f over some abstract domain A, say graphs, is normally measured in terms of resources required by its best implementation on some particular model of computation, most commonly the random access machine (RAM). This implementation, however, computes a function $\widehat{f}$ over some concrete domain C, say binary strings, rather than A. So, prior to considering the cost of running $\widehat{f}$, one should first establish that $\widehat{f}$ does actually implement f.

However, there is little meaning to a claim that a single function $\widehat{f}$ over some domain C implements the intended function f over domain A without also specifying how the two domains are related. The following definition is common.[1]

Definition 6.1 (*Simulation [single-valued representation]*) A concrete (partial) function $\widehat{f} : C \to C$ *simulates* an abstract (partial) function $f : A \to A$ *with respect to* a particular injective representation $\rho : A \xrightarrow{\text{1-1}} C$ if $\rho(f(x)) = \widehat{f}(\rho(x))$ for all $x \in A$. [In the case of partial functions, we also demand that $\widehat{f}(\rho(x))$ be undefined whenever $f(x)$ is.]

One clearly needs to require, as we have, that a representation be injective. Otherwise, any and all functions could be simulated by the identity function with respect to a representation that maps the whole abstract domain to a single constant.

The above definition will be extended to functions with arity wider than 1 and multivalued representations in Definitions 6.2 and 6.4 below. Figure 6.1 depicts the multivalued case.

[1] Allowing different representations for input and output, as in "A more general notion of simulation is obtained if we let drop the requirement that $\mathfrak{R}^{(1)}$ and $\mathfrak{R}^{(2)}$ have the same input and output sets.... $\mathfrak{R}^{(1)}$ can be *weakly simulated* on $\mathfrak{R}^{(2)}$ if there exist such $\mathscr{E}$ and $\mathscr{D}$ with the property that for each program π_1 for $\mathfrak{R}^{(1)}$ there exists a program π_2 on $\mathfrak{R}^{(2)}$ such that $\mathscr{D}\mathfrak{R}^{(2)}_{\pi_2}\mathscr{E} = \mathfrak{R}^{(1)}_{\pi_1}$" [14, p. 21], can lead—if one is not careful—to the same kinds of problems we will encounter in Sect. 6.9.

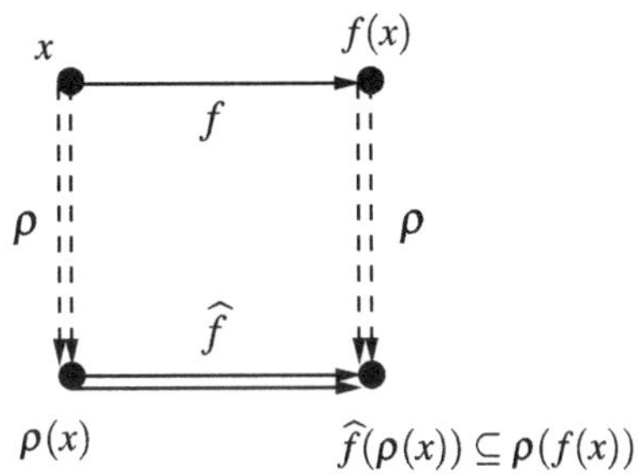

Fig. 6.1 The function $\widehat{f}$ simulates the function f via a representation ρ between their domains

The choice of representation can make all the difference in the world. If one is not honest, then a computable function can end up implementing an incomputable one by getting the representation itself to do the bulk of the work.

Example 6.1 Consider any standard enumeration TM_m, $m = 0, 1, \ldots$, of Turing machines, and define the following incomputable functions over the natural numbers $\mathbb{N}$:

- $h : \mathbb{N} \to \mathbb{N}$ enumerates (the numerical "codes" of) those machines that halt on the empty tape;
- $\bar{h} : \mathbb{N} \to \mathbb{N}$ enumerates those that do not.

So $h(\mathbb{N}) \uplus \bar{h}(\mathbb{N}) = \mathbb{N}$, where $h(\mathbb{N})$ is the image $\{h(n) \mid n \in \mathbb{N}\}$ of h and $\bar{h}(\mathbb{N})$ is the image of $\bar{h}$. Then the *incomputable* function

$$H(m) := \begin{cases} \min h(\mathbb{N}) & \text{if } \mathsf{TM}_m \text{ halts} \\ \min \bar{h}(\mathbb{N}) & \text{if } \mathsf{TM}_m \text{ does not} \end{cases}$$

is implemented by the *computable* parity function ($n \mod 2$) under the following bijective representation:

$$\rho(m) := \begin{cases} 2h^{-1}(m) + 1 & \text{if } \mathsf{TM}_m \text{ halts} \\ 2\bar{h}^{-1}(m) & \text{if } \mathsf{TM}_m \text{ does not}. \end{cases}$$

We have

$$\rho(m) \mod 2 = \rho(H(m)) = \begin{cases} 1 & \text{if } \mathsf{TM}_m \text{ halts} \\ 0 & \text{if } \mathsf{TM}_m \text{ does not}, \end{cases}$$

as required. □

The problem with the above "implementation" of the halting function obviously lies in the representation, which clearly gives the impression of itself doing the (computationally) impossible.

6.3 Honest Representations

The cause of the problem we just saw with the "dishonest" representation is not the (mathematically well-defined) mapping itself but rather the lack of suitable context for it. In particular, the integer successor function, for instance, cannot be implemented by any computable function under the nefarious representation ρ of the previous section, though it is part and parcel of our normal view of the naturals. As we will see, we can allow an honest representation to be any arbitrary injective (multivalued) function as long as we also pay attention to the internal structure of the abstract domain.

Imagine that Abe, the person posing instances of a problem, thinks in terms of an abstract domain A, such as integers, graphs, or pictures. Abe must have some means of describing for himself each of the elements of A, most commonly by means of a finite set G of "generators" of A (cf. [4, 9, 22, 23]). These generators give structure to A and meaning to its elements as described by ground terms H over G. For generators to do their job, every element of A must be equal to the value of at least one term in H; so at least one generator must be a scalar constant (of arity 0).[2]

Examples of generators for the natural numbers include:

$$\mathsf{0} \text{ (nullary zero)}$$
$$\blacksquare' \text{ (postfix successor } \lambda n.\, n+1),$$

(in unary "caveperson" style), as well as

$$\mathsf{0} \text{ (nullary zero, } \lambda.0, \text{ usually suppressed)}$$
$$\blacksquare\mathsf{0} \text{ (postfix doubling, } \lambda n.\, 2n)$$
$$\blacksquare\mathsf{1} \text{ (postfix doubling plus one, } \lambda n.\, 2n+1)$$

for the commonplace binary representation, and

$$\mathsf{0} \text{ (nullary zero, } \lambda.0, \text{ usually suppressed)}$$
$$\blacksquare\mathsf{0} \text{ (postfix tripling, } \lambda n.\, 3n)$$
$$\blacksquare\mathsf{1} \text{ (postfix tripling plus one, } \lambda n.\, 3n+1)$$
$$\blacksquare\mathsf{2} \text{ (postfix tripling plus two, } \lambda n.\, 3n+2)$$

for ternary. With the latter two, there are infinitely many representations of the number zero.

---

[2]Unlike the development in [4], where effectiveness of an algorithm was at issue, here we are not insisting that the generators form a *free* term algebra (a Herbrand universe): more than one term may designate the same abstract element.

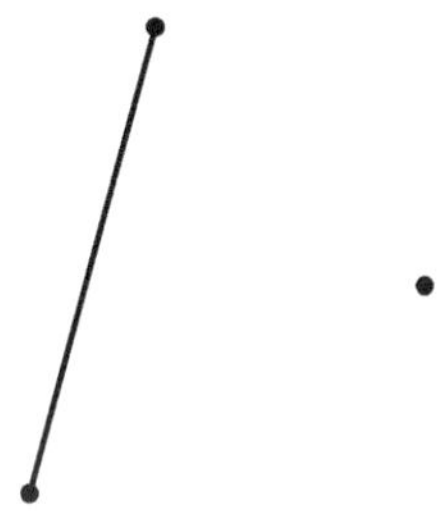

Fig. 6.2 An abstract undirected, unlabeled graph

For undirected, unlabeled graphs, G, with vertices V (G denotes the set of graphs whose vertices are taken from the set V of vertices), an example of a set of generators is

$$\begin{aligned}
&\square : V (\text{nullary first-vertex}) \\
&\blacksquare' : V \to V (\text{postfix next-vertex}) \\
&\varnothing : G \text{ (nullary empty-graph)} \\
&\bullet\,;\blacksquare\ : G \times V \to G \text{ (binary add-vertex to graph)} \\
&(\bullet) + \blacksquare \frown \blacksquare\ : G \times V \times V \to G \text{ (ternary add-edge to graph)}\ .
\end{aligned}$$

Over these generators, the graph depicted in Fig. 6.2 is the value of the ground term

$$(\varnothing\,;\square\,;\square'\,;\square'') + \square \frown \square'\ ,$$

wherein there is an edge between the "first" and "second" vertices. It is also the value of the term

$$(\varnothing\,;\square''\,;\square'\,;\square) + \square'' \frown \square'\ ,$$

wherein there is an edge between the "third" and "second" vertices. [In general, generators can be partial.]

Accordingly, we formalize the notion of (honest) representation as any injective multivalued function from an abstract domain that is structured by generators. Recall that a multivalued function $\rho : A \rightrightarrows C$ (or set-valued function $\rho : A \to \mathscr{P}(C)$) is *injective* if $\rho(x) \cap \rho(y) = \varnothing$ for all distinct $x, y \in A$.

Definition 6.2 (*Representation*)

- An *abstract domain* is a set A of elements, including (always) Boolean values TRUE and FALSE, equipped with a finite set G of generators for the whole domain, which also includes the binary equality relation $=$. Every element of A must be equal to at least one ground term over G.
- A *representation* of A in a "concrete" domain C is an *injective* multivalued function $\rho : A \overset{\text{1-1}}{\rightrightarrows} C$. We will insist that $\rho(\text{TRUE})$ is finite.

- The representation $\rho(\langle a_1, \ldots, a_n \rangle)$ of a tuple of abstract elements a_i is the set $\rho(a_1) \times \rho(a_2) \times \ldots \times \rho(a_n)$, the set of all tuples $\langle c_1, \ldots, c_n \rangle$, such that $c_i \in \rho(a_i)$.

The equality relation and Boolean constants are required for interpreting the output, as we will see. Having only finitely many representations of TRUE will allow Abe to understand and compare results of Cay's computations.

Having representations as multi-valued, rather than single-valued, functions gives the freedom to have many representations for the same abstract element, as is very commonly done in practice. For example, one may represent the rational number one-half by $^1/_2$, $^7/_{14}$, etc., and the unordered set $\{7, 2, 3\}$ by the sequences $\langle 2, 3, 7 \rangle$, $\langle 7, 3, 2 \rangle$, etc.

The choice of what is "abstract" and what is "concrete" is in "the eyes of the beholder"; it is in the final analysis an arbitrary formal choice. One may view a number as an abstract entity, represented by a concrete string over the symbols 0 and 1, while another views the symbol 1 as an abstract entity represented by some ink dots or electric pulse, etc. Likewise, the equality relation $=$ depends on the choice of what an abstract entity is. For example, if the abstract domain is graphs, then the graph of Fig. 6.2 is a single entity and all of its different generating terms yield equal entities, while in the case that the abstract domain consists of graphs with numbered vertices, the different generating terms yield isomorphic, but unequal, entities.

An alternative to the proposed generator-based approach for describing abstract elements would be to define them by means of a set of relations. For graphs, this might be the relation telling whether an edge is present between two given vertices. It is well known that using such relations, rather than generating functions, increases the complexity of many procedures. (For example, exhaustively checking all vertex combinations for getting an adjacent vertex.) Furthermore, we argue in Sect. 6.9 that this alternative does not at all fit the bill. Intuitively, a set of functions allows one to also generate the representations, while a set of relations does not.

6.4 Honest Implementation

A function $\widehat{f}$ over some concrete domain C honestly implements a function f over an abstract domain A if it preserves the functionality of f under the representation, while also preserving the meaning of the domain elements as given by the domain generators.

We formally consider a *(computational) family* F over a domain A to be an algebra (in the universal-algebra sense), consisting of the domain (universe) A and operations F over A (of any arity), along with a matching vocabulary. Our definition of honest implementation will require the simulation of the desired function together with a set of generators. The implementation notion is then really about an algebra as a whole.

When we have cause to care about the intensionality (internal workings) of a computational mechanism, we will talk about a *(computational) model* comprising a set of *algorithms*, each of which involves a set of states, a subset of which are initial states, and a (partial and/or multivalued) transition function over states [18].

Definition 6.3 (*Simulation [multivalued representation]*)

- A function $\widehat{f}$ over a domain C *simulates* a function f of arity ℓ over a domain A via representation $\rho : A \overset{1\text{-}1}{\rightrightarrows} C$ if, for every $\bar{x} \in A^\ell$, we have $\widehat{f}(\rho(\bar{x})) \subseteq \rho(f(\bar{x}))$.
- Likewise, a family of functions $\widehat{F}$ over C *simulates* a family F over A (via representation ρ) if each $f \in F$ is simulated via the same ρ by some $\widehat{f} \in \widehat{F}$.

As usual, functions are extended to operate over sets by letting $f(S) := \{f(\bar{x}) \mid \bar{x} \in S\}$.

Definition 6.4 (*Honest Implementation*) Consider an abstract domain A with generators G.

- A family of functions $\widehat{F}$ over C provides an *implementation* of a family F over A if F is simulated by $\widehat{F}$.
- We will refer to the implementation as *close* if the simulation is via a bijection.
- An implementation $\widehat{F}$ over C is *honest* as long as F includes the generators G as well as equality.
- We will say that a function $\widehat{f}$ over C *honestly implements* a single function f over A if the implementation also supplies simulations $\widehat{g}$ of each generator $g \in \mathsf{G}$ including equality over A. In other words, we require that $\{\widehat{f}\} \cup \widehat{\mathsf{G}}$ implement $\{f\} \cup \mathsf{G}$, for some set $\widehat{\mathsf{G}}$ of concrete generators and implementation $\widehat{=}$ of abstract equality.

See the illustration in Fig. 6.1.

The point is that $\widehat{f}$ implements a function f with respect to a specific set of generators. If Abe considers an abstract domain with generators that are natural, computable, and trackable for him, but completely useless for his sister, Sal, then $\widehat{f}$ is an honest implementation of f for Abe but not for Sal.

We give next an example of an honest implementation of an abstract function over the rationals $\mathbb{Q}$ by means of a concrete function over strings.

Example 6.2 The task is to implement rational multiplication, $\mathsf{m} : \mathbb{Q} \times \mathbb{Q} \to \mathbb{Q}$, by means of a string-based model of computation.

- The abstract domain (A in the definition) is the set of rational numbers $\mathbb{Q}$ (with the subset of integers $\mathbb{Z}$ and its subsets the positive integers $\mathbb{Z}^+$ and negative integers $\mathbb{Z}^-$, plus the truth values), along with the following generators:

$$
\begin{aligned}
&\mathsf{o} : \mathbb{Z} \text{ (nullary zero)}\\
&\mathsf{I} : \mathbb{Z}^+ \text{ (nullary one)}\\
&\mathsf{s} : \mathbb{Z}^+ \to \mathbb{Z}^+ \text{ (unary successor)}\\
&\mathsf{n} : \mathbb{Z}^+ \to \mathbb{Z}^- \text{ (unary negation)}\\
&\mathsf{q} : \mathbb{Z} \times \mathbb{Z}^+ \to \mathbb{Q} \text{ (quotient of an integer by a positive integer)}
\end{aligned}
$$

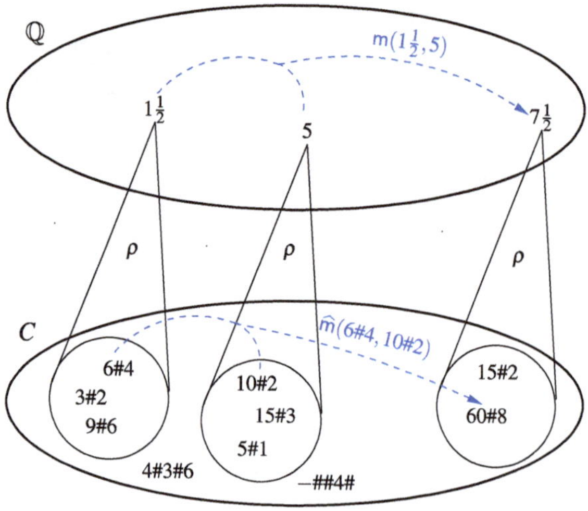

Fig. 6.3 Representing abstract rational numbers by concrete strings, and implementing the multiplication function

- The concrete domain (C) is the set Σ^* of finite strings over the symbols $\{0, 1, 2, 3, 4, 5, 6, 7, 8, 9, -, \#\}$, plus TRUE and FALSE.
- Let $\underline{x} \in \Sigma^*$ denote the number $x \in \mathbb{Z}$ in decimal notation.
- The representation $\boldsymbol{\rho} : \mathbb{Q} \overset{\text{1-1}}{\rightrightarrows} \Sigma^*$ is defined by $\boldsymbol{\rho}(r) := \{\underline{x} \;\#\; \underline{y} \mid r = x/y,\ x \in \mathbb{Z},\ y \in \mathbb{Z}^+\}$.
- The implementations of the generators and equality are as follows:

$\widehat{\mathsf{0}}$ returns the string $\mathtt{0}$

$\widehat{\mathsf{1}}$ returns the string $\mathtt{1}$

$\widehat{\mathsf{s}}(w)$: If w is the decimal representation $\underline{x}$ of $x \in \mathbb{Z}^+$, then $\widehat{\mathsf{s}}(w)$ returns the decimal representation $\underline{x+1}$ of $x + 1$. Otherwise it is undefined.

$\widehat{\mathsf{n}}(w)$ returns the string $-w$

$\widehat{\mathsf{q}}(u, v)$ returns the string $u \;\#\; v$

$\widehat{=}(u, v)$ returns TRUE iff $u = \underline{x}_1 \;\#\; \underline{x}_2$, $v = \underline{y}_1 \;\#\; \underline{y}_2$, and $x_1/x_2 = y_1/y_2$, for some $x_1, y_1 \in \mathbb{Z}$ and $x_2, y_2 \in \mathbb{Z}^+$.

- The implementation $\widehat{\mathsf{m}}(u, v)$ of multiplication is the following: If $u = \underline{x}_1 \;\#\; \underline{x}_2$ and $v = \underline{y}_1 \;\#\; \underline{y}_2$, where $x_1, y_1 \in \mathbb{Z}$ and $x_2, y_2 \in \mathbb{Z}^+$, then the implementation returns the string $\underline{x}_1 \cdot \underline{y}_1 \;\#\; \underline{x}_2 \cdot \underline{y}_2$. Otherwise, multiplication is not defined.

See Fig. 6.3.

An implementation of multiplication that reduces the resulting fraction is as honest an implementation as the above one. □

6.5 Honest Computability

We first demonstrate the reasonableness of our demands on implementations by taking a careful look at honest "effective" computations.

Since we believe the Church-Turing Thesis in light of the arguments and proofs in [4, 12], we shall use the term *effective computation* to stand for the functionality of a Turing machine over strings or of a recursive function over the natural numbers. Let TM denote the family of Turing-machine computable string functions and REC, the family of recursive numerical functions.

Armed with our definition of honest implementation, we are prompted to define honest computability over arbitrary abstract domains as follows:

Definition 6.5 (*Honest Computability*) A function over an abstract domain is *honestly computable* if it, and generators of its domain, can be honestly implemented by the recursive functions REC over the natural numbers (or by the Turing-computable functions TM over strings).

This definition guarantees that concrete representations for all the elements of the abstract domain can also be effectively generated.[3] It follows that, if the Turing family TM implements a family consisting of an arbitrary function f over a domain A and a finite set of generators for A, then f is—by definition—honestly computable.

Lemma 6.1 *If the recursive functions simulate a set of generators via some representation, then that representation—restricted to be univalued—can be effectively defined by structural induction.*

For example, consider these generators G for the naturals $\mathbb{N}$: zero, o, and successor, s. Suppose they are mapped to the constant $\widehat{\mathsf{o}}$ and the recursive function $\widehat{\mathsf{s}}$, respectively, under a (multivalued) representation $\eta : \mathbb{N} \rightrightarrows \mathbb{N}$. Define ρ, a single-valued restriction of η, by (structural) induction (over H, the ground terms of G) as follows:

$$\begin{aligned}\rho(\mathsf{o}) &:= \widehat{\mathsf{o}}\\ \rho(\mathsf{s}(n)) &:= \widehat{\mathsf{s}}(\rho(n))\,.\end{aligned}$$

We have by induction that $\rho(n) \in \eta(n)$ for all n.

We get also that every function $f : \mathbb{N} \to \mathbb{N}$ that is implemented by a recursive $\widehat{f} : \mathbb{N} \to \mathbb{N}$ under η must also be recursive, since

$$f(n) = \rho^{-1}(\widehat{f}(\rho(n))) = \eta^{-1}(\widehat{f}(\eta(n)))$$

is the composition of computable functions. (The inverse ρ^{-1} of a single-valued computable representation is computable by search.)

A similar argument applies to other sets of generators for other abstract domains.

[3]The developments in [4, 5, 12] do not directly address the issue of honesty.

Specifically, under the above (lax) assumptions, if a concrete function $\widehat{f}$ is computable, then the implemented function f can in fact be programmed effectively as an abstract state machine (ASM) [8, 16]. ASMs are a framework providing a most general programming paradigm in which one can precisely express (ineffective as well as effective) algorithms over arbitrary domains. In our case, f may be effectively computed over the combined domain $A \uplus \mathbb{N}$ by programming:

$$f(g(x_1, \ldots, x_\ell)) := \rho^{-1}(\widehat{f}(\widehat{g}(\rho(x_1), \ldots, \rho(x_\ell))))$$

for each $g \in \mathsf{G}$. For this to work, we presume the availability of an effective equality test for A.

To summarize the development so far, we would say that a function f over Abe's abstract domain A is honestly computed by Cay's implementation iff Cay can evaluate terms of the form $f(t_1, \ldots, t_\ell)$—or more generally terms over f and generators G of A—where the $t_i \in \mathsf{H}$ are terms over G, and Abe, the querier, can check the results (using the equality predicate). Furthermore, f is deemed effective if Cay's implementation uses effective means (such as those provided by a Turing machine).

6.6 Honest Comparisons

The fact that honest implementations effectively generate representations for all abstract domain elements guarantees the "completeness" of the recursive functions and of Turing machines in the sense that no representation can enlarge its computational power.

Definition 6.6 (*Completeness* [3]) A family F is *complete* if it cannot simulate any strict superset of itself.

Theorem 6.1 *Consider a computational family F over the natural numbers $\mathbb{N}$, and suppose that the recursive functions* REC *simulate F and, furthermore, that* REC $\subseteq$ *F. Then $F =$* REC.

Proof We show that every function $h \in F$ is also in REC. Since REC $\subseteq F$, we know the successor function s over $\mathbb{N}$ in also in F. Because REC implements F, there must be functions $\widehat{h}, \widehat{\mathsf{s}} \in$ REC, such that for every $\bar{n} \in \mathbb{N}^*$, $\widehat{h}(\rho(\bar{n})) \subseteq \rho(h(\bar{n}))$, and for every $n \in \mathbb{N}$, $\widehat{\mathsf{s}}(\rho(n)) \subseteq \rho(\mathsf{s}(n))$. (The notation S^* is used here and later for all finite tuples of elements of S.)

Given a vector $\bar{n} := \langle n_1, \ldots, n_\ell \rangle \in \mathbb{N}^*$, we can compute $h(\bar{n})$ by the following recursive procedure:

- Construct the vector

$$\widehat{n} = \langle \widehat{n}_1, \ldots, \widehat{n}_\ell \rangle = \langle \widehat{\mathsf{s}}^{n_1}(x_0), \ldots, \widehat{\mathsf{s}}^{n_\ell}(x_0) \rangle ,$$

which represents $\bar{n}$, by choosing any $x_0 \in \boldsymbol{\rho}(0)$ and applying $\widehat{s}$ to that value n_j times for the jth component.

- Compute
$$\widehat{k} := \widehat{h}(\widehat{n}) \; .$$
[If h is partial and diverges on $\bar{n}$, then the simulating function $\widehat{h}$ will likewise diverge on $\widehat{n}$.]
- Search for the number k that is represented by $\widehat{k}$, by computing
$$\min_{i \in \mathbb{N}}[\widehat{\mathsf{s}}^{\,i}(x_0) \mathrel{\widehat{=}} \widehat{k}].$$
This search is guaranteed to terminate after an iteration k, such that $\widehat{k} \in \boldsymbol{\rho}(k)$, since there is a fixed finite set of truth values $\widehat{\text{TRUE}}$ to test for.
- At this moment, $\widehat{k}$ represents $k = h(\bar{n})$.

□

It follows that the recursive functions **REC** are complete in the defined sense. By the same token, Turing machines are complete (see [3]), whereas two-counter machines and the lambda calculus are not. Two-counter machines can neither square nor exponentiate [21], but famously implement all recursive functions via the (expansive) representation $\boldsymbol{\rho} : n \mapsto 2^n$ (as shown by the late Marvin Minsky [19]).

We now know what it means for families to have the same computational power:

Definition 6.7 (*Equipotence*) Families $\widehat{F}$ and F are *equipotent* if they simulate each other.

The representations by means of which $\widehat{F}$ implements F and vice-versa need not be the same, even if they both operate over the same domain.

For example, **REC** and **TM** are equipotent via representations like Gödel numbers and tally numbers.

By this definition, 2-counter machines are equipotent with 3-counter machines (though the former model includes strictly fewer functions), but not with 1-counter ones.[4]

If follows that the honest (and self-consistent) way to compare computational power (when representations are allowed), is to say that a family F' is *strictly more powerful* than another family F if F' simulates F but not vice-versa. This is in fact true for F', the recursive functions and F, the primitive recursive ones, but this popular claim requires showing that there is no (injective) representation whatsoever via which the primitive recursive functions can simulate all the recursive ones. See [3].

If an abstract function is computable whenever it has an honest recursive implementation, how can one show that an abstract function h is incomputable, short of

[4]"Combining these simulations, we see that two-counter machines are as powerful as arbitrary Turing machines (one-counter machines are strictly less powerful)" [17, p. 33]. But who says that one-counter machines cannot also simulate more than they can compute? They cannot [2, Thm. 40].

trying all possible representations? The answer is that h is *incomputable* whenever there is at least one *bijective* representation that provides recursive implementations of the generators but a non-recursive implementation of h.

Theorem 6.2 *Every function h over an abstract domain with generators* G *that is honestly computable is also recursively implemented by each and every close numerical implementation of h (that is, an implementation in* $\mathbb{N}$ *via a bijective representation) that implements the operations in* G *by means of recursive functions.*

The reason we need a *bijective* counterexample to establish incomputability is that one can always have the part of the implementation that works with numbers outside the image $\rho(A)$ of a non-surjective representation ρ (like $\rho(n) = 2n$) do something outlandish (to odd numbers).

Proof Let $\pi : A \leftrightarrow \mathbb{N}$ be a bijection from the abstract domain A of h, and let $\widetilde{h} : \mathbb{N} \to \mathbb{N}$ implement h under π. If $\widehat{h}$, the function that simulates h under some ρ, is recursive, then $\widetilde{h}$ must also be recursive. This is because

$$\widetilde{h}(n) = \pi(h(\pi^{-1}(n))) = \pi(\rho^{-1}(\widehat{h}(\rho(\pi^{-1}(n)))))$$

and, by Lemma 6.1, both ρ and π are computable from any standard numerical encoding of generator terms H over G. Hence, $\widetilde{h}(n)$ is recursive. □

6.7 Honest Universality

A (partial) function ω is said to be "universal" for a whole family F of (partial) functions (such as all the recursive functions, for instance) if it computes the whole family by being supplied with the code $\ulcorner f \urcorner$ of the desired $f \in F$ as an extra argument. If ω works with a concrete domain C, whereas the functions in F operate on an abstract domain A, then representations $\rho : A \to C$ are in order once again. Then we would say that varyadic ω is universal for F (with respect to encoding $\ulcorner \cdot \urcorner : F \to C$ and representation $\rho : A \overset{1\text{-}1}{\rightrightarrows} C$) if

$$\omega(\ulcorner f \urcorner, \rho(\bar{x})) \subseteq \rho(f(\bar{x}))$$

for all $f \in F$ and $\bar{x} \in A^*$ of the right length for the arity of f.

Another potential problem with the notion of universal function is that some models of computation—like Turing machines—do not take their inputs separately, but, rather, all functions are unary (string-to-string for Turing machines). In such cases, one needs to be able to represent pairs (and tuples) as single elements. One standard pairing function for the naturals is the injection $\langle i, j \rangle := 2^i 3^j$. For strings, one usually uses an injection like $\langle u, w \rangle := u \,;\, w$, where ";" is some symbol not in the original string alphabet.

There are several ways to go. The pairing function could reside in the abstract domain A, or in the concrete domain C, or in the representation of A as C. Regardless, this need raises a critical issue. Unless we demand that pairing be effective, there could be an implementation of the universal function that does too much, computing even non-effective functions. For example, a naïve definition might simply ask that pairing be injective and say that ω is universal for some set F of functions if $f(x) = \omega(\langle \ulcorner f \urcorner, x \rangle)$ for all $f \in F$ and $x \in C$, for some arbitrary encoding $\ulcorner \cdot \urcorner : F \to C$ of functions. The problem is that an injective pairing could cheat and include the answer in the "pair". For Turing machines, say, the pair $\langle u, w \rangle$ might be represented as $u\ ;\ w$ when machine u halts on input w and as $u : w$ when it doesn't. Better yet, one could map $\langle \ulcorner f \urcorner, y \rangle \mapsto [f(y), \ulcorner f \urcorner, y]$, where the square brackets are some ordinary tupling function for the domain. Then a putative universal machine could effortlessly "compute" virtually anything, computable or otherwise, just by reading the encoded input pair.

Davis [7] and, later, Rogers [20] proposed general definitions of universality for Turing machines and for partial-recursive functions, respectively. Both insist that pairing be effectively computable. But we are talking about models in which no function takes two arguments, so we might not have an appropriate notion of computable binary function at our disposal. To capture effectiveness of pairing in such circumstances, we demand the existence of component-wise successor functions. Given a "successor" function s for domain C (that is, $C = \{s^n(x_0)\}$ for some $x_0 \in C$) and a pairing function $\langle \cdot, \cdot \rangle : C \times C \xrightarrow{\text{1-1}} C$, the component-wise successor functions operate as follows: $s_1 : \langle a, b \rangle \mapsto \langle s(a), b \rangle$ and $s_2 : \langle a, b \rangle \mapsto \langle a, s(b) \rangle$. If s, s_1, and s_2 are all computable, then we will say that *pairing is effective*. This is because one can program pairing so that $\langle z, y \rangle := s_1^i(s_2^j(\langle x_0, x_0 \rangle))$, where $z = s^i(x_0)$ and $y = s^j(x_0)$. And if pairing is effective, then its two projections (inverses), $\text{1ST} : \langle a, b \rangle \mapsto a$ and $\text{2ND} : \langle a, b \rangle \mapsto b$, are likewise effective. (Generate all representations of pairs in a dovetailed, zig-zag fashion, until the desired one is located. What the projections do with non-pairs is left up in the air.)

Another concern is that requiring that pairing be computable is too liberal for the purpose. One does not really want the pairing function to do all the hard real work itself. For example, the mapping could include $f(x)$ in the pair, even if it only can do that for f that are known to be total (like, for the primitive recursive functions, of which there are infinitely many), or all functions that halt within some recursive bound. That would make it a trivial matter to be universal for those functions—just transcribe the answer from the input.

Definition 6.8 (*Honest Pairing*) A pairing function is *honest* if it is effective and bijective.

This way, there is no room for hiding information.

For bijective pairing with computable projections, there is an effective means of forming a pair $\langle a, b \rangle$ by enumerating all of C until the two projections give a and b, respectively. With bijectivity alone, sans computability, one could still hide a fair amount of incomputable information in a bijective mapping. For instance, imagine that 0 is the code of the totality predicate and that the rest of the naturals code the partial-recursive functions in a standard order. Map pairs $(i+1, z)$ to $3\langle i, z\rangle$, where $\langle\cdot,\cdot\rangle$ is a standard pairing; map $(0, z)$ to $3j+1$ when z is the (code of the) jth total (recursive) function; and map $(0, z)$ to $3k+2$ when z is the kth non-total (partial recursive) function. Now, let U be some standard computable universal function. Then, for y divisible by 3, $\omega(y) := U(y/3)$ would compute all the partial-recursive functions, whereas $\omega(y) := y \equiv 1 \pmod 3$ would compute the incomputable totality predicate when $y = (0, z)$ is not divisible by 3.

Definition 6.9 (*Honest Universality*) Let F be some family of unary functions over an abstract domain A. Unary function ω over concrete domain C is *universal* for F, via pairing function $\langle\cdot,\cdot\rangle$ over C, if $\{\lambda y.\, \omega\langle a, y\rangle \mid a \in C\}$ implements F. If, in addition, pairing is bijective, then we call the universal function *honest*.

That is, ω is universal if, for fixed representation $\rho : A \to C$ and encoding $\ulcorner\cdot\urcorner : F \to C$, we have $\omega\langle\ulcorner f\urcorner, \rho(x)\rangle = \rho(f(x))$, for $f \in F$ and $x \in A$. Of course, we are interested in the case where both pairing and the universal function are effectively computable.

Theorem 6.3 ([10]) *Let F be some family of unary functions over a domain A, including generators and equality. Then, if there is a computable unary universal function (over any domain C) for F, via an effective pairing, then all the implemented functions in F are also computable.*

Suppose $F = \{f_z\}_z$ is some standard enumeration of (the definitions of) the partial-recursive functions. Based on Davis's (second) definition of a universal Turing machine, which relies on a notion of effective mappings between strings and numbers, namely, recursive in Gödel numberings, Rogers defines (in his third definition) what we may refer to as the *universal property* of a unary numerical function ω, namely, that $f_z(x) = \pi(\omega\langle z, x\rangle)$ for some recursive bijection π and effective (but perhaps dishonest) pairing $\langle\cdot,\cdot\rangle$.

The following follows from the definitions:

Theorem 6.4 ([10]) *If a function has the universal property, then it is honestly universal. Furthermore, there must exist an honest computable universal function.*

6.8 Honest Complexity

We turn now to the question of complexity of problems over abstract domains. But before one can measure complexity, one needs a measure of input size and a measure of the cost of a computation as a function of that size.

A size measure is associated with each (ground) term over the generators. This provides the flexibility of considering each of the various views of the same abstract element differently. Since problems often involve more than one input value, we need to measure the size of tuples of terms.

Definition 6.10 (*Size*) A *size* measure for an abstract domain is a function $|\cdot|: \mathsf{H}^* \to \mathbb{N}$, where H^* is the set of tuples of (ground) terms over the generators of the domain.

Complexity is measured with respect to this size, whatever it may be.

Examples of size measures for terms denoting graphs are tree height of the term, as well as the number of vertices or number of edges in a graph. Note that the two latter measures assign the same size to all terms of the same graph. Usually the size of a tuple is the sum of the sizes of its individual components.

One might argue that a size measure should not be this arbitrary, but should enforce a compact representation of the abstract elements, as Garey and Johnson demanded of the representation of numbers in the paragraphs quoted at the outset, namely, that the size of a natural number n should be order $\log n$. In many cases, however, this is too demanding. For instance, a set of n elements taken from some unordered set may have $n!$ reasonable representations. Checking equality between two such representations, in order to choose a single canonical representation for each set, might require a quadratic number of element comparisons. Even more involved is the case of graphs. If we are asked to decide the existence of a Hamiltonian path in an unlabeled graph, we should not demand that there be a unique or almost-unique way of constructing each graph, considering that graph isomorphism is a difficult problem. But there are exponentially many isomorphic graphs, so the standard representations of graphs are as wasteful as is the unary encoding of numbers. It is also standard practice to store data in compressed form, and it can easily take exponential time and space to reconstruct before manipulating.

The cost assigned to a computation over the concrete domain C depends on the relevant aspects of the computational model in question. For example, the cost can be the number of steps of a RAM model or the number of tape cells used by a Turing machine. (RAMs are in fact nearly optimal for time and space [11].) As with the size measure, cost is also in "the eyes of the beholder". Given a cost measure for computation in the model, we define the cost of terms, as follows:

Definition 6.11 (*Cost*) The *cost* $\kappa(\widehat{h}(t_1, \ldots, t_\ell))$ of a concrete term $\widehat{h}(t_1, \ldots, t_\ell)$ is the cost of a computation that constructs the concrete values $c_i \in C$ arguments t_i and then computes $\widehat{h}(c_1, \ldots, c_\ell)$, the value of $\widehat{h}$ for the concrete values thus obtained.

In some cases the cost of a computation might be the sum of the costs of its steps, as is natural for time complexity, while in other cases a different aggregation, such as maximum, is appropriate, as is done for space complexity. Often, the declared size of the input is approximately the cost of constructing it, so the impact on complexity of including the cost of construction is negligible.

Equipped with size and cost measures, we are ready to formalize our intuition of when a complexity measure is honest. The complexity of an implementation must take the specific means of representation into account. We have demanded that an honest implementation of an abstract function also provide implementations of the abstract domain's equality and generators (Definition 6.4). We may assume that every generator has a unique implementation. (Different implementations should have different names, thus refer to different, but possibly equivalent, generators.)

Our definition of the complexity of a function resembles the standard one; it is just that our notion of the cost of computing a function includes the cost of generating the representation of the input.

Definition 6.12 (*Honest Complexity*) Consider an abstract domain A with ground generator terms H and an honest implementation $\widehat{h} : C^\ell \to C$ over concrete domain C, implementing a function $h : A^\ell \to A$ over A. Let $m : \mathbb{N} \to \mathbb{N}$ be a complexity measure. Then we say that $\widehat{h}$ has *honest (worst-case) complexity* of at most m if $\kappa(\widehat{h}(\bar{t})) \leqslant m(|\bar{t}|)$, for all tuples $\bar{t} \in \mathsf{H}^\ell$ of terms.

Average and probabilistic complexities can be defined analogously.

While the complexity of implementing generators influences the complexity of implementing f, the complexity of implementing the equality relation need not affect it. For example with abstract graphs, equality checks are very involved, yet many graph operations need not check for graph equality (isomorphism). We do insist, however, that every implementation also implements the equality check in order to enforce a correct interpretation of the abstract domain—having the ability to use the equality implementation, Abe, the person posing instances of the function f, can verify whether the result of f's implementation is indeed proper. (Cf. Sect. 6.5 and the proof of Theorem 6.1.)

To sum up, to preclude dishonest measures of complexity, we require that the implementor Cay charge not only for calculating the answer to Abe's query, but also for building its native representation of the query from Abe's language of generators. That way, any new information hidden in the representation is put there by Cay and the costs incurred are charged for.

6.9 Dishonest Decisions

It is standard to classify the difficulty of a problem according to its membership in a set of functions or relations, for example, whether it is Turing-computable, in polynomial time, or in polynomial space. *Computational models*, such as Turing machines with arbitrary outputs, compute sets of (partial) functions, whereas *decision models*, such as finite automata or Turing machines with only "yes" or "no" outputs, compute sets of relations.

We argue that computational families, which implement functions, capture the essence of computational power more accurately than do decision families, which implement relations. For that reason, we based the notion of honest implementation and complexity, even for decision problems, on functions rather than on decision procedures. The underlying reason for the better adequacy of functions than relations is that the former can also comprise the means to generate the representations of objects.

As defined in Sect. 6.5, a computational family over a domain A is a set of functions $F \subseteq \{f : A^* \to A\}$; likewise a *decision family* over A is a set of relations $R \subseteq \{r \subseteq A^n \mid n \in \mathbb{N}\}$. Specific computational or decision families are defined via some internal mechanism, a *model of computation*, a point that will play a rôle in our arguments later.

Decision families are inherently incomplete, in that one can readily "increase" their power via a representation that adds some information on top of the represented element [3]. For example, let h be an incomputable decision problem over Σ^*, and consider the representation $\boldsymbol{\rho} : \Sigma^* \to \Sigma^*$, where $\boldsymbol{\rho}(w) = h(w)w$. (The representation just adds the incomputable bit $h(w)$ before the word w.) Then, Turing machines can "decide", via the representation $\boldsymbol{\rho}$, both h and all of the ordinary Turing-decidable problems.

Surprisingly, the weak computational model of finite automata (FSAs) is already powerful enough to decide, via a suitable representation, *any* countable set of (decidable or undecidable) relations [13]. The representation hides with each domain element a finite amount of data of relevance to finitely-many relations, such that each decision procedure gets all the data it needs from the represented inputs.

Let Σ be the binary alphabet $\{0, 1\}$ throughout the remainder of this section.

Lemma 6.2 ([13]) *For every countable set R of relations over the natural numbers $\mathbb{N}$, there is an injection $\boldsymbol{\rho} : \mathbb{N} \to \Sigma^*$, such that the set* FSA *of finite automata simulates R via $\boldsymbol{\rho}$, viewing a relation $r \subseteq \mathbb{N}$ as a Boolean function $r : \mathbb{N} \to \Sigma$.*

Accordingly, a FSA a computes the function $a : \Sigma^* \to \{\boldsymbol{\rho}(0), \boldsymbol{\rho}(1)\}$, returning $\boldsymbol{\rho}(0)$ when the input word is rejected and $\boldsymbol{\rho}(1)$ when accepted.

Proof Let $r_1, r_2, \ldots$ be any enumeration of the relations in R. Define the representation $\boldsymbol{\rho} : n \mapsto r_1(n)\, r_2(n) \cdots r_n(n)$. For every n, the length of $\boldsymbol{\rho}(n)$ is n, and it gives explicit answers to the first n relational questions r_i. Now, for every relation $r_i \in R$, consider the FSA a_i depicted in Fig. 6.4. One can easily verify that for each $m \in \mathbb{N}$, the automaton a_i accepts $\boldsymbol{\rho}(m)$ iff $m \in r_i$. For an input word w of length $m \geqslant i$, a_i finds the answer whether $m \in r_i$ at the i th digit of w. For the finitely-many inputs of

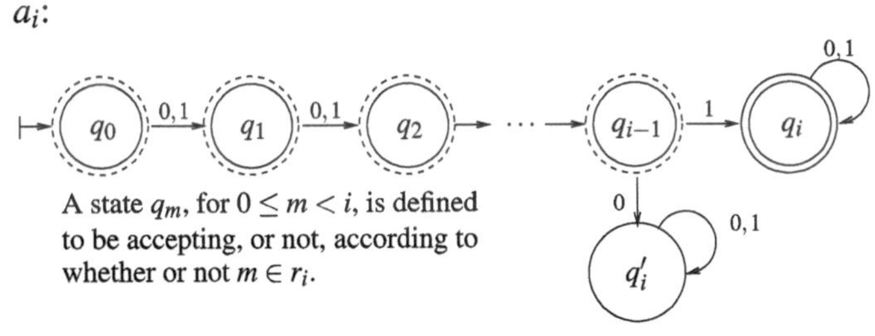

Fig. 6.4 The finite automaton a_i, which implements an arbitrary relation r_i via the representation ρ of the proof of Lemma 6.2

length $m < i$, representing numbers up to (but not including) i, the first i states of a_i are fixed to accept (and "return" $\rho(1)$) or reject (returning $\rho(0)$) the input word $\rho(m)$ according as to whether $m \in r_i$. □

One might have presumed that this disturbing sensitivity to representations would be resolved by limiting representations to bijections, but this is unfortunately not the case, as shown in [13].

Theorem 6.5 ([13]) *For every countable set R of relations over $\mathbb{N}$ there is a bijection $\pi : \mathbb{N} \leftrightarrow \Sigma^*$, such that the set* FSA *of finite automata closely implements R via π.*

The above-described inherent incompleteness of decision families, that they can easily be enlarged by representing input differently, stems from their inability to generate the representation of the input. On the other hand, as shown in Theorem 6.1, the set of recursive functions is complete, in the sense that it cannot honestly implement an incomputable function, regardless of the choice of representation.

A question naturally arises considering the completeness of recursive functions (Theorem 6.1) and the inherent incompleteness of decision families (Lemma 6.2 and Theorem 6.5): Where does the proof of Lemma 6.2 break down if we try to modify it to demonstrate that every set of functions can be computed, via a suitable representation, by finite-state transducers (input-output automata)?

The answer is that, with the aim of computing a countable set of functions $\{f_1, f_2, \ldots\}$, the representation that is used in the proof of Lemma 6.2 may be generalized to something like $\rho(n) = f_1(n)\ \$\ f_2(n)\ \$ \cdots \$\ f_n(n)$. Then, for every function f_i, there is indeed a transducer a_i, such that, for every n, we have $a_i(\rho(n)) = f_i(n)$. This, however, doesn't fit the bill. To properly represent f_i, we need for a_i to return $\rho(f_i(n))$, not $f_i(n)$. One might be tempted to suggest instead a representation η that already provides the represented values, as in $\eta(n) = \eta(f_1(n))\ \$\ \eta(f_2(n))\ \$ \cdots \$\ \eta(f_n(n))$. This is, however, a circular definition: Let f_1 be the successor function s. Representing 1, we have $\eta(1) = \eta(\mathsf{s}(1)) = \eta(2) = \eta(\mathsf{s}(1))\ \$\ \eta(\mathsf{s}(2)) = \cdots$.

Finally, it may be worthwhile noting that Turing's halting problem is immune to the particular representation of programs [1], as are similar problems, though—as we have seen—decision procedures are quite sensitive to the representation of input data. Here the problem is to decide whether machine TM_m halts on input string w. Problem instances are pairs $\langle \ulcorner m \urcorner, w \rangle$ consisting of an encoding $\ulcorner m \urcorner$ of the machine along with the input w or an encoding $\ulcorner w \urcorner$ thereof. However, the pairing function itself must be honest, as explained in Sect. 6.7. In that situation, the encoding of any given machine (or computer program) can only hide a finite amount of information, not enough to answer the halting problem for all inputs to the machine, though the representation of those inputs themselves could hide the answers.

6.10 Discussion

We have proposed to regard an abstract function as honestly and effectively implemented if it can be effectively computed given its arguments as constructor terms.

Analogously, we suggest that the cost of generating concrete representations of queries be included in the honestly considered cost of deciding problems regarding abstract objects.

Demanding of an implementation that it also generate its internal representations of the input from an abstract term description of that input precludes the hiding of incomputability in the representation used for concrete implementations and, likewise, obviates cheating on complexity problems by giving away the answer in the representation. It also means that checking parity of a binary string should be considered linear-time (in input length), not constant-time. Put another way, presenting a number with least-significant digit first is just as dishonest as ordering the nodes of a graph by its Hamiltonian path. (In general, the sublinearity of various deterministic algorithms, ignoring the cost of constructing the input, strongly depends on how the input is presented.)

Often, one analyzes alternative representations with respect to the complexity of a set of basic functions. Considering graphs, for example, it is common to compare the adjacency-list representation with adjacency matrices. While the former provides greater efficiency for adding a vertex, it has a steeper edge removal cost. In these cases, the complexity of generating the input representation might be considered another aspect of the complexity tradeoffs.

> Many persons who are not conversant with mathematical studies, imagine that because the business of the [Analytical] engine is to give its results in *numerical notation*, the *nature of its processes* must consequently be *arithmetical* and *numerical*, rather than algebraical and analytical. This is an error. The engine can arrange and combine its numerical quantities exactly as if they were *letters* or any other *general* symbols; and in fact it might bring out its results in algebraical *notation*, were provisions made accordingly. It might develope three sets of results simultaneously, viz. *symbolic* results; *numerical* results (its chief and primary object); and *algebraical* results in *literal* notation. This latter however has not been deemed a necessary or desirable addition to its powers, partly because the necessary arrangements for effecting it would increase the complexity and extent of the mechanism to a degree that would not be commensurate with the advantages, where the main object of the invention is to translate into *numerical* language general formulae of analysis already known to us, or whose laws of formation are known to us. But it would be a mistake to suppose that because its *results* are given in the *notation* of a more restricted science, its *processes* are therefore restricted to those of that science. The object of the engine is in fact to give the *utmost practical efficiency* to the resources of *numerical interpretations* of the higher science of analysis, while it uses the processes and combinations of this latter.
>
> —Augusta Ada Lovelace, Notes to "On Babbage's Analytical Engine" (1843)
> [emphasis in the original]

References

1. Abramsky, S. (2011). Undecidability of the halting problem: A self-contained pedagogical presentation. Unpublished note.
2. Boker, U. (2008). The influence of domain interpretations on computational models. Ph.D. thesis, Tel Aviv University, School of Computer Science.

3. Boker, U., & Dershowitz, N. (2006). Comparing computational power. *Logic Journal of the IGPL, 14*, 633–648. Retrieved February 9, 2016, from http://nachum.org/papers/ComparingComputationalPower.pdf.
4. Boker, U., & Dershowitz, N. (2008). The Church-Turing Thesis over arbitrary domains. In: A. Avron, N. Dershowitz and A. Rabinovich (Eds.), *Pillars of computer science, essays dedicated to Boris (Boaz) Trakhtenbrot on the occasion of his 85th birthday*. Lecture Notes in Computer Science (Vol. 4800, pp. 199–229). Berlin, Germany: Springer. Retrieved February 9, 2016, from http://nachum.org/papers/ArbitraryDomains.pdf.
5. Boker, U., & Dershowitz, N. (2010). Three paths to effectiveness. In A. Blass, N. Dershowitz and W. Reisig (Eds.), *Fields of logic and computation: Essays dedicated to Yuri Gurevich on the occasion of his 70th birthday*. Lecture Notes in Computer Science (Vol. 6300, pp. 36–47). Berlin, Germany: Springer. Retrieved February 9, 2016, from http://nachum.org/papers/ThreePathsToEffectiveness.pdf.
6. Cai, J. -Y. (1991). Computations over infinite groups. In L. Budach (Ed.), *Proceedings of the 8th International Symposium on Fundamentals of Computation Theory (FCT)*, Gosen, Germany. Lecture Notes in Computer Science (Vol. 529, pp. 22–32). Springer.
7. Davis, M. (1957). The definition of universal Turing machine. *Proceedings of the American Mathematical Society, 8*, 1125–1126.
8. Dershowitz, N. (2012). The generic model of computation. In *Proceedings of the Seventh International Workshop on Developments in Computational Models (DCM 2011, July 2012, Zürich, Switzerland). Electronic Proceedings in Theoretical Computer Science* (pp. 59–71). Retrieved February 9, 2016, from http://nachum.org/papers/Generic.pdf.
9. Dershowitz, N., & Falkovich, E. (2011). A formalization and proof of the Extended Church-Turing Thesis. In *Proceedings of the Seventh International Workshop on Developments in Computational Models (DCM 2011). Electronic Proceedings in Theoretical Computer Science* (Vol. 88, pp. 72–78). Zürich, Switzerland. Retrieved February 9, 2016, from http://nachum.org/papers/ECTT_EPTCS.pdf.
10. Dershowitz, N., & Falkovich, E. (2012) Honest universality. *Special issue of the Philosophical Transactions of the Royal Society A, 370*, 3340–3348. Retrieved February 9, 2016, from http://nachum.org/papers/HonestUniversality.pdf.
11. Dershowitz, N., & Falkovich, E. (2017). The invariance thesis. Logical Methods in Computer Science. Retrieved February 9, 2016, from http://nachum.org/papers/InvarianceThesis.pdf. To appear.
12. Dershowitz, N., & Gurevich, Y. (2008). A natural axiomatization of computability and proof of Church's Thesis. *Bulletin of Symbolic Logic, 14*, 299–350. Retrieved February 9, 2016, from http://nachum.org/papers/Church.pdf.
13. Endrullis, J., Grabmayer, C., & Hendriks, D. (2015). Regularity preserving but not reflecting encodings. In *Proceedings of the 30th Annual ACM/IEEE Symposium on Logic in Computer Science (LICS)*, Kyoto, Japan (pp. 535–546). IEEE. Retrieved February 9, 2016, from http://arxiv.org/pdf/1501.04835v1.pdf.
14. Engeler, E. (1968). *Formal languages: Automata and structures. Lectures in Advanced Mathematics*. Chicago, IL: Markham Publishing Company.
15. Garey, M. R., & Johnson, D. S. (1979). *Computers and intractability: A guide to the theory of NP-completeness*. New York, NY: W. H. Freeman.
16. Gurevich, Y. (2000). Sequential abstract state machines capture sequential algorithms. *ACM Transactions on Computational Logic, 1*, 77–111.
17. Harel, D., Kozen, D., & Tiuryn, J. (2000). *Dynamic logic*. Cambridge, MA: MIT Press.
18. Knuth, D. E. (1968). Algorithm and program; information and data. *Communications of the ACM, 9*, 654.
19. Minsky, M. L. (1967). *Computation: Finite and infinite machines*. Englewood Cliffs, NJ: Prentice-Hall.
20. Rogers, Jr., H. (1965). On universal functions. *Proceedings of the American Mathematical Society, 16*, 39–44. Retrieved February 9, 2016, from http://www.jstor.org/stable/2033997.

21. Schroeppel, R. (1972). *A two counter machine cannot calculate* 2^N. Artificial Intelligence Memo #257, Massachusetts Institute of Technology, A. I. Laboratory. Retrieved February 9, 2016, from ftp://publications.ai.mit.edu/ai-publications/pdf/AIM-257.pdf.
22. Shapiro, S. (1982). Acceptable notation. *Notre Dame Journal of Formal Logic*, *23*, 14–20.
23. Weihrauch, K. (1987). *Computability*. EATCS Monographs on Theoretical Computer Science (Vol. 9). Berlin, Germany: Springer.

Chapter 7
Why Post Did [Not] Have Turing's Thesis

Wilfried Sieg, Máté Szabó and Dawn McLaughlin

> ...I study Mathematics as a product of the human mind and not as absolute...
>
> (Post in Anticipation, i.e., (Post [47], p. 64)).

Abstract The *conceptual confluence* of Post's and Turing's analysis of combinatory processes, respectively of mechanical procedures, is the central topic in Davis and Sieg's [14]. Where Turing argued convincingly for the adequacy of his notion of machine computation in 1936, Post viewed his *identical* notion in the same year as being tied to a working hypothesis in need of "continual verification". Post gave novel and informative arguments for his thesis or, as he put it, *generalization*. He insisted, however, that ultimately a psychological analysis "of mental processes involved in combinatory mathematical processes" has to be given. In this way, he hoped to obtain a *natural law* and thus the basis for the claim that the undecidability and incompleteness theorems constitute "a fundamental discovery in the limitations of the mathematizing power of Homo Sapiens". Our detailed analysis of (the background for) his work on the issues leads to an unambiguous answer to the question *Did Post have Turing's Thesis?*: He did [not].

This paper was written for Martin Davis, brilliant mathematical logician, expert computer scientist, and dedicated student of Post; his humanity and devotion to logic are transcendent.— For Wilfried Sieg, he has been a mentor and friend for more than thirty years; during the last few years we have been collaborating, e.g., organizing a session for the *Turing Centenary Conference* in Cambridge in 2012 and writing a joint paper, our [14]. It has been a pleasure and privilege to do so.

W. Sieg (✉) · M. Szabó · D. McLaughlin
Carnegie Mellon University, Pittsburgh, USA
e-mail: sieg@cmu.edu

M. Szabó
e-mail: mszabo@andrew.cmu.edu

D. McLaughlin
e-mail: dawnm@andrew.cmu.edu

E.G. Omodeo and A. Policriti (eds.), *Martin Davis on Computability, Computational Logic, and Mathematical Foundations*,
Outstanding Contributions to Logic 10, DOI 10.1007/978-3-319-41842-1_7

Keyword Turing's thesis

7.1 Introduction

In 1936, Post and Turing presented two models of computation that were essentially identical. Turing, in his long paper, argued for the adequacy of his model and proved that the *Entscheidungsproblem* for predicate logic is unsolvable by procedures that can be carried out by his machines; Post, in his very short paper, described his model and only conjectured that it is equivalent to Gödel's *(general) recursive functions*.[1] The *Entscheidungsproblem* for subsystems of *Principia Mathematica*, including one corresponding to predicate logic, had motivated Post's work already in the early 1920s. Post's and Turing's focus on this particular *finiteness problem* concerning syntactic configurations was perhaps the reason for developing a common perspective. Undoubtedly, theirs is a uniquely direct approach to combinatory procedures and stands in stark contrast to Gödel's indirect way via (effectively) calculable number theoretic functions. The latter way was, however, straightforwardly and strongly rooted in mathematical practice.

Indeed, the experience with number theoretic calculation procedures had crystallized, in the late 19th and early 20th century, into the concept of *primitive recursive functions*. That concept is found in Dedekind's [15]. Skolem as well as Hilbert and Bernays used it extensively throughout the 1920s. Gödel [22] in 1931 put this class of functions to work for the arithmetization of syntax in his proof of the incompleteness theorems. In the 1934 Princeton lectures, he defined the larger class of *(general) recursive functions*. The precise mathematical notions introduced by Gödel, Church, and Turing were eventually shown to be equivalent. However, Church had already in late 1935 suggested identifying the informal concept of calculable functions with that of recursive functions. The history of and the arguments surrounding *Church's Thesis* will not be reviewed here; there are, after all, comprehensive accounts in the literature, e.g., (Kleene [32], Sects. 62 and 70), (Gandy [21]), (Sieg [48, 51]).

Turing's model of machine computation has been recognized as special, in part because it vividly captures the "mechanical" aspect of algorithmic procedures. More important is Turing's argument from 1936, given in Sect. 9 of his [56], that any mechanical procedure carried out by a *human computer* can be executed by one of his machines, a claim now labeled *Turing's Thesis*. His analysis has great intuitive appeal and articulates restrictive *boundedness* and *locality conditions*, but it does not constitute a proof of the general claim. Turing's argument does prove a suitably restricted claim, if one focuses attention on finite strings of letters from a finite alphabet and accepts the above conditions for strings and operations on them. We will come back to this argument later on, as it is precisely here that Turing's and Post's approaches diverge.

[1]Church [2] had obtained, also in 1936, the same unsolvability result as Turing. He used Gödel's recursive functions, which he knew to be equivalent to the λ-definable ones, as the precise notion of computability.

Post does not argue in his 1936 paper that the worker in his model, when operating in the given symbol space, can carry out all combinatory processes, but suggests contemplating "wider and wider formulations" and reducing them to his so-called *formulation 1*. Nevertheless, he makes the enigmatic claim that the incompleteness and undecidability results brought to light "that a fundamental discovery in the limitations of the mathematizing power of Homo Sapiens has been made". (Post [41], p. 105) That claim has been frequently contrasted with, if not opposed to Gödel's view that these results do not imply "any bounds for the powers of human reason, but rather for the potentialities of pure formalism in mathematics". (Gödel [26], p. 370) The dramatic fashion in which Post presents his claim almost begs for such an interpretation; it suggests, furthermore, he would have agreed with Church and Kleene judging Gödel's search for "humanly effective but non-mechanical procedures" as a fruitless pursuit.

If the "limitations" position were the only one Post had on the matter, then the above interpretation would stand without question. However, Post's position is more complex and evolving in his work. In the *first part* of this paper we describe Post's 1921 solution of the finiteness problem for the sentential logical part of *Principia Mathematica*. That was presented in (Post [38]) indicating also a generalization of the project with the explicit goal of obtaining a decision procedure for all of *Principia Mathematica*. In the *second part*, we use Post's *Anticipation* paper [47] to follow his work in the early 1920s in pursuit of this goal: different *canonical systems* are formulated and shown to be equivalent to normal ones. The results of this work and his experience with the *tag problem* led Post to a complete reversal of his project. However, that required accepting what Davis called, *Post's Thesis*; the path to that thesis is described in the *third part*. The *fourth part* gives an account of Post's anticipation of the undecidability and incompleteness results due to Church, Turing, and Gödel. Post concludes from his version of the incompleteness theorem that mathematical thinking is, and must be, *creative*; we explore here how Post's notion can be understood. Finally, Post's sustained reflections on *inescapable limitations* of mathematical thinking and natural laws, as well as on symbolic logics and finitary constraints allow us to find convincing reasons why he did [not] have Turing's Thesis. That is the topic of the *final three parts*.

7.2 The Finiteness Problem for *Principia Mathematica*

Emil Post was a graduate student at Columbia University from 1917 to 1920 and attended during those years a seminar by his advisor Cassius Keyser on Whitehead and Russell's [61] *Principia Mathematica*.[2] Nevertheless, the main influence on his *programmatic* approach to *Principia Mathematica* and symbolic logic more gener-

[2]The source of all biographical data concerning Post is Martin Davis [11].—Keyser wrote a review of the first two volumes of *Principia Mathematica*, see (Keyser [29]). In Sect. 7.8 we will discuss some of his views that seem to have influenced Post.

ally seems to have been C. I. Lewis.[3] Lewis's Survey of Symbolic Logic from 1918 [34] distinguishes, in Chapter 6, Sects. 2 and 3, between the *heterodox* and *orthodox* views of "the nature of mathematics and of logistic". On the orthodox view Lewis writes:

> "Logistic" may be taken to denote any development of scientific matter, which is expressed exclusively in ideographic language and uses predominantly the operations of symbolic logic. (Lewis [34], p. 340)

From that viewpoint mathematical systems are "nothing more nor less than ... complex logical structure[s]". Their subject matter is "freed from ... appeal to intuition or perception"; after all, in proofs that proceed in accord with logical principles, "nothing depends upon the fact that the terms denote certain ... entities". (Lewis [34], pp. 340–341).

Lewis views the treatment of *Principia Mathematica* as orthodox and remarks on its logistic:

> One might characterize the logistic of *Principia Mathematica* roughly by saying that the order of logistic is assumed, and the order of the other branches then follows from the meaning of their terms. (Lewis [34], p. 354)

From the heterodox view the individual steps in proofs are not logical operations, but "fundamentally arbitrary" and definitely *pre-logical*, since "they underlie the proofs of logic as well as of other branches". Lewis explains, "The assumption of these operations—substitution, etc.—is the most fundamental of all the assumptions of logistic." Thus, a quite different view of the subject emerges:

> It is possible to view the subject in a way, which makes such pre-logical principles the fundamentally important thing, and does not regard as essential the use of symbolic logic as foundation. (Lewis [34], p. 355)

As a consequence of the non-foundational view of logic, a mathematical system "is any set of strings of recognizable marks in which some of the strings are taken initially and the remainder derived from these by operations performed according to rules which are independent of any meaning assigned to the marks." (Lewis [34], p. 355).

Post takes up Lewis's heterodox view and, in his Columbia dissertation, investigates the sentential subsystem of *Principia Mathematica* from a strictly metamathematical perspective. The thesis is identical with the article *Introduction to a General Theory of Elementary Propositions* that was published in 1921 as [38]. The work aimed for the "highest generality" that, according to Post, had not been reached by Whitehead and Russell:

> ...owing to the particular purpose the authors [Whitehead and Russell] had in view they decided not to burden their work with more than was absolutely necessary for its achievement, and so gave up the generality of outlook which characterized symbolic logic.
> (Post [38], p. 163)

---

[3] C. I. Lewis's influence on Post is discussed also in (Urquhart [60], pp. 619–620).

In the first part of the paper, Post attempts to recover this generality of outlook. The resulting developments are viewed, in accord with the heterodox perspective, as purely formal ones; thus, any *useful* "instruments of logic or mathematics" will be employed for their study. In the *Introduction*, Post emphasizes the central point that the theorems of the first part of his paper are *about* the system of sentential logic; they are *not included* in it.

To underline the significance of this central point, Post states that his "most important theorem" gives a "uniform method for testing the truth [i.e., provability] of any proposition of the system" and allows establishing relations between propositions. Such relations show definitely "that the postulates of 'Principia' are capable of developing the complete system of the logic of propositions without ever introducing results extraneous to that system—a conclusion that could hardly been arrived at by the particular processes used in that work." (Post [38], p. 164) This remark appeals to the *completeness theorem* for the calculus that can be extracted from the corollary Post formulated on page 171: The set of true formulae [i.e., provable ones] is identical with the set of positive ones [i.e., tautologies].[4]

How is the heterodox view of the system realized? Through, what is for us today, a standard presentation of the syntax of a logical system. The well-formed formulae (*enunciations*) are inductively defined from variables $p, p_1, p_2, \ldots, q, q_1, q_2, \ldots, r, r_1, r_2, \ldots$ using only the connectives (*primitive functions*) $\sim$ and $\vee$; the definition is labeled by **I**. Then the rule of substitution is stated and labeled by **II**: "The assertion of a function involving a variable p produces the assertion of any function found from the given one by substituting for p any other variable q, or $\sim q$, or $(q \vee r)$." The only inference (*production*) rule for deducing theorems (*assertions*) is modus ponens. It takes the form **III**:

"$\vdash P$" and "$\vdash :\sim P . \vee . Q$" produce "$\vdash Q$.",

where P and Q are meta-variables. Finally, the axioms (*primitive assertions*) are taken from *Principia Mathematica* and presented under **IV**:

$\vdash :\sim (p \vee p). \vee .p,$
$\vdash :\sim q. \vee .p \vee q,$
$\vdash :\sim (p \vee q). \vee .q \vee p.$
$\vdash :\sim [p \vee (q \vee r)]. \vee .q \vee (p \vee r),$
$\vdash : . \sim (\sim q \vee r). \vee :\sim (p \vee q). \vee .p \vee r,$

Truth-values and tables are then introduced. The truth-value of a proposition is denoted by $+$, if it is true, and by $-$, if it is false. Post emphasizes that "this meaning of $+$ and $-$ is convenient to bear in mind as a guide to thought, but in the actual

[4]The completeness theorem for sentential logic had been established in its modern formulation in Bernays's [2] *Habilitationsschrift* of 1918, for the same calculus of *Principia Mathematica*. The independence investigations that are also contained in that work were published in (Bernays [3]).—Post's [43] book on *Iterative Systems* originated as a "companion piece" to his dissertation. There he determined "all the non-equivalent sub-languages of the language of the complete two-valued propositional calculus". (Post [43], p. 3) For a contemporary view of this highly interesting work, see (Urquhart [60], Sect. 5).

development that follows they are to be considered merely as symbols which we manipulate in a certain way". (Post [38], p. 166) Thus, even truth-values and tables are treated as syntactic objects that can be manipulated on the basis of rules. The "most important theorem", called by Post *Fundamental Theorem*, is formulated as follows:

> **Theorem.** *A necessary and sufficient condition that a function* [formula] of F be asserted [proved] as a result of the postulates **II, III, IV** is that all its truth-values be $+$.
>
> (Post [38], p. 169)

Post states, the (proof of the) Fundamental Theorem solves the finiteness problem of the sentential subsystem of *Principia Mathematica* as it obviously "gives a direct method for testing whether that function can or cannot be asserted"; moreover, he claims:

> ...if the test shows that the function can be asserted the above proof [of the Fundamental Theorem] will give us an actual method for immediately writing down a formal derivation of its assertion by means of the postulates of *Principia.* (Post [38], p. 171)

There are many further informative observations, theorems, and corollaries; for example, on page 172, the "Post completeness" of the calculus is formulated as a theorem in this way:

> **Theorem.** *Every function of the system can either be asserted by means of the postulates or else is inconsistent with them.*

However, here we are interested in the *generalizations* of the heterodox outlook on systems Post suggests developing. According to Post, two directions of further development can be pursued. The first direction concerns the extension of the decision procedure for sentential logic to other parts of *Principia Mathematica.* In [39], an abstract from 1921 entitled *On a Simple Class of Deductive Systems*, Post reports to have solved the decision problem for what we now call monadic predicate logic; the full paper was never published. More ambitious, but as we will see closely related, is the second direction:

> ...we might take cognizance of the fact that the system of "Principia" is but one particular development of the theory—particular in the primitive functions it employs and in the postulates it imposes on those functions—and so [we] might construct a general theory of such developments.
>
> (Post [38], p. 164)

The "syntactic" and "semantic" parts of the theory concerning sentential logic are both generalized. As to the latter and the *truth-tables* involved, Post considers functions with m truth-values, where m is greater than 2. His pioneering work on *m-valued Truth-Systems* proved to be influential; together with Łukasiewicz, Post is considered as one of the founders of many-valued logic.[5] As to the former and the *method*

---

[5]There is an interesting difference between their motivations. Łukasiewicz mentions in his [35, 36] from 1918 and 1920, respectively, that he was led to the idea of a third truth-value while working on antinomies and "the principle of contradiction in Aristotle's work" (Łukasiewicz [35], p. 86). In order to avoid antinomies, the "third logical value may be interpreted as *possibility*" (Łukasiewicz [35], p. 87). Post, in contrast, does

of postulation, Post considers a more general syntax of the subsystem he investigated. Instead of assuming only $\sim$ and $\vee$ as primitive functions, he introduces μ functions $f_1(p_1, p_2, \ldots p_{m_1})$, $f_2(p_1, p_2, \ldots p_{m_2})$, $\ldots$, $f_\mu(p_1, p_2, \ldots p_{m_\mu})$ with arbitrary numbers of arguments in order to obtain "complete generality". The rules for constructing formulae under **I** are standard, and substitution **II** is modified from the earlier discussion to suit the wider syntactic context. Finally, modus ponens is replaced in **III** by production rules of a much more general form, and so are the axioms in **IV**. The resulting system is later considered to be in *canonical form A*, and the details are discussed at the beginning of Sect. 7.3 below.

The remainder of this section of Post's paper, entitled *Generalization by Postulation*, is viewed as "merely an introduction to the general theory". (Post [38], p. 177) Post defines (in)consistency of a system in a new way, as one cannot make use of the function $\sim$, which might not be among the primitive ones:

> *Definition.* A system will be said to be inconsistent if it yields the assertion of the unmodified variable p. [Thus, the system asserts every formula.] (Post [38], p. 177)

Post's further definitions and considerations reveal his expectations at this time. For a consistent system, a *true* formula is defined to be one "that can be asserted as a result of the postulates"; i.e., all the theorems of such a system are true by definition. If adding a formula to the postulates of the system makes the system inconsistent, then that formula is defined to be *false*.

Without any further explanation, Post asserts after the definition of *false*, "We can then state that in the system 'Principia' every function is true or false." That suggests to him a further definition for an arbitrary consistent system; namely, such a system is called *closed* if every function is either true or false. In a footnote, he remarks that he would have called such a system *categorical*, if that name had not been used "in a different connection".[6] Justifying his terminology Post writes:

> ...the postulates of such a [closed] system automatically exclude the possibility of any added assertion—a state of affairs we believe to be highly desirable in the final form of a logical theory. (Post [38], p. 177)

Post clearly expected Principia Mathematica to be closed or syntactically complete, in the modern sense of this notion. Furthermore, Post emphasized already in the introduction the virtue of such a broadened outlook, namely, that it will "serve to prepare us for a similar analysis of that complete system [i.e., of *Principia*], and so ultimately of mathematics". (Post [38], p. 164) It is against the background of these expectations that Post's work in the early 1920s should be seen.

(Footnote 5 continued)
not give any interpretation of the multiple truth-values except for the purely mathematical one; he treats the corresponding truth tables as syntactic objects similar to their treatment in the two-valued case.

[6]The emergence of *categoricity* and *completeness* is described in (Awodey and Reck [1]).

7.3 Canonical Systems and Their Reductions

Post indeed attempted "a similar analysis" of *Principia Mathematica* while he held a Procter fellowship at Princeton in 1920–21. This work together with additional work from the first half of the 1920s is reported in *Absolutely unsolvable problems and relatively undecidable propositions—account of an anticipation.*[7] That paper, briefly referred to as *Anticipation*, was published posthumously in Davis's 1965 [12]. In §2 of *Anticipation* Post went beyond the purely sentential logic of *Principia Mathematica* and analyzed parts $*10$ and $*11$. That turns out to be a presentation of predicate logic, as the variables are taken to have a "common range", i.e., they are not restricted to types. Post describes this logical system as a system in *canonical form B*.[8]

The work in *Anticipation* establishes the equivalence of canonical systems of forms A, B, and C with systems in *normal form*.[9] The canonical form A in *Anticipation* is given in exactly the form it was given in (Post [38]): **I** is formulated as above, the substitution rule **II** and the production rules **III** are presented in appropriately more general forms. We quote Post's formulation:

> **II.** The assertion of a function involving a variable p produces the assertion of any function found from the given one by substituting for p any other variable q, or $f_1(q_1, \ldots, q_{m_1})$,..., or $f_\mu(q_1, \ldots, q_{m_\mu})$.
>
> **III.** "$\vdash g_{11}(P_1, P_2, \ldots, P_{k_1})$" ... "$\vdash g_{\kappa 1}(P_1, P_2, \ldots, P_{k_\kappa})$"
>
> ....
>
> "$\vdash g_{1\kappa_1}(P_1, P_2, \ldots, P_{k_1})$" ... "$\vdash g_{\kappa\kappa_\kappa}(P_1, P_2, \ldots, P_{k_\kappa})$"
>
> produce produce
>
> "$\vdash g_1(P_1, P_2, \ldots, P_{k_1})$" ... "$\vdash g_\kappa(P_1, P_2, \ldots, P_{k_\kappa})$",
>
> where the P's are any combination of the f's including the special case of the unmodified variable, while the g's are particular combinations of this kind which need not have all the indicated arguments.
>
> **IV.** $\vdash h_1(p_1, p_2, \ldots, p_{l_1})$,
>
> $\vdash h_2(p_1, p_2, \ldots, p_{l_2})$,
>
> ...
>
> $\vdash h_\lambda(p_1, p_2, \ldots, p_{l_\lambda})$,

---

[7]The purely mathematical part reducing a combinatorial system (in form C) to a normal one had already been published in 1943 under the title *Formal reductions of the general combinatorial decision problem*. The long final note of this paper gives a "brief résumé" of the larger developments of which this mathematical investigation was a part and which are fully described in (Post [47]).

[8]The motivation for canonical form B, but also the familiar concepts of *reduction* and *equivalence* are discussed in detail on pages 7–8 of *Anticipation* and, similarly, in (Post [45], p. 289).

[9]Indeed, full equivalence of $*10$ and systems in canonical form seems to have been a goal; see footnotes 26, 79, and 90 of *Anticipation*. Thus, the solvability of the finiteness problem of any of the systems would imply solvability for all the others, in particular for predicate logic.—Footnote 79 indicates that the reduction of systems in normal form to those of form A, the so-called *closing of the circle*, was not done in 1920–21, but only later.

> where the h's are particular combinations of the f's.
>
> (*Anticipation*, p. 5)

Canonical form B shares I and IV with A, but II and III are modified as follows:

> **$\mathbf{II}_0$:** II restricted to the *replacing of a variable by any other variable, and that not present in the given assertion.*
>
> **$\mathbf{III}_0$:** III with the added restriction that *each capital P of a conclusion is present in at least one premise of the corresponding production.*
>
> (*Anticipation*, pp. 7–8)

The reduction of canonical form A to B is relatively straightforward. The basic idea is to add a new primitive propositional function e to the system in form B to those available in the A system, such that $e(P)$ is an assertion of the B system, just in case P is an enunciation of the A system, including variables. That is achieved by taking "$\vdash e(p)$" as the **IV** of the B system and by adding a set of production rules to $\mathbf{III}_0$ that mimics the syntax of the A system. More productions are then added to duplicate the effect of **IV** of the A system, ensuring that the primitive assertions of the A system are assertions of the B system as well. The final step is to reproduce the production rules of the A system in $\mathbf{III}_0$. The system thus obtained is in canonical form B and has all the assertions of the system in form A plus the assertions concerning the enunciations of the A system.

The systems in canonical form C radically move away from the quasi-logical form of the A and B systems; they have as their *basis* nothing more than a finite number of distinct symbols $a_1, a_2, \ldots, a_\mu$. The formulae of such a system are all the finite sequences of these symbols, repetitions of the same symbol are allowed. That is, an arbitrary enunciation of the system is a sequence $a_{i_1}, a_{i_2}, \ldots, a_{i_n}$ of symbols from the basis. A finite set of such enunciations constitutes the *primitive assertions* of the system, i.e., its axioms. Finally, a finite set of productions allows the generation of new assertions from old ones; they are of this general form:

$$
\begin{array}{c}
g_{11}\ P_{i'_1}\ g_{12}\ P_{i'_2} \ldots g_{1m_1}\ P_{i'_{m_1}}\ g_{1(m_1+1)} \\
g_{21}\ P_{i''_1}\ g_{22}\ P_{i''_2} \ldots g_{2m_2}\ P_{i''_{m_2}}\ g_{2(m_2+1)} \\
\cdots\cdots\cdots\cdots\cdots\cdots\cdots\cdots \\
g_{k1}\ P_{i^{(k)}_1}\ g_{k2}\ P_{i^{(k)}_2} \ldots g_{km_k}\ P_{i^{(k)}_{m_k}}\ g_{k(m_k+1)} \\
\text{produce} \\
g_1\ P_{i_1}\ g_2\ P_{i_2} \ldots g_m\ P_{i_m}\ g_{m+1}
\end{array}
$$

The rules and their use are described by Post as follows:

> …the g's are specified sequences of the primitive a's, including the null sequence, and each P of the conclusion is present in at least one premise. In the application of these productions the P's may be identified with arbitrary sequences of the above type, it being understood however, that the conclusion may not be null. (*Anticipation*, p. 25)

The systems of canonical form C are today better known as *Post production systems*.

The reduction from canonical form B to C is more complicated than that from A to B, but the basic methodology is similar. Two additional goals have to be met: the elimination of the use of the parenthesis notation employed in form B and the restriction of the alphabet to a finite one, in contrast with the infinite number of variables allowed in B. The latter goal is accomplished by means of a single primitive assertion $\alpha_0 a_0$, where α_0 can be interpreted as asserting that the following string of a_0's represents a variable of the B system; the rule "$\alpha_0 P$ produces $\alpha_0 a_0 P$" guarantees that infinitely many variables are recognized. (*Anticipation*, p. 26) The former goal is accomplished via a translation of the parenthesis notation into a dot notation, and a representation of that notation in the C system. Once the variables of the B system are suitably represented by $\alpha_0 a_0 \ldots a_0$ the full reduction of a system in form B to one in form C consists in mimicking the syntax of the B system, as well as reproducing the axioms and production rules. So we end up with all the assertions of the B system being assertions of the C system, plus a set of assertions of the C system concerning the enunciations and variables of the B system. In contrast to the previous reduction where the system in form B had but a single primitive assertion, we end up with more primitive assertions than those of the reduced system.

Finally, these combinatorial C systems are reduced to systems in *normal form*. The latter systems have a single primitive assertion. Their production rules have just one premise, each involving a single variable; they are of a remarkably simple form:

$$gP \text{ produces } Pg'$$

The transition from many primitive assertions to one is effected by taking the *logical product* of the primitive assertions of the C system as the primitive assertion of the normal system. This is accomplished using two new letters in the alphabet of the normal system; see (*Anticipation*, p. 31). The production rules of the C system are similarly translated into rules taking the logical product of the premises of the original rule as the single premise of the normal rule. Additional rules allow the manipulation of such products; these modified production rules are then recast in normal form by introducing (finitely many) additional letters and (finitely many) new "helper" productions.[10]

[10] A version of this reduction, much easier to understand than Post's, is given in (Minsky [37]) and was further refined in (Szabó [55]).—For Post, as pointed out in (De Mol [18], p. 53), this result supported his conjecture that all of *Principia Mathematica* could be reduced to a normal system; he wrote in (*Anticipation*, p. 45): "…for if the meager formal apparatus of our final normal systems can wipe out all the additional greater complexities of canonical form B, the more complicated machinery of the latter should clearly be able to handle formulations correspondingly more complicated than itself."—To emphasize the significance of this result we quote a pregnant remark from Minsky ([37], p. 240): "We have long felt that this result is one of the most beautiful theorems in mathematics. The fact that any formal system can be reduced to Post canonical systems with a single axiom and productions of the restricted [normal] form is in itself a remarkable discovery, and even more so when we learn that this was found in 1921, long before the formalization of metamathematics became so popular.".

Let us briefly note that many of the informal observations Turing uses to restrict calculations of a human computer to computations of his machines are reflected in Post's reductions. Turing shifts, in Sect. 9 of his [56], calculations from two-dimensional paper to a linear tape divided into squares; that is justified by the assertion that the two-dimensional character of the paper is "no essential of computation" (p. 135). The production rules of forms A, B, and C are two-dimensional, as each of them takes multiple premises to a single conclusion. In the final normal form, however, all productions take only a single premise, and these rules are written on a single line. Perhaps the two-dimensional character of the original production rules is an artifact of the presentation, but even were it not, the reduction to normal form would be proof that the two-dimensional character was indeed not essential. Turing required the operations of a human computer to be analyzed into such simple ones that, in his own terminology, it is difficult to imagine them further divided; this requirement is certainly satisfied by the normal production rules. In any case, the (computation) steps in both Turing's and Post's case are carried out on strings, one-dimensionally. That is an expression of the conceptual confluence in Post's and Turing's work diagnosed by (Davis and Sieg [14]). Given this confluence it is perhaps not surprising that Post eventually arrived at the same conclusion as Turing—against his deeply held initial expectations. We see part of the circuitous route to this conclusion in the next part.

7.4 Reasons for a Thesis

In §§7–10 of *Anticipation*, Post asserts and sketches proofs of results he had obtained in the early 1920s. These results are analogues of the undecidability and incompleteness theorems Turing and Gödel presented in their papers of 1936 [56] and 1931 [22], respectively. For the undecidability of predicate logic in particular, an adequate concept of computability or mechanical procedure is needed relative to which unsolvability is shown. The adequacy of such a concept is usually expressed as a *Thesis*, most often as Turing's Thesis, Church's Thesis, or the Church-Turing Thesis. Post formulated an assertion that is equivalent to those theses; he called it simply *generalization*, but it was dubbed *Post's Thesis* by Davis in his [10]. We describe now the considerations that led up to it and made it ultimately plausible to Post. Let us start out by quoting Post's own formulation at the end of §7 of *Anticipation*:

> *Every generated set of sequences on a given set of letters* $a_1, a_2, \ldots, a_\mu$ *is a subset of the set of assertions of a system in normal form with primitive letters* $a_1, a_2, \ldots, a_\mu, a'_1, a'_2, \ldots, a'_\mu$, *i.e., the subset consisting of those assertions of the normal system involving only the letters* $a_1, a_2, \ldots, a_\mu$. (*Anticipation*, p. 46)

Post had shown in §2 of *Anticipation*, as we discussed earlier, that parts *10 and *11 of *Principia Mathematica* are reducible to a single system in canonical form B. He made an even stronger claim now:

> From this experience, and the knowledge of the kind of forms and kind of operations appearing in the whole *Principia Mathematica* [...] it becomes reasonably certain that all of *Principia Mathematica* can in similar fashion be reduced to a system in canonical form B.
> (*Anticipation*, p. 44)

As canonical form B is reducible to canonical form C and the latter to normal form, it is reasonable to believe that *the whole system of Principia Mathematica* is reducible to normal form.

> If then we think of *canonical form C as a method of generating a set of (finite) sequences* [...], we see that the generated sets of sequences yielded by all systems in canonical form C are the same as those yielded by the formally simpler normal systems.
> (*Anticipation*, p. 45, our emphasis)

However, to make the Thesis plausible, one has to argue that normal systems indeed yield *all* generated sets of sequences (over a finite alphabet). Post advances two arguments for this claim: the first establishes the reducibility of very general kinds of wider formulations to normal systems; the second shows that the familiar method of diagonalization does not lead out of the sets generated by normal systems.

The first argument starts out, in the last paragraph on page 45 of *Anticipation*, with the observation that the premises and conclusion of any production rule of a system in canonical form C can "completely be described in logical terms and the primitive relation of precedence in a sequence". The meaning of "logical terms" is not specified, but the later part of §7 provides the basis for a suitable interpretation: the properties of production rules of systems in canonical form C and their operations can be expressed in the language of *Principia Mathematica*. That is, one can express that a given string is obtainable from a given premise by substituting certain primitive letters into the operational variables, that a sequence of strings contains instances of all the premises of a given production rule and that its consequence can be produced; thus, the generation of sequences can be represented in *Principia Mathematica* together with some postulates for sequences.

Such an embedding of systems in canonical form C in an expanded system of *Principia Mathematica* "suggests the possibility of describing more complicated operations for the purpose of generating sets of sequences." (*Anticipation*, p. 45) For it might be that the whole apparatus of *Principia Mathematica* is capable of expressing more complex operations than those that can be given by production rules of canonical form C. Thus, these possibly more complex operations have to be accounted for by arguing that the sets of sequences generated by them can be generated by systems in form C and, in turn, by normal systems. Post asks us to *suppose* that each operation is of the form, "a certain number of premises, described in logical terms, gives rise to a certain conclusion, likewise described". (*Anticipation*, p. 45) They may consequently be written in the form: $P_1, P_2, \ldots, P_k$ produces P, where $P_1, P_2, \ldots, P_k, P$ have certain properties $f_1(P_1), f_2(P_2), \ldots, f_k(P_k), f(P)$. In addition, *suppose* that a set of postulates on the letters $a_1, a_2, \ldots, a_\mu$ is fixed and "*Principia Mathematica* is used as the logic of the resulting mathematical system", which Post calls *sequence-Principia Mathematica system*. (*Anticipation*, p. 46) Granting the generality of *Principia Mathematica*, sequences $P_1, P_2, \ldots,$

P_k, P will have the properties $f_1(P_1)$, $f_2(P_2)$, ..., $f_k(P_k)$, $f(P)$ if the latter are assertions in the *sequence-Principia Mathematica system*. (*Anticipation*, p. 46) Then, the operations can be written in the form: $f_1(P_1)$, $f_2(P_2)$, ..., $f_k(P_k)$, $f(P)$, P_1, P_2, ..., P_k produce P. As a result, the system for generating sequences on $a_1, a_2, \ldots, a_\mu$ is the system of *Principia Mathematica* supplemented "by certain postulates and operations of the same general type". (*Anticipation*, p. 46) Notice the condition that postulates and operations are to be of the *same general type* as those of *Principia Mathematica*. This is essential in the following argument.

Principia Mathematica can be reduced to a system in normal form and, with the supposition emphasized in the previous paragraph, reducibility can be expected of the expanded system as well. (*Anticipation*, p. 46) That is, even after *Principia Mathematica* has been expanded to a system for generating sets of sequences using complex operations, it is plausible that a system in normal form can be set up, such that the set of its assertions involving only the letters a_1, a_2, ..., a_μ is the same as the set of sequences generated by the expanded *Principia Mathematica* system. This conclusion led Post to the *generalization* we quoted earlier and called, following Davis, *Post's Thesis*. The reasons are for Post "the generality of the system of *Principia Mathematica*, and its seeming inability to lead to any other generated sets of sequences on a given set of letters than those given by our normal systems". (*Anticipation*, p. 46) The recognition of the significance of this web of reductions led, as Post put it at the end of §6, "to a reversal of our entire program" (*Anticipation*, p. 44) that had been aiming for a positive solution of the general finiteness problem.

Let us come, finally, to the second argument we mentioned at the beginning of this part. Using a diagonal argument an apparent counterexample to *Post's Thesis* can be constructed from specially generated sets of sequences. According to the Thesis, the generated sets of sequences that involve only the letter a are subsets of assertions of normal systems with letters $a, a_1, a_2, \ldots, a_\mu$, where $\mu = 0, 1, 2, \ldots$ Given the definition of normal systems, it is clear that there are only enumerably many of them. For a fixed enumeration, we can define a set D_a of a-sequences: D_a contains the sequence a^m of m a's if and only if a^m is *not* an assertion of the m-th normal system. D_a differs from the set of assertions of each normal system, as it disagrees with the m-th such system on a^m. Post remarks that this set is ***not*** a counterexample to the thesis; after all:

> [We] have merely *defined* a set of a-sequences, whereas to yield a true counter-example we must show how to *generate* that set, i.e., set up a system of "combinatory iteration" whose operations would at some time yield each and every a-sequence in that set, but would never yield an a-sequence not in the set. (*Anticipation*, p. 47)

However, were the finiteness problem solvable for normal systems, the above set could be generated, as any solution would decide, in particular, whether or not the a-sequence of length m is an assertion of the m-th normal system.

> That is, a solution of the finiteness problem for all normal systems would yield a counter-example disproving the correctness of our proposed generalization. (*Anticipation*, p. 47)

Thus, we are led to the following biconditional: Post's Thesis holds *if and only if* the finiteness problem for normal systems is *unsolvable*.

Post points out, "nothing in the above argument weakens the reasoning that led us to our generalization". (*Anticipation*, p. 47)[11] It should be remarked that Post's unsuccessful attempts to solve the finiteness problem for tag systems made the unsolvability of normal systems plausible to him and committed him to his generalization even more strongly[12]:

> We therefore hold on to that generalization and conclude that *the finiteness problem for the class of all normal systems is unsolvable, that is, that there is no finite method which would uniformly enable us to tell of an arbitrary normal system and arbitrary sequence on the letters thereof whether that sequence is or is not generated by the operations of the system from the primitive sequence of the system.* (*Anticipation*, pp. 47–8)

Thus, we have an informal *sketch* of an argument for the undecidability of the *sequence-Principia Mathematica* system, on account of the unsolvability of the finiteness problem for normal systems and their connection to *sequence-Principia Mathematica* via canonical systems of form B and C. The unsolvability of the finiteness problem rests on the argument leading to the *generalization* (i.e., *Post's Thesis*), which in turn depends crucially on the supposition that the operations and postulates of the full *sequence-Principia Mathematica* system are all of the same general type. Thus, the methodological issue revolves around the question of why only extensions of this restricted form should be admitted.[13] In the next part we take the Thesis for granted and, with it in the background, look at the proof sketches Post provides for his analogues of Turing's undecidability result and Gödel's first incompleteness theorem. In Sect. 7.6 we will come back to the "central methodological issue".

---

[11]Kleene was convinced by the same argument; see (Kleene [33], p. 59). In (Post [45], p. 285) "overwhelming evidence" is adduced for Church's Thesis by reference to footnote 2 in (Kleene [31]). There one finds a concise and masterful summary of the evidence, as Kleene saw it, for the "identification" of effective calculability and recursiveness; Kleene's remarks are quoted in footnote 34, below.

[12]The tag systems are special normal systems: the g's are all of the same length; each g' depends only on the first letter of its corresponding g.—The important role of the problem of "tags" was emphasized in §3 of *Anticipation*; see also the very careful analysis in (De Mol's [16, 17]). Indeed, De Mol argues that the tag systems were crucial for Post for two main reasons: (1) they prompted his belief that there might be absolutely unsolvable problems, and (2) they inspired the formulation of normal systems.

[13]The character of this argument and its similarity to "absoluteness" considerations of Gödel, Church, Hilbert & Bernays, and Turing will be discussed briefly in footnote 29 of Sect. 7.7 below.—Martin Davis pointed out that the above considerations of course do not establish the undecidability of predicate logic. Post is very cautious on what he claims to have established with respect to that problem; see footnotes 79 and 90 of *Anticipation* and also footnote 10 above.

7.5 Anticipating Turing's and Gödel's Results

Post's considerations that are presented in this part take his Thesis for granted. It is perfectly clear to Post that the significance of his results depends on it. For example, Post writes, referring to the unsolvability of the finiteness problem for the class of all normal systems:

> The correctness of this result is clearly entirely dependent on the trustworthiness of the analysis leading to the above generalization. (*Anticipation*, p. 48)

In order to obtain this very result, he outlines in §9 of *Anticipation* "a minimum mathematical development". He sharpens the informal proof he sketched in §8, beginning with the definition of a particular ordering of all the bases of normal systems, called the *σ-ordering*. The diagonal set D_a relative to this ordering is called the *N-set*. The definition is followed by the "almost trivial theorem" that no normal system has the N-set as the set of assertions involving only the letter a. Post asserts that this theorem provides the "mathematical basis for the no finite method theorem" and emphasizes that it would be trivial, "were it not for the all embraciveness of normal systems" (*Anticipation*, p. 50), i.e., the correctness of his Thesis.

In the remainder of §9 proofs of statements are only sketched, although "a complete mathematical proof thereof clearly can be given". (*Anticipation*, p. 50)[14] Post labels the statements cautiously as (Theorem)-s. Among them is the formulation of an "important intermediate" assertion.

> **(Theorem).** *There exists a normal system K and a correspondence C such that for each normal system and enunciation thereof there is one and only one enunciation in K by correspondence C, and such that such an enunciation in K is asserted when and only when the corresponding normal system versus enunciation is such that the enunciation is an assertion in that normal system.* (*Anticipation*, p. 51)

Post refers to the normal system K as "the *complete normal system* because, in a way, it contains all normal systems". A footnote attached to this remark states, "the 'complete normal system' would thus correspond to Turing's 'universal computing machine'".[15] (*Anticipation*, fn. 95)

The normal system M has a *finite-normal-test* if there exists a normal system M' on the primitive letters of M supplemented by at least the letter b and such that the following correspondence between their enunciations holds: P is an assertion in M' when and only when it is an assertion in M; bP is an assertion in M' when and only when P is not an assertion in M.[16] Thus, for each enunciation P of M exactly one of the two sequences P or bP is an assertion of M', and which is the assertion depends entirely on whether P is or is not an assertion of M. We have

[14]The full sentence is this: "Our remaining 'theorems' deserve that name only in the sense that a complete mathematical proof thereof clearly can be given—as contrasted with our generalization of §7".

[15]The footnotes were added by Post at the time of writing the *Anticipation* paper in the late 1930s and early 1940s. They were not part of his original notes from the 1920s.

[16]That is, Post remarks in note 96, b serves as a negation symbol.

consequently a (Theorem) that can be proved by a *reductio ad absurdum* argument: if there were such a system constituting a finite-normal-test for K, then a system could be constructed that would generate exactly the N-set; but that is impossible.

> **(Theorem).** *There exists no finite-normal-test for the complete normal system K.* (*Anticipation*, p. 52)

Following the argument for this (Theorem), Post describes "the positive content" of its proof. In conjunction with the previous (Theorem) it allows us to construct from a finite-normal-test L for K a normal system L' such that L will give a *wrong* answer to the question, whether the a-sequence of length m' is an assertion of L', which is the m'-th normal system.

The above requirements on a finite-normal-test for K are weakened in §10 or rather, as Post puts it in his letter to Gödel of 30 October 1938, his examination of the "source of the contradiction" in the above argument led to a particular statement in the extending logic "such that neither it nor its negative was asserted" in that logic. (Gödel [28], p. 170) So let K be again the complete normal system and assume that L is a normal system whose alphabet includes that of K. Now L is not *always* to give an answer to the question, whether or not an enunciation of K is an assertion of K; rather, when L gives an answer, it has to be correct. More precisely, let L have at least the primitive letter b in addition to the primitive letters of K. If S is a normal system and P one of its enunciations, we require (S, P) to be an assertion of L if and only if P is an assertion of S; whereas, if $b(S, P)$ is an assertion of L, then P must not be an assertion of S. In terms of the complete system K this is equivalent to the following: (S, P) is an assertion of L if and only if it is an assertion of K, and if $b(S, P)$ is an assertion of L, then (S, P) is not an assertion of K. Post calls such an L a *normal-deductive-system* adjoined to K and remarks, if b is added to K, the resulting system is normal-deductive.[17]

Observe an important property of any normal-deductive-system L: such a system cannot prove both (S, P) and $b(S, P)$. This property is similar to Gödel's consistency requirement, at least with respect to the set of enunciations of the form (S, P) and $b(S, P)$. Via a diagonal argument an analogue of Gödel's First Theorem[18] can be obtained, showing that a particular enunciation is undecidable:

> **(Theorem).** *No normal-deductive-system* $[L]$ is complete, there always existing a normal system S and enunciation P thereof such that P is not in S [thus, (S, P) is not in L], while $b(S, P)$ is not in the normal-deductive-system $[L]$. (*Anticipation*, p. 54)

[17]This explains why the requirement on L was not weakened in case (S, P) is an assertion of L: "There is no reason for doing so since by suitably adjoining K to such a weak L the stronger L would result." (*Anticipation*, p. 53).

[18]Gödel formulated the first incompleteness theorem in its full generality as pertaining to *all* consistent formal systems containing some elementary number theory most strikingly in 1964 in his [26], the Postscriptum to his Princeton Lecture Notes. For the "precise and unquestionably adequate" characterization of *formal* systems he appealed to Turing's and Post's work. He wrote there: "A formal system can simply be defined by any mechanical procedure for producing formulas, called provable formulas.".

Thus every normal-deductive system L is incomplete at least with respect to the set of enunciations of the form (S, P). The restriction to such enunciations does not diminish the significance of the result; after all, Gödel's Theorem similarly states "the incompleteness of any symbolic logic with respect to the class of arithmetical propositions". (*Anticipation*, fn. 101)

After the above incompleteness (Theorem) has been asserted, Post adds a "still more important" one, stating that normal-deductive systems can always be extended. According to him, the following statement "rules out the possibility of a completed symbolic logic. That is, any symbolic logic can be made more complete." (*Anticipation*, fn. 101) In the letter to Gödel, mentioned above, he makes the reason for this extendibility clearer; L does not decide the enunciation (S, P), but the proof of the above (Theorem) and an appeal to the meaning of (S, P), "S asserts P", do decide it. The statement is formulated in the following way:

> **(Theorem).** *No normal-deductive-system is equivalent to the complete logical system (if such there be); better, given any normal-deductive-system there exists another which second proves more theorems (to put it roughly) than the first.* (Anticipation, *p. 54*)

These results taken together with those of §9 of *Anticipation* lead Post to the conclusion, *"A complete logic is impossible."* This is for Post "an iconoclastic result from a logician's point of view", as it means "logic must be informal not only in some parts of its description, but also in its very operation". (*Anticipation*, pp. 54–5)

"Better still", Post writes, *"The Logical Process is Essentially Creative."* This creative aspect of the logical process is the crucial conclusion for Post.[19] In the very introduction to *Anticipation* he had emphasized already:

> [P]erhaps the greatest service the present account could render would stem from its stressing of its final conclusion that mathematical thinking is, and must be, essentially creative. (*Anticipation*, p. 4)

He immediately adds in a footnote a deeply puzzling remark that seems to provide a reason for the creativeness of mathematical thinking:

> Yet, as this account emphasizes, the creativeness of human mathematics has a counterpart inescapable limitation thereof—witness the absolutely unsolvable (combinatory) problems. (*Anticipation*, fn. 12)

A similar remark is found in his [45] after Post establishes what he called there *Gödel's theorem in miniature*:

> The conclusion is unescapable [sic] that even for such a fixed, well defined body of mathematical propositions, *mathematical thinking is, and must remain, essentially creative*. To the writer's mind, this conclusion must inevitably result in at least a partial reversal of the entire axiomatic trend of the late nineteenth and early twentieth centuries, with a return to meaning and truth as being of the essence in mathematics. (Post [45], p. 295)

---

[19] At this very spot Post remarks (*Anticipation*, p. 55) that his conclusion goes contrary to the viewpoint of C. I. Lewis as it was reported above in Sect. 7.2. Furthermore, he mentions that it is "not so much contrary to Russell's viewpoint (since he does not fully express himself)" and that it is "in line with Bergson's *Creative Evolution*". Post is not correct with his remark on Lewis, as the latter had a more sophisticated understanding of mathematics than expressed through the programmatic heterodox view; see (Lewis [34], pp. 359–361).

To understand the first claim, we not only have to clarify the ordinary meaning of *counterpart*, we also have to gain a deeper insight into the meaning Post associates with creativeness.

Here is, first of all, an attempt to rephrase Post's claim, guided by the Oxford English Dictionary as to the meaning of counterpart[20]: *the creativeness of human mathematics* is a natural complement to *its inescapable limitation*. Indeed, in the twice-mentioned letter to Gödel, Post characterizes his incompleteness theorem "as a corollary of the existence of absolutely unsolvable problems". (Gödel [28], p. 170) He describes then also how the detailed analysis of his proof of the unsolvability result as sketched above, once the consistency of the logic and the meaning of the relevant *Entscheidungsproblem* have been granted, leads "to a definite yes or no answer to the enunciation that the assumed logic failed to decide". He therefore concluded:

> that mathematical proof was *essentially creative* in that once having set up a formal system relative to say the above Entscheidungsproblem we could then always transcend that system, i.e. add to the set of assertions relative to the same body of enunciations—a conclusion I believe also reached in your work. (Gödel [28], pp. 170–71, our emphasis)

The enunciation to be added can actually be effectively constructed, given the appropriate technical set up.[21] A mathematical and restrictive understanding of creativeness is clear, but let us also point out that our parallel reading of the informal concept is supported in *Anticipation*: footnote 7 consists of just the following sequence of words: "Produced, created—in practice, written down"; it is attached to the word "generated" (according to rules) and is to make that word's meaning clear. In sum, we have reached a restrictive understanding of creativeness and how it can be viewed as a natural complement to the limitation of mathematical thinking.

Before discussing this issue further, we may ask, why generated sets should play such a central and exclusive role as instruments for solving combinatory problems. In *Anticipation*, pages 2–3, the issue is described in its proper conceptual context, and this context gives also a first hint as to the restrictive character of the generative process.[22] According to Post, recursiveness, λ-definability and even Turing-computability were introduced, to capture effective calculability. His own work, in contrast, attempts to capture the informal notion of generated sets for the following reason:

---

[20]One particularly fitting meaning for *counterpart* is articulated as follows: "One of two parts which fit and complete each other; a person or a thing forming a natural complement to another.".

[21]The various features mentioned in the informal discussion found their way into the definition of a "creative set" of natural numbers in (Post [45], p. 295). Post envisions on page 296 not only a finite, but indeed transfinite iteration of this extending process. The iteration along Kleene's constructive ordinals is actually carried out in (Davis [9], p. 190) continuing work in (Davis [8]). The statement added to a particular system S is the Gödel sentence G for S. Assuming that S satisfies the standard representability and derivability conditions, G is equivalent to the consistency statement for S. The results, for this type of extension, from Feferman's [20] *progressions of theories* (and Turing's [57] *ordinal logics*) can be directly transferred to the Post-Davis construction.

[22]Clearly, Post's discussion excludes sets definable by generalized inductive definitions, like Kleene's O, as generated sets, as they require a "rule" with infinitely many premises.

> This [notion of generated sets] derives from the idea of a *symbolic logic* rather than that of an algorithm, and may be described by saying that each member of the set is at some time generated by the continued application of a given method, while that method will at no time yield an individual ... not in the set. (*Anticipation*, p. 3, our emphasis)

We saw in Sect. 7.3, how canonical systems of forms A and B were used to give precise, rule-based descriptions of significant fragments of *Principia Mathematica*, namely, sentential and predicate logic. Post assumed that the full system of *Principia Mathematica* could be described in a similar way and, furthermore, that it would be syntactically complete. Thus, the finiteness problem would be solved, as in the case of sentential logic, if an obviously decidable criterion for provability could be found. This line of thought made it strategically appropriate to focus on generating theorems and to work on simplifying the production rules. In this way, Post was led first to normal systems and then to the reversal of his whole program, having discovered the unsolvability of the decision problem for all normal systems. For the sake of its broader significance, the reversal required the generalization for generated sets.

The *inescapable limitation of human mathematics* is exemplified, as we saw, by the absolute unsolvability of a particular combinatorial problem. That is the rock bottom of Post's analysis of the incompleteness phenomenon and the ultimate grounding for his appeal to return to *meaning* and *truth*, as well as to use an open concept of *proof*. These three notions, freed from a reliance on formal procedures, are involved in the argument for the creativeness of human mathematics, and Post believes that these developments will effect "a reversal of the entire axiomatic trend of the late 19th and early 20th centuries". So it is crucial to grasp Post's reasons for viewing the combinatorial problem as *absolutely* unsolvable—for *Homo Sapiens Mathematicus*. In Sect. 7.6 we explore Post's way of securing a natural law and thus the basis for the claim of *absolute* unsolvability; it was expressed by Post repeatedly, but most directly in his letter to Gödel: "... the absolute unsolvability of that problem [the above combinatorial problem] has but a basis in the nature of physical induction at least in my work and I still think in any work." (Gödel [28], p. 171) We assume that the adjective "physical" is simply meant to contrast this form of ("ordinary" scientific) induction from mathematical induction; it is the physical induction that is used to support a natural law.

Note to the reader. The next two parts take on the difficult and genuinely challenging task to disentangle two strands in Post's thinking about the central methodological issue. There is, on the one hand, the *investigation of symbol complexes and mechanical operations* on them; this seems to be motivated by the finitary character of symbolic logics. There is, on the other hand, the *ambition to ground the finiteness and discreteness assumptions* concerning such logics in structural features of the human mind or, rather, self-consciousness. The task is made even more difficult by the fact that the considerations concerning the ambitious grounding are presented through the fragmentary excerpts in the Appendix to *Anticipation*; see footnote 26 below. Our analysis is just a beginning.

7.6 A Natural Law: Inescapable Limitations

In the introductory remarks to the Appendix of *Anticipation*, Post reemphasizes that the undecidability and incompleteness results are "evidences of limitations in man's mathematical powers". (*Anticipation*, p. 56) In a similar vein Post had noted in ([41], p. 105, fn. 8), as a consequence of these results, "that a fundamental discovery in the limitations of the mathematizing power of Homo Sapiens has been made". The limitations are explained more comprehensively in the first footnote of *Anticipation*:

> [...] The writer cannot overemphasize the fundamental importance to mathematics of the existence of absolutely unsolvable combinatory problems. True, with a specific criterion of solvability under consideration, say recursiveness, the unsolvability in question, as in the case of the famous problems of antiquity, becomes merely unsolvability by a given set of instruments. And, indeed, the corresponding proofs for combinatory problems are almost trivial in comparison with the classic unsolvability proofs. The fundamental new thing is that for the combinatory problems the given set of instruments is in effect the only humanly possible set. (*Anticipation*, p. 1, fn. 1)

The last sentence, with its claim that "the given set of instruments is in effect the only humanly possible set" for solving combinatory problems, receives more concrete content in a footnote that points to the central methodological problem that has to be resolved in order to justify the claim:

> Since the earlier formal work made it seem obvious that the actual details of the outline [for the proofs of the above results] could be supplied, the further efforts of the writer were directed towards establishing *the universal validity of the basic identification of generated set with normal set.* (Post [44], p. 215, fn. 18, our emphasis)

Post clearly does not see this identification as a definition; after all, he intends to *establish* its universal validity.

One direction of the identification, i.e., normal sets are generated ones, is taken to be correct; thus, the inclusion of generated sets among the normal ones is at issue and is seen by Post as a "partially verified conclusion". (*Anticipation*, p. 3) This was articulated also in his [41] from 1936, where Post conjectured that wider and wider formulations (of generating systems) would all be logically reducible to his "formulation 1". He considered this conjecture as a "working hypothesis" which would be changed by the successful pursuit of the reductive program "not so much to a definition or axiom but to a *natural law*". (*Anticipation*, p. 105)[23] Indeed, Post observed, "the work done by Church and others [establishing equivalences of

[23]Turing, in his illuminating and informal paper from 1954, entitled *Solvable and unsolvable problems*, formulates the thesis not for mechanical procedures or generated sets, but rather for *puzzles* as follows: "The normal form for puzzles is the *substitution type of puzzle* [i.e., a particular kind of Post canonical system]." He remarks then, "The statement is moreover one which one does not attempt to prove. ... for its status is something between a theorem and a definition. In so far as we know a priori what is a puzzle and what is not, the statement is a theorem. In so far as we do not know what puzzles are, the statement is a definition which tells us something about what they are." (Turing [59], p. 15) As puzzles can be given "finite coordinates", they are more general syntactic configurations. It should be mentioned that Post also considered broader classes of syntactic configurations; see (Urquhart [60], p. 643).

various formulations] carries this identification considerably beyond the working hypothesis stage". The natural-law-perspective is expressed again in a footnote to his ([45], p. 286). Post mentions there that Kleene had used "Thesis" as a label for the identification. However, in contrast to Kleene, Post feels "that, ultimately, 'Law' will best describe the situation" and points out, via his [41] paper, that this law is in need of "continual verification". ([41], p. 105, fn. 8)

How can such a natural law be verified, continually? The short answer to the question is, by the kind of "physical induction" Post appealed to in his letter to Gödel, written on 30 October 1938. We quoted from that letter at the end of Sect. 7.5, namely, that the *absolute* unsolvability of the decision problem for all normal systems "has but a basis in the nature of physical induction...". Post then claims with respect to Gödel's own logical system, a version of *Principia Mathematica*, "that [physical] induction could have gone far enough to include your particular system theorematically". (Gödel [28], p. 171) That means he could have proved, as he had done for (subsystems of) *Principia Mathematica*, that Gödel's system is also reducible to a normal system. So it seems that the reduction of particular systems, or of wider formulations, to one of his canonical systems has the point of inductively strengthening the evidence for the problematic half of the identification.

Post asserted forcefully in 1936 that the identification should not be masked under a definition; this is directed against Church who had proposed in his [4] to definitionally identify effective calculability with recursiveness. Church challenged this assertion in his review [6] of (Post [41]) by saying, "effectiveness in the ordinary sense has not been given an exact definition, and hence the working hypothesis in question has not an exact meaning". Church's remark is correct, but does not undermine Post's program of inductively strengthening the connection between the informal concept of generated sets and the mathematically defined normal ones. Whatever is done in support of Post's reductive program is useful, and indeed necessary, to justify the use of "effectively calculable" in Church's definition.[24] We seem to be at a standstill of an almost purely terminological kind.

At this point a substantive question should be raised. Assume that the generalization has indeed been confirmed as a natural law; does it support Post's claim concerning the limitation of human mathematical powers? An attempt to answer this question reveals that much more than a terminological choice is at stake. Indeed, a crucial turn in argumentation is required that brings in not only the human mind and its way of understanding mathematics, but also the mediating role of symbolic logics. For syntactically complete theories the connection between (informal) mathematics and its representation in symbolic logics is unproblematic and direct. That was taken

[24]Kleene, straddling Post's and Church's positions, wisely remarked in his [32], the classical *Introduction to metamathematics*, "While we cannot prove Church's thesis, since its role is to delimit precisely an hitherto vaguely conceived totality, we require evidence that it cannot conflict with the intuitive notion which it is supposed to complete; i.e. we require evidence that every particular function which our intuitive notion would authenticate as effectively calculable is general recursive. The thesis may be considered a hypothesis about the intuitive notion of effective calculability, or a mathematical definition of effective calculability; in the latter case, the evidence is required to give the theory based on the definition its intended significance." (Kleene [32], pp. 318–9).

for granted by Post in his [38] and was discussed in Part 1.[25] If all symbolic logics are incomplete, then the connection has to be anchored in some other way.

Having restated that the development in §§9–10 of *Anticipation* concerning undecidability and incompleteness "owes its significance entirely to the universal character of our characterization of an arbitrary generated set of sequences as given in §7", Post points to this new direction at the beginning of the Appendix to *Anticipation*. He disowns the idea that the considerations of §7, as described in Sect. 7.4, were intended as a proof-like argument. Instead, he claims famously:

> Establishing this universality is not a matter of mathematical proof, but of psychological analysis of the mental processes involved in combinatory mathematical processes.
> (*Anticipation*, p. 55)

What role such a psychological analysis might play is further clarified by a distinction Post makes in footnote 6 of *Anticipation*. There he separates "a formulation which includes an equivalent for every possible 'finite process' ", from "a description which will cover every possible method for setting up finite processes".[26]

The psychological analysis aims for a suitable description covering "every possible method for setting up finite processes"; that theme had already been alluded to at the end of §8:

> But for full generality a complete analysis would have to be made of all the possible ways in which the human mind could set up finite processes for generating sequences. The beginning of such an attempt will be found in the Appendix. (*Anticipation*, p. 48)

We will now point to the crucial stages of Post's attempt to arrive at *full generality*. The beginning of the needed *complete analysis* is described as follows:

> We begin here a *derivation* of the logic of *finite operations* and ultimately of all of the *logic of mathematics* from *first principles*. These principles are supposed to be a digest[27] of our experience of the logico-mathematical activity... (*Anticipation*, p. 56)

---

[25]In footnote 12 of *Anticipation*, Post asserts that "the bubble of symbolic logic as universal logical machine finally [has] burst" on account of the undecidability and incompleteness results; he adds, "Actually, the old dream of symbolic logic is finding partial realization in Tarski's recent work on decision problems."

[26]In the very footnote in which Post articulates this difference, he also asserts that the first goal has been achieved by the work in §7. The contributions to the proposed complete analysis, needed to achieve the second goal, are fragmentary. They are sometimes quite obscure and difficult to grasp, in particular, those related to the "analysis of proof" with the goal of finding an absolutely undecidable proposition. See *Anticipation* footnotes 4 and 6 as well as the remarks on the "process of proof" starting on page 59. Post writes, the limitations in man's mathematical powers "suggest that in the realms of proof ... a problem may be posed whose difficulties we can never overcome; that is that we may be able to find a definite proposition which can never be proved or disproved." (*Anticipation*, p. 56) Then he refers back to footnote 1 in which he describes, more expansively, a "fundamental problem", namely, the question of "the existence of absolutely undecidable propositions which in some a-priori fashion can be said to have a determined truth-value, and yet cannot be proved or disproved by any valid logic." (*Anticipation*, p. 1) That is, of course, in striking opposition to the rationalist optimism of Hilbert and Gödel that is beautifully expressed in (Gödel [25], p. 164).

[27]From Latin 'digesta' (n., pl.) meaning 'Matters methodically arranged'.

The logico-mathematical activity is, of course, an activity of the human mind "as situated in the universe", and its objects "may be anything in the universe". Its method "seems to be essentially that of *symbolization*". The use of language is for Post a central symbolizing activity:

> It may be noted that language, the essential means of human communication is just symbolization. (*Anticipation*, p. 57)

Leaving aside a detailed discussion of one feature of this activity Post considers as important (and we don't fully understand), namely, its self-consciousness, we point to one central effect through this fundamental remark of Post's:

> ... we shall here *not* consider the *original* objects which are symbolized, but only the relations and operations upon these resulting *symbols*... (*Anticipation*, p. 57)

It is essential "that these symbols enter into certain spatial relations". The "result of logical thought" is conceived of as a "spatial configuration of symbols". For the study at hand, "We are to regard our symbols as without properties except that of permanence, distinguishability and that of being part of certain symbol-complexes." (*Anticipation*, p. 57) Consequently, the core of the project is now "an analysis of these spatial relations..."

After a long and complex discussion of "the creative germ of the thinking process" and the nature of proof, Post returns on page 62 of *Anticipation* to the analysis of symbolisms in connection with finite processes. He presents a summary of the "method" for obtaining a description and indicates which of its elements, in addition to mere symbols, constitute the sought-after description:

> We return here to a more complete discussion and analysis of the very first part of the present research i.e., in connection with finite methods. We shall here generalize to finite methods for obtaining any results not just *test* for truth and falsity...[28]
>
> We shall here first give what is at least a first approximation to a definitive solution of finding a *natural normal* form for symbolic representation.
>
> There are three stages in the analysis we give. In the first stage we have the things symbolized. ...
>
> ......
>
> This then gives us our second stage in our analysis, namely a system of symbolizations for corresponding mathematical states. (*Anticipation*, p. 62)

The subsequent reflections are concerned with the symbolizations that are now *assumed* to be *finite* and *discrete* (and we will come back to them in the next Part.) As to the correspondence between symbolizations and mathematical states, Post asserts:

> Now the system of symbolizations in question is essentially to be a human product and each symbolization [is to be] a human way of describing the original mathematical state. (*Anticipation*, p. 63)

A discussion of the "third and last stage in this analysis" follows. The symbolizations "represent the original mathematical states" and, given the finiteness and discreteness

[28]Recall from Sect. 7.2 the non-semantic understanding of truth and falsity.

assumptions, they can be "completely described". Post finally concludes, "Hence these descriptions can be considered to represent or symbolize those mathematical states." (*Anticipation*, p. 63)

The above remarks are all taken from the main text of the Appendix, i.e., from Post's notes and diary from the 1920s. In footnote 120, the very last of the footnotes that comment on the early work and were written 15 to 20 years later, Post quotes a remark from the 1920s that makes explicit, how he thought of his work at that time: "The main outline of the work is completed and we really have a case of Filling In." Post then continues the footnote with this devastating judgment:

> Actually, but the surface of the problem was thus, perhaps, barely scratched, the problem, that is, of describing "all the finite processes of the human mind," at least in so far as they might concern the generalization of §7. (*Anticipation*, p. 67)

So it seems that the hoped-for description, "which will cover every possible method for setting up finite processes", had not been achieved in Post's own judgment. However, given the basic assumptions, he may very well be seen as having arrived at a formulation "which includes an equivalent for every possible 'finite process'"; that is indeed Post's considered judgment. (See footnote 26 above.) We will examine this issue in Sect. 7.7.

7.7 Symbolic Logics: Finitary Constraints

What Post did *not* argue for explicitly at all, except "dogmatically", is the correspondence between *mathematical states* and *finite and discrete symbol complexes*, with the latter in some sense *representing* the former. The nature of that representation is crucial for the claim that the undecidability and incompleteness results constitute a discovery of limitations of the *mathematizing* power of Homo Sapiens. After all, those results concern straightforwardly only symbolic logics in which mathematics can be developed. Post writes in footnote 12 of *Anticipation*, "Symbolic Logic may be said to be Mathematics become self-conscious." And the former is according to Post by necessity *finitary*.

The finitary feature of symbolic logics is crucial not only for Post but also for Church and logicians in general, when they insist on an *effective* concept of deduction. In *Anticipation* Post asserts:

> Where we say "symbolic logic" the tendency now is to say "finitary symbolic logic". However, it seems to the writer that logic should be considered essentially a human enterprise, and that when this is departed from, it is then incumbent on such a writer to add a qualifying "non-finitary". (*Anticipation*, fn. 10)

In his ([45], p. 288) Post leaves out the qualifying "finitary" when writing, "The assertions [theorems] of an arbitrary symbolic logic constitute a generated set A of what may be called symbol-complexes or formulas." He justifies the omission in a footnote referring to (Church [7]), where Church defends an *effectiveness criterion* that "necessarily applies" to inference steps or rules of procedure of *any* "formal system of logic":

> There is some current tendency to apply the name "logic" to schemes which are similar to accepted systems of logic, but involve one or more rules of procedure which lack this characteristic of effectiveness. Such schemes may perhaps be of interest as abstract definitions of classes of formulas, but they cannot in my opinion be called "logics" except by a drastic (and possibly misleading) change of the usual meaning of that word. For they do not provide an *applicable* criterion as to what constitutes a valid proof. (*Anticipation*, p. 225)

This remark not only reflects the normative requirement that logics should have an *applicable*, i.e., *decidable* proof predicate, but also Church's considerations in his classical 1936 paper [5]. There Church argued that all functions "computable in a logic" are recursive by his well known *step-by-step argument*; see (Gandy [21]) or (Sieg [49]). For Post this would be of interest, but certainly not conclusive, as the identification of *decidable* and *recursive* is taken for granted.

When discussing §7 of *Anticipation* in Sect. 7.4 above, we saw an attempt to give a convincing argument for a reducibility claim that is analogous to Church's: given *any* expansion of *Principia Mathematica* by "postulates and operations of the same general type" one can set up an equivalent normal system.[29] In his [44], Post formulates a broader claim. He writes that the methods developed for the reduction of systems of form C to normal systems led him to conclude,

> that not only *Principia Mathematica*, but any symbolic logic whose operations could effectively be reproduced in *Principia Mathematica*, and hence probably any (finitary) symbolic logic could be reduced to a system in canonical form, and consequently to a system in normal form. (Post [44], p. 215, fn. 18)

In his letter to Gödel we quoted earlier, Post states straightforwardly that *any* symbolic logic is reducible to normal form. How this can be achieved and given a philosophical grounding was indicated already in Part 5, when we examined his analysis of *symbolizations*. We quoted there an extended passage from *Anticipation* that ended with the sentence "This gives us our second stage in our analysis, namely a system of symbolizations for corresponding mathematical states." As we saw, symbolizations are taken to be spatial configurations and are assumed to be finite and discrete. They are constituted from parts that are unanalyzable in a given discussion, but have certain properties and stand in particular relations. They are, as we discussed, a "human

---

[29] Post's argument for this assertion resembles Gödel's for the *absoluteness* of the notion of computable functions in his [24, 26]. A similar argument for identifying the notion of calculable functions with recursiveness is found in Church's letter to Pepis from June 8, 1937, which is reprinted in the Appendix of (Sieg [49]). Each argument shows that, as long as broad informal conditions are satisfied, the extensions of particular kinds don't allow for more computations than the restricted frameworks. Considerations of the same kind are found in Supplement II of Hilbert & Bernays's *Grundlagen der Mathematik II* as well as in (Turing [56], Sect. 9, II).—This notion of "absoluteness", obviously quite different from Post's, is discussed in (Sieg [51], 572–7). We should point out that Post's argument suffers from the same kind of subtle circularity as Gödel's and Church's, because it is required that the extensions have to have postulates and rules of the same general form as those of *Principia Mathematica*.

product" and a "human way of describing the original mathematical state[s]". Thus, Post asserts, the need for the following assumption is readily seen, namely,

> that the number of these elementary properties and relations used is finite and that there is a certain specific finite number of elements in each relation. (*Anticipation*, p. 63)

This is the basis for the third and last stage of Post's analysis, namely, the complete "finite and normal" description of the symbol complexes.

Corresponding to the three stages in the analysis of symbolizations, Post considers three stages in the analysis of *methods*. The first and rather inchoate stage consists of "any and **all** methods" [emphasis in the text]. "To allow for the most wonderful creations", Post elaborates, "my image of such methods involved dark clouds pierced by flashes of lightning accompanied by rolling thunder." Some of these methods are symbolized at the second stage, and their application results in a finite sequence of symbol-complexes "due to discreteness and finiteness". The third stage then consists in "reducing the method of passing from symbol complex" to an operation of "normal type". Such normal operations apply to the complete description of symbol complexes obtained at the third stage of the analysis of symbolizations. The underlying process is *iteration* and is viewed as mechanical, "merely machine like" (cf. also footnote 87).

The philosophical grounding of these third stages of Post's analyses is in both cases given through only vaguely conceived conscious experience of mathematical states, respectively in similarly unstructured methods; it is obviously problematic and, in the end, quite dogmatic. The dogmatic aspect relates in particular to consciousness, concerning which Post simply remarks in (*Anticipation*, p. 65): "…what we are *conscious of* is not mathematics, but a symbolization of it …" [emphasis in the text]. Recall that symbolizations are assumed to be finite and discrete—on account of the fact that their system is essentially a "human product" and "each symbolization a human way of describing the original mathematical state". Is it in this sense that Post made the remark we chose as the motto for our paper "…I study Mathematics as a product of the human mind and not as absolute…"?—In any event, the presumed grounding of finiteness and discreteness assumptions in features of the human mind or consciousness is rather ineffectual. And yet, only such a grounding, together with connections to mathematical states, would give support for the sweeping claims of *absolute unsolvability* and, as a consequence, of the *inescapable limitation* of human mathematical thinking.

Let us try now to turn Post's analysis on its head to see the similarity and radical difference to Turing's argument for his thesis. When reflecting on the reduction of the logic of finite operations to finite and discrete symbol complexes and operations of a normal type, Post articulates most clearly, that "only elementary recognition seems to be needed for the logical aspect of the *operational description* of mathematics". The nature of such elementary recognition is indicated, for example, by Post's remarks concerning judgments about symbol complexes: "a single undivided act of judgment" is required for recognizing the crucial properties of and relationships between parts of a symbol complex (*Anticipation*, p. 63). So, if we disregard the connection to mathematical states and focus purely on symbolizations (without meaning), then the requirement to appeal only to elementary recognition would seem to justify the finiteness and discreteness assumptions for symbol complexes—quite in Turing's way; the restricted mechanical, machine-like character of operations is taken for granted anyway. In this way Post's analysis is cut off not only from stage one (as Post himself does), but also from stage two. When Post's views are articulated in this way, his *reversed argument* that starts with (the demand for) elementary recognition and ends with finiteness conditions is strikingly similar to that Turing gives in 1936 in his [56] for the claim that the mechanical operations of a human computer can be carried out by one of his machines.[30]

Relating Turing's analytic work to his own, Post notes the parallelism, but retains the connection of the last two stages and remarks:

> Fundamental is the distinction between the static outer symbol-space with its assumed capacity of unbounded complexity, and the dynamic mental world with, however, its obvious limitations. This has been fully captured by Turing in his finite number of mental states hypothesis. (Anticipation, fn. 118)

States of mind of human computers do play a critical role for Turing's analysis in Part I of Sect. 9 of his paper. In analogy to the finiteness constraint on immediately recognizable sequences, Turing asserts also that human computers have only a finite number of states of mind. However, Turing replaces states of mind by "a more physical and definite counterpart" (in Part III of the very same section):

> It is always possible for the computer to break off from his work, to go away and forget all about it, and later come back and go on with it. If he does this he must leave a note of instructions (written in some standard form) explaining how the work is to be continued. This note is the counterpart to "state of mind". (Turing [56], p. 253)

Thus, it is not clear in what sense Turing's final analysis uses a "finite number of mental states hypothesis"; it is also not clear at all in what sense it would capture fully "the dynamic mental world ... with its obvious limitations".[31] Turing, as is clear from the last quote, describes mechanical procedures that are completely external to the computing agent; these externalized computations are the objects of his analysis, not mental states or the human mind. Some cognitive capacities are of course taken for granted, but for his restrictive analysis Turing appeals to only one obviously psychological fact, namely, the limitation of the human sensory apparatus; it and only it provides the basis for the fundamental finiteness and locality conditions.

---

[30]Turing's argument was analyzed in (Sieg [48]); it is put into a broader systematic framework in (Sieg [51]).

[31]Turing is discussed in notes 6, 9, 112, 118, and 120 of *Anticipation*. Post most strongly emphasizes the role of the "finite number of mental states hypothesis". However, why would its correctness make Post's position (as he remarks in note 9) "largely academic"? And, why would it make (as he says in note 6) "the detailed development envisioned in the Appendix unnecessary"? - Post is grappling with a different problem; in addition, it is not just the *number* of mental states that is important for Post's considerations, but also their internal elementary, discrete *structure*. (Wider formulations, as stressed in [41], are to achieve greater *psychological fidelity*.).

These considerations are directly and beautifully mirrored in Post's description of Turing machines and their computations in his [46] from 1947.[32] His mathematically precise description via a form of canonical system is used to prove the undecidability of the word problem for semi-groups by reducing it to the halting problem for Turing machines. Turing was impressed by the result and the method of proof; in his [58] he extended the undecidability result to semi-groups with cancelation. The discussion of the methodological and mathematical issues surrounding computability in his [59] does not mention machines; the issues are rather articulated exclusively in terms of puzzles and substitution puzzles, i.e., particular canonical systems.[33] That naturally underlines the conceptual confluence in their work: it is also, it seems to us, an expression of deep appreciation. The latter is, of course, also expressed in the abstract of (Turing [58]), when he describes the word problem as being reduced to an unsolvable problem that is "connected with the *logical computing machines introduced by Post and the author*", referring to (Post [41]) from 1936 and his own classical paper from the same year.

7.8 Yes, No, and Broader Contexts

Post's *generalization* asserts: *sequences generated (by a finite and mechanical process) can be generated by a normal system.* In Sect. 7.4 we discussed some of Post's reasons for accepting this assertion. They include an absoluteness argument that is similar to Gödel's; see footnotes 13 and 29. In his 1936 paper, Post emphasized the significance of proofs of the equivalence between different concepts. He claimed in (Post [45], p. 285) that there is an "overwhelming evidence" for the coextensiveness of the concept of recursive function "with the intuitive concept effectively calculable function". As to the substance of the *overwhelming evidence*, Post referred to footnote 2 of (Kleene [31]) where Kleene summarized *the* considerations in favor of coextensiveness.[34] All of this fits perfectly well into the intellectual patterns of

[32]Note in particular that the internal configurations of machines or m-configurations, which correspond to the states of mind of the human computer, are directly incorporated into the symbolic configurations on which the (Post) production rules operate.

[33]See (Sieg [52]) for the discussion of this variant of Turing's Thesis introducing a normal forms for puzzles; see also footnote 23.

[34]For the reader's convenience, we quote the essential points of Kleene's footnote, which is attached to the statement, "A function of natural numbers, with natural numbers as values, is taken to be effective if it is Herbrand-Gödel recursive." Here is the (partial) quote: "...This notion of effectiveness appears, on the following evidence, to be general. A variety of particular effective functions and classes of effective functions (selected with the intention of exhausting known types) have been found to be recursive. Two other notions, with the same heuristic property, have been proved to be equivalent to the present one, viz., Church-Kleene λ-definability and Turing computability. Turing's formulation comprises the functions computable by machines. ... Functions determined by algorithms and by the derivation in symbolic logics of equations giving their values (provided the individual steps have an effectiveness property which may be expressed in terms of recursiveness) are recursive.".

the early developments of computability theory and receives further strong support from the conceptual confluence of Post's own work with Turing's. In this sense then, *Post did have Turing's Thesis* and accepted the standard arguments for it.[35]

However, in these considerations it is taken for granted that a notion of mechanical, algorithmic procedure is being analyzed. Gödel emphasized in his [26] that *finite procedure*—in the context of undecidability and incompleteness results—must be understood as *mechanical procedure*. "This meaning, however," he asserted, "is required by the concept of formal system, whose essence it is that reasoning is completely replaced by mechanical operations on formulas." He added parenthetically:

> Note that the question of whether there exist finite *non-mechanical* procedures, not equivalent with any algorithm, has nothing whatsoever to do with the adequacy of the definition of "formal system" and of "mechanical procedure". (Gödel [26], p. 72)

Post's project had greater ambitions, arguing for the *absolute* unsolvability of combinatorial problems and concluding that the mathematical powers of Homo Sapiens are *inescapably* limited.

For Post to reach his ambitious goal, he needed a prima facie stronger generalization: *sequences generated (by a finite process of the human mind) can be generated by a normal system*. In addition, in order to obtain the connection to mathematics, the representation of mathematical states by symbol systems is required, as we discussed extensively in Sect. 7.6. That Gordian Knot is tied in a single step by appealing to a feature of the human mind that is epitomized by the remark "...what we are *conscious of* is not mathematics, but a formalization of it..." When drawing the conclusions concerning the limitations of the human mind, Post refers to "mental processes involved in combinatory mathematical processes". (*Anticipation*, p. 55) On account of the structure of our mind, then, the combinatory mathematical processes involve for humans only finite mental processes that ultimately are captured by mechanical ones. Thus, a thesis or rather a sequence of theses has been formulated that is of a dramatically different character from Turing's. And in that sense, *Post did not have Turing's Thesis*, but rather a stronger "mental Thesis". It supports, in particular, the claim made in *Anticipation*, "The fundamental new thing is that for the combinatory problems the given set of instruments is in effect the only humanly possible one." This "mental Thesis" is subject to precisely the criticism that was incorrectly raised against Turing, when Gödel in his [27] tried to pinpoint a "philosophical error in Turing's work"; see (Sieg [53]).

Why did Post focus attention on finite processes of the human mind and insist on their *psychological* analysis? We saw in Sect. 7.7 that finite processes have to be considered because of the finitary character of symbolic logics; and that is necessitated in turn by the fact that they are a *human product*. The psychological analysis is to anchor the human way of doing logic and mathematics in features of the mind.[36]

---

[35]We contended in Sect. 7.7, turning "Post's analysis on its head", that one can extract from his analysis, when stripped of philosophical preconceptions and reversed, an argument that is strikingly similar to Turing's.

[36]This has, however, significant repercussions on the very mathematical powers: they are limited, but also creative (in the restrictive sense we pointed out at the end of Sect. 7.5). That was made clear

A related explanation could point to the influence of Post's advisor Keyser. In his book *Mathematical Philosophy*, Keyser has a chapter devoted to the psychology of mathematics.[37] Mathematics is not only viewed as "an enterprise of the mind", but it is also claimed, "mathematical phenomena *represent* mental phenomena" (Keyser [30], p. 412; emphasis in the text). If we take, as Post certainly does, the structure and development of symbolic logics as mathematical phenomena, then the latter can be taken, following Keyser, as representing mental phenomena.

There might be a second point of influence, as Keyser ascribes great importance to the *phenomenon of generalization*, i.e., "the process by which the human mind from time to time enlarges the empire of its rational activity". (Keyser [30], p. 413) It is, after all, distinctive that Post calls his version of the identification claim *generalization*. Keyser discusses the generalization of the number concept as a significant example; indeed, he considers it to be one of "the probably best [specimens of generalization] to be found in the history of thought". Through a succession of generalizations the domain of the number concept was extended from at first containing just the integers to "embracing", as he puts it, "positives and negatives, rationals and irrationals, reals and imaginaries, cardinals and ordinals, including the transfinite numbers of Georg Cantor". (Keyser [30], p. 413) Not to allow suitable generalizations, according to Keyser, hampers progress in mathematics; he attributes the delay, by two thousand years, of "the advent of the concept of hyperspace and n-dimensional geometry" to a *backward psychology of mathematics*. (Keyser [30], p. 407) Relating these observations back to Post, it was "the m-dimensional space analogy" that led him to introduce in Sect. 12 of his [38] a generalization of the classification of functions; generalization is also the broad theme of two of Post's mathematical papers, his [40, 42].

Generalization and the attendant conceptual innovation were important to Post also in his [45], where he starts out with the observation, "Recent developments of symbolic logic have considerable importance for mathematics both with respect to its philosophy and practice." The undecidability and incompleteness results support claims in the *philosophy of mathematics*, as they lead "to far-reaching conclusions on the nature of logical activity, and hence of mathematics". (*Anticipation*, p. 48) The concept of recursive function or its proved equivalents is viewed as significant also for the *practice of mathematics*: Post wants to demonstrate "that this concept [of recursive function] admits development into a mathematical theory much as the group concept has been developed into a theory of groups". He remarks, however,

(Footnote 36 continued)
by Post, when asserting, "mathematicians are better than machines", as they can prove theorems machines cannot. Such a proof requires not only an argument outside the system, but for the creative extension of the system it also requires, that the extending statement be recognized as true. See *Anticipation*, footnotes 12 and 100.

[37]Post finished his graduate education in 1920; Keyser's book was published only in 1922. Yet it is most likely that Post was familiar with Keyser's views expressed in the book. In the Preface (page vii) Keyser mentions that the book is the result of more than forty years of reflection on the nature of mathematics.

that only "a very limited portion of a sub-theory of the hoped-for general theory" is being developed in his own paper.

We can of course ask, what such a general theory *in analogy* to group theory might be. The development in Post's [45] cannot be viewed as being analogous to group theory, as there is no abstract concept of computability under which particular notions of effectiveness fall: many different notions can be shown to be *equivalent*, but they don't share the defining characteristics of such a general concept. What one might be looking for is perhaps best exemplified by the unifying, axiomatic treatment of different conceptions of "real numbers" that had been introduced in the second half of the 19th century, e.g., Dedekind cuts, Cauchy sequences, or Cantor fundamental sequences. In this case the relevant abstract concept is that of a *complete ordered field*; all the mentioned examples fall under it. In addition, if we take Dedekind cuts as our canonical reals, then every other complete ordered field is isomorphic to the system of cuts. (Thus, they are all isomorphic.)[38]

If one were to pursue a structural-axiomatic approach for computability, one would try to find an abstract concept under which the various models of computation fall. In analogy to proving that all complete ordered fields are isomorphic to the system of Dedekind cuts, one would try to prove in the computability case a *representation theorem* stating, the models falling under the abstract concept are all reducible to Turing machines. That approach has actually been pursued, see (Sieg [50, 51]); the abstract concept is that of a *computable (discrete) dynamical system*. The approach is thoroughly motivated by the confluent work of Post and Turing and their attempts to consider "wider formulations", i.e., configurations and local operations on them that are more general than those permitted in canonical systems or substitution puzzles. Their mathematical work and their methodological reflections have inspired the definition of a *computable dynamical system* and the proof that this abstract concept is equivalent to that of a Turing machine.

The general point is then this: the characteristic conditions of computable dynamical systems articulate minimal abstract conditions that a combination of *finite configurations* and *mechanical operations* have to satisfy to still count as a "wider formulation" or as a "puzzle". Thus, we don't have to face a mysterious thesis for the concept of computability; we rather have to face the ordinary and very hard issues for judging the adequacy of a mathematical concept to capture aspects of our intellectual or physical experience.

Acknowledgements We thank Martin Davis, Liesbeth De Mol, and Ulrik Buchholtz for encouraging, but also critical remarks. Liesbeth's detailed comments prompted real changes in our presentation. We are also very grateful to Alberto Policriti and Eugenio Omodeo, who as editors of this volume were extremely patient with our difficulties completing this essay.

[38]The underlying distinction between structural and formal axiomatics is discussed in Sieg [54].

References

1. Awodey, Steve and Reck, Erich. 2002. "Completeness and Categoricity, Part II: 20th Century Metalogic to 21st Century Semantics." *History and Philosophy of Logic* 23, no. 2: 77–94.
2. Bernays, Paul. 1918. *Beiträge zur axiomatischen Behandlung des Logik-Kalküls*. Habilitationsschrift. Göttingen. Reprinted in [20], 222–68.
3. Bernays, Paul. 1926. "Axiomatische Untersuchung des Aussagen-Kalküls der 'Principia Mathematica'." *Mathematische Zeitschrift* 25, 305–320.
4. Church, Alonzo. 1935. "An Unsolvable Problem of Elementary Number Theory. Preliminary Report (Abstract)." *Bulletin of the American Mathematical Society* 41: 332–33.
5. Church, Alonzo. 1936. "An Unsolvable Problem of Elementary Number Theory." *The American Journal of Mathematics* 58, no. 2: 345–63.
6. Church, Alonzo. 1937. "Review of (Post, 1936)." *The Journal of Symbolic Logic* 2, no. 1: 43.
7. Church, Alonzo. 1938. "The Constructive Second Number Class." *Bulletin of the American Mathematical Society* 44, 224–32.
8. Davis, Martin. 1950. *On the Theory of Recursive Unsolvability*. Dissertation, Princeton University.
9. Davis, Martin. 1958. *Computability and Unsolvability*. New York: McGraw-Hill.
10. Davis, Martin. 1982. "Why Gödel Didn't Have Church's Thesis." *Information and Control* 54, no. 1–2: 3–24.
11. Davis, Martin. 1994. "Emil L. Post: His Life and Work." In [14], xi–xxviii.
12. Davis, Martin, ed. 1965. *The Undecidable*. Hewlett, New York: Raven Press. Reprinted by Dover Publications, Mineola, N.Y., 2004.
13. Davis, Martin. 1994. *Solvability, Provability, Definability: The Collected Works of Emil L. Post*. Basel: Birkhäuser.
14. Davis, Martin and Wilfried Sieg. 2016. "Conceptual Confluence in 1936: Post & Turing." In *Turing Centenary Volume* edited by Thomas Strahm and Giovanni Sommaruga. Basel: Birkhäuser, 3–27.
15. Dedekind, Richard. 1888. *Was sind und was sollen die Zahlen?* Braunschweig: Vieweg.
16. De Mol, Liesbeth. 2006. "Closing the Circle. An Analysis of Emil Post's Early Work." *The Bulletin of Symbolic Logic* 12, no. 2: 267–289.
17. De Mol, Liesbeth. 2011. "On the complex behaviour of simple tag systems. An experimental Approach." *Theoretical Computer Science* 412, no. 1–2: 97–112.
18. De Mol, Liesbeth. 2013. "Generating, Solving and the Mathematics of Homo Sapiens. Emil Post's Views on Computation." In *A Computable Universe – Understanding and Exploring Nature as Computation*, edited by Hector Zenil. New Jersey: World Scientific, 45–62.
19. Ewald, William and Wilfried Sieg, eds. 2013. *David Hilbert's Lectures on the Foundations of Arithmetic and Logic, 1917–1933*. Heidelberg: Springer.
20. Feferman, Solomon. 1962. "Transfinite recursive progressions of axiomatic theories." *The Journal of Symbolic Logic 27*: 259–316.
21. Gandy, Robin. 1988. "The confluence of ideas in 1936." In *The Universal Turing Machine – A Half-century Survey*, edited by Rolf Herken. Oxford: Oxford University Press, 55–111.
22. Gödel, Kurt. 1931. "On Formally Undecidable Propositions of Principia Mathematica and Related Systems I." Translated by Elliott Mendelson. In [13], 5–38.
23. Gödel, Kurt. 1934. "On Undecidable Propositions of Formal Mathematical Systems." In [13], 41–71.
24. Gödel, Kurt. 1936. "On the Length of Proofs." In [13], 82–3.
25. Gödel, Kurt. 193?. "[Undecidable Diophantine Propositions]." In *Kurt Gödel. Collected Works*, edited by Solomon Feferman, et. al. Vol. 3. Oxford: Oxford University Press, 164–75.
26. Gödel, Kurt. 1964. "Postscriptum to (Gödel, 1934)." In [13], 71–3.
27. Gödel, Kurt. 1972. "Some remarks on the undecidability results." In *Kurt Gödel. Collected Works*, edited by Solomon Feferman, et. al. Vol. 2. Oxford: Oxford University Press, 305–6.
28. Gödel, Kurt. 2003. "Correspondence with Emil Post." In *Kurt Gödel. Collected Works*, edited by Solomon Feferman, et. al. Vol. 5. Oxford: Oxford University Press, 169–74.

29. Keyser, Cassius. 1912. "Principia Mathematica." *Science* ser. II, vol. 35, no. 890: 106–10.
30. Keyser, Cassius. 1922. *Mathematical philosophy, a study of fate and freedom; lectures for educated laymen*. New York: E. P. Dutton & Company.
31. Kleene, Stephen Cole. 1938. "On Notation for Ordinal Numbers." *Journal of Symbolic Logic* 3, no. 4: 150–55.
32. Kleene, Stephen Cole. 1952. *Introduction to Metamathematics*. Groningen: Elsevier.
33. Kleene, Stephen Cole. 1981. "Origins of Recursive Function Theory." *Annals of the History of Computing* 3, no. 1: 52–67.
34. Lewis, Clarence Irving. 1918. *A Survey of Symbolic Logic*. Berkeley: University of California Press.
35. Łukasiewicz, Jan. 1918. "Farewell Lecture." In *Selected Works of Jan Łukasiewicz*, edited by Ludwik Borkowski. Amsterdam: North-Holland Publishing Company. 1970. 84–6.
36. Łukasiewicz. 1920. "On Three-Valued Logic." In *Selected Works of Jan Łukasiewicz*, edited by Ludwik Borkowski. Amsterdam: North-Holland Publishing Company. 1970. 87–8.
37. Minsky, Marvin. 1967. *Computation: Finite and Infinite Machines*. London: Prentice-Hall.
38. Post, Emil. 1921. "Introduction to a General Theory of Elementary Propositions." *American Journal of Mathematics* 43, no. 3: 163–85.
39. Post, Emil. 1921. "On a Simple Class of Deductive Systems. (Abstract)" In *Solvability, Provability, Definability: The Collected Works of Emil L. Post*. Basel: Birkhäuser. 545.
40. Post, Emil. 1930. "Generalized differentiation." *Transactions of the American Mathematical Society* 32: 723–81.
41. Post, Emil. 1936. "Finite Combinatory Processes – Formulation 1." *The Journal of Symbolic Logic* 1, no. 3: 103–5.
42. Post, Emil. 1940. "Polyadic groups." *Transactions of the American Mathematical Society* 48: 208–350.
43. Post, Emil. 1941. *The Two-Valued Iterative Systems of Mathematical Logic*. Princeton. Princeton University Press.
44. Post, Emil. 1943. "Formal Reductions of the General Combinatorial Decision Problem." *American Journal of Mathematics* 65, no. 2: 197–215.
45. Post, Emil. 1944. "Recursively Enumerable Sets of Positive Integers and Their Decision Problems." *American Mathematical Society* 50, no. 5: 284–316.
46. Post, Emil. 1947. "Recursive unsolvability of a problem of Thue." *Journal of Symbolic Logic* 12: 1–11.
47. Post, Emil. 1965. "Absolutely Unsolvable Problems and Relatively Undecidable Propositions - Account of an Anticipation." In [12], 375–441.
48. Sieg, Wilfried. 1994. "Mechanical procedures and mathematical experience." In *Mathematics and Mind*, edited by Alexander George. Oxford: Oxford University Press, 71–117.
49. Sieg, Wilfried. 1997. "Step by Recursive Step: Church's Analysis of Effective Calculability". *The Bulletin of Symbolic Logic* 3, no. 2: 154–180. Reprinted in 2014. (with a Postscriptum) in *Turing's Legacy: Developments from Turing's ideas in logic*, edited by Rod Downey. Cambridge: Cambridge University Press, 434–66.
50. Sieg, Wilfried. 2002. "Calculations by man and machine: conceptual analysis." In *Reflections on the foundations of mathematics*, edited by Wilfried Sieg, Richard Sommer, and Carolyn Talcott. Natick: A. K. Peters, 390–409.
51. Sieg, Wilfried. 2009. "On Computability." In *Philosophy of Mathematics (Handbook of the Philosophy of Science)* edited by Andrew Irvine. Amsterdam: North-Holland Publishing Company, 535–630.
52. Sieg, Wilfried. 2012. "Normal forms for puzzles: A variant of Turing's Thesis." In *Alan Turing – His work and Impact*, edited by S. Barry Cooper and Jan van Leeuwen. Amsterdam: Elsevier, 332–8.
53. Sieg, Wilfried. 2013. "Gödel's Philosophical Challenge (to Turing)." In *Computability*, edited by Jack Copeland, Carl Posy, and Oron Shagrir. Cambridge: MIT Press, 183–202.
54. Sieg, Wilfried. 2014. "The ways of Hilbert's axiomatics: structural and formal." *Perspectives on Science* 22, no. 1: 133–57.

55. Szabó, Máté. 2014. *Post and Kalmár on Turing and Church.* MS Thesis, Carnegie Mellon University, Department of Philosophy.
56. Turing, Alan. 1936. "On computable numbers with an application to the Entscheidungsproblem." *Proceedings of the London Mathematical Society*, 2, vol. 42: 230–65.
57. Turing, Alan. 1939. "Systems of logic based on ordinals." *Proceedings of the London Mathematical Society*, ser. 2, 45: 161–228.
58. Turing, Alan. 1950. "The word problem in semi-groups with cancellation." *The Annals of Mathematics*, Ser. 2, Vol. 52, no. 2: 491–505.
59. Turing, Alan. 1954. "Solvable and unsolvable problems." *Science News* 31: 7–23.
60. Urquhart, Alasdair. 2009. "Emil Post." In *Handbook of the History of Logic. Logic from Russell to Church*, edited by Dov M. Gabbay and John Woods. Vol. 5. Amsterdam: North–Holland Publishing Company. 429–78.
61. Whitehead, Alfred and Bertrand Russell. 1910. *Principia Mathematica.* Vol. 1. Cambridge: Cambridge University Press.

Chapter 8
On Quantum Computation, Anyons, and Categories

Andreas Blass and Yuri Gurevich

Abstract We explain the use of category theory in describing certain sorts of anyons. Yoneda's lemma leads to a simplification of that description. For the particular case of Fibonacci anyons, we also exhibit some calculations that seem to be known to the experts but not explicit in the literature.

Keywords Anyon model · Fibonacci anyons · Quantum computing · Categories · Yoneda Lemma · Mathematical foundations

8.1 Introduction

This paper attempts to explain the use of category theory in describing certain sorts of *anyons*. These are rather mysterious physical phenomena which, one hopes, will provide a basis for quantum computing needing far less error correction than other approaches.

The first author of this paper has long been a fan of category theory; even as a graduate student, he was described by one of his professors as "functorized". The second author has been far more skeptical about the value of category theory in computer science, because of its distance from applications and because of the peril of potential (and in some cases actual) over-abstraction. In 2012, both authors began working with the Quantum Architectures and Computing (QuArC) Group at Microsoft Research and found anyons to be near the top of the group's agenda. Seeing calculations and applications that use unitary matrices to represent braiding of anyons, we naturally wondered what Hilbert space these matrices are intended to operate on. We made rather a nuisance of ourselves by asking different people, on

---

A. Blass (✉)
Mathematics Department, University of Michigan, Ann Arbor, MI 48109–1043, USA
e-mail: ablass@umich.edu

Y. Gurevich
Microsoft Research, One Microsoft Way, Redmond, WA 98052, USA
e-mail: gurevich@microsoft.com

E.G. Omodeo and A. Policriti (eds.), *Martin Davis on Computability, Computational Logic, and Mathematical Foundations*,
Outstanding Contributions to Logic 10, DOI 10.1007/978-3-319-41842-1_8

different occasions, what anyons actually are, from a mathematical point of view. Are they Hilbert spaces? Are they vectors in a Hilbert space? Are they something else? It turned out that the only mathematically sound answer in the literature involved a special sort of categories, *modular tensor categories*.[1] So the second author agreed that categories can be quite relevant to important applications in computer science.

Our purpose in this paper is to describe some of the ideas surrounding categories and anyons in general and the special case of Fibonacci anyons and their category description. We hope that our presentation will be accessible and useful for mathematicians and computer scientists who have some acquaintance with the basics of category theory. Where we need to go beyond the basics, we explain, albeit briefly, the concepts from category theory that we use. We have also included a section describing the physical background that this mathematics is intended to formalize.

To describe more of our motivation for studying anyons, we need to presuppose some general information that will be explained in later sections of this paper. In particular, we shall refer to the fusion rule $\tau \otimes \tau = \tau \oplus 1$ for Fibonacci anyons τ (and the vacuum 1). We hope that the following paragraphs will give the reader a rough idea of what we are looking at, and that re-reading them after the rest of the paper will provide a less rough idea.

In contrast to what occurs elsewhere in quantum theory, the states (represented, as usual, by vectors in Hilbert spaces, up to scalar multiples) in the modular tensor category picture are ways in which one configuration can fuse to form another configuration.[2] They are not the configurations themselves. For example, in the Fibonacci case, there is a 2-dimensional Hilbert space of ways for three anyons to be regarded as (or to fuse into) one anyon; this is the Hilbert space $\mathrm{Hom}(\tau \otimes \tau \otimes \tau, \tau)$.

When we first heard about Fibonacci anyons, we thought that the fusion rule $\tau \otimes \tau = \tau \oplus 1$ meant that, if we put two τ anyons together, then the result might look like one τ anyon or like the vacuum (this much is true in the modular tensor category model) and that the general result would be a superposition of these two alternatives. But the model doesn't allow such superpositions. Nor does the model say anything about the probabilities of the two possible outcomes.

Instead, we get superpositions of the following sort. Start with three τ's. Fuse the first two to get one τ or vacuum. If you got vacuum, then the overall result is one τ, namely the third of the original ones, which you haven't yet fused. If, on the other hand, fusing the first two τ's gives a τ, then fusing that with the third τ might produce a τ. (It might also produce vacuum, but that's irrelevant for the present discussion.) So we have two ways to end up with one τ, according to whether the first two τ's fused to vacuum or to τ. And it is these two ways that the model allows superpositions of. Another possibility for getting two ways here is to fuse the last two τ's first and then fuse the result with the first τ. These two form another basis of the same 2-dimensional Hilbert space of "ways". The relation between the two bases is

[1] Other answers explained the physics, in terms of excitations, but these matters are not the subject of this paper, which is specifically about mathematics except for the introductory material summarized in Sect. 8.2.

[2] For more on the notion of fusion, see Remark 1 at the end of this introduction.

(part of) the associativity isomorphism of the modular tensor category. Yet another possibility would begin by fusing the first and third τ's. The modular tensor category representation of this possibility would use a braiding isomorphism to move the first anyon to be adjacent to the third (or vice versa), and it would depend on the path along which that anyon is moved around the second one.

In Sect. 8.2, we give a general introduction to anyons from the point of view of physics and quantum computation. That section is intended to give the reader a rough idea of what anyons are and why researchers in quantum computation would be interested in them. The treatment here is quite superficial, and we give references for more detailed treatments.

In Sect. 8.3, we gradually introduce modular tensor categories, and we explain how they are intended to be used to describe anyons. This section borrows heavily from the axiomatization given in [9], but with some modifications and rearrangements.

Section 8.4 is devoted to an application of one of the central theorems of category theory, known as Yoneda's Lemma, to producing a simplified view of modular tensor categories.

Finally, in Sect. 8.5, we consider the special case of Fibonacci anyons. This special case is unusually simple in some respects. Nevertheless (or perhaps therefore) it occupies a prominent place in quantum computing research. Section 8.5 begins with a general description of Fibonacci anyons and then exhibits some calculations, whose results seem to be well known to some in the quantum computing community but which we have not been able to find written down in the literature.

More detailed treatments of modular tensor categories are available in the papers [9] of Panangaden and Paquette and [11] of Wang. Much of our exposition is based on the former. For other aspects of anyons and topological quantum computation, see, for example, [5] and the references there.

Remark 1 We encountered numerous explanations of the notion of *fusion* of anyons, and they seemed to contradict each other. At one extreme was the picture of fusion as a physical process in which anyons are brought into spatial proximity with each other and energy is released as they form a new anyon (or perhaps annihilate each other). A minor modification of this picture is that energy need not be released; it might actually be consumed in the process. Another picture, however, did not insist that the anyons be brought together. They could remain far apart, and a suitable global measurement of the system's quantum numbers could reveal how they "fused". A path to reconciling these apparently contradictory pictures is suggested by a comment at the end of Sect. II.A of [8]; the idea is as follows. Consider several anyons, which we intend to fuse. As long as they are far apart, the various possible results of their fusion have energies that are very close together. (In technical terms, the ground state of the system is very nearly degenerate.) So the different fusion results can be distinguished in principle but not practically. When the anyons are brought closer together, though, the energy differences between the fusion possibilities become larger, and so it becomes practical to distinguish these possibilities. Thus, the discrepancy between various views of fusion seems to be largely a discrepancy between what can be observed in principle (or what is "really" happening) and what can be detected in practice.

8.2 Quantum Theory and Anyons

This section is a superficial summary of a small part of quantum theory and some basic information about anyons. The physics described here is intended merely to provide an orientation for understanding the mathematics in the rest of the paper.

8.2.1 Quantum Mechanics

In quantum theory, the state of a physical system is typically represented by a non-zero vector in a complex Hilbert space $\mathcal{H}$, but all non-zero scalar multiples of a vector represent the same state. Thus, the states constitute the projective space associated to $\mathcal{H}$. Because of the freedom to adjust scalar factors, one often imposes the normalization that the vectors representing a state should have norm 1; there still remains a freedom to adjust the phase, i.e., a scalar factor of absolute value 1.

If a system has an observable property with infinitely many possible values, for example position or momentum, then the Hilbert space of its states must be infinite-dimensional. In quantum computing, however, one usually ignores many such properties and concentrates on only a small number (often only one) of properties with only finitely many possible values. As a result, one deals with finite-dimensional Hilbert spaces. (This simplification is analogous to modeling a classical computer by a configuration of bits, not taking account of its other physical properties, like position or momentum or temperature, unless these threaten to interfere with the bits of interest.)

The automorphisms of a Hilbert space $\mathcal{H}$ are the *unitary* transformations, i.e., the linear bijections that preserve the inner product structure. These play several important roles, both in physics and in quantum computation. First, they provide the dynamics of isolated quantum systems. That is, the state of an isolated system will evolve in time by the action of a one-parameter group (the parameter being time) of unitary operators.[3] Second, if a system has symmetries, i.e., if it is invariant under some transformations, then these transformations are usually modeled by unitary operators.[4] Finally, the design of quantum algorithms is based on unitary operators. We want the system to evolve from a state that we know how to produce to another state from which we can extract useful information by a measurement. That evolution is described by a unitary operator. So an algorithm designer wants to find unitary operators that represent a useful evolution of a state. In addition to finding such operators, we want to represent them as compositions of simpler ones, called *gates*, that we know how to implement.

---

[3]Here we use the so-called Schrödinger picture of quantum mechanics. A physically equivalent alternative view, the Heisenberg picture, has the states remaining constant in time, while the operators modeling properties of the state evolve by conjugation with a one-parameter group of unitary operators.

[4]A few discrete symmetries can be modeled by anti-unitary transformations.

Where classical computation uses bits, whose possible values are denoted by 0 and 1, quantum computation uses *qubits*. A measurement of a qubit produces two possible values; the qubit itself is represented by a 2-dimensional Hilbert space, in which a certain orthonormal basis, usually written $\{|0\rangle, |1\rangle\}$, corresponds to the two values. In contrast to the classical case, though, the Hilbert space structure provides many other states in addition to these two basic ones. Any non-zero linear combination of $|0\rangle$ and $|1\rangle$ represents a possible state of the system. If the state is represented by the unit vector $x|0\rangle + y|1\rangle$, then measuring the qubit in the $\{|0\rangle, |1\rangle\}$ basis will produce the outcome 0 with probability $|x|^2$ and the outcome 1 with probability $|y|^2$. Such a state is a *superposition* of the two basic states. More precisely, this state vector is the superposition, with coefficients x and y, of the vectors $|0\rangle$ and $|1\rangle$, respectively.

It is more accurate to speak of superposition of vectors than of superposition of states. The reason is that, although phase factors don't affect the state represented by a vector, *relative* phases do affect superpositions. Thus, for example, although $|1\rangle$ and $-|1\rangle$ represent the same state of a qubit, the superpositions $(|0\rangle + |1\rangle)/\sqrt{2}$ and $(|0\rangle - |1\rangle)/\sqrt{2}$ represent quite different states.

It is almost true in general that, for any two states of any quantum system, any superposition of the associated vectors also represents a possible state of that system. The word "almost" in the preceding sentence refers to the possibility of *superselection rules*. These rules specify that, for certain quantities, like electric charge, it is impossible to superpose two states with different values of those quantities. Thus, when discussing a system for which several values of the electric charge can occur, we are, in effect, dealing with several separate Hilbert spaces, called *superselection sectors*, one for each value of the charge. One can, and sometimes one does, form the direct sum of these Hilbert spaces to obtain a Hilbert space containing all the possible states of that system, but most of the vectors in that direct sum, involving superpositions with different charges, do not represent physically possible states. We prefer, in this paper, to deal with superselection sectors as separate Hilbert spaces and forgo their direct sum. For more information about superselection rules, see [4].

In reality, there are very few superselection rules—arising from certain conserved quantities like electric charge, baryon number, and parity—but in the study and application of anyons one often artificially adds superselection rules, and we shall encounter such rules in the category-theoretic treatment below. This amounts to deciding not to consider superpositions of vectors from certain Hilbert spaces, i.e., to consider those superselection sectors separately rather than considering their direct sum.

In the presence of superselection rules, the operators that one considers are operators acting on each of the superselection sectors separately. In the case of true superselection rules, the dynamics of the system and any gates that one could construct are given by unitary operators acting on each sector separately. In the case of artificial superselection rules, nature may not cooperate with our artificial rules, and states in one sector may evolve out of that sector. Such evolution interferes with our understanding and intentions; it is often called "leakage" and one strives to avoid it.

In addition to the unitary operators mentioned above, Hermitian (or self-adjoint) operators on the Hilbert space of states also play an important role in quantum mechanics, because they model observable properties of a system. The connection between Hermitian operators and (real-valued) observables is easy to describe in the case of finite-dimensional Hilbert spaces $\mathscr{H}$.[5] Let the Hermitian operator A have (distinct) eigenvalues $a_1, \ldots, a_k$, with associated eigenspaces $S_1, \ldots, S_k$. (Some of these eigenvalues may have multiplicity greater than 1, but they are to be listed only once among the a_i's. The associated S_i will then have dimension greater than 1.) These eigenspaces are orthogonal to each other, and their sum is all of $\mathscr{H}$. Any unit vector $|\psi\rangle \in \mathscr{H}$ can be expressed as the sum of its projections $|\phi_i\rangle$ to the subspaces S_i. Measuring A on a system in state $|\psi\rangle$ produces one of the eigenvalues a_i; the probability of getting the result a_i is the squared norm of the projection, $\| |\phi_i\rangle \|^2$. Note that the dimension of $\mathscr{H}$ is an upper bound for the number of distinct eigenvalues a_i of any Hermitian operator on $\mathscr{H}$. In particular, any measurement performed on a qubit will have at most two possible outcomes. It is in this sense that a qubit is the quantum analog of a classical bit.

8.2.2 Anyons

To understand anyons, it is useful to recall first that ordinary particles are of two sorts, *bosons* and *fermions*. These differ in several respects, beginning with the action of spatial rotations on the corresponding Hilbert spaces. For particles in ordinary 3-dimensional space, the group $SO(3)$ of Euclidean rotations of that space acts on the states of the particle. (More precisely, the group of all Euclidean motions acts, but we abstract from the particle's position and consider only its orientation in space; thus we ignore translations and consider only the group of rotations.) Because the vector representing a state is defined only up to a phase factor, the action of the rotation group is not a representation in the usual sense but a *projective representation*. This means that each rotation g of physical 3-dimensional space is represented by a unitary operator $\rho(g)$ on the Hilbert space, but this $\rho(g)$ is unique only up to a phase factor. It is customary to make some arbitrary choice of these phase factors, so that we can speak unambiguously of $\rho(g)$. The arbitrariness of the choice is, however, reflected in the fact that $\rho(gh)$ and $\rho(g)\rho(h)$ need not be equal but can differ by a phase factor. Furthermore, ρ and ρ' are considered equivalent representations if they differ only by these arbitrary phase factors. It is reasonable to ask, in this connection, why the operators $\rho(g)$ need to be unitary or even linear, rather than only linear up to phase factors. The reason is that, unlike absolute phases, relative phases are relevant in superpositions, so physical symmetries must preserve them.

It turns out that any projective representation ρ of $SO(3)$ is given by a genuine unitary representation $\tilde{\rho}$ of the universal covering group of $SO(3)$, namely $SU(2)$

[5]In the infinite-dimensional case, the description is similar but one must take into account the possibility of a continuous spectrum of the operator, in addition to or instead of discrete eigenvalues.

(see for example [1] and [10]). That is, if $p : SU(2) \to SO(3)$ denotes the 2-to-1 projection map, we have $\rho \circ p$ equivalent to $\tilde{\rho}$. More concretely, it means that there are two sorts of projective representations of $SO(3)$, up to equivalence. One sort is the ordinary unitary representations of $SO(3)$; the other is given by unitary representations of $SU(2)$ that send the non-trivial element $-I$ of the kernel of p to the operator $-I$. (Throughout this paper, we use I, sometimes with subscripts, to denote identity transformations, functions, morphisms, etc.) The first sort of representation corresponds to bosons, whose state vectors (not merely their states) are unchanged when rotated gradually through a full revolution. The second sort corresponds to fermions, where a rotation through 2π changes the state vector by a sign.

A second distinction between bosons and fermions, even more important for our purposes, is the behavior of systems of several identical particles. Because the particles are identical, any permutation of the particles leaves the state unchanged and therefore changes the state vector by at most a phase factor. As a result, we have a one-dimensional projective representation of the symmetric group. Again, it turns out that there are just two possibilities (both of which are actual unitary representations of the symmetric group). Either all permutations leave the state vectors unchanged, or the even permutations leave the state vectors unchanged while the odd permutations reverse the vectors' signs.

A deep theorem of relativistic quantum field theory, the spin-statistics theorem, says that these two behaviors of multi-particle states under permutations exactly match the two behaviors of single-particle states under rotations. Interchanging two identical bosons leaves the state vector of the pair unchanged; interchanging two identical fermions reverses the sign of the state vector.

The preceding discussion of bosons and fermions depends crucially on the fact that the particles are in ordinary 3-dimensional space. If particles were confined to a 2-dimensional space, more possibilities would arise.

Specifically, the rotation group in two dimensions, $SO(2)$, has more sorts of projective representations than $SO(3)$ does; the reason is ultimately that the universal covering group of the circle group $SO(2)$ is the additive group of real numbers, and the covering projection is not 2-to-1 but ∞-to-1. The result is that a gradual rotation of a particle through 2π can multiply its state vector by an arbitrary phase factor, not just ± 1. The possibility of getting *any* phase here led to the name *anyon*.

Reducing the dimensionality of space from 3 to 2 also affects the possibilities for permuting identical particles. For simplicity, consider the case where there are just two particles, and we interchange them. We can perform the interchange gradually, in the plane, by rotating the 2-particle system counterclockwise by π around the midpoint between the particles. Alternatively, we can achieve the same interchange by a clockwise rotation. In 3-dimensional space, these two options are equivalent in the sense that they can be gradually deformed into each other, by rotating the plane of the particles' motion about the line through the particles' initial positions. In 2-dimensional space, there is no such deformation without making the particles collide. Winding one particle around the other any number of times, we get infinitely many ways to achieve one and the same permutation. With more than two particles, there are more complicated ways to achieve the same permutation by moving the particles around in the plane. As a result, in place of (projective) representations of symmetric groups, we have representations of braid groups. For example, in the case of two particles, in place of the group of two possible permutations of the particles, we have the group of all integers, with integer n representing a counterclockwise rotation by $n\pi$ (and negative n representing clockwise rotations).

The preceding discussion was oversimplified in that (among other things) when moving particles around each other, we ignored any rotation that the individual particles might have undergone during the motion. A more accurate presentation would need to suitably combine the braid and rotation groups.

8.2.3 Anyons in Reality

As explained above, anyons do not occur in 3-dimensional space; it is necessary to reduce the number of spatial dimensions to 2. Since we live in a 3-dimensional space, will we ever find anyons? It turns out that anyon-like behavior occurs for certain excitations in materials that are so thin as to be effectively two-dimensional. A detailed discussion of this would take us too far from the purpose of this paper, so we refer the reader to Sect. 1.1 of [9].

We emphasize, however, that the anyons are not what one would ordinarily think of as "particles" but rather excitations in some medium, which exhibit particle-like behavior. It should be noted in this connection that it is not unusual, in other contexts, for excitations to behave like particles and thus to be analyzed mathematically as if they were particles. For example, vibrational excitations in crystal lattices are treated as particles called phonons. Similarly, photons are excitations of the electromagnetic field. In quantum field theory, all particles are excitations of the corresponding fields.

8.2.4 Anyons in Quantum Computation

Quantum computation is unpleasantly susceptible to environmental disturbances. Its advantages over classical computation depend on maintaining superpositions of state vectors, with high precision in the coefficients of those vectors. Small disturbances can easily modify those coefficients or, indeed, destroy superpositions altogether. Significant effort must therefore be devoted to error correction, and this makes algorithms slower and harder to design.

It has been suggested [6] that qubits could be more robust, i.e., less susceptible to disturbances, if they were implemented using certain sorts of anyons. For example, if qubits were encoded in the way two anyons wind around each other, then this winding, being a topological property of the system, would be robust. A small disturbance in the actual motion of the anyons would leave the winding number intact. This hope of reducing the error correction needs of quantum computing has motivated much of the current interest in anyons.

In this approach to quantum computation, braiding of anyons serves not only to store information but also to process it. In general, as mentioned above, quantum computation proceeds by initializing a quantum state, then applying a unitary transformation to it, and finally measuring some observable in the resulting transformed state. The unitary transformation used here must be designed so that a feasible measurement produces a useful result. In addition, there must be a way to implement the unitary transformation as the composition of a sequence of simpler unitary transformations, usually called gates in this context. In the anyon approach to quantum computation, the most basic unitary gates arise from the braiding of anyons around each other, and a crucial question is whether these gates are *universal* in the sense that arbitrary gates can be approximated by composing the basic ones.

It is worth noting explicitly that, in this picture, a qubit is not encoded in the state of a single anyon but rather in a whole system of several anyons. This feature will be quite prominent in the category picture described in the rest of this paper.

8.3 Modular Tensor Categories

In this section we describe the category-theoretic structure that has been developed to support a mathematical theory of anyons. Much of what we describe here is in [9], though we have modified some aspects and rearranged others.

Throughout this section, we let $\mathscr{A}$ be a category, intended to describe the quantum-mechanical behavior of a system of anyons. $\mathscr{A}$ will carry several sorts of additional structure, roughly classified as "additive" and "multiplicative" structure, all subject to various axioms. We describe the structures and the axioms a little at a time. We begin with the additive structure, because this is where Hilbert spaces enter the picture, so it is the basis for the connection with the usual formalism of quantum theory.

The vectors in our Hilbert spaces will be the morphisms of $\mathscr{A}$. Specifically, for each pair of objects X, Y of $\mathscr{A}$, the set $\mathrm{Hom}(X, Y)$ of morphisms from X to Y will have the structure of a Hilbert space. So we have many Hilbert spaces, one for each pair X, Y of objects. Some of these Hilbert spaces will be mere combinations of others, but there will still be several different "basic" Hilbert spaces. This means physically that we regard the system as being subject to superselection rules, which keep these Hilbert spaces separate.

We assume familiarity with some basic notions of category theory, specifically, the notions of product (including terminal object, which is the product of the empty family), coproduct (including initial object), equalizer, coequalizer, monomorphism, epimorphism, isomorphism, functor, and natural transformation. Definitions and examples can be found in [7] or [3, Chap. 1].

8.3.1 Additive Structure

We begin by requiring $\mathscr{A}$ to be an abelian category. This requirement, formulated in detail below, provides a well-behaved addition operation on each of the sets $\mathrm{Hom}(X, Y)$, although the requirement is formulated in purely category-theoretic terms and does not explicitly mention this addition operation.

Axiom 1 (*Abelian*) $\mathscr{A}$ is an abelian category. That is

1. There is an object 0 that is both initial and terminal. A morphism that factors through this zero object will be called a zero morphism and denoted by 0. Note that each $\mathrm{Hom}(X, Y)$ contains a unique zero morphism.
2. Every two objects have a product and a coproduct.
3. For every morphism $\alpha : X \to Y$, the pair α, 0 has an equalizer and a coequalizer. These are called the *kernel* and *cokernel* of α.
4. Every monomorphism is the kernel of some morphism, and every epimorphism is the cokernel of some morphism.

This axiom has a surprisingly rich collection of consequences, developed in detail in Chap. 2 of [3]. We list here only some of the highlights, which will be important for this paper, and we refer the reader to [3] for the proofs and additional information.

Proposition 1 ([3], Theorem 2.12) *Any morphism that is both monic and epic is an isomorphism.*

(More generally, as one can easily check, in any category, any equalizer that is an epimorphism is an isomorphism.)

Proposition 2 ([3], Theorem 2.35) *The product and coproduct of any two objects coincide.*

That is, given two objects X and Y, there is an object $X \oplus Y$ that serves simultaneously as the product of X and Y, with projections $p_X : X \oplus Y \to X$ and $p_Y : X \oplus Y \to Y$, and as the coproduct of X and Y, with injections $u_X : X \to X \oplus Y$ and $u_Y : Y \to X \oplus Y$. (If $X = Y$, then our notations for the projections and injections become ambiguous, and we use p_1, p_2, u_1, u_2 instead.) For brevity, we often refer to $X \oplus Y$ as the *sum* of X and Y, rather than as the product or coproduct.

As a product, $X \oplus X$ admits a diagonal morphism $\Delta_X : X \to X \oplus X$, namely the unique morphism whose composites with both projections are the identity morphism I_X of X. Dually, as a coproduct, it admits the folding morphism $\nabla_X : X \oplus X \to X$, whose composites with both of the injections are I_X. Using the diagonal and folding morphisms, we can define a binary operation, called addition, on $\mathrm{Hom}(X, Y)$ for any objects X and Y. Given $f, g : X \to Y$, we define $f + g : X \to Y$ to be the composite

$$X \xrightarrow{\Delta_X} X \oplus X \xrightarrow{f \oplus g} Y \oplus Y \xrightarrow{\nabla_Y} Y,$$

where $f \oplus g$ is obtained from the functoriality of products (or of coproducts—they yield the same result).

Proposition 3 ([3], Theorems 2.37 and 2.39) *This addition operation makes each* $\mathrm{Hom}(X, Y)$ *an abelian group, with the zero morphism serving as the identity of the group. Composition of morphisms is additive with respect to both factors; that is, when either factor is fixed, the composite* $f \circ g$ *is an additive function of the other factor.*

Axiom 2 (*Vectors*) Each of these abelian groups $\mathrm{Hom}(X, Y)$ carries an operation of multiplication by complex numbers, making $\mathrm{Hom}(X, Y)$ a vector space over $\mathbb{C}$, and making composition of morphisms bilinear over $\mathbb{C}$.

The complex vector spaces $\mathrm{Hom}(X, Y)$ will play the role of quantum-mechanical state spaces. For this purpose, they should also be equipped with inner products, making them Hilbert spaces, but, following [9], we refrain from assuming an inner product structure at this stage of the development.[6] It turns out that much of what we shall do later does not depend on the availability of inner products in the vector spaces $\mathrm{Hom}(X, Y)$.

An object S in the abelian category $\mathscr{A}$ is called *simple* if $S \not\cong 0$ and every monomorphism into S is either a zero morphism or an isomorphism. In other words, S is a non-zero object with no non-trivial subobjects. Because of the abelian structure of $\mathscr{A}$, this definition can be shown (using [3, Theorem 2.11]) to be equivalent to its dual: A non-zero object is simple if and only if it has no non-trivial quotients, i.e., every epimorphism out of S is either a zero morphism or an isomorphism.

Axiom 3 (*Semisimple*) Every object in $\mathscr{A}$ is a finite sum of simple objects.

This axiom considerably simplifies the structure of the vector spaces $\mathrm{Hom}(X, Y)$. In the first place, as shown in [3, Sect. 2.3], morphisms from a sum $\bigoplus_j S_j$ to another sum $\bigoplus_k S'_k$ are given by matrices of morphisms between the summands. Specifically, the matrix associated to $f : \bigoplus_j S_j \to \bigoplus_k S'_k$ has as its a, b entry the composite

$$S_b \xrightarrow{u_b} \bigoplus_j S_j \xrightarrow{f} \bigoplus_k S'_k \xrightarrow{p'_a} S'_a.$$

Composition of morphisms in $\mathscr{A}$ corresponds to the usual multiplication of matrices.

Furthermore, when the summands are simple, we have the following additional information about the matrix entries, a generalization of Schur's Lemma in group representation theory.

Proposition 4 *If* $f : S \to S'$ *is a morphism between two simple objects, then* f *is either the zero morphism or an isomorphism.*

Proof The kernel of f is a monomorphism into S, and if it is an isomorphism then f is zero. So, by simplicity of S, we may assume that the kernel of f is zero and therefore (by [3, Theorem 2.17*]) f is a monomorphism. Similarly, by considering

[6] In fact, inner products are never explicitly assumed in [9]. They are, however, implicit in the statement, in Sect. 5.1 of [9], that certain bases "are – of course – related by a unitary transformation".

the cokernel of f and invoking the simplicity of S', we may assume that f is an epimorphism. But then, by Proposition 1, f is an isomorphism. □

The last axiom in this subsection combines two finiteness assumptions.

Axiom 4 (*Finiteness*)

1. There are only finitely many non-isomorphic simple objects.
2. Each of the vector spaces $\mathrm{Hom}(X, Y)$ is finite-dimensional over $\mathbb{C}$.

The first of these two finiteness requirements is merely a technical convenience. The second, however, gives the following important information about the endomorphisms of simple objects.

Proposition 5 *If S is a simple object, then* $\mathrm{Hom}(S, S) \cong \mathbb{C}$.

Proof The operation of composition of morphisms is a multiplication operation that makes the vector space $\mathrm{Hom}(S, S)$ into an algebra over $\mathbb{C}$. Since S is simple, Proposition 4 says that every non-zero element of this algebra is invertible. That is, $\mathrm{Hom}(S, S)$ is a division algebra over $\mathbb{C}$. But $\mathbb{C}$ is algebraically closed, so the only finite-dimensional division algebra over it is $\mathbb{C}$ itself. □

Note that the isomorphism $\mathrm{Hom}(S, S) \cong \mathbb{C}$ in this proposition can be taken, as the proof shows, to be an isomorphism of algebras, not just of vector spaces. In particular, the identity morphism of S corresponds to the number 1.

Combining this proposition with our earlier observations about matrices, we find that any morphism $f : \bigoplus_j S_j \to \bigoplus_k S'_k$ between any two objects in $\mathscr{A}$ is given by a matrix whose entries are complex numbers. Moreover, the a, b entry is 0 unless $S_b \cong S'_a$. From this observation, it easily follows that, when an object X of $\mathscr{A}$ is expressed as a sum $\bigoplus_j S_j$ of simple objects, the isomorphism types of the summands S_j and their multiplicities are completely determined by X. That is, the representation of X as a sum of simple objects is essentially unique.

8.3.2 Multiplicative Structure

In this subsection, we introduce the multiplicative structure that makes $\mathscr{A}$ a braided monoidal category. The central idea is that, if objects X and Y represent certain anyons, then $X \otimes Y$ should represent a system consisting of both of these anyons. We must, however, remember that the Hilbert spaces that occur in this context are not the objects of $\mathscr{A}$ but the vector spaces of morphisms between the objects.

A system consisting of two anyons of types X and Y would, if measured as a whole, appear as another anyon, whose type might not be entirely determined by the types X and Y. Formally, this means that $X \otimes Y$ is a sum of several simple objects. Furthermore, there might be several "ways" for a composite system to appear as having a particular type Z, modeled as several morphisms from $X \otimes Y$ to Z, and our Hilbert spaces will also contain superpositions of these.

The multiplicative structure will also include a unit object 1; its intended interpretation is the vacuum. Thus, $1 \otimes X$ and $X \otimes 1$ amount to just X because a system consisting of X and nothing is the same as X.

The first aspect of multiplicative structure can be stated rather briefly as the following axiom, but we expand it afterward because we shall need the details later.

Axiom 5 (*Multiplication*) $\mathscr{A}$ is a monoidal category.

This means that it is equipped with a "multiplication" functor $\otimes : \mathscr{A} \times \mathscr{A} \to \mathscr{A}$ and a "unit object" 1 that satisfy the usual associative and unit laws up to coherent isomorphism. Let us first explain "satisfying the laws up to isomorphism" and then discuss "coherent".

Associativity would mean that $A \otimes (B \otimes C)$ is the same as $(A \otimes B) \otimes C$ for any objects A, B, C (and similarly for morphisms). Associativity up to isomorphism means that these objects need not be equal but they are isomorphic, and we are given specific isomorphisms

$$\alpha_{A,B,C} : (A \otimes B) \otimes C \to A \otimes (B \otimes C)$$

for all A, B, C, and furthermore these isomorphisms constitute a natural transformation between functors $\mathscr{A} \times \mathscr{A} \times \mathscr{A} \to \mathscr{A}$.

Similarly, the requirement that the object 1 be a unit up to isomorphism means that we are given natural isomorphisms

$$\lambda_A : 1 \otimes A \to A \quad \text{and} \quad \rho_A : A \otimes 1 \to A.$$

As is well-known from classical algebra, the associative law implies associative identities for more than three factors at a time; for example, if $*$ is an associative operation, then all five of the possible parenthesizations of $a * b * c * d$ give the same result. The analogous result for categories is that any natural isomorphism α as above produces natural isomorphisms between any two parenthesizations of $A \otimes B \otimes C \otimes D$. There is, however, an embarrassment of riches, as we can build, from α (and its inverse), several isomorphisms between such parenthesizations of four factors. Specifically, the "extreme left" and "extreme right" parenthesizations are connected by a product of three α's:

$$((A \otimes B) \otimes C) \otimes D \xrightarrow{\alpha_{A,B,C} \otimes I_D} (A \otimes (B \otimes C)) \otimes D \xrightarrow{\alpha_{A,B\otimes C,D}} A \otimes ((B \otimes C) \otimes D) \xrightarrow{I_A \otimes \alpha_{B,C,D}} A \otimes (B \otimes (C \otimes D)).$$

The same two parenthesizations are connected by a product of two other α's:

$$((A \otimes B) \otimes C) \otimes D \xrightarrow{\alpha_{A\otimes B,C,D}} (A \otimes B) \otimes (C \otimes D) \xrightarrow{\alpha_{A,B,C\otimes D}} A \otimes (B \otimes (C \otimes D)).$$

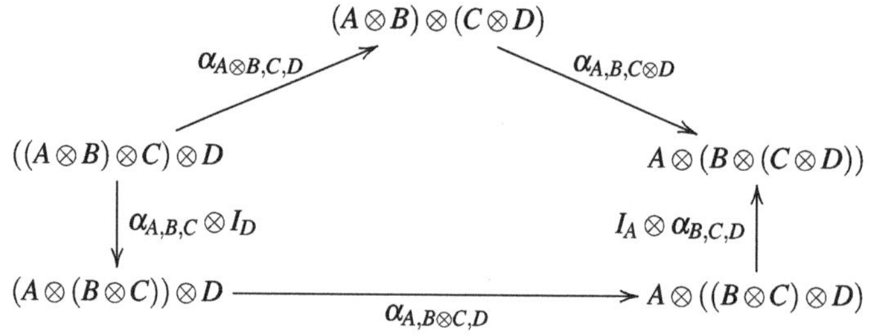

Fig. 8.1 The pentagon condition

One aspect of "coherence" is that these two transformations must agree, so that there is a single, well-defined way of shifting the parentheses from the left to the right. This requirement is often called the *pentagon condition*, because the diagram exhibiting these two transformations together has the shape of a pentagon. In this connection, the first composition, involving three morphisms, is sometimes called the "long side" of the pentagon, and the second composition is the "short side". In Fig. 8.1, the short side is the top of the pentagon while the long side contains the vertical sides and the bottom.

Another aspect of coherence is that two ways of simplifying $(A \otimes 1) \otimes B$ should agree, namely $\rho_A \otimes I_B$ and

$$(A \otimes 1) \otimes B \stackrel{\alpha_{A,1,B}}{\longrightarrow} A \otimes (1 \otimes B) \stackrel{I_A \otimes \lambda_B}{\longrightarrow} A \otimes B.$$

It is easy to think of other compositions of α's, λ's, and ρ's that should agree, for example the many ways of connecting different parenthesizations of five or more factors. Fortunately, all of these requirements can be deduced from the two that we have exhibited here. This is Mac Lane's coherence theorem, and we refer to Chap. VII of [7] for its precise statement, its proof, and additional information about monoidal categories.

The pentagon condition will play a major role in the rest of this paper, because the associativity isomorphism α is often nontrivial and of considerable interest. The unit isomorphisms λ and ρ, on the other hand, will play essentially no role, because one can safely identify $1 \otimes X$ and $X \otimes 1$ with X and take $\lambda_X = \rho_X = I_X$ for all X. From now on, we will make these simplifying identifications.

The idea that $\otimes$ represents combining two anyons (or two systems of anyons) into a single system suggests that this operation should be commutative, i.e., that $X \otimes Y$ should be naturally isomorphic to $Y \otimes X$. The next axiom postulates the existence of such an isomorphism, with good behavior in connection with the associativity isomorphism α.

Axiom 6 (*Braiding*) The monoidal structure on $\mathscr{A}$ is equipped with a *braiding*, i.e., a natural isomorphism $\sigma_{X,Y} : X \otimes Y \to Y \otimes X$ subject to two requirements, first that the following two composite isomorphisms be equal:

$$(A \otimes B) \otimes C \stackrel{\alpha_{A,B,C}}{\longrightarrow} A \otimes (B \otimes C) \stackrel{\sigma_{A,B\otimes C}}{\longrightarrow} (B \otimes C) \otimes A \stackrel{\alpha_{B,C,A}}{\longrightarrow} B \otimes (C \otimes A)$$

and

$$(A \otimes B) \otimes C \stackrel{\sigma_{A,B}\otimes I_C}{\longrightarrow} (B \otimes A) \otimes C \stackrel{\alpha_{B,A,C}}{\longrightarrow} B \otimes (A \otimes C) \stackrel{I_B \otimes \sigma_{A,C}}{\longrightarrow} B \otimes (C \otimes A),$$

and, second, the analogous equality with each $\sigma_{X,Y}$ replaced with ${\sigma_{Y,X}}^{-1}$.

Recall, from Sect. 8.2.2, that anyons inhabit two-dimensional space and therefore, when two of them are interchanged, it is necessary to keep track of how they move around each other. A clockwise rotation by π around the midpoint between them is not the same as, nor even deformable to, a counterclockwise rotation. So we should describe $\sigma_{X,Y}$ not merely as switching X with Y but as doing so in a counterclockwise direction. The choice of direction here is a matter of convention; ${\sigma_{Y,X}}^{-1}$ is then the clockwise rotation achieving the same interchange. Thus, we expect that, in general, $\sigma_{X,Y} \neq {\sigma_{Y,X}}^{-1}$. (If these two were always equal, then we would have a *symmetric monoidal* category rather than a braided one.)

A useful picture, often used in connection with braiding, is to imagine the factors in a $\otimes$-product as being lined up from left to right. Then the counterclockwise interchange $\sigma_{X,Y}$ amounts to moving X from the left of Y to the right of Y by passing X in front of Y. ${\sigma_{Y,X}}^{-1}$ also moves X from the left to the right of Y, but it does so by passing X behind Y.

The equality of the two composite morphisms in the definition of braiding is called the *hexagon condition* (Fig. 8.2). In terms of moving anyons around each other, it expresses the fact that moving A past $B \otimes C$ by passing A in front of $B \otimes C$ is equivalent to first passing A in front of B and then passing A in front of C. The hexagon condition for ${\sigma_{Y,X}}^{-1}$ has a similar pictorial description with "in front of" replaced with "behind".

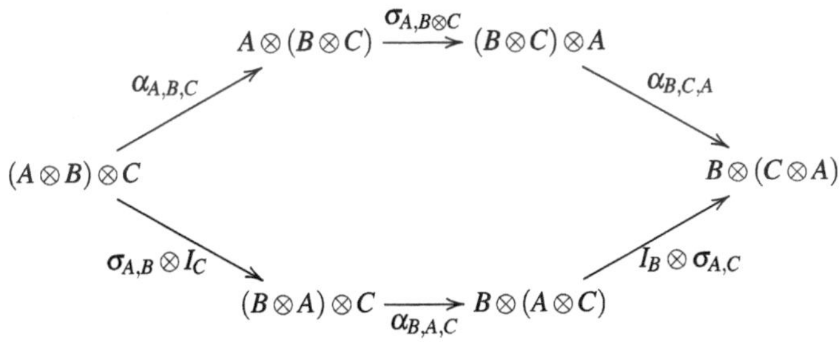

Fig. 8.2 The hexagon condition

The last axiom in this subsection relates the multiplicative structure discussed here with the additive structure from the preceding subsection.

Axiom 7 (*Additive-Multiplicative*)

1. The monoidal unit 1 is simple.
2. The product operation $\otimes$ is bilinear on morphisms.

In more detail, item (2) here means that the function

$$\mathrm{Hom}(A, B) \times \mathrm{Hom}(C, D) \to \mathrm{Hom}(A \otimes C, B \otimes D)$$

given by the functoriality of $\otimes$ is bilinear with respect to the $\mathbb{C}$-vector space structures of the Hom-sets. It follows from this, via results in [3, Sect. 2.4], that $\otimes$ distributes over $\oplus$ on objects, i.e., that $X \otimes (Y \oplus Z)$ is canonically isomorphic to $(X \otimes Y) \oplus (X \otimes Z)$.

8.3.3 Duals, Twists, and Modularity

In this subsection, we collect some additional axioms to complete the definition of modular tensor categories. These axioms will not play a role in the computations we do later. We list them for the sake of completeness, but we make only a few comments about them and refer the reader to [9, Sects. 4.3, 4.5, and 4.7] for more thorough explanations.

Axiom 8 (*Antiparticles*) For each object X of $\mathscr{A}$, there is a *dual* object X^*, and there are two morphisms $i_X : 1 \to X \otimes X^*$ and $e_X : X^* \otimes X \to 1$, such that the compositions

$$X^* \xrightarrow{I_{X^*} \otimes i_X} X^* \otimes X \otimes X^* \xrightarrow{e_X \otimes I_{X^*}} X^*$$

and

$$X \xrightarrow{i_X \otimes I_X} X \otimes X^* \otimes X \xrightarrow{I_X \otimes e_X} X$$

are equal to the identity morphisms I_{X^*} and I_X, respectively. Furthermore, dualization commutes with $\otimes$ and $\oplus$ and preserves 1 and 0.

For the sake of readability, we have exhibited the compositions in this axiom without the parentheses and associativity isomorphisms that technically should be there. We follow the same convention for iterated $\otimes$ below.

The intention behind this axiom is that, if X represents some particle, then X^* represents its antiparticle. The morphism i_X represents creation of a particle-antiparticle pair from the vacuum, and e_X represents annihilation of such a pair.

The operation of dualization becomes a contravariant functor from $\mathscr{A}$ to itself if one defines the dual f^* of a morphism $f : X \to Y$ to be the composite

$$Y^* \xrightarrow{I_{Y^*} \otimes i_X} Y^* \otimes X \otimes X^* \xrightarrow{I_{Y^*} \otimes f \otimes I_{X^*}} Y^* \otimes Y \otimes X^* \xrightarrow{e_Y \otimes I_{X^*}} X^*.$$

Axiom 9 (*Rotations*) There is a natural isomorphism δ with components $\delta_X : X \to X^{**}$ respecting the monoidal structure and duality in the sense that

$$\delta_1 = I_1, \quad \delta_{X \otimes Y} = \delta_X \otimes \delta_Y, \quad \text{and} \quad \delta_{X^*} = ({\delta_X}^*)^{-1}.$$

By combining these δ isomorphisms with the morphisms i and e from duality, one can obtain isomorphisms $X \to X$ that represent twisting an anyon by 2π; see [9, Sect. 4.5] for details.

Monoidal categories satisfying the "Antiparticles" axiom are called *rigid*, and those that also satisfy the "Rotations" axiom are called *ribbon* categories.

Axiom 10 (*Modularity*) For any two simple objects X and Y, let $s_{X,Y} : 1 \to 1$ be the morphism

$$\begin{aligned} 1 = 1 \otimes 1 &\xrightarrow{i_X \otimes i_Y} X \otimes X^* \otimes Y \otimes Y^* \xrightarrow{I_X \otimes \sigma_{X^*,Y} \otimes I_Y} X \otimes Y \otimes X^* \otimes Y^* \\ &\xrightarrow{I_X \otimes \sigma_{Y,X^*} \otimes I_Y} X \otimes X^* \otimes Y \otimes Y^* \xrightarrow{\delta_X \otimes I_{X^*} \otimes \delta_Y \otimes I_{Y^*}} X^{**} \otimes X^* \otimes Y^{**} \otimes Y^* \\ &\xrightarrow{e_{X^*} \otimes e_{Y^*}} 1 \otimes 1 = 1. \end{aligned}$$

Since $\mathrm{Hom}(1, 1) = \mathbb{C}$, these morphisms $s_{X,Y}$ constitute a matrix of complex numbers, with rows and columns indexed by the isomorphism classes of simple objects. This matrix is required to be invertible.

Notice that, if $\mathscr{A}$ were not merely braided but symmetric, then the σ's and the σ^{-1}'s in this composite would cancel out, and we would have $s_{X,Y} = t_X t_Y$ where t_X is the composite

$$1 \xrightarrow{i_X} X \otimes X^* \xrightarrow{\delta_X \otimes I_{X^*}} X^{**} \otimes X^* \xrightarrow{e_{X^*}} 1,$$

and similarly for t_Y. Thus, the matrix s described in the modularity axiom would be the product of a column vector by a row vector (in this order). Such a matrix has rank at most 1. By requiring this matrix to be invertible, the axiom says that, as far as the rank of this matrix is concerned, the braiding is as far as possible from being symmetric.

8.4 Yoneda Simplification

In this section, we point out a simplification of the additive structure of $\mathscr{A}$, based on Yoneda's Lemma. That lemma (see [7, Sect. 3.2]) says roughly that an object in any category is determined, up to isomorphism, by the morphisms into it. More precisely, any category $\mathscr{C}$ is equivalent to a full subcategory of the category $\hat{\mathscr{C}}$ of contravariant functors from $\mathscr{C}$ to the category of sets.[7] Under this equivalence, any object X of $\mathscr{C}$ corresponds to the functor $\mathrm{Hom}(-, X)$, i.e., the functor sending each object U of $\mathscr{C}$ to the set of morphisms $U \to X$ and sending each morphism $f : U \to V$ to the operation $\mathrm{Hom}(V, X) \to \mathrm{Hom}(U, X)$ of composition with f.

In the case of our category $\mathscr{A}$, we can greatly simplify $\hat{\mathscr{A}}$ while still maintaining the Yoneda equivalence. In the first place, since every object U of $\mathscr{A}$ is a finite sum, and thus in particular a coproduct, of simple objects, $U = \bigoplus_{j \in F} S_j$, morphisms $U \to X$ amount to F-indexed families of morphisms $S_j \to X$. More precisely, any $f : U \to X$ is determined by the composite morphisms $f \circ u_j : S_j \to X$, and, conversely, any family of morphisms $g_j : S_j \to X$ arises in this way from a unique morphism $U \to X$. Thus, $\mathscr{A}$ is equivalent to a full subcategory of the category $\hat{\mathscr{S}}$ of set-valued functors on the category $\mathscr{S}$ of simple objects in $\mathscr{A}$.

Up to equivalence, we need not use all the simple objects; it suffices to have at least one representative from each isomorphism class of simple objects. So we can replace the $\mathscr{S}$ of the preceding paragraph by a skeleton of it, i.e., a full subcategory $\mathscr{S}_0$ consisting of just one representative per isomorphism class.

The structure of this new, skeletal $\mathscr{S}_0$ admits, thanks to the finiteness axiom and Proposition 4 the following description. There are finitely many objects. The morphisms from any object to itself form a copy of $\mathbb{C}$. If U and V are distinct objects, then the only morphism from U to V is zero.

As a result, the Yoneda embedding, simplified as above, sends each object X of $\mathscr{A}$ to a finite family of vector spaces, indexed by the simple objects U in $\mathscr{S}_0$, namely the vector spaces $\mathrm{Hom}(U, X)$. Furthermore, the morphisms $X \to Y$ in $\mathscr{A}$ are given by *arbitrary* families of linear maps $g_U : \mathrm{Hom}(U, X) \to \mathrm{Hom}(U, Y)$ between corresponding vector spaces. The reason for "arbitrary" is that, because of the paucity of morphisms in $\mathscr{S}_0$, all such families automatically satisfy the commutativity conditions required in order to be natural transformations and thus to be morphisms in the functor category $\hat{\mathscr{S}_0}$.

Summarizing, we have that, up to equivalence of categories, $\mathscr{A}$ can be described as the category whose objects (resp. morphisms) are families of finite-dimensional vector spaces (resp. linear maps), indexed by the objects of $\mathscr{S}_0$. Furthermore, it is easy to check that sums in $\mathscr{A}$ are given, via this equivalence, by direct sums of vector spaces.

In other words, the additive structure of $\mathscr{A}$ is trivial. The interesting structure is the monoidal structure, and this can be quite complicated. In particular, the associativity isomorphisms α and the braiding isomorphisms σ, though given (like any morphisms) by linear maps, need not have a particularly simple structure.

The analysis of the multiplicative structure of $\mathscr{A}$ can be facilitated by taking advantage of the semisimplicity of $\mathscr{A}$ and the fact that $\otimes$ distributes over $\oplus$. If we know how $\otimes$ acts on simple objects, distributivity determines how it acts on sums of simple objects, and, by semisimplicity, those are all the objects. Moreover,

---

[7]There are set-theoretic issues if $\mathscr{C}$ is a proper class rather than a set, but these issues need not concern us here. The finiteness conditions imposed on our anyon category $\mathscr{A}$ ensure that it is equivalent to a small, i.e., set-sized, category.

because the associativity and braiding isomorphisms are natural, and thus in particular commute with the injection and projection morphisms of sums, the behavior of these isomorphisms on arbitrary objects is determined by their behavior on simple objects. Better yet, the pentagon and hexagon conditions will be satisfied in general as soon as they are satisfied for simple objects.

Thus, the additive and multiplicative structure of $\mathscr{A}$ can be completely described by giving

1. a complete list of non-isomorphic simple objects (including the unit or vacuum 1),
2. for each pair of objects in this list, their $\otimes$-product, expressed as a sum of objects from the list,
3. the associativity isomorphisms $\alpha_{X,Y,Z}$ for all X, Y, Z in the list, and
4. the braiding isomorphisms $\sigma_{X,Y}$ for all X, Y in the list,

subject to the pentagon and hexagon conditions.

We shall not be concerned here with duality and ribbon structure, but it could also be reduced to a consideration of the simple cases.

Often, items (1) and (2) here determine or at least greatly constrain items (3) and (4) via the pentagon and hexagon conditions. One such situation is the subject of the next section. Other examples, both of strong constraints on (3) and (4) and of weak constraints can be found in [2].

8.5 Fibonacci Anyons

8.5.1 *Definition and Additive Structure*

In this section, we consider the special case of *Fibonacci anyons*. These are defined by specifying the category $\mathscr{A}$ as follows. There are just two simple objects, 1 (the vacuum, the unit for $\otimes$) and τ. Each is its own dual. (Recall that Axiom 8 requires each object to have a dual; dualization is additive, so we need only specify the duals of the simple objects.) The monoidal structure is given by $\tau \otimes \tau = 1 \oplus \tau$ (plus the fact that 1 is the unit, so $1 \otimes \tau = \tau \otimes 1 = \tau$ and $1 \otimes 1 = 1$).

The terminology "Fibonacci anyon" comes from the fact, easily verified using the distributivity of $\otimes$ over $\oplus$, that iteration of $\otimes$ gives $\tau^{\otimes n} = f_{n-1} \cdot 1 \oplus f_n \cdot \tau$, where the f's are the Fibonacci numbers defined by the recursion $f_{-1} = 1$, $f_0 = 0$, and $f_{n+1} = f_n + f_{n-1}$. Here and below, we use the notation $k \cdot S$ to mean the sum of k copies of the object S of $\mathscr{A}$. (The notation makes sense for arbitrary objects S, but we shall need it only for simple S.)

As explained in Sect. 8.4, we can identify the category $\mathscr{A}$ with the category of pairs (V_1, V_τ) of finite-dimensional complex vector spaces. Explicitly, an object X is identified with the pair $(\mathrm{Hom}(1, X), \mathrm{Hom}(\tau, X))$. In particular, the unit 1 in $\mathscr{A}$

is identified with $(\mathbb{C}, 0)$, and τ is identified with $(0, \mathbb{C})$. This identification respects the additive structure: $\oplus$ in $\mathscr{A}$ corresponds to componentwise direct sum of pairs of vector spaces.

8.5.2 Tensor Products

The multiplicative structure of $\mathscr{A}$, on the other hand, is quite far from componentwise tensor product of vector spaces, as the latter would make $\tau \otimes \tau = \tau$ (because $\mathbb{C} \otimes \mathbb{C} \cong \mathbb{C}$). Our goal in the rest of this paper is to determine the multiplicative structure in terms of pairs of vector spaces.

The equation $\tau^n = f_{n-1} \cdot 1 \oplus f_n \cdot \tau$ mentioned above already determines that structure as far as the objects are concerned, but there remains much to be said about the morphisms.

A morphism from one pair of vector spaces (V_1, V_τ) to another such pair (W_1, W_τ) is a pair of linear transformations $(m_1 : V_1 \to W_1, m_\tau : V_\tau \to W_\tau)$. We can think of it as a pair of matrices, provided we fix bases for all the vector spaces involved here.

The choice of bases involves considerable arbitrariness, but there is a (somewhat) helpful guiding principle, namely that, if we have already chosen bases for two vector spaces, then the union of those bases serves naturally as a basis for the direct sum of those vector spaces. Some caution is needed, though, because the same vector space can arise as a direct sum in several ways and can thus have several equally natural bases. Indeed, much of our work below will be finding the transformations that relate such bases.

The guiding principle tells us nothing about choosing bases for the one-dimensional spaces V_1 and V_τ in the pairs $1 = (V_1, 0)$ and $\tau = (0, V_\tau)$. There isn't even any non-zero morphism between these simple objects to suggest a correlation between the choice of bases. Nor do we get canonical bases here by evaluating compound expressions that fuse to τ or to 1 or to a sum of these. So we might as well identify these one-dimensional spaces with $\mathbb{C}$ and use the number 1 as the basis vector in both of them.

Then $\tau \otimes \tau = 1 \oplus \tau = (\mathbb{C}, \mathbb{C})$ already has a basis for each of the two vector spaces. Let us turn to the triple product

$$\tau \otimes (\tau \otimes \tau) = \tau \otimes (1 \oplus \tau) = (\tau \otimes 1) \oplus (\tau \otimes \tau) = \tau \oplus (1 \oplus \tau) = 1 \cdot 1 \oplus 2 \cdot \tau.$$

As a pair of vector spaces, it is isomorphic to $(\mathbb{C}, \mathbb{C}^2)$, but we have some additional information about it, namely that it was obtained as the sum of $\tau \otimes 1 = \tau$ and $\tau \otimes \tau = 1 \oplus \tau$. Our guiding principle thus suggests choosing a basis in $\mathbb{C}^2$ that respects this sum decomposition. That is, one of the basis vectors in $\mathbb{C}^2$ should come from the first τ and the other should come from the second summand, $1 \oplus \tau$.

Consider, however, the analogous computation with the other way of parenthesizing the triple product:

$$(\tau \otimes \tau) \otimes \tau = (1 \oplus \tau) \otimes \tau = (1 \otimes \tau) \oplus (\tau \otimes \tau) = \tau \oplus (1 \oplus \tau) = 1 \cdot 1 \oplus 2 \cdot \tau.$$

It also leads to the pair of vector spaces $(\mathbb{C}, \mathbb{C}^2)$, and it also provides a suggestion for a basis of $\mathbb{C}^2$. There is, however, no guarantee that this suggestion agrees with the one in the preceding paragraph. We shall see below that the two suggestions are actually guaranteed to *disagree*. We have two bases for $\mathbb{C}^2$, and there will be a non-trivial matrix transforming the one into the other. We shall find that this matrix is almost uniquely determined.

There could, a priori, have also been two different natural bases for the first component $\mathbb{C}$ in $\tau^{\otimes 3}$, although we shall see that, in this particular situation, they coincide.

These basis transformation matrices, relating the bases that arise from $\tau \otimes (\tau \otimes \tau)$ and from $(\tau \otimes \tau) \otimes \tau$, amount to the associativity isomorphism $\alpha_{\tau,\tau,\tau}$ in the definition of the monoidal category $\mathscr{A}$.

Recall from Sect. 8.4 that all the associativity isomorphisms of $\mathscr{A}$ are determined by those with simple objects as subscripts. One of these is the $\alpha_{\tau,\tau,\tau}$ mentioned just above; the others involve one or more 1's in the subscript. Fortunately, all those others are identity maps, thanks to the identification of $1 \otimes X$ and $X \otimes 1$ with X. So the entire associativity structure of $\mathscr{A}$ comes down to two matrices, a 2×2 matrix relating the two bases for $\mathbb{C}^2$ and a number (a 1×1 matrix) relating the two bases for $\mathbb{C}$. These matrices are subject to the constraint given by the pentagon condition (Fig. 8.1). Below, we shall calculate that constraint explicitly. It will almost uniquely determine α.

We shall also calculate the constraint imposed by the hexagon condition on the braiding isomorphisms σ (Fig. 8.2). Again, the only component that needs to be computed is $\sigma_{\tau,\tau}$. The components where at least one subscript is 1 are trivial, and the components with non-simple objects as subscripts reduce, by distributivity, to ones with simple subscripts.

8.5.3 *Notation for Basis Vectors*

In order to compute the isomorphisms $\alpha_{\tau,\tau,\tau}$ and $\sigma_{\tau,\tau}$ for Fibonacci anyons, we shall view them as matrices, using suitable bases for the relevant vector spaces, and we shall calculate the constraints imposed on those matrices by the pentagon and hexagon conditions. We begin by setting up a convenient notation for those bases.

The domains and codomains of the morphisms under consideration are obtained from τ and 1 by iterated $\otimes$. We must, of course, be careful about the parenthesization of such $\otimes$-products because, as we saw above, different parenthesizations can lead to different bases; indeed, $\alpha_{\tau,\tau,\tau}$ contains exactly the information about how two such bases are related.

In general, given a parenthesized $\otimes$-product of τ's and 1's, we can use the defining equations for Fibonacci anyons, particularly $\tau \otimes \tau = 1 \oplus \tau$, and the distributivity of $\otimes$ over $\oplus$, to convert the given product into a sum of τ's and 1's. Each summand in that sum arises from the original product as a result of certain choices of 1 or τ when expanding some occurrences of $\tau \otimes \tau$.

For example, in the equation

$$\tau \otimes (\tau \otimes \tau) = \tau \otimes (1 \oplus \tau) = (\tau \otimes 1) \oplus (\tau \otimes \tau) = \tau \oplus (1 \oplus \tau) = 1 \cdot 1 \oplus 2 \cdot \tau$$

considered above, the summand 1 at the right end of the equation arose from the $\tau \otimes (\tau \otimes \tau)$ at the left end by first choosing the summand τ in the evaluation of $(\tau \otimes \tau)$ at the first step in the equation, and then, after applying the distributive law at the second step, choosing the summand 1 in the evaluation of $\tau \otimes \tau$ at the third step. These choices can be visualized as the tree

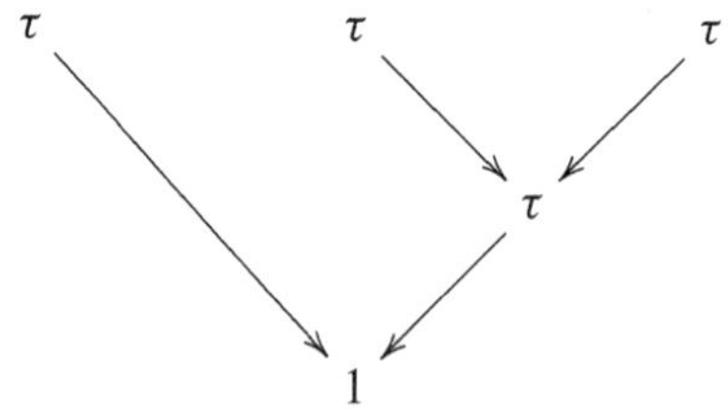

or, in a more compressed notation,

$$(\tau \underset{1}{\cdot} (\tau \underset{\tau}{\cdot} \tau)).$$

Here the three τ's and the parentheses describe the $\otimes$-product $\tau \otimes (\tau \otimes \tau)$ that we began with, and the symbols under the dots indicate the choice of summand at each step. The inner $\underset{\tau}{\cdot}$ indicates that, from the evaluation of the inner $\tau \otimes \tau = 1 \oplus \tau$, we chose the τ summand. After applying distributivity, that leads us to $\tau \otimes \tau$, from which, as indicated by the outer $\underset{1}{\cdot}$, we chose the summand 1.

The other possible choices during the same evaluation would be written

$$(\tau \underset{\tau}{\cdot} (\tau \underset{\tau}{\cdot} \tau)) \quad \text{and} \quad (\tau \underset{\tau}{\cdot} (\tau \underset{1}{\cdot} \tau))$$

and depicted by the trees

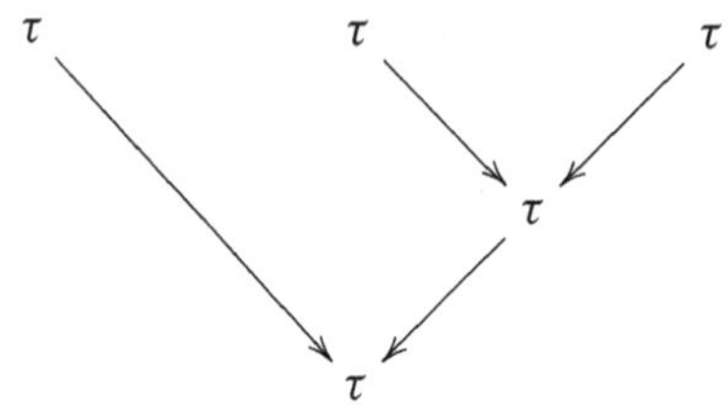

and

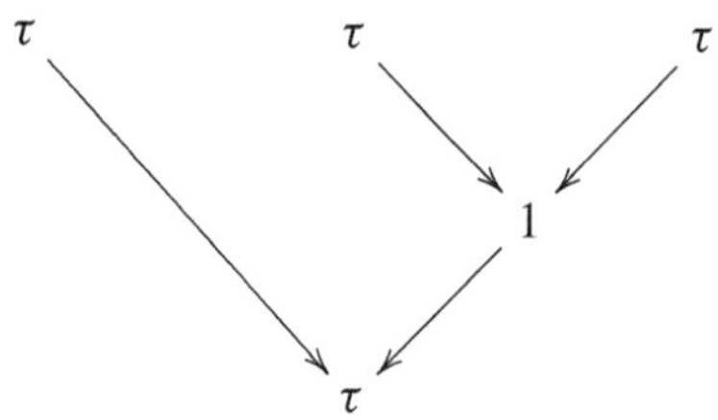

.

The first of these indicates that, as before, we chose the τ summand when evaluating the inner $\otimes$, obtaining, when distributivity is applied, the summand $\tau \otimes \tau = 1 \oplus \tau$, but then we chose the τ rather than the 1. The second indicates that, when evaluating the inner $\tau \otimes \tau$, we chose the summand 1, so that, after applying distributivity, we got $\tau \otimes 1$. Here, there is no choice remaining to be made; $\tau \otimes 1$ is simply τ. Nevertheless, we write τ under the outer dot and at the root of the tree, to make it obvious that the final result here is τ.

In what follows, we shall systematically use the compressed notation, but the reader can easily draw the tree diagrams. Indeed, these diagrams are just the parse trees of the compressed notations. The trees can also be viewed as a sort of Feynman diagrams, depicting how the anyons at the leaves of the tree fuse on their way to the root.

In our notation, we write a product of τ's or 1's, with τ's or 1's also under the dots, to represent specific summands (1 or τ) in the fully distributed expansion of a $\otimes$-product of τ's and 1's. To evaluate $(X \underset{1}{\cdot} Y)$, first evaluate X and Y; then apply $\otimes$ to them; and then take the 1 summand in the result. To evaluate $(X \underset{\tau}{\cdot} Y)$ do the same except that you take the τ summand in the result. These notations will never be used in situations where they would be meaningless because the required summand is not present in the result; that is, we never write $(X \underset{1}{\cdot} Y)$ when one of X, Y evaluates to 1 and the other to τ, for then $\otimes$ yields only τ; and we never write $(X \underset{\tau}{\cdot} Y)$ when both of X, Y evaluate to 1. As in one of the examples above, we include the subscript under the dot even when that subscript is forced because one of the factors evaluates to 1.

Notice that our notation provides symbols, like the three examples above, that denote not only an object 1 or τ (which can be read off by just looking under the outermost dot in the notation) but also a particular occurrence of that $1 = (\mathbb{C}, 0)$ or $\tau = (0, \mathbb{C})$ as a subspace (direct summand) of a specific $\otimes$-product, namely the product with the same factors and the same parentheses as in our notation.

In other words, if we are given a parenthesized $\otimes$-product of 1's and τ's, representing the pair of vector spaces (V_1, V_τ), then by replacing each $\otimes$ by either $\underset{1}{\cdot}$ or $\underset{\tau}{\cdot}$, we obtain (either a meaningless expression because some required summand is absent or) a notation for a subspace of V_1 or V_τ. It denotes a subspace of V_1 (resp. V_τ) just in case the outermost $\otimes$ was replaced by $\underset{1}{\cdot}$ (resp. $\underset{\tau}{\cdot}$).

Our notation provides names for certain summands $1 = (\mathbb{C}, 0)$ or $\tau = (0, \mathbb{C})$ of certain objects (V_1, V_τ) of the Fibonacci category $\mathscr{A}$. We shall also use the same notation for the resulting basis vectors. That is, once we have a copy of, say, $(\mathbb{C}, 0)$ in (V_1, V_τ), the number 1 in $\mathbb{C}$ corresponds to some vector in V_1, and we shall use the same notation for this vector as for the summand. The same goes for the case of copies of $(0, \mathbb{C})$ in (V_1, V_τ); they provide vectors in V_τ.

Notice that, if we begin with some parenthesized $\otimes$-product of 1's and τ's, with value (V_1, V_τ) in $\mathscr{A}$, and if we form all possible (meaningful) notations by replacing $\otimes$ by $\underset{1}{\cdot}$ or $\underset{\tau}{\cdot}$, then the resulting vectors, as described in the preceding paragraph, constitute bases for the vector spaces V_1 and V_τ. This observation is just a restatement of the fact that the original parenthesized $\otimes$-product is the direct sum of all the simple objects obtainable by making the choices indicated by the subscripts in our notation.

8.5.4 Associativity

Now that we have a general notation system for the basis vectors in parenthesized $\otimes$-products, we turn to the specific cases involved in associativity and the pentagon condition.

The unique "interesting" component of associativity, $\alpha_{\tau,\tau,\tau}$, which we sometimes abbreviate as simply α, is an isomorphism from $(\tau \otimes \tau) \otimes \tau$ to $\tau \otimes (\tau \otimes \tau)$, both of which are, as pairs of vector spaces, a 1-dimensional V_1 and a 2-dimensional V_τ. The first parenthesization gives a basis vector

$$((\tau \underset{\tau}{\cdot} \tau) \underset{1}{\cdot} \tau) \quad \text{for } V_1$$

and two basis vectors

$$((\tau \underset{1}{\cdot} \tau) \underset{\tau}{\cdot} \tau) \quad \text{and} \quad ((\tau \underset{\tau}{\cdot} \tau) \underset{\tau}{\cdot} \tau) \quad \text{for } V_\tau.$$

The second parenthesization similarly gives a basis vector

$$(\tau \underset{1}{\cdot} (\tau \underset{\tau}{\cdot} \tau)) \quad \text{for } V_1$$

and two basis vectors

$$(\tau \underset{\tau}{\cdot} (\tau \underset{1}{\cdot} \tau)) \quad \text{and} \quad (\tau \underset{\tau}{\cdot} (\tau \underset{\tau}{\cdot} \tau)) \quad \text{for } V_\tau.$$

Our task is to compute the transformation α between these bases.[8] This α has two components, the first relating two bases of the one-dimensional space V_1 and the second relating two bases of the two-dimensional space V_τ. These are given, respectively, by a non-zero number p such that

$$((\tau \underset{\tau}{\cdot} \tau) \underset{1}{\cdot} \tau) = p(\tau \underset{1}{\cdot} (\tau \underset{\tau}{\cdot} \tau))$$

[8] We have chosen to regard V_1 and V_τ as each being a single space, independent of the parenthesization. The different parenthesizations give (possibly) different bases for these spaces. An alternative view is that each parenthesization gives its own V_1 and V_τ, isomorphic to $\mathbb{C}$ and $\mathbb{C}^2$ respectively, with their standard bases, while α gives an isomorphism between the two V_1's and an isomorphism between the two V_τ's. The two viewpoints are easily intertranslatable and the computations that follow would be the same in either picture.

and a non-singular matrix $\begin{pmatrix} q & r \\ s & t \end{pmatrix}$ such that

$$((\tau \underset{1}{\cdot} \tau) \underset{\tau}{\cdot} \tau) = q(\tau \underset{\tau}{\cdot} (\tau \underset{1}{\cdot} \tau)) + r(\tau \underset{\tau}{\cdot} (\tau \underset{\tau}{\cdot} \tau))$$
$$((\tau \underset{\tau}{\cdot} \tau) \underset{\tau}{\cdot} \tau) = s(\tau \underset{\tau}{\cdot} (\tau \underset{1}{\cdot} \tau)) + t(\tau \underset{\tau}{\cdot} (\tau \underset{\tau}{\cdot} \tau)).$$

Here "non-zero" for p and "non-singular" for the matrix embody the requirement that α is an isomorphism.

We shall now investigate the constraints imposed on p, q, r, s, t by the pentagon condition.

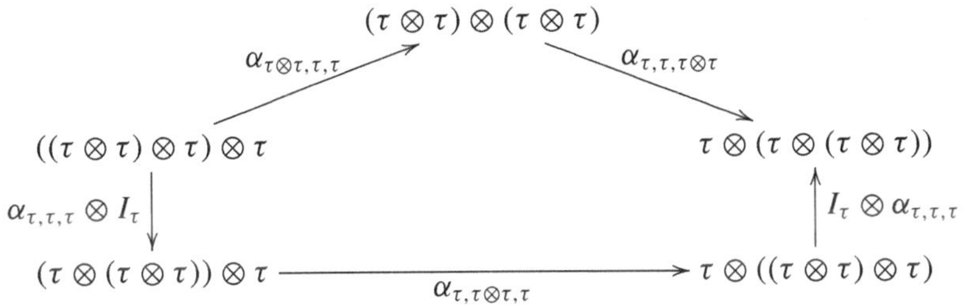

That condition involves the $\otimes$-product of four τ's, parenthesized in five ways, and we shall need to consider the natural bases for all five parenthesizations. Since $\tau^{\otimes 4} = (\mathbb{C}^2, \mathbb{C}^3)$, each parenthesization will give two vectors as a basis for the 1 component and three as a basis for the τ component. We begin by considering the τ components, whose bases are displayed below. (There is no significance to the chosen ordering of the five bases, nor the ordering of the three vectors within each basis.)

$$\begin{array}{lll}
(((\tau \underset{\tau}{\cdot} \tau) \underset{1}{\cdot} \tau) \underset{\tau}{\cdot} \tau) & (((\tau \underset{1}{\cdot} \tau) \underset{\tau}{\cdot} \tau) \underset{\tau}{\cdot} \tau) & (((\tau \underset{\tau}{\cdot} \tau) \underset{\tau}{\cdot} \tau) \underset{\tau}{\cdot} \tau) \\
((\tau \underset{1}{\cdot} \tau) \underset{\tau}{\cdot} (\tau \underset{\tau}{\cdot} \tau)) & ((\tau \underset{\tau}{\cdot} \tau) \underset{\tau}{\cdot} (\tau \underset{1}{\cdot} \tau)) & ((\tau \underset{\tau}{\cdot} \tau) \underset{\tau}{\cdot} (\tau \underset{\tau}{\cdot} \tau)) \\
((\tau \underset{1}{\cdot} (\tau \underset{\tau}{\cdot} \tau)) \underset{\tau}{\cdot} \tau) & ((\tau \underset{\tau}{\cdot} (\tau \underset{1}{\cdot} \tau)) \underset{\tau}{\cdot} \tau) & ((\tau \underset{\tau}{\cdot} (\tau \underset{\tau}{\cdot} \tau)) \underset{\tau}{\cdot} \tau) \\
(\tau \underset{\tau}{\cdot} ((\tau \underset{\tau}{\cdot} \tau) \underset{1}{\cdot} \tau)) & (\tau \underset{\tau}{\cdot} ((\tau \underset{1}{\cdot} \tau) \underset{\tau}{\cdot} \tau)) & (\tau \underset{\tau}{\cdot} ((\tau \underset{\tau}{\cdot} \tau) \underset{\tau}{\cdot} \tau)) \\
(\tau \underset{\tau}{\cdot} (\tau \underset{1}{\cdot} (\tau \underset{\tau}{\cdot} \tau))) & (\tau \underset{\tau}{\cdot} (\tau \underset{\tau}{\cdot} (\tau \underset{1}{\cdot} \tau))) & (\tau \underset{\tau}{\cdot} (\tau \underset{\tau}{\cdot} (\tau \underset{\tau}{\cdot} \tau)))
\end{array}$$

Each row in this picture is a basis for the 3-dimensional V_τ; specifically, it is the basis arising from the same parenthesization of $\tau \otimes \tau \otimes \tau \otimes \tau$ as the parenthesization in our notation.

When writing transformation matrices between these bases, we must regard each basis as given in a specific order, because rows of a matrix come in an order. We (arbitrarily) choose the orders in which the bases are displayed above.

The five isomorphisms that appear in the pentagon condition amount to five transformations between these bases. Let us consider these one at a time, beginning with the one connecting the first two bases in the table. Here we are dealing with the isomorphism

$$\alpha_{\tau\otimes\tau,\tau,\tau} : (((\tau \otimes \tau)\otimes)\tau \otimes \tau) \to ((\tau \otimes \tau) \otimes (\tau \otimes \tau)).$$

The first subscript of this α, namely $\tau \otimes \tau$, can be decomposed as the sum $1 \oplus \tau$, and the naturality of α then implies that $\alpha_{\tau\otimes\tau,\tau,\tau}$ is the direct sum of $\alpha_{1,\tau,\tau}$ and $\alpha_{\tau,\tau,\tau}$. The first of these two summands is the identity, like all associativity isomorphisms where one of the three factors is 1. The second summand is given by our matrix $\begin{pmatrix} q & r \\ s & t \end{pmatrix}$. As a result, we find that the transformation $\alpha_{\tau\otimes\tau,\tau,\tau}$ connecting the first two bases in our list is (taking into account the order in which the basis vectors are listed)

$$\alpha_{\tau\otimes\tau,\tau,\tau} = \begin{pmatrix} 0 & q & r \\ 1 & 0 & 0 \\ 0 & s & t \end{pmatrix}.$$

In this matrix, the 1 in the (2,1) position and the two zeros in its row arise from the fact that the identity map $\alpha_{1,\tau,\tau}$ sends the second vector in our first basis to the first vector in the second basis. Had we listed $(((\tau \underset{1}{\cdot} \tau) \underset{\tau}{\cdot} \tau) \underset{\tau}{\cdot} \tau)$ first rather than second in our first basis, the matrix for $\alpha_{\tau\otimes\tau,\tau,\tau}$ would have been a block diagonal matrix with 1 in the upper left corner.

An exactly analogous computation gives the isomorphism between the second and the last bases in our list:

$$\alpha_{\tau,\tau,\tau\otimes\tau} = \begin{pmatrix} q & 0 & r \\ 0 & 1 & 0 \\ s & 0 & t \end{pmatrix}$$

Multipying these two matrices, we get the transformation from the first basis (parenthesized to the left) to the last (parenthesized to the right) that corresponds to the “short” side of the pentagon (two morphisms, across the top of the diagram). This product is

$$\begin{pmatrix} rs & q & rt \\ q & 0 & r \\ st & s & t^2 \end{pmatrix}.$$

Turning to the long side of the pentagon (three morphisms), we find that the middle one, corresponding to rows 3 and 4 in our list of bases and to the bottom of the diagram, is quite analogous to the two that we have already computed. It is

$$\alpha_{\tau,\tau\otimes\tau,\tau} = \begin{pmatrix} q & 0 & r \\ 0 & 1 & 0 \\ s & 0 & t \end{pmatrix}.$$

The remaining two isomorphisms for the long side of the pentagon (the vertical arrows in the diagram) are a bit different, as they involve α's on three of the four factors and an identity map on the remaining factor. Let us consider $\alpha_{\tau,\tau,\tau} \otimes I_\tau$, which connects the first basis in our list to the third. In effect, this ignores the rightmost factor and acts like α on the first three factors. In other words, it is given by the same matrix as the transformation from the basis

$$((\tau \underset{\tau}{\cdot} \tau) \underset{1}{\cdot} \tau) \qquad ((\tau \underset{1}{\cdot} \tau) \underset{\tau}{\cdot} \tau) \qquad ((\tau \underset{\tau}{\cdot} \tau) \underset{\tau}{\cdot} \tau)$$

to the basis

$$(\tau \underset{1}{\cdot} (\tau \underset{\tau}{\cdot} \tau)) \qquad (\tau \underset{\tau}{\cdot} (\tau \underset{1}{\cdot} \tau)) \qquad (\tau \underset{\tau}{\cdot} (\tau \underset{\tau}{\cdot} \tau)).$$

Notice that, in each of these bases the first element is in the V_1 component, so that component of α, namely p, enters the picture. Indeed, the matrix connecting these bases is

$$\alpha_{\tau,\tau,\tau} \otimes I_\tau = \begin{pmatrix} p & 0 & 0 \\ 0 & q & r \\ 0 & s & t \end{pmatrix}.$$

Similarly, the remaining isomorphism on the long side of the pentagon is also

$$I_\tau \otimes \alpha_{\tau,\tau,\tau} = \begin{pmatrix} p & 0 & 0 \\ 0 & q & r \\ 0 & s & t \end{pmatrix}.$$

Multiplying the three matrices for the long side of the pentagon, and equating, as the pentagon condition requires, the resulting product to the product that we obtained for the short side of the pentagon, we have

$$\begin{pmatrix} p^2q & prs & prt \\ prs & q^2+rst & qr+rt^2 \\ pst & qs+st^2 & rs+t^3 \end{pmatrix} = \begin{pmatrix} rs & q & rt \\ q & 0 & r \\ st & s & t^2 \end{pmatrix}.$$

This is the V_τ part of the pentagon condition. Before turning to the V_1 part, let us extract as much information as possible from the matrix equation that we have just derived.

Suppose, toward a contradiction, that $p \neq 1$. Then the (1,3) and (3,1) components of our matrix equation give $rt = st = 0$, so either $r = s = 0$ or $t = 0$. If $r = s = 0$, then the (1,2) component of the matrix equation gives that $q = 0$ also, but this contradicts the fact that $\begin{pmatrix} q & r \\ s & t \end{pmatrix}$ is non-singular. There remains the case that $t = 0$. Then the (2,2) component says $q = 0$, the (2,3) component says $r = 0$, and we again contradict the non-singularity of $\begin{pmatrix} q & r \\ s & t \end{pmatrix}$. So we have contradictions in all cases if $p \neq 1$.

So $p = 1$. Now the (1,1) entry of the matrix equation gives $q = rs$. Substituting that into the (2,2) component, we get $q(q+t) = 0$, so either $q = 0$ or $q = -t$. The first of these options leads, via the (1,2) entry, to $rs = 0$ and thus to a contradiction to non-singularity, as before. Therefore $q = -t$.

From the (2,3) and (3,2) entries, we get that $(q + t^2)r = r$ and $(q + t^2)s = s$. We cannot have both $r = 0$ and $s = 0$, as that would give $q = 0$ in the (1,2) entry and contradict non-singularity. So we must have $q + t^2 = 1$. In view of $q = -t$, this means $q^2 + q - 1 = 0$ and therefore

$$q = -t = \frac{-1 \pm \sqrt{5}}{2}.$$

This evaluation of q and t, together with the earlier results

$$p = 1 \quad \text{and} \quad rs = q,$$

satisfy, as one easily checks, the entire matrix equation above. The least trivial item to check is the (3,3) entry, $rs + t^3 = t^2$, which, in view of the equations above, becomes $q - q^3 = q^2$, i.e., $0 = q(q^2 + q - 1)$, and this is true because q was obtained as a solution of $q^2 + q - 1 = 0$.

All of the preceding calculation was based on the V_τ component of $\tau^{\otimes 4}$; we still have the V_1 component of the pentagon equation to work out. Again, we have a list of five bases, now for a 2-dimensional space, as follows.

$$\begin{array}{ll}
(((\tau \underset{1}{\cdot} \tau) \underset{\tau}{\cdot} \tau) \underset{1}{\cdot} \tau) & (((\tau \underset{\tau}{\cdot} \tau) \underset{\tau}{\cdot} \tau) \underset{1}{\cdot} \tau) \\
((\tau \underset{1}{\cdot} \tau) \underset{1}{\cdot} (\tau \underset{1}{\cdot} \tau)) & ((\tau \underset{\tau}{\cdot} \tau) \underset{1}{\cdot} (\tau \underset{\tau}{\cdot} \tau)) \\
((\tau \underset{\tau}{\cdot} (\tau \underset{1}{\cdot} \tau)) \underset{1}{\cdot} \tau) & ((\tau \underset{\tau}{\cdot} (\tau \underset{\tau}{\cdot} \tau)) \underset{1}{\cdot} \tau) \\
(\tau \underset{1}{\cdot} ((\tau \underset{1}{\cdot} \tau) \underset{\tau}{\cdot} \tau)) & (\tau \underset{1}{\cdot} ((\tau \underset{\tau}{\cdot} \tau) \underset{\tau}{\cdot} \tau)) \\
(\tau \underset{1}{\cdot} (\tau \underset{\tau}{\cdot} (\tau \underset{1}{\cdot} \tau))) & (\tau \underset{1}{\cdot} (\tau \underset{\tau}{\cdot} (\tau \underset{\tau}{\cdot} \tau)))
\end{array}$$

Computations analogous to (but shorter than) the earlier ones give, for the short side of the pentagon,

$$\alpha_{\tau\otimes\tau,\tau,\tau} = \begin{pmatrix} 1 & 0 \\ 0 & p \end{pmatrix} \text{ and } \alpha_{\tau,\tau,\tau\otimes\tau} = \begin{pmatrix} 1 & 0 \\ 0 & p \end{pmatrix}.$$

So the product for the short side is simply $\begin{pmatrix} 1 & 0 \\ 0 & p^2 \end{pmatrix}$. For the long side, we get

$$\alpha_{\tau,\tau\otimes\tau,\tau} = \begin{pmatrix} 1 & 0 \\ 0 & p \end{pmatrix}$$

and

$$\alpha_{\tau,\tau,\tau} \otimes I_\tau = I_\tau \otimes \alpha_{\tau,\tau,\tau} = \begin{pmatrix} q & r \\ s & t \end{pmatrix}.$$

Equating the product of the long side and the product of the short side, we get

$$\begin{pmatrix} 1 & 0 \\ 0 & p^2 \end{pmatrix} = \begin{pmatrix} q^2 + prs & qr + ptr \\ qs + pts & rs + pt^2 \end{pmatrix}.$$

This matrix equation is automatically satisfied because of the equations that we had already derived from the V_τ component of the pentagon condition. So there is no new information in the V_1 component.

We can, however, get some additional information if we impose the requirement that the associativity isomorphisms be unitary transformations. This amounts to requiring the vector spaces of morphisms $\mathrm{Hom}(X, Y)$ to be Hilbert spaces and requiring our natural bases for them to be orthonormal.

Unitarity tells us nothing new about p, since we already know $p = 1$, but unitarity of $\begin{pmatrix} q & r \\ s & t \end{pmatrix}$ gives the equations

$$q^2 + |r|^2 = q^2 + |s|^2 = 1 \quad \text{and} \quad q(\bar{s} - r) = q(s - \bar{r}) = 0,$$

where bars denote complex conjugation and where we used the fact that q is real. So $s = \bar{r}$ and, since $rs = q$, we get first that q has to be positive,

$$q = \frac{-1 + \sqrt{5}}{2},$$

and second that

$$r = \sqrt{q}e^{i\theta} \quad \text{and} \quad s = \sqrt{q}e^{-i\theta}$$

for some real θ. Thus, we finally have, under the assumption of unitarity,

$$\alpha_{\tau,\tau,\tau} = \begin{pmatrix} 1 & 0 & 0 \\ 0 & q & \sqrt{q}e^{i\theta} \\ 0 & \sqrt{q}e^{-i\theta} & -q \end{pmatrix}$$

with $q = \frac{-1+\sqrt{5}}{2}$ and θ arbitrary. The presence of θ here is an artifact of our choice of bases. If we modified the final vector in each of our bases, $((\tau \underset{\tau}{\cdot} \tau) \underset{\tau}{\cdot} \tau)$ in the domain of $\alpha_{\tau,\tau,\tau}$ and $(\tau \underset{\tau}{\cdot} (\tau \underset{\tau}{\cdot} \tau))$ in the codomain, by a phase factor $e^{-i\theta}$, then, with respect to the new bases, we would have

$$\alpha_{\tau,\tau,\tau} = \begin{pmatrix} 1 & 0 & 0 \\ 0 & q & \sqrt{q} \\ 0 & \sqrt{q} & -q \end{pmatrix}.$$

8.5.5 Braiding

We now turn to the task of computing the braiding σ in the Fibonacci anyon category $\mathscr{A}$. The only nontrivial component of the natural isomorphism σ is $\sigma_{\tau,\tau}$, because components with a subscript 1 are identity morphisms and components with non-simple subscripts reduce to direct sums of components with simple subscripts.

The nontrivial component $\sigma_{\tau,\tau}$ is an isomorphism from $\tau \otimes \tau = 1 \oplus \tau$ to itself. Representing objects of $\mathscr{A}$ by pairs of vector spaces, we have that $\sigma_{\tau,\tau}$ is an automorphism of $(\mathbb{C}, \mathbb{C})$, so it amounts to two non-zero scalars, a multiplying vectors in the first (1) component and b multiplying vectors in the second (τ) component. These are subject to the hexagon identity, which equates the composites

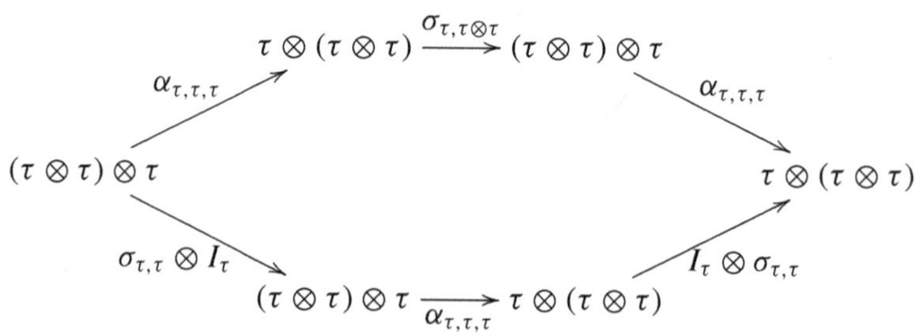

as well as the analogous identity with σ^{-1} in place of σ.

Consider the first (1) component of this equation. In the bottom composition, the $\sigma_{\tau,\tau}$ factors in the first and third morphisms must act on the τ components so that the $\otimes$-product with I_τ has a 1 component. So both of these are b. The α between them, acting on the 1 component, is an identity map, because our previous calculation gave $p = 1$. So the bottom of the hexagon is b^2. In the top, both of the α's are again just 1. The σ in the middle of that row is $\sigma_{\tau,1\oplus\tau}$, i.e., the direct sum of $\sigma_{\tau,1}$ and $\sigma_{\tau,\tau}$. The first of these two summands has no 1 component; the second does, and it is a. So the top of the hexagon is just a, and the hexagon condition reads $a = b^2$. (The corresponding calculation for σ^{-1} gives only $a^{-1} = b^{-2}$, which is no new information.)

Now consider the second (τ) component of the hexagon equation. We do the calculation in matrix form, using the natural bases

$$((\tau \underset{1}{\cdot} \tau) \underset{\tau}{\cdot} \tau) \quad \text{and} \quad ((\tau \underset{\tau}{\cdot} \tau) \underset{\tau}{\cdot} \tau) \quad \text{for } (\tau \otimes \tau) \otimes \tau$$

and

$$(\tau \underset{\tau}{\cdot} (\tau \underset{1}{\cdot} \tau)) \quad \text{and} \quad (\tau \underset{\tau}{\cdot} (\tau \underset{\tau}{\cdot} \tau)) \quad \text{for } \tau \otimes (\tau \otimes \tau).$$

With respect to these bases, $\alpha_{\tau,\tau,\tau}$ is given by $\begin{pmatrix} q & r \\ s & t \end{pmatrix}$ as computed earlier. Both $\sigma_{\tau,\tau} \otimes I_\tau$ and $I_\tau \otimes \sigma_{\tau,\tau}$ are given by

$$\begin{pmatrix} a & 0 \\ 0 & b \end{pmatrix} = \begin{pmatrix} b^2 & 0 \\ 0 & b \end{pmatrix},$$

because in each case, $\sigma_{\tau,\tau}$ acts as a on the first basis vector (where it interchanges two τ's that were combined to 1) and as b on the second (where it interchanges two τ's that were combined to τ). Finally, $\sigma_{\tau,\tau\otimes\tau}$ is the direct sum of $\sigma_{\tau,1}$, which is 1, and $\sigma_{\tau,\tau}$ acting on the τ component, which is b; since that direct sum decomposition matches our choice of bases, $\sigma_{\tau,\tau\otimes\tau}$ is given by the matrix $\begin{pmatrix} 1 & 0 \\ 0 & b \end{pmatrix}$. Multiplying the matrices for each of the rows, we find that the hexagon identity, in the τ component, reads

$$\begin{pmatrix} q^2 + brs & (q+bt)r \\ (q+bt)s & rs + bt^2 \end{pmatrix} = \begin{pmatrix} b^4 q & b^3 r \\ b^3 s & b^2 t \end{pmatrix}.$$

Since we know, from our associativity calculation, that r and s are not zero, the (1,2) and (2,1) entries of this matrix equation reduce to $q + bt - b^3 = 0$, or, since $t = -q$,

$$b^3 = q(1-b).$$

The (1,1) and (2,2) entries give, after we remember that $rs = q$ and cancel a common factor q,

$$q + b = b^4 \quad \text{and} \quad 1 + bq + b^2 = 0.$$

The last of these equations, being quadratic in b, can be solved explicitly:

$$b = \frac{-q \pm \sqrt{q^2 - 4}}{2}.$$

We note that, since $q = \frac{\sqrt{5}-1}{2}$ is between 0 and 1, the square root in the formula for b is imaginary, so the two values of b are each other's complex conjugates. The product of the two values for b is 1, so b is a complex number of absolute value 1 with real part $\frac{-q}{2}$.

The ambiguity in the choice of b is unavoidable in this situation. Replacing one choice by the other just replaces σ by its inverse (since $|b| = 1$), and there is nothing in the algebra of $\mathscr{A}$ that distinguishes the counterclockwise motion defining σ from the clockwise motion defining σ^{-1}. To put it another way, the change from one value of b to the other can be exactly compensated by reflecting the orientation of the (2-dimensional) space in which the anyons live.

Although we have now computed b and thus also $a = b^2$, we can get a more useful view of these numbers by manipulating the three equations above that relate b to q. Solving the last one for q in terms of b, and substituting the result, $q = \frac{-b^2-1}{b}$ into the other two equations, we obtain from the first equation that

$$b^3 = \frac{b^3 - b^2 + b - 1}{b}, \text{ i.e., } b^4 - b^3 + b^2 - b + 1 = 0,$$

which means that $-b$ is a primitive fifth root of unity and therefore b is a primitive tenth root of unity. The third equation above confirms that by reducing to $b^5 = -1$.

Among the four primitive tenth roots of unity only two, $e^{\pm 3\pi i/5}$, have negative real parts, as b does (recall that its real part is $-q/2$). So we conclude that, up to complex conjugations,

$$b = e^{3\pi i/5} \text{ and therefore } a = e^{6\pi i/5}.$$

This completes the calculation of the braiding σ for Fibonacci anyons.

Remark 2 The multiplicative structure for Fibonacci anyons, summarized by the fusion rule $\tau \otimes \tau = 1 \oplus \tau$, is perhaps the simplest nontrivial fusion rule. Other fusion rules have been analyzed, either by hand as we have done here or with computer support. The appendix of [2] summarizes much of what is known about specific examples. There does not, however, seem to be any general theory for arbitrary fusion rules.

8.5.6 Fibonacci Anyons and Quantum Computation

In Sect. 8.2, we mentioned the hope that, by using anyons to encode qubits, one could use braiding to transform anyon states in various ways, thereby enabling quantum computation. Two anyons are not sufficient for this purpose, because the braid group on two strands is abelian, whereas quantum computation needs non-commuting unitary transformations. In the case of Fibonacci anyons, the computation in the preceding subsection shows that the braiding transformation $\sigma_{\tau,\tau}$ is diagonal in a suitable basis, so it splits into one-dimensional representations; this again shows its inadequacy for quantum computation.

With three Fibonacci anyons, the situation improves dramatically. In a suitable basis, the transformation that braids the first two of the three anyons, $\sigma_{\tau,\tau} \otimes I_\tau$, is still diagonal. The same goes for the transformation that braids the second and third anyons, but the suitable bases in these two cases are not the same. They differ by an associativity isomorphism α. More precisely, one is the conjugate of the other by $\alpha_{\tau,\tau,\tau}$. They do not commute.

In fact, such braiding transformations suffice to approximate arbitrary unitary transformations of the two-dimensional Hilbert space V_τ for $\tau^{\otimes 3}$. Furthermore, using six Fibonacci anyons to code two qubits, one can approximate, by braiding, the so-called "controlled not" gate, which, in combination with one-qubit gates, is sufficient to produce all unitary gates for an arbitrary number of qubits; that is, it is sufficient for quantum computation. We refer to [9, Sect. 6] for these combinations of Fibonacci braidings.

References

1. Bagchi, B., & Misra, G. (2000). A note on the multipliers and projective representations of semi-simple Lie groups. *Sankhyā: The Indian Journal of Statistics*, *62*, 425–432.
2. Bonderson, P. H. (2007). Non-abelian anyons and interferometry. Ph.D. thesis, California Institute of Technology.
3. Freyd, P. (1964). *Abelian categories: An introduction to the theory of functors*. Harper & Row.
4. Giulini, D. (2007). Superselection rules. http://arxiv.org/pdf/0710.1516.pdf.
5. Kauffman, L., & Lomonaco, S. (2010). Topological quantum information theory. In S. Lomonaco (Ed.), *Quantum information science and its contribution to mathematics. Proceedings of symposia in pure mathematics* (Vol. 68, pp. 103–177). American Mathematical Society.
6. Kitaev, A. (2003). Fault-tolerant quantum computation by anyons. *Annals of Physics*, *303*, 2–30. http://arxiv.org/abs/quant-ph/9707021.
7. Mac Lane, S. (1971). *Categories for the working mathematician*. Graduate Texts in Mathematics 5. Springer.
8. Nayak, C., Simon, S., Stern, A., Freedman, M., & Das Sarma, S. (2008). Non-abelian anyons and topological quantum computation. *Reviews of Modern Physics*, *80*, 1083–1159.
9. Panangaden, P., & Paquette, É. O. (2011). A categorical presentation of quantum computation with anyons, Chapter 15. In B. Coecke (Ed.), *New structures for physics*. Lecture Notes in Physics (Vol. 813, pp. 983–1025). Springer.
10. Raghunathan, M. S. (1994). Universal central extensions. *Reviews of Modern Physics*, *6*, 207–225.
11. Wang, Z. (2010). *Topological quantum computation*. CBMS Regional Conference Series in Mathematics (Vol. 112). American Mathematical Society.

Chapter 9
Taking Physical Infinity Seriously

Don Perlis

Abstract The concept of infinity took centuries to achieve recognized status in the field of mathematics, despite the fact that it was implicitly present in nearly all mathematical endeavors. Here I explore the idea that a similar development might be warranted in physics. Several threads will be speculatively examined, including some involving nonstandard analysis. While there are intriguing possibilities, there also are noteworthy difficulties.

Keywords Non-standard analysis · Infinity · Physics

9.1 Introduction

Infinity plays a central role in mathematics, and arguably always has—despite occasional negative characterizations (even by some of the most esteemed practitioners). Today surely there is little question about its importance in the minds of the vast majority of mathematicians.[1] There is also very wide appreciation of the idea that whither goes mathematics, there also goes physics (and often the other way around). And yet in physics the notion of infinity plays a rather curious "fix-it-up" role, rather like duct tape, that is brought out for use whenever needed but then put firmly back

[1] In [8] Martin Davis includes a discussion of infinity in mathematics in terms of imaginative powers of our minds (my words, not his), and (partly) justifies this by analogy with physics—somewhat the reverse of my point here, but one I am equally sympathetic to.

My thanks for helpful comments and clarifications from: Paulo Bedaque, Juston Brodie, Jeff Bub, Jean Dickason, Sam Gralla, Dan Lathrop, Carlo Rovelli, Ray Sarraga, and two anonymous reviewers—none of whom however is to be blamed for any errors or outrageousnesses that remain.

D. Perlis (✉)
University of Maryland, College Park, USA
e-mail: perlis@cs.umd.edu

E.G. Omodeo and A. Policriti (eds.), *Martin Davis on Computability, Computational Logic, and Mathematical Foundations*,
Outstanding Contributions to Logic 10, DOI 10.1007/978-3-319-41842-1_9

in the tool box again. Thus it is not kept front and center in actual physical models, quite unlike its now central and fundamental role in mathematics.[2]

This is part of a much larger issue: how mathematics relates to physical reality. This involves many aspects that we will not touch on here, other than some brief comments. For instance, Wigner [26] regards it as "unreasonable" that there is such a strong connection between math and physics. And Kreisel [14] has considered whether quantities that are physically observable (according to a given physical theory) can be generated by a Turing machine; such a theory he calls "mechanical". See also [1, 16, 22], all of whom discuss cosmological issues such as whether space is infinite in extent; Rovelli [22] in particular distinguishes—similarly to a distinction we shall draw—between infinite divisibility and infinite extent.

A related question is: what sort of universe is needed in order for there to be a possibility of mathematics at all? That is, not actual mathematical practice, but simply the possibility of "stuff" sufficient to allow, for instance, such things as sequences, records, relations. There would seem to be a requisite minimum level of temporality and spatiality even for natural numbers to have any meaningfulness. And, perhaps deeper: what counts as stuff, and what is it for stuff to "be"? But we will leave these questions aside, and return to our main theme.[3]

Here I will describe a number of examples in which infinity is used explicitly in physics, and possible developments that these might suggest, including a few detours along the way.[4] Yet I must add that, as a non-physicist, I also approach the broader topic with some trepidation; and while I have consulted a number of physicists in the writing of this paper, still any misconceptions are completely my own. I trust the reader will pardon any sense that I am throwing in the kitchen sink; this essay represents some possibly far-flung imaginings that perhaps do not fall altogether within customary styles in scientific writing.

The rest of this paper is organized as follows: We describe the examples just referred to above, to distinguish several modes of use of infinities in physics; next I review some ideas due to Jose Benardete on a Zeno-like puzzle about infinity, and some related issues concerning particles, densities, and spin; we then turn to non-standard analysis as one methodology that appears to shed some light (in connection with Dirac delta functions), but has difficulties of its own.

---

[2]One prominent example that will *not* be discussed at any length here are the divergent Feynman integrals (among others) of quantum field theory (QFT). See for instance the excellent Wikipedia entry for Renormalization [28].

[3]I can't resist noting that in roughly 1968-9 Martin Davis mentioned to me that in his estimation a huge unclarity underlay foundational issues in mathematics and in particular set theory: what counts as a thing?

[4]That the topic is appropriate to a volume dedicated to Martin Davis, I justify with the observations that (i) Martin helped instill in me a general love for ideas on topics far and wide; and (ii) at least two of Martin's writings bear on related themes: nonstandard analysis [7] and quantum physics [6]. I note that Rovelli [21] entertains an idea already present in [6], namely that of observer-dependent reference frames in quantum mechanics; and (personal note from Rovelli) this also apparently has come up in writings of Kochen and Isham as well, all after Martin's contribution appeared. See also [24] for more on this theme.

9.2 Multiple Uses of Infinity in physics

Quantum mechanics provides us with many intriguing examples of our subject; I give three here. First, Schrödinger's solution of his wave equation for the energy levels of the hydrogen atom involves an argument in which infinity plays the role of a kind of *reductio*, or proof by contradiction, leading to the rejection of the infinity. Second, that same solution results in an infinite set of energy levels, which are pointedly *not* rejected. Third, Dirac introduced the (infinite-valued on an infinitesimal interval) delta function because it provided a highly simplifying and intuitively satisfying notation for his vastly influential treatment of quantum mechanics. I briefly summarize each of these uses of infinity below.

In a 1926 paper, Schrödinger solved his famous wave equation for the special case of the hydrogen atom. Along the way he had to set to zero certain series terms, since otherwise they would lead to variables with infinite values. (The remaining terms provide solutions for energy levels of the hydrogen atom that are the familiar Bohr ones that closely match experiment[5]—but not quite close enough; later refinements were needed, including spin and relativistic effects.) So in this case, a variable taking on an infinite value is used as a reason to reject it and instead consider only alternative lines of argument. This of course is not new to Schrödinger but in fact is a common form of argument, applicable whenever the variable in question is something one has reason to think should be finite. I provide this particular example of such a *reductio* use of infinity here (as opposed to any number of others) simply because it is curious that it arises in the same setting in which the next example occurs. We may refer to this first as a *dense* physical infinity: a physical variable (that in principle might be measured by means of instruments within certain physical confines) taking on (but perhaps should not do so) an infinite value. This is employed via a *reductio* to eliminate the infinity (sometimes easily as above, sometimes with enormous effort and controversy as in QFT).

Yet a result of Schrödinger's argument is that the distinct possible energy levels of the hydrogen atom alluded to above are infinite in number, and in fact a specific formula is derived for the possible energies, E_n where $n = 1, 2, ...$ *This* infinitude is not shrugged off as unphysical; each and every E_n is taken as representing an in-principle possible physical energy for the atom.[6] Indeed, it is the excellent match-up with experiment that makes the Schrödinger result so convincing.[7] Of course, it is similar in kind to the infinitude of possible heights (or potential energies) of a projectile above ground level, which is also not seen as unusual. These perhaps

[5]E.g., when associated to the spectral lines found by Balmer in 1886.

[6]A very recent result [9] even derives the famous centuries-old Wallis formula for π from the very same infinite sequence of hydrogen's energy levels, something no one had the faintest idea could happen, suggesting that the infinitude has yet further significance—although just what that may be is unclear.

[7]For instance, had Schrödinger's calculation led instead to a sequence of values for E_n that stopped after $n = 20$, surely there would have been a frenzied attempt by experimentalists to find twenty-one energy levels to test the theoretical result.

amount, in the end, to little more than the fact that the infinite (unbounded) set of real numbers, **R**, is taken as the possible range of values for many physical variables (with some limitations as dictated by a given situation—but the infinitude is not in general ruled out). This is a *range-of-values* physical infinity: a mere listing of possible values, of which there may be infinitely many. Yet it is a possibility that, in some sense, describes (a working picture of) the universe: the universe has in it an unbounded range of allowable values for certain variables.[8]

One way to make these two standard physical uses of infinity more intuitive may be this: if a variable represents a measurable quantity, something that one might detect in an experiment, then the measured value must be finite: we have no means to measure an actual infinity; whereas any—even an infinite—*number* of finite values might be measured (given enough time). Or: there may be an infinite amount of space, matter, or energy, in the universe; but not right where the measuring instruments are located. Note that we are not taking a stand on such a view; in fact, we are exploring alternative possibilities!

Indeed, one can reason: there may be things physically present that we cannot measure. One such that comes to mind is the wavefunction itself; this is sometimes[9] characterized as the fundamental "reality" of which our measurements ferret out (and even modify) some features but never reveal the full thing in itself. If the wavefunction is really there, yet never fully revealed, why not also infinite energies and other quantities? Or consider space and time (or spacetime) themselves: we never measure all of space or time, by any means. Yet in measuring bits and pieces, we convince ourselves that there is a great deal more, and in the case of some theories even that the universe has an infinitude of such pieces, either extended (range-of-values) or densely packed.

Our third example is Dirac's delta function. This is in wide use by physicists (and not only in quantum mechanics). Yet the delta function is routinely viewed as a useful fiction, not something to take seriously except as a convenient shorthand for a much more cumbersome and less intuitive set of tools. This mode we then call the *useful fiction* infinity: we use it but we don't believe it corresponds to anything physical.[10] Nonetheless, it seems to fall also into the *dense* mode of infinity.

Thus we have cases where a dense infinity is outlawed (by reductio), and others where it is accepted as a useful fiction; and there are also cases (range-of-values)

---

[8]The chapter by Blass and Gurevich in this volume similarly comments on "infinitely many possible values, for example of position or momentum" and the corresponding infinite-dimensional Hilbert space of such a system's states. This is closely realted to the idea of an infinite extent of space, which may or may not be the case—but such is not seen as a reason to reject a model outright. Similarly, the infinitely-many possible reference frames in quantum mechanics suggested in [6] is not suspect on the basis of the infinity involved.

[9]More so some decades ago; it seems now a minority view.

[10]This is reminiscent of the early uses of imaginary numbers: they were clearly useful, but it was far less clear that such a number could be a *thing* in any sense available back then. Eventually two developments helped: (i) the observation that imaginary numbers can be interpreted as rotations, and (ii) formal/abstract methodology: if something has a consistent mathematical use, that is all that is needed in order for it to *be* an object of mathematical study.

where infinity is accepted as quite physically sensible. Much of what we are considering here is whether some of the "fiction" cases should perhaps be considered as less fiction and more real physics. Delta functions are one case in point (we shall return to them below) but not the only one.

9.3 Benardete's Challenge

Benardete [4] discusses novel variants of a paradox of Zeno. Here is a version that suits our purposes: Imagine that an impenetrable barrier is erected at each point $x = 1/2^n$ for n $= 1, 2, \ldots$; we suppose the barriers to be of zero thickness (or of decreasing thickness as they close in on $x = 0$, so that they do not overlap or touch each other, and so that they do not overlap or touch $x = 0$). Moreover, imagine that each barrier is immovable once so placed. Finally, imagine that a projectile is aimed at the barriers from a point to the left, i.e., from some $x < 0$.

Let us first of all note that this appears to be a case of dense infinity. There is an infinitude of physical entities in a finite region. To be sure, this particular setup is highly implausible; we are bringing it into the discussion as an easy warmup case, before proceeding to more physically plausible cases.

Now, what will happen as the projectile moves rightward? Since there is nothing apparent to impede the projectile at negative positions ($x < 0$), it would seem that it should continue its rightward motion until it strikes a barrier. But before it can strike a barrier at $x = 1/2^n$ it must first strike (and pass through) all those to its left (at $x = 1/2^m$ for all $m > n$). This is impossible by the conditions of the problem. So it cannot strike any barrier at all! Hence it must stop its rightward motion, never passing zero, yet without touching anything that would be a cause for its rightward motion to cease.

This has been debated in various philosophical papers; see [13, 19, 30]. In [18] standard physics is brought to bear on the puzzle in the forms of classical mechanics, quantum mechanics, and relativity, showing for instance in the classical case that a field effect in the form of a repulsive force is mandated by Newton's Laws, so that the projectile is bounced back to the left before passing zero. But the lesson for us here is that even a dense infinity need not be paradoxical when seen from within standard physical theory. (Of course, one can resurrect a paradox by insisting the barriers produce no forces outside their own immediate locations; and the lesson then would be that this is inconsistent with standard physics.)

Another version of the puzzle involves a continuous barrier-wall extending from some point $b > 0$ all the way back to, but not including, $x = 0$. That is, this is a wall of width b but with its left face missing. While a seeming bit of physical nonsense (at least in terms of materials made of atoms) it is a familiar enough entity in mathematics, essentially a half-open half-closed interval. And the same form of argument applies as in the earlier Benardete example. It would seem that physical entities cannot be isolated quite as well as our imaginations might like: physical interactions will occur and cannot be dismissed by mere stipulation.

Thus the Benardete examples provide a kind of dense infinity, but not apparently one that "breaks" anything. Perhaps this is because it does not directly involve an infinite density of standard physical quantities like mass or charge or energy. (A closer analysis might turn up an infinite sort of potential energy, however.) In any event, when we turn to something "real" such as an electron, the situation presents itself more starkly.

9.4 The Electron—Getting to the Point

An electron presents a somewhat related challenge. An electromagnetic field exists around any charged particle. If the particle is not in motion, then it is simply an electric field, $\boldsymbol{E}$, given by Coulomb's Law. But the same law mandates that the field's magnitude E increases at locations closer to the particle, approaching infinity in the limit. In addition, the charge density is zero outside the immediate location of the electron, and infinity at that location. Finally, the mass density is also infinite at the location of the electron, and zero elsewhere. These claims are based on the not uncommon assumption that an electron has no spatial extent and is located at a literal mathematical point; experimentally, the electron's radius is less than 10^{-22} m [15].[11] A similar situation arises in the case of a black hole, where the mass density becomes infinite at the mathematical point (singularity) of the hole itself.[12]

One way to mathematically represent the situation of an infinite point density is via a Dirac delta function, namely one that is infinite at the point in question, and zero elsewhere. This—usually taken as a convenient fiction as already noted—does the trick really well and surprisingly often, and is now a standard item in the physics toolbox. However, delta functions can quickly turn from convenience to headache, due to the nonlinearity of many applications. That is, the usual way to "precisify" a delta function is as a Schwartz distribution: a linear functional on a space of functions. However—as Wald [25] points out—in many applications (nonlinear ones) delta functions (when viewed as distributions) cannot be sensibly multiplied, and this poses significant difficulties for their use where there are point sources of fields. This is a bit outrageous: why cannot one multiply two functions? The answer is that the Schwartz representation really groups these "fiction-functions" into equivalence classes (ones that provide the same results for certain special integration properties[13]), and integration does not always respect some of the desired characteristics

[11] But see for instance [23].

[12] See [27] for an interesting discussion of electrons as black holes. A related set of issues involve the self-force and self-energy of an electron (or any point charge): the field created by a charge affects not only space surrounding the charge but also at the charge location(s) as well. Thus an electron's field influences it's own behavior. Similar considerations apply to any particle with non-zero mass: the associated gravitational field should affect the particle itself; see [25].

[13] Namely: $\int_{-\infty}^{+\infty} f_1(x)g(x) = \int_{-\infty}^{+\infty} f_2(x)g(x)$ for all "test" functions g.

needed for non-linear applications. Yet once ungrouped from each other and treated as genuine functions, delta functions can indeed by multiplied, as we will see in the next section.[14]

Summarizing a bit, one way that infinity arises in physics is as follows: a vector field (such as gravitational or electrostatic force) depends on the spatial separation between one body and another, in a way that increases without bound as that distance decreases to zero. In particular, in these two instances, the force is proportional to the reciprocal of the square of the distance. When that distance is zero, the expression for the force becomes one divided by zero: $1/0$.

Now, division by zero is extremely problematic; it is not simply that it is not defined, but that it is both overdetermined and underdetermined. $0/0$ can be set equal to any number ($0/0 = x$) with impunity, since $0 = 0x$. And $1/0$ cannot be set equal to any number at all, since $1 \neq 0x$. So there is no non-arbitrary nor even consistent way to define division by zero that respects the basic concept of division: $(a/b)b = a$, that is, as the inverse of multiplication.

It is tempting to say that this is because the real numbers are too restrictive, and that $1/0 = \infty$. But then what is $2/0$? And do we allow $1 = 0 \times \infty$? These notions contain hints of a possible solution. In fact, mathematical physics often employs such intuitions, in the form of infinitesimals and infinities; again think of the standard delta function, that is zero at all non-zero reals, yet when infinitesimally close to zero it rises up to infinity.

But mathematicians have invented many sorts of numbers, going well beyond the familiar real and complex fields, including some that explicitly contain infinities as first-class objects. Which fits the physical situation best? We shall not attempt to answer this here, nor even to survey the existing options. Instead, we shall discuss just one such option, with particular application to delta functions and—possibly—to point particles.

9.5 NSA

One well-known approach to making sense of infinite and infinitesimal quantities is nonstandard analysis (NSA), where the real number system **R** is extended to ***R**, which includes "numbers" that are larger than every real, and also ones that are smaller than every positive real and yet are themselves larger than 0. The latter (small ones) and their negatives become the infinitesimals in common use in physical reasoning. This was the aim of Robinson [20]: to develop ***R** and to show that in fact the familiar intuitive arguments using infinitesimals then become quite rigorous.

[14] This is not to say that successful application to non-linear differential equations is an automatic benefit; as noted, it is not the product per se but rather integration properties of products that is at issue.

But infinitesimals are not the same thing as zero; they are simply very very close to zero; one might say that they form a kind of fuzzy zero—and more generally, that each real r has about it a band of new numbers (r plus any infinitesimal) that "coat" r so closely that for ordinary purposes r and its coat are indistinguishable.[15]

A key point is that, while being in zero's coat, an infinitesimal ε nonetheless has a well-defined reciprocal $1/\varepsilon$, which is infinite (larger than every real). We still do not have a reciprocal for zero itself, but perhaps we can dispense with that, and when a variable "approaches" zero we may try to regard it as being in zero's coat rather than being zero itself. More generally, the coat of a real r then provides stand-ins for r, which are r-ish in more or less degree (but all of them are r-ish and not s-ish for any other real s).

As Robinson has shown, ***R** can be given a very rigorous definition, so that it remains an algebraic field and respects the "usual" mathematical properties of **R**. These properties are given sharp characterization, roughly as follows: for any sentence S that can be expressed in a particular formal language L (including much of standard math notations, for instance +, $\times$, constants, =, $<$, $\forall$, set-membership, etc.—but NOT using a symbol for **R** itself), S is true when interpreted as being about elements in **R** iff it is true about ***R**.[16] Now this "transfer principle" between **R** and ***R** is the basis for a great many applications of NSA.[17] But results of such applications—at least when those results are interpreted as being about **R** (or more precisely about the "set-theoretic superstructure for **R**")—generally are theorems that can also be proven (though maybe less easily or intuitively) without NSA. One of the suggestions we are raising here is this: perhaps ***R** (or its superstructure) can be taken seriously as a model of physical reality, to see whether this sheds light on infinities that arise in physics.[18]

One very nice (traditional) application of NSA is the delta function, which now can be defined an as actual (non-fictional) function from ***R** to ***R**. For instance, given an infinitesimal ε, let $\delta(x) = 0$ for all numbers (in ***R**) that lie outside $[-\varepsilon/2, \varepsilon/2]$, and let $\delta(x) = 1/\varepsilon$ for numbers in that interval. The graph of such a function then is an infinitesimally thin, infinitely high rectangle, and the area under it is exactly $\varepsilon \times 1/\varepsilon = 1$. And then the integral of $\delta(x)$ times any function *f from ***R** to ***R** (that is an appropriate extension of an integrable function f on the reals), gives f(0)—or more precisely, gives the average value of *f in that interval, which is itself in the coat of—and so normally indistinguishable from—f(0).

---

[15]I apologize for introducing the term *coat* for this; already in use are: monad, haze, cloud, halo. My excuse is that a *coat of paint* is thin, hugs close to its target, and is not to be touched by other entities (at least while wet).

[16] Details can get a bit complicated; see [7].

[17]There are by now dozens of books and hundreds or articles on the subject of NSA in general and applications of the transfer principle in particular. See for instance [2, 5].

[18]See [12] for a rare exceptional—but alas all too preliminary—treatment of NSA's nonstandard universe itself as having physical significance, in this case to QFT.

But now the *product* of any two such delta functions from *R to *R is unproblematically another function from *R to *R. There is a tradeoff, however. For we must *choose* a particular delta function to use in a given application, rather than opt for the distributional approach that lumps many such together.[19]

9.6 Back to the Real World

Now we return to physics, and in particular to the electron. We regard it as being a point, or rather, we take its radius to be in the coat of 0 (or whatever point it is centered on). That is, we will postulate it to be a ball of infinitesimal radius. In particular, let some ε_e be that radius, and assume its mass m is uniformly distributed.[20] Now we will attempt to characterize its spin ($\hbar/2$) as a physical angular momentum L of actual rotation, namely with an angular frequency ω so that we get the usual classical formula:

$$L = \hbar/2 = I\omega = (2/5)m\varepsilon_e^2\omega$$

Since ε_e is infinitesimal then ω must be infinite, since the LHS is finite.

The idea of treating spin as a possible rotational phenomenon was considered long ago (see below), but taking the radius to be a positive real r; this led to trouble with special relativity (SR). A point on the surface of the electron ball would—in order that the rotation provide the proper angular momentum of spin, have to travel at speeds in excess of the speed of light. But to reach such a speed would require infinite energy, according to SR, and that traditionally is taboo. Here then is a possible advantage of NSA: suppose we allow physical quantities to be infinite.

Let's calculate the speed v of a point on the surface of an "electron ball" with (initially real) radius r that is rotating with angular momentum $\hbar/2$. From the above equation, we get

$v = \omega(1/2\pi)(2\pi r) = \omega r = 5\hbar/(4mr)$

If we insist that $v < c$ then we find

$c > 5\hbar/(4mr)$

or

$r > 5\hbar/(4mc) = 0.5 \times 10^{-12}$ m.

---

[19]Further investigation (I am unaware of any work on this topic) may reveal advantages to particular "natural" choices for a delta function in particular applications. For now I simply point out one from Robinson's book (p. 138): $\frac{1}{\sqrt{\varepsilon\pi}}\exp(-\frac{x^2}{\varepsilon})$. For real values of ε this is just an ordinary Gaussian, which arises quite naturally in many situations, and has very nice mathematical properties. Possibly in the nonstandard realm it will also play a helpful role. Note that this is not claimed to resolve issues about non-linear applications where integration properties of products arise.

[20]Note that this means the ball will be a proper subset of the coat, since coats have no boundary; if they did, then for instance $2\varepsilon_e$ would be outside the coat, which makes no sense for it too is infinitesimal.

This is essentially the negative result found by Goudsmit and Uhlenbeck [11] that made them (and others) give up the idea of spin as deriving from an actual physical rotation, since it was known even then that r is less than 3×10^{-15} m.[21]

There is an alternative: allowing $v \geq c$, and also allowing infinite energies, as well as replacing r by ε_e. But why insist that r be infinitesimal? This is not strictly necessary. But since as already noted, it is commonly thought that $r = 0$ (an electron is an actual point with no extent, no volume)[22] and since we are allowing infinities anyway, it is tempting to go "all the way" (at least all the way to infinitesimals, if not literally to zero).

Back to our calculations: if ε_e is infinitesimal then as noted above, the angular frequency ω is infinite. But what then is the speed of a point in the electron coat, at distance ε_e from the origin of rotation? It will be as above, but replacing r with ε_e, hence infinite:

$$v = \omega\varepsilon_e = 5\hbar/(4m\varepsilon_e)$$

This infinite speed precisely produces the finite angular momentum $\hbar/2$. That is, the infinite speed of a point within the electron coat (which itself is at infinitesimal distance ε_e from the origin of rotation), works together with that infinitesimal distance to produce the needed finite angular momentum of spin.

However, not everything works out so nicely. The kinetic energy of mass m with speed v, in SR, is

$$T = mc^2(\gamma - 1)$$

where $\gamma = 1/\sqrt{1-(v/c)^2)}$

When $v = c$, γ is infinite, hence T would seem to be infinite. This is well-known, of course, and is a primary reason that c is regarded as an unreachable upper limit on all speeds of massive objects. But it is now even worse: for *this* infinity (of γ) seems to be of the totally impossible kind: $1/0$.[23]

There is however another interpretation: multiplying through by $\sqrt{1-(v/c)^2)}$, we get

$$\sqrt{1-(v/c)^2)}T = mc^2(1 - \sqrt{1-(v/c)^2)})$$

and for $v = c$ this becomes $0 \times T = mc^2$. A reasonable conclusion now is that $m = 0$: a particle traveling at light-speed has no mass.[24] And T is not further constrained here, infinite or otherwise. Presumably it can (for $v = c$) be taken as the energy of

[21] But see for instance [10, 17], for this is still a topic of dispute.

[22] For many purposes; but in QFT for instance this is not quite right.

[23] It is no good trying to wriggle out of this by supposing T is an NSA sort of infinity; that would correspond to v being "almost" the same as c (in the same coat, so that v/c is in the coat of 1). For in fact we need—for the Goudsmit/Uhlenbeck model—that v be even greater than c. And then γ actually has an imaginary value! This leads into the even stranger physics of tachyons.

[24] This can actually be given a positive spin (pun intended). The Higgs field endows particles with mass according to whether they are retarded by it—retarded from traveling at light-speed, that is. Particles that are not so retarded are by definition massless!

an appropriate light-speed particle. Whether this is physical nonsense or not, at least we are getting some sort of "results" from such an approach.

9.7 Summary and Discussion

We have isolated three uses of infinities in physics: dense, range-of-values, and useful-fiction. Range-of-values seems generally unproblematic, but illustrative of the idea that our understanding of the universe can involve infinities of some sort. These are not directly measured, but rather are supported by a mix of inductive reasoning and evidence; and they do not seem to present major difficulties.

The infinities in the Benardete example perhaps lie in between range-of-values and dense: many location values are posited, yet they come close to representing an infinite density of something—but it is not clear just what. And there is no outright paradox if we apply ordinary physics and a little commonsense.

But when we replace imagined barriers with actual physical entities such as fields, things can quickly get bizarre, as in infinite values for charge and force and mass densities. While our discussion focused on a point-model of the electron, any point-source field will do. There are standard tools for representing this—for instance the delta function—but these are usually seen as merely useful calculational devices and not as possible models of what the universe is like. I am arguing that the great success of such tools speaks to the strong possibility of an underlying phenomenon well-worth trying to model.

I am not urging that NSA need be the mathematical physics of the future. There are certainly other directions to consider, such as the surreal numbers studied by Conway, Kruskal and others (see Wikipedia entry [29]). In addition, Bell [3] presents an approach to infinitesimals (but not infinities) based on "smooth worlds" where logic (and geometry) gets even stranger than in NSA yet where physics again comes into play. And indeed infinity (of the dense kind) might happen not to be physically sensible at all. But the idea should not be discarded out of hand.

References

1. Aguirre, A. (2011). Cosmological intimations of infinity. In M. Heller & W. H. Woodin (Eds.), *Infinity: New research frontiers* (pp. 176–192). Cambridge University Press.
2. Albeverio, S. (1988). Nonstandard analysis in mathematical physics. In N. Cutland (Ed.), *Nonstandard analysis and its applications* (pp. 182–220). Cambridge Univeristy Press.
3. Bell, J. L. (2008). *A primer of infinitesimal analysis* (2nd ed.). Cambridge University Press.
4. Benardete, J. (1964). *Infinity: An essay in metaphysics*. Oxford: Clarendon Press.
5. Cutland, N., Di Nasso, M., & Ross, D. A. (Eds.). (2006). *Nonstandard methods and applications in mathematics* (Vol. 25). AK Peters Ltd.
6. Davis, M. (1977). A relativity principle in quantum mechanics. *International Journal of Theoretical Physics*, *16*(11), 867–874.
7. Davis, M. (2005). *Applied nonstandard analysis*. Dover. (Reprinted from 1977 Wiley edition).

8. Davis, M. (2014). Pragmatic platonism. In N. Tenant (Ed.), *Foundational adventures*. College Publications.
9. Friedmann, T., & Hagen, C. R. (2015). Quantum mechanical derivation of the Wallis formula for π. *Journal of Mathematical Physics, 56*.
10. Giulini, D. (2008). Electron spin or "classically non-describable two-valuedness". *Studies In History and Philosophy of Science Part B: Studies In History and Philosophy of Modern Physics, 39*(3), 557–578.
11. Goudsmit, S., & Uhlenbeck, G. (1925). Unpublished manuscript.
12. Gudder, S. (1994). Toward a rigorous quantum field theory. *Foundations of Physics, 24*(9), 1205–1225.
13. Hansen, C. S. (2011). New Zeno and actual infinity. *Open Journal of Philosophy, 1*(02), 57.
14. Kreisel, G. (1974). A notion of mechanistic theory. *Synthese, 29*(1), 11–26.
15. Meschede, D. (2007). *Optics, light and lasers: The practical approach to modern aspects of photonics and laser physics* (2nd ed.). Wiley-VCH.
16. Misner, C. W. (1981). Infinity in physics and cosmology. In *Proceedings of the American Catholic Philosophical Association* (Vol. 55, pp. 59–72).
17. Ohanian, H. C. (1986). What is spin. *American Journal of Physics, 54*(6), 500–505.
18. Perlis, D., & Sarraga, R. (1976). Physical theory and the divisibility of space and matter. Technical report, Math Dept, Univ of Puerto Rico, Mayaguez.
19. Priest, G. (1999). On a version of one of Zeno's paradoxes. *Analysis, 59*(261), 1–2.
20. Robinson, A. (1996). *Non-standard analysis*. Princeton University Press. (Reprint of 1974 2nd edition; first published in 1966 by North-Holland).
21. Rovelli, C. (1996). Relational quantum mechanics. *International Journal of Theoretical Physics, 35*(8), 1637–1678.
22. Rovelli, C. (2011). Some considerations on infinity in physics. In M. Heller & W. H. Woodin (Eds.), *Infinity: New research frontiers* (p. 167). Cambridge University Press.
23. Sasabe, S. (1992). Virtual size of electron caused by its self-field. *Journal of the Physical Society of Japan, 61*(8), 2606–2609.
24. Van Fraassen, B. C. (2010). Rovelli's world. *Foundations of Physics, 40*(4), 390–417.
25. Wald, R. M. (2011). Introduction to gravitational self-force. In L. Blanchet, A. Spallicci & B. Whiting (Eds.), *Mass and motion in general relativity* (pp. 253–262). Springer.
26. Wigner, E. P. (1960). The unreasonable effectiveness of mathematics in the natural sciences. *Communications on Pure and Applied Mathematics, 13*(1), 1–14.
27. Wikipedia (2015). Black hole electron—Wikipedia, the free encyclopedia. Retrieved September 29, 2015.
28. Wikipedia (2015). Renormalization—Wikipedia, the free encyclopedia. Retrieved September 29, 2015.
29. Wikipedia (2015). Surreal number—Wikipedia, the free encyclopedia. Retrieved September 29, 2015.
30. Yablo, S. (2000). A reply to new Zeno. *Analysis, 60*(2), 148–151.

Chapter 10
Banishing Ultrafilters from Our Consciousness

Domenico Cantone, Eugenio G. Omodeo and Alberto Policriti

> *The reader who remembers these key points will do well in what follows. In particular, it is now quite all right to entirely forget how the nonstandard universe was defined and to banish ultrafilters from our consciousness.*
> (Martin Davis, *Applied Nonstandard Analysis*, 1977)

Abstract The way in which Martin Davis conceived the first chapter of his book "*Applied nonstandard analysis*" is a brilliant example of information hiding as a guiding principle for the design of widely applicable constructions and methods of proof. We discuss here a common trait that we see between that book and another writing of the year 1977, "*Metamathematical extensibility for theorem provers and proof-checkers*", which Martin coauthored with Jacob T. Schwartz. To tie the said part of Martin's study on nonstandard analysis to proof technology, we undertake a verification, by means of a proof-checker based on set theory, of key results of the non-standard approach to analysis.

Keywords Proof checking · Proof engineering · Nonstandard analysis · Foundations of infinitesimal calculus

D. Cantone (✉)
DMI, Università di Catania, Viale A. Doria 6, 95125 Catania, Italy
e-mail: cantone@dmi.unict.it

E.G. Omodeo
DMG/DMI, Università di Trieste, Via Valerio 12/1, 34127 Trieste, Italy
e-mail: eomodeo@units.it

A. Policriti
DMIF, Università di Udine, Via Delle Scienze 206, 33100 Udine, Italy
e-mail: alberto.policriti@uniud.it

E.G. Omodeo and A. Policriti (eds.), *Martin Davis on Computability, Computational Logic, and Mathematical Foundations*,
Outstanding Contributions to Logic 10, DOI 10.1007/978-3-319-41842-1_10

10.1 Introduction

Year 1977: Martin Davis appears in print with "*Applied nonstandard analysis*" [14], whose subject is less close to computability and computational logic than the various areas to which Martin has contributed before. Nevertheless we will tie that book—appropriately, we believe—to another publication of the same year, "*Metamathematical extensibility for theorem provers and proof-checkers*" [23, pp. 120–146], jointly authored by Martin and his friend and colleague "Jack" (namely Jacob T. Schwartz).[1] We aim at unveiling an affinity between some of the matter which that book treats in preparation for analysis proper and the field of automated reasoning of which Martin has been a trailblazer since its early days,[2] and at taking advantage of that link for a proof-checking undertaking which we see as promising.

Martin's book is dedicated to the memory of Abraham Robinson, the creator of nonstandard analysis. At the Summer Institute for Symbolic Logic held at Cornell University, a scientific gathering that both had attended in 1957, Robinson gave a talk in which he "made the provocative remark that the auxiliary points, lines, or circles 'constructed' as part of the solution to a geometry problem can be thought of as being elements of what is now called the Herbrand universe for the problem" [15, pp. 7–8].[3] At the same meeting Martin reported on his own implementation, three years earlier on a JOHNNIAC machine, of Presburger's decision procedure for elementary additive number theory [12]. This proximity of interests between the two distinguished scholars about automating proofs was, presumably, coincidental.

In 1977, on the other hand, disappointment is beginning to take place in the automated deduction community (see [5]), as researchers experience the combinatorial explosion plaguing the automatic search for mathematical proofs even if pruned by the best available techniques. More emphasis is now placed on comfortable interaction between man and computerized proof assistants, and on proof checkers (see, e.g., [38]) as opposed to fully automatic theorem provers. Specific knowledge pertaining to diverse branches of mathematics begins to be perceived as essential for an advancement of the proof techniques; Ballantyne and Bledsoe [3] (see also [2]) succeed in automating the proofs of hard theorems in analysis using methods which rely on the nonstandard viewpoint.

The new context brings to the fore issues related to correct-program technology and proof engineering. An emblem of the times is the Clear specification

[1] See [22] and, therein, the enjoyable [16]; see also [21] and [1, pp. 478–480]. The above-cited [23] led to the sole joint publication by Martin and Jack, namely [24].

[2] Landmark contributions of Martin to automatic theorem-proving in 1st-order predicate logic have been [10, 13, 19, 20, 25], historically occurring between Paul C. Gilmore's and Dag Prawitz' methods, on the one hand, and J. Alan Robinson's resolution principle on the other. Concerning the *linked conjunct* method then proposed by Martin and his team at Bell Labs, see [29, 39].

[3] The term '*Herbrand universe*', today widely used, appeared for the first time in the influential paper [13] (reviewed in [34]); but [17, p. 432] contends that it would be more historically correct to credit the construction of that universe to Thoralf Skolem.

language [7], paving the way to the OBJ family of languages, which will integrate specification, prototyping, and verification into a system with a single underlying logic: theorem-proving is now aimed at providing mechanical assistance for proofs that are needed in the development of software and hardware. This is the scene encountered by the joint work [24] of Martin and Jack, at the dawn of large-scale proof technology.

Here is the issue they raised. "For use of mechanized proof verifier systems to remain comfortable over a wide range of applications, ⋯ it should be possible to augment the system by adding new symbols, schemes of notation, and extended rules of inference of various kinds" [24, p. 217]. A stringent requirement is that the envisioned changes to a system do not disrupt its soundness: a proof verifier should, therefore, be furnished with the metamathematical capability of justifying its progressive augmentations.

Use of a metamathematical extension mechanism, [24] points out, leads to the common acceptance of algebraic calculations in lieu of detailed predicate calculus proofs. Although recourse to the methods of nonstandard analysis in lieu of the ε-δ methods is not mentioned in that paper, we see that less familiar but expedient detour as being in accord with the matter under discussion.

As an arena for experimenting with this circle of ideas, we have undertaken a merciless formal remake of [14, Chap. 1] with Jack's proof checker Ref, see [35, Chap. 4], which embodies a variant of the Zermelo-Fraenkel set theory. This task, which has hardly anything to do with analysis *per se*, is an essential prerequisite if we are to bring the methods of nonstandard analysis within the scope of Ref. As a result of the "mathematical simplicity, elegance, and beauty of these methods"—and of "enthusiasm ⋯ not unrelated to the well-known pleasures of the illicit"—, we expect to eventually get the reward of "their far-reaching applications" (see [14, p. viii]).

Our effort will also suggest changes to Ref's current implementation which can improve its metamathematical extensibility.

We have set up substantial ground for specifying and proving, by means of the Ref verifier (very succinctly described in Sect. 10.6), the two consequences of Łoś's theorem which we need (namely, Theorems 10.1 and 10.2 in Sect. 10.3): once we will have fully achieved those goals, we will move on to work on Robinson's concurrence theorem and on a few other crucial propositions (Theorems 10.3, 10.4, 10.5, and 10.6 of Sect. 10.4). To complete our job we must then introduce "schemes of notation and extended rules of inference of various kinds" that properly assist Ref's users in exploiting nonstandard methods.

In order to reach the goals of our experiment, we must express in set-theoretic terms metalevel notions such as the evaluation of a sentence in a universe; another not entirely trivial task concerns the representation of individuals (thought of as 'non-sets') within a formal system which deals with sets whose construction ultimately relies on nothing but the null set $\emptyset$. For these two matters, to be discussed in Sect. 10.9 and in Sects. 10.7 and 10.8 respectively, our experiment is innovative, at least as regards the Ref proof checker. In other respects, we can benefit from work previously done: among other things we found, already adequately formalized, a theory of ordinal numbers conceived *à la* Raphael M. Robinson, and the ultrafilter theorem obtained using Zorn's lemma.

Concerning proof checkers, issues of reuse have an even greater relevance than for theorem provers. Such issues pertain more to proof engineering than to computational logic:[4] rather than going through the same proof pattern several times, one should abstract a common method to be recalled over and over again, with all the conveniences offered by technology.

Reuse is supported in Ref by a construct named 'THEORY' (see [31] and [35, pp. 19–25]), similar to—although of a less algebraic nature—a mechanism for parameterized specifications of the aforementioned Clear specification language. This paper will discuss how to organize THEORYs that enable one to tackle without reiteration of techniques the foundations of nonstandard analysis; hopefully, it will stimulate reflections on good "proof hiding" practices, of the kind which Martin's passage [14, p. 42], quoted in the epigraph to this paper, seems eager to suggest.

10.2 Basic Construction for Nonstandard Analysis

> **Why nonstandard analysis?** Nonstandard analysis is a technique rather than a subject ··· The subject can be claimed to be of importance insofar as it leads to simpler, more accessible expositions, or (more important) to mathematical discoveries. [14, p. 1]

The initial part of [14] dwells on how to enlarge a standard universe into a nonstandard one. While taking stock at the end of the first chapter, Martin stresses that much of the machinery developed up to there is not used in the remainder of the book; then, in recapitulating which key points the reader should remember, he underlines the three main tools of nonstandard analysis: *transfer principle*, *concurrence*, and *internality*.

We will now give a quick account of the elaborate ultrapower construction whose details Martin deems "quite all right", after that turning point, "to banish from our consciousness". We thereby undertake a formal recasting of that construction with Ref, in order to encapsulate it within Ref's THEORYs.

The STANDARD UNIVERSE is the SUPERSTRUCTURE

$$\widehat{\mathsf{s}} = \overbrace{\underbrace{\overbrace{\mathsf{s} \cup \mathscr{P}(\mathsf{s})}^{\mathsf{s}_1} \cup \mathscr{P}(\mathsf{s} \cup \mathscr{P}(\mathsf{s}))}_{\mathsf{s}_2} \cup \mathscr{P}(\mathsf{s} \cup \mathscr{P}(\mathsf{s}) \cup \mathscr{P}(\mathsf{s} \cup \mathscr{P}(\mathsf{s})))}^{\mathsf{s}_3} \cup \cdots$$

[4]See [8, pp. 5–6]. In a recent personal web-page, David Aspinall (Univ. of Edinburgh) defines *Proof Engineering* to mean the activity on construction, maintenance, documentation and presentation of large formal proof developments. Within Proof Engineering, according to Aspinall, "Software Engineering provides the techniques to develop large, structured and well-specified repositories of computer code; proof checking provides the mechanisms to provide a complete semantics and formal correctness as an absolute quality criterion.".

built on a set $\mathsf{s} = \mathsf{s}_0$, whose $(n+1)$-st stage is $\mathsf{s}_{n+1} = \mathsf{s}_n \cup \mathscr{P}(\mathsf{s}_n)$ for each $n \in \mathbb{N} = \{0, 1, 2, \dots\}$ (as customary, $\mathscr{P}$ designates the powerset operator). It is essential that s consists of *individuals*; namely that $\emptyset \notin \mathsf{s}$ and that no element of any element of s pops up at any stage, viz., $\widehat{\mathsf{s}} \cap \bigcup \mathsf{s} = \emptyset$. Every set w of individuals generates a superstructure $\widehat{w}$, much as we have just indicated for s; e.g., $\widehat{\emptyset}$ consists of the entities known as *hereditarily finite sets*.

The superstructure $\widehat{\mathsf{s}}$ gets embedded into another one, $\widehat{\mathsf{w}}$, built on a specific set $\mathsf{w} \supset \mathsf{s}$ of individuals, by means of a function $x \mapsto {}^*x$; in particular ${}^*\mathsf{s} = \mathsf{w}$. A set $\widetilde{\mathsf{w}} \subset \widehat{\mathsf{w}}$ is cut out of the wider superstructure: this $\widetilde{\mathsf{w}}$, satisfying the revealing equality $\widetilde{\mathsf{w}} = \bigcup_{i \in \mathbb{N}} {}^*\mathsf{s}_i$, will be the NONSTANDARD UNIVERSE paired with $\widehat{\mathsf{s}}$.

Such companions $\widehat{\mathsf{s}}$, $\widetilde{\mathsf{w}}$ will—in a sense—have the same properties. An unrestrained formulation of this principle would have paradoxical consequences, though, and we must postpone to Sect. 10.3 the precise formulation of criteria enabling the transferability of properties. In a major instance studied in [14, Chap. 2], s includes an Archimedean ordered field D, e.g., the field $\mathbb{Q}$ of rational numbers or the field $\mathbb{R}$ of real numbers; then ${}^*\mathsf{D}$, included in w, will still satisfy the laws of an ordered field but will violate the Archimedean property which—roughly speaking—rejects infinitely large or infinitely small elements.[5]

Before showing how to construct $\widetilde{\mathsf{w}}$, let us make it clear which are the sets which qualify as universes:[6]

Definition 10.1 A set $\mathscr{U}$ is called a UNIVERSE if $\emptyset \in \mathscr{U}$ and the following properties hold for all x, y:

Upward closure: If $x, y \in \mathscr{U}$, then $\{x, y\} \in \mathscr{U}$.
Downward closure: If $x \in \mathscr{U}$ and $x \cap \mathscr{U} \neq \emptyset$, then $x \subseteq \mathscr{U}$
(this says that each element x of $\mathscr{U}$ is either an individual, hence has no element in $\mathscr{U}$, or is included in $\mathscr{U}$). ⊣

The upward closure property readily yields that a universe $\mathscr{U}$ is always closed with respect to the Kuratowski ordered pair formation $\langle x, y\rangle =_{\mathrm{Def}} \{\{x\}, \{x, y\}\}$; by also exploiting downward closure we get, for each function $g \in \mathscr{U}$ such that $g \cup \mathsf{dom}(g) \subseteq \mathscr{U}$, that the result $g\lceil x$ of applying g to a set x belongs to $\mathscr{U}$. (By function we mean here a single-valued set of ordered pairs; moreover, when g fails to be a function or x does not belong to its domain, $g\lceil x$ is meant to designate $\emptyset$.) Every superstructure based on a set of individuals is a universe, so it is closed with respect to pair formation and to function application.

[5]In particular, when $\mathsf{D} = \mathbb{R}$, we get a field, ${}^*\mathbb{R}$, of entities called *hyperreal* numbers. In ${}^*\mathbb{R}$ there are positive numbers lying infinitely close to zero; the reciprocals of such infinitesimals must, of course, exceed any positive integer.

[6]Our definition of universe marginally differs from the one given in [14, p. 15] in that we are not assuming individuals to be given beforehand. Certain proper classes can also be regarded as universes, according to a plain generalization of this definition to be seen in Fig. 10.5.

The construction of $\widetilde{\mathsf{W}}$ relies on a pair æ, ï such that

(1) $\text{æ} \subseteq \mathscr{P}(\text{ï}) \setminus \{\emptyset\}$;
(2) $x \cap y \in \text{æ}$ for all $x, y \in \text{æ}$;
(3) $y \in \text{æ}$ whenever $x \in \text{æ}$ and $x \subseteq y \subseteq \text{ï}$;
(4) no strict superset of æ meets the *filter* conditions (1)–(3).

(Consequently, see [14, p. 10], $\{x \cap \text{ï}\,,\ \text{ï} \setminus x\} \cap \text{æ} \neq \emptyset$ holds for every set x.) By well-established terminology, æ is an *ultrafilter*[7] over the *index set* $\text{ï} = \bigcup \text{æ}$. Momentarily we do not commit our choice of æ and ï in any way; this choice is most relevant, though, for the applicability of the nonstandard techniques.

We say that a property $C(j)$ of elements of ï holds *a.e.* ('almost everywhere') if $\{j \in \text{ï} \mid C(j)\} \in \text{æ}$, that is, if the indices satisfying C form a set which belongs to æ. Thus, for example, the condition $gj = hj$ *a.e.* defines an equivalence relation over $\mathsf{S}^{\text{ï}}$, the set of all functions from the index set into standard individuals; we can then pick a representative element ρg out of each equivalence class $\left\{h \in \mathsf{S}^{\text{ï}} \mid gj = hj \text{ } a.e.\right\}$, and finally get the set

$$\mathsf{W} = \left\{\rho g : g \in \mathsf{S}^{\text{ï}}\right\}$$

of *nonstandard individuals*. This is an enlargement of S, whose elements can in fact be put in natural correspondence with the representatives of *a.e.* constant functions (the injection of S into W is $x \mapsto \rho\, g_x$, where $g_x \in \{x\}^{\text{ï}}$, i.e. $g_x = \text{ï} \times \{x\}$). We will manage to enforce the strict inclusion $\mathsf{W} \supsetneq \mathsf{S}$ in Sect. 10.4; our present assumptions only suffice to ensure that $\mathsf{W} \supseteq \mathsf{S}$.

The construction at issue continues with the specification of a function, $\bar{\ }$, whose set of values will be the universe $\widetilde{\mathsf{W}}$ we are after and whose domain is layered in a way mimicking the hierarchical organization

$$\widehat{\mathsf{S}} = \bigcup_{n \in \mathbb{N}} \mathsf{S}_n = \mathsf{S}_0 \uplus \biguplus_{n \in \mathbb{N}} \left(\mathscr{P}(\mathsf{S}_n) \setminus \mathsf{S}_n\right)$$

(where $\uplus$ and $\biguplus$ designate disjoint unions) of the standard universe:

$$\bar{\ }: \bigcup_{n \in \mathbb{N}} \left\{f \in \widehat{\mathsf{S}}^{\text{ï}} \mid fj \in \mathsf{S}_n \text{ } a.e.\right\} \longrightarrow \widehat{\mathsf{W}}\,.$$

For each $f \in \widehat{\mathsf{S}}^{\text{ï}}$ such that $fj \in \mathsf{S}_0$ *a.e.*, we put $\bar{f} = \rho g$, where $g \in \mathsf{S}^{\text{ï}}$ is such that $fj = gj$ *a.e.* Next, for successive numbers $n \in \mathbb{N}$, we define *à la* Mostowski the image $\bar{f}$ of each function f such that $fj \in \mathscr{P}(\mathsf{S}_n) \setminus \mathsf{S}_n$ *a.e.*, by putting

$$\bar{f} = \left\{\bar{g} : g \in \widehat{\mathsf{S}}^{\text{ï}} \mid gj \in \mathsf{S}_n \cap (fj) \text{ } a.e.\right\}.$$

[7] A slicker characterization of ultrafilters will be shown in Fig. 10.7.

The following facts admit straightforward proofs:

- $\widetilde{\mathsf{W}}$, the set of all images $\bar{f}$, is a universe;
- $\bar{f} \in \bar{g}$ if and only if $fj \in gj$ *a.e.*;
- $\bar{f} = \bar{g}$ if and only if $fj = gj$ *a.e.*

Much as before, there is a natural one-one correspondence between $\widehat{\mathsf{S}}$ and those functions, in the domain of $\bar{\ }$, which are *a.e.* constant; hence the embedding *of $\widehat{\mathsf{S}}$ into $\widetilde{\mathsf{W}}$ announced at the beginning of this section is plainly induced by $\bar{\ }$. This function *will soon be extended by bringing into its domain many subsets of $\widehat{\mathsf{S}}$ which do not belong to $\widehat{\mathsf{S}}$.

Before going any further, let us pause to recall that Martin works under the assumption that "we have available some given sufficiently large set $\mathscr{I}$ of true individuals (sometimes called *urelemente*), about which we assume nothing except that they are not sets" [14, p. 11], and he repeatedly stresses that questions as to the true 'nature' of such entities are irrelevant to mathematical practice.[8] Anyway, we will have to face this issue (see Sect. 10.8) while carrying out our formalization task, because our framework will be a set theory devoid of individuals proper: our 'individuals' will simply be sets whose elements are 'inaccessible' from within the superstructure.

10.3 Bounded Formulae and the Transfer Principle

> The link between logic and computing is to a great extent the notion of a formal language, which is the kind of language machines understand. [18, p. 83]
>
> Formulas of $\mathscr{L}_U$ can be used not only to make assertions about U, but also to define subsets of U. [14, p. 23]

In order to make assertions about a universe $\mathscr{U}$ and to introduce its definable subsets, [14, pp. 20–21] specifies a language $\mathscr{L}_{\mathscr{U}}$ endowed with:

T0. constants $\boldsymbol{c}$, which are in one-one correspondence with the elements of $\mathscr{U}$ (each $\boldsymbol{c}$ is meant to designate the corresponding element c of $\mathscr{U}$);
T1. a countable infinitude $\boldsymbol{x}_1, \boldsymbol{x}_2, \boldsymbol{x}_3, \ldots$ of variables (each ranging over $\mathscr{U}$);
T2. dyadic function symbols $\langle s, t \rangle$ and $(s \restriction t)$ (which are meant to designate, respectively, ordered pair formation and function application);

F0. dyadic relation symbols $(s = t)$ and $(s \in t)$ (designating $=$ and $\in$);
F1. propositional connectives $\neg$ (monadic) and **&** (dyadic);
F2. bounded quantifiers of the form $(\exists \boldsymbol{x}_n \in t)$, where t stands for a term where $\boldsymbol{x}_n$ does not appear.

---

[8] In a similar attitude, [11, p. 54] states that "one possible view is that the integers are atoms and should not be viewed as sets. Even in this case, one might still wish to prevent the existence of unrestricted atoms. In any case, for the 'genuine' sets, Extensionality holds and the other sets are merely harmless curiosities.".

More detailed syntactic rules about terms and formulae of $\mathscr{L}_{\mathscr{U}}$, as well as the semantics of $\mathscr{L}_{\mathscr{U}}$, follow the pattern familiar to anyone who has encountered first-order predicate languages; we leave them as understood for the time being and will belabor this point when arriving at our formalization task (see Sect. 10.9). Anyway, it will best suit our purposes to handle only formulae in *negative normal form*: hence we admit as primitive constructs also the propositional connective $\vee$ and bounded universal quantifiers $(\forall \boldsymbol{x}_n \in t)$; moreover, we confine $\neg$ inside contexts of the forms $\neg(s = t)$ and $\neg(s \in t)$, shortened as usual to $(s \neq t)$ and $(s \notin t)$.

If exactly one variable, say $\boldsymbol{x}_n$, occurs free in a formula α of $\mathscr{L}_{\mathscr{U}}$, then we indicate by $\alpha(\boldsymbol{c})$ the sentence[9] resulting from α when all free occurrences of $\boldsymbol{x}_n$ get replaced by a constant, $\boldsymbol{c}$, that designates some $c \in \mathscr{U}$.

Definition 10.2 A set $d \subseteq \mathscr{U}$ is called DEFINABLE if there is a formula α of $\mathscr{L}_{\mathscr{U}}$ with one free variable such that $d = \{c \in \mathscr{U} \mid \alpha(\boldsymbol{c}) \text{ is true in } \mathscr{U}\}$. $\dashv$

Consider, now, the languages $\mathscr{L}_{\widehat{\mathsf{S}}}$ and $\mathscr{L}_{\widetilde{\mathsf{W}}}$ of the standard universe and of its nonstandard counterpart. Let the notation $\models \alpha$ express the fact that α, a sentence of $\mathscr{L}_{\widehat{\mathsf{S}}}$, is true in $\widehat{\mathsf{S}}$; similarly, indicate by ${}^*{\models}\beta$ the fact that β, a sentence of $\mathscr{L}_{\widetilde{\mathsf{W}}}$, is true in $\widetilde{\mathsf{W}}$.

A translation $\lambda \mapsto {}^*\lambda$ of terms and formulae from $\mathscr{L}_{\widehat{\mathsf{S}}}$ into $\mathscr{L}_{\widetilde{\mathsf{W}}}$ can be specified as follows: to get ${}^*\lambda$, replace every constant $\boldsymbol{c}$ occurring in λ by the constant ${}^*\boldsymbol{c}$ that designates the image ${}^*c$ of c.

We can now state two propositions, both easily obtainable from Łoś's theorem, a fundamental result of model theory which we underplay here:

Theorem 10.1 *If α, β are formulae of $\mathscr{L}_{\widehat{\mathsf{S}}}$ where the only free variable is $\boldsymbol{x}_1$ and*

$$\{c \in \widehat{\mathsf{S}} \mid \models \alpha(\boldsymbol{c})\} = \{c \in \widehat{\mathsf{S}} \mid \models \beta(\boldsymbol{c})\}$$

holds, then

$$\{c \in \widetilde{\mathsf{W}} \mid {}^*{\models} {}^*\alpha(\boldsymbol{c})\} = \{c \in \widetilde{\mathsf{W}} \mid {}^*{\models} {}^*\beta(\boldsymbol{c})\} .$$

Theorem 10.2 (Transfer principle) *For every sentence α of $\mathscr{L}_{\widehat{\mathsf{S}}}$,*

$${}^*{\models} {}^*\alpha \textit{ if and only if } \models \alpha .$$

Thanks to Theorem 10.1, we can add to the domain of the function * every definable subset d of $\widehat{\mathsf{S}}$, via the unambiguous stipulation that

$${}^*d = \{c \in \widetilde{\mathsf{W}} \mid {}^*{\models} {}^*\alpha(\boldsymbol{c})\} \text{ when } d = \{c \in \widehat{\mathsf{S}} \mid \models \alpha(\boldsymbol{c})\} .$$

---

[9]When the need will arise, we will adjust this notation also to terms, indicating by $t(\boldsymbol{c})$ a term devoid of variables resulting from replacement of a variable of t by a constant $\boldsymbol{c}$.

Davis thus briefly conveys the significance of the transfer principle:

> There is a formal language that can be used to make assertions that are ambiguous in that they can refer to either of the two structures. ··· The *transfer principle* roughly states that the same assertions of the formal language are true in the standard universe as in the nonstandard universe. It is typically used by proving a desired result in the nonstandard universe, and then, noting that the result is expressible in the language, concluding that it holds in the standard universe as well. [14, pp. 2–3]

Let us pause again, to observe that the task of formalizing within set theory such model-theoretic propositions as the above Theorems 10.1 and 10.2 presupposes that we encode terms and formulae via sets: we will display a technique for that purpose in Sect. 10.9. Similar tasks arise frequently in logic, when it comes to investigate inside a formal system some meta-theoretical issues regarding the system itself. E.g., in preparation for the proof that an axiomatic theory of sets is essentially undecidable one will encode its formulae, inside $\widehat{\emptyset}$ (see [33]) or even by means of natural numbers. Our encoding cannot be carried out with the same parsimony of means, due to the tight interplay between syntax and intended semantics in our languages (see the formation rule T0 of each $\mathscr{L}_{\mathscr{U}}$); we will manage, nonetheless, to encode the formulae of $\mathscr{L}_{\widehat{\mathsf{s}}}$ inside $\widehat{\mathsf{s}}$ and the ones of $\mathscr{L}_{\widehat{\mathsf{w}}}$ inside $\widehat{\mathsf{w}}$.

10.4 A Kind of 'All-at-Once Compactification'

> Another technique is *concurrence*. This is a logical technique that guarantees that the extended structure contains all possible completions, compactifications and so forth. [14, p. 3]

Suppose that s is infinite. If ï is also infinite and an injection g of ï into s exists, it will suffice to require that no finite set belongs to the ultrafilter æ in order that $gj \neq x$ *a.e.* for any $x \in \mathsf{s}$; thus g must differ from any function h from ï to s which is *a.e.* constant, and *nonstandard individuals exist*! This is one way of making the nonstandard enlargement non-trivial (see [28, p. 52]).

Preliminary to the construction of a much richer nonstandard universe, [14, p. 34] defines concurrence. In our own, slightly readjusted terms:

Definition 10.3 Relative to a universe $\mathscr{U}$, a dyadic relation r such that $r \in \mathscr{U}$ and $r \cup \mathsf{dom}(r) \subseteq \mathscr{U}$ is said to be CONCURRENT if to every *finite* $d \subseteq \mathsf{dom}(r)$ there corresponds some $b \in \mathscr{U}$ s.t. $d \times \{b\} \subseteq r$. ⊣

Now let ï be the set of functions ϕ such that $\mathsf{dom}(\phi)$ is the set of all concurrent relations $r \in \widehat{\mathsf{s}}$ and $\phi\, r$ is a finite subset of $\mathsf{dom}(r)$ for each such r. The ultrafilter æ will then be chosen so that ï $= \bigcup$ æ holds and the membership relation

$$\{\phi \in \text{ï} \mid \psi\, r \subseteq \phi\, r \text{ for each concurrent } r \in \widehat{\mathsf{s}}\} \in \text{æ}$$

also holds, for each $\psi \in$ ï. Here comes a key theorem, due to Abraham Robinson:

Theorem 10.3 (Concurrence theorem) *To every concurrent relation* $r \in \widehat{\mathsf{S}}$ *there corresponds some* $\ell \in \widetilde{\mathsf{W}}$ *such that* $\{{}^*a : a \in \mathsf{dom}(r)\} \times \{\ell\} \subseteq {}^*r$.

From this claim, [14, p. 36] draws the conclusion that nonstandard individuals exist: for, assuming $\mathbb{N} \subseteq \mathsf{S}$ in order to slightly simplify the argument, one such is the 'limit' element ℓ corresponding to the concurrent relation

$$\{\langle n, m\rangle : n \in \mathbb{N},\ m \in \mathbb{N} \mid n < m\} ;$$

in fact, $\ell \in {}^*\mathbb{N} \setminus \mathsf{S}$.

> The third technique is *internality*. A set s of elements of the nonstandard universe is *internal* if s itself is an element of the nonstandard universe; otherwise, s is *external*. A surprisingly useful method of proof is one by reductio ad absurdum in which the contradiction is that some set one knows to be *external* would in fact be *internal* under the assumption being refuted. [14, p. 3]

Definition 10.4 We call

EXTERNAL SET: every element of $\widehat{\mathsf{W}} \setminus \widetilde{\mathsf{W}}$;
INTERNAL SET: every element of $\widetilde{\mathsf{W}} \setminus \mathsf{W}$. ⊣

After showing, with the aid of the transfer principle, that ${}^*\mathbb{N} \setminus \mathbb{N}$ is an external set, [14, pp. 39–41] provides criteria for demonstrating the internality of specific sets:

Theorem 10.4 (Internality theorem) *If* $d \subseteq \widetilde{\mathsf{W}}$ *is definable in* $\widetilde{\mathsf{W}}$ *and* a *is an internal set, then* $a \cap d$ *is an internal set.*

Theorem 10.5 *If* a *and* b *are internal sets, then so is* $a \times b$.

Theorem 10.6 (Internal function theorem) *If* $f \in b^a$, *where* a *and* b *are internal sets, and for a suitable term* t *of* $\mathscr{L}_{\widetilde{\mathsf{W}}}$ *involving one free variable*

$$fc \ \textit{ is the value of } \ t(\mathbf{c}) \ \textit{ in } \widetilde{\mathsf{W}}\,,\ \textit{ for each } \ c \in a\,,$$

then f *is internal.*

Along the way, [14, pp. 39–41] shows $\mathbb{N}$ to be an external set.

10.5 Key Application of the Nonstandard Methods

In [14, Chap. 2] the construction of the nonstandard universe is used twice: first to obtain $\mathbb{R}$, the field of real numbers, from the field $\mathbb{Q}$ of the rationals; on second application, to work out the structure of ${}^*\mathbb{R}$ from $\mathbb{R}$. The first use can supersede such classical constructions as the ones devised by George Cantor and Richard Dedekind.

The second use brings infinitesimals into play, along with their inverses, which are infinite numbers: one is thus led into the realm of HYPERREAL numbers.

To briefly see how these embeddings work, consider first an *ordered field* D (in the customary sense). For any such field, we can assume w.l.o.g. that $\mathbb{Q} \subseteq D$.

Definition 10.5 Put

$$F = \bigcup_{n \in \mathbb{N}} \{ x \in D \mid 0 \leq x \leq n \ \vee \ 0 < -x \leq n \},$$
$$I = \{ x \in D \mid x = 0 \ \vee \ (1/x) \in D \setminus F \}.$$

An element x of D is said to be FINITE, INFINITE, or INFINITESIMAL, depending as whether $x \in F$, $x \in D \setminus F$, or $x \in I$. For x, y in D, we say that x IS NEAR y if $x - y \in I$; if so, we write $x \approx y$.

D is called ARCHIMEDEAN if $F = D$; otherwise stated, if $I = \{0\}$. ⊣

As is plain, I is an ideal in the subring F of D; moreover, $\approx$ is an equivalence relation on D, whose restriction to F equals the equivalence relation induced by I. Consequently, the quotient $F/\approx = F/I$ is a ring; actually, it is an Archimedean ordered field.

Suppose next that D is an *Archimedean* ordered field and that $\mathsf{D} \subseteq \mathsf{S}$, where S is as in Sects. 10.2, 10.3, and 10.4. By virtue of the transfer principle, the ${}^*\mathsf{D}$ resulting from D through the ultrapower construction is, in its turn, an ordered field (of which D is a subfield). It is no longer Archimedean, though; for, its nonnull subset ${}^*\mathbb{N} \setminus \mathsf{S}$ consists of elements which are infinite. If we now designate by F and I the set of all finite, respectively infinitesimal, elements of ${}^*\mathsf{D}$, then it readily turns out that the canonical homomorphism $^\circ$ of F onto F/I acts as a monomorphism of D into F/I. After so embedding D in the Archimedean field F/I, [14, p. 51] goes on to prove that F, D, I, and ${}^*\mathsf{D} \setminus \mathsf{F}$ are all external subsets of ${}^*\mathsf{D}$; then, by resorting to the concurrence theorem, [14] obtains the following:

Theorem 10.7 (Dedekind's Theorem) *If A, B are nonnull subsets of D such that $a < b$ holds for all $a \in A$ and $b \in B$, then there is a $c \in \mathsf{F}/\mathsf{I}$ such that $a \leq c \leq b$ holds for all $a \in A$ and $b \in B$.*

From this, [14] gets that

Theorem 10.8 *F/I is a complete ordered field,*

after noting that between two elements x, y of an Archimedean ordered field such that $x < y$ there always lies a $q \in \mathbb{Q}$ such that $x < q < y$. Archimedean ordered fields exist (one such is, of course, $\mathbb{Q}$); therefore, a complete ordered field exists as well. Up to isomorphism, this must be *unique* (owing, in particular, to the fact that any complete ordered field is Archimedean): by definition, $\mathbb{R}$ is taken to be this field.

If we go over the same construction again, now taking $\mathsf{D} = \mathbb{R} \subseteq \mathsf{S}$, we can naturally identify F/I with $\mathbb{R}$ and, accordingly, think of $^\circ$ as being the field homomorphism that sends each finite hyperreal number to its *standard part*, namely to the sole real number which lies near it. It can also be shown (see [14, pp. 53, 56]) that infinitesimally near each real number there is a $q \in {}^*\mathbb{Q}$.

Typical notions of elementary real analysis can be captured in new terms from the nonstandard viewpoint, after which classical theorems can be obtained by nonstandard methods. Various illustrations of this are provided in [14, pp. 56–74], e.g.:

Theorem 10.9 *Consider a sequence* $\{ s_n : n \in \mathbb{N} \setminus \{0\} \}$ *of real numbers* s_n *and a real number* ℓ. *Then*

- *the sequence converges to* ℓ *if and only if* $({}^*s)_n \approx \ell$ *holds for* all *infinite* $n \in {}^*\mathbb{N}$;
- $({}^*s)_n \approx \ell$ *holds for* some *infinite* $n \in {}^*\mathbb{N}$ *if and only if, for each* $\varepsilon > 0$ *in* $\mathbb{R}$, *the inequality* $|s_n - \ell| < \varepsilon$ *is satisfied for infinitely many* $n \in \mathbb{N}$.

Theorem 10.10 *Let* f *be a real-valued function on the closed interval* $[a, b] =_{Def} \{x \in \mathbb{R} \mid a \leq x \leq b\}$, *where* $a, b \in \mathbb{R}$ *and* $a < b$. *Then* f *is continuous at* $x_0 \in [a, b]$ *if and only if, for all* $x \in {}^*[a, b]$, $\quad x \approx x_0$ *implies* ${}^*f(x) \approx {}^*f(x_0)$.

Theorem 10.11 *Let* f *be a continuous real-valued function on the closed interval* $[a, b]$. *If* $f(a) < 0 < f(b)$, *then* $f(c) = 0$ *holds for some* $c \in [a, b]$.

Proof 1 (Sketch) Consider the function $t : \mathbb{N} \times \mathbb{N} \longrightarrow \mathbb{R}$ defined as follows:

$$t(n, i) = \begin{cases} a + i\,(b - a)/n & \text{if } n \in \mathbb{N} \setminus \{0\} \text{ and } 0 \leq i \leq n\,, \\ 0 & \text{otherwise,} \end{cases}$$

so that ${}^*t : {}^*\mathbb{N} \times {}^*\mathbb{N} \longrightarrow {}^*\mathbb{R}$ meets an analogous condition, by the transfer principle.

Choose $\nu \in {}^*\mathbb{N} \setminus \mathbb{N}$. Since $L = \{ i \in {}^*\mathbb{N} \mid f({}^*t(\nu, i)) > 0 \text{ and } i \leq \nu \}$ is a definable subset of ${}^*\mathsf{S}$, L is also internal by Theorem 10.4; and since $\nu \in L$, there is a least element $j > 0$ in L. If we take c to be the standard part of ${}^*t(\nu, j)$, it turns out that $c \approx {}^*t(\nu, j) \approx {}^*t(\nu, j-1)$; therefore $f(c) \approx f({}^*t(\nu, j)) \approx f({}^*t(\nu, j-1))$, and hence $f(c) = {}^\circ(f({}^*t(\nu, j))) = {}^\circ(f({}^*t(\nu, j-1)))$, where the inequalities ${}^\circ(f({}^*t(\nu, j))) \geq 0$ and ${}^\circ(f({}^*t(\nu, j-1))) \leq 0$ hold. We conclude that $f(c) = 0$, as desired. □

10.6 Basic Features of Our Proof Checker

Our proof-checker Ref, a.k.a. ÆtnaNova or Referee, processes script files, named SCENARIOS, which consist of definitions, theorems, and detailed proofs of the theorems. After checking a scenario for syntactic validity, Ref verifies that the proofs are compliant with the version of set theory built into it. The language in which scenarios are written extends the usual language of first-order predicate logic with constructs reflecting the theory which underlies Ref: we can for example, as shown by most of the abbreviating definitions in Fig. 10.1,[10] exploit a very flexible set abstraction construct of the form

---

[10]About Ref's built-in operator ***arb*** (X) that occurs thrice in Fig. 10.1, suffice it to say for the time being that it selects an element of its operand X when $X \neq \emptyset$, and that ***arb*** $(\emptyset) = \emptyset$.

DEF $\bigcup$: [El'ts of el'ts]	$\bigcup \mathsf{S}$	$=_{\text{Def}} \{\mathsf{u} : \mathsf{v} \in \mathsf{S}, \mathsf{u} \in \mathsf{v}\}$
DEF $\mathscr{P}$: [All subsets]	$\mathscr{P}(\mathsf{S})$	$=_{\text{Def}} \{\mathsf{x} : \mathsf{x} \subseteq \mathsf{S}\}$
DEF pair_0: [Ord'd pair]	$\langle \mathsf{X}, \mathsf{Y} \rangle$	$=_{\text{Def}} \{\{\mathsf{X}\}, \{\mathsf{X}, \mathsf{Y}\}\}$
DEF pair_1: [Left proj.]	$\mathsf{Q}^{[1]}$	$=_{\text{Def}} \boldsymbol{arb}(\{\mathsf{x} : \mathsf{s} \in \mathsf{Q}, \mathsf{x} \in \mathsf{s} \mid \mathsf{s} = \{\mathsf{x}\}\})$
DEF pair_2: [Right proj.]	$\mathsf{Q}^{[2]}$	$=_{\text{Def}} \boldsymbol{arb}(\{\mathsf{y} : \mathsf{d} \in \mathsf{Q}, \mathsf{y} \in \mathsf{d} \mid \mathsf{Q} = \{\{\mathsf{y}\}\} \vee \mathsf{d} \backslash \{\mathsf{y}\} \in \mathsf{Q}\})$
DEF map_1: [Domain]	$\mathbf{dom}(\mathsf{F})$	$=_{\text{Def}} \left\{\mathsf{p}^{[1]} : \mathsf{p} \in \mathsf{F}\right\}$
DEF map_2: [Restriction]	$\mathsf{F}_{\restriction \mathsf{A}}$	$=_{\text{Def}} \left\{\mathsf{p} \in \mathsf{F} \mid \mathsf{p}^{[1]} \in \mathsf{A}\right\}$
DEF map_3: [Image]	$\mathsf{F} \restriction \mathsf{x}$	$=_{\text{Def}} \boldsymbol{arb}\left(\mathsf{F}_{\restriction \{\mathsf{x}\}}\right)^{[2]}$
DEF map_4: [Is a map]	$\mathsf{Is\_map}(\mathsf{F})$	$\leftrightarrow_{\text{Def}} \left\langle \forall \mathsf{p} \in \mathsf{F} \mid \mathsf{p} = \langle \mathsf{p}^{[1]}, \mathsf{p}^{[2]} \rangle \right\rangle$
DEF Fin: [Finitude]	$\mathsf{Finite}(\mathsf{F})$	$\leftrightarrow_{\text{Def}} \left\langle \forall \mathsf{g} \in \mathscr{P}(\mathscr{P}(\mathsf{F})) \backslash \{\emptyset\}, \exists \mathsf{m} \mid \mathsf{g} \cap \mathscr{P}(\mathsf{m}) = \{\mathsf{m}\} \right\rangle$

Fig. 10.1 A few basic operations over sets and maps; two special properties

$$\left\{ \textit{term} : \textit{iterators} \mid \textit{condition} \right\}$$

to specify many familiar operations and relations over sets.

Ref's second-order construct named THEORY enables one to package definitions and theorems into reusable proofware components. Besides providing theorems of which it holds the proofs, a THEORY has the ability to bring into a mathematical discourse decisive clues.[11] Like procedures of a programming language, Ref's THEORYs have input formal parameters, in exchange for whose actualization they supply useful information. Actual input parameters must satisfy a conjunction of statements, called the ASSUMPTIONS of the THEORY. A THEORY usually encapsulates the definitions of entities related to the input parameters and it supplies, along with some consequences of the assumptions, theorems talking about those internally defined entities that the THEORY returns as output parameters. After having been derived by the user once and for all inside the THEORY, the consequences of the assumptions, as well as the claims involving the output parameters, are available to be exploited repeatedly.

Two THEORY interfaces are shown in Fig. 10.2. The THEORY finiteImage awaits as input parameters a set f_0, assumed to be finite, and a *global* function g, namely one that sends every set x to a value $\mathsf{g}\, x$; whenever applied to fitting actual parameters, this finiteImage will simply produce a claim of the form $\mathsf{Finite}(\{\mathsf{g}\, x : x \in \mathsf{f}_0\})$. The other one, reachGlob,[12] only expects a global function g; it will return the global function glob_Θ sending every set b to the smallest superset $\{b, \mathsf{g}\, b, \mathsf{g}(\mathsf{g}\, b), \ldots\}$ of

[11] In a passage echoing Abraham Robinson's 'provocative remark' which we have recalled in the Introduction through Martin's words, Jack says about this ability of THEORYs [35, p. 9]: "··· definitions serve to 'instantiate', that is, to introduce the objects whose special properties are crucial to an intended argument. Like the selection of crucial lines, points, and circles from the infinity of geometric elements that might be considered in a Euclidean argument, definitions of this kind often carry a proof's most vital ideas". A typical case of this kind is, in arithmetic, the selection of the least natural number that meets some key property.

[12] This is a specialized variant of the THEORY reachability presented in [35, Sect. 7.3]. As seen here, the formal output parameters of a THEORY always carry a subscript Θ.

THEORY finiteImage$(f_0, g(X))$
Finite(f_0)
$\Rightarrow$
Finite$(\{g(x) : x \in f_0\})$
END finiteImage

THEORY reachGlob$(g(X))$
$\Rightarrow$ (glob$_\Theta$)
$\langle \forall y,x,z \mid y \in glob_\Theta(x) \,\&\, z \in glob_\Theta(y) \rightarrow z \in glob_\Theta(x) \rangle$
$\langle \forall b,x,y \mid b \in glob_\Theta(b) \,\&\, (x \in glob_\Theta(b) \,\&\, y = g(x) \rightarrow y \in glob_\Theta(b)) \rangle$
$\langle \forall b,t \mid b \in t \,\&\, \langle \forall x \in t \mid g(x) \in t \rangle \rightarrow glob_\Theta(b) \subseteq t \rangle$
$\langle \forall b \mid glob_\Theta(b) = \{b\} \cup \{g(u) : u \in glob_\Theta(b)\} \rangle$
$\langle \forall b \mid \{ f \subseteq glob_\Theta(b) \mid \langle \forall x \in glob_\Theta(b) \mid x \in f \leftrightarrow g(x) \in f \rangle \} = \{\emptyset, glob_\Theta(b)\} \rangle$
END reachGlob

Fig. 10.2 Interfaces of two Ref THEORYS

$\{b\}$ which is closed under application of g to its own elements, as precisely stated by the claims which this THEORY will supply.

An example of the use of reachability through a global function is the construction of the set of all natural numbers intended *à la* von Neumann, which can be carried out in two steps:[13]

APPLY $\langle$glob$_\Theta$: count$\rangle$reachGlob$(g(X) \mapsto X \cup \{X\})$$\Rightarrow$THM nats$_a$: [*Upwardcounting*]
$\langle \forall y,x,z \mid y \in count(x) \,\&\, z \in count(y) \rightarrow z \in count(x) \rangle$ &
$\langle \forall b,x,y \mid b \in count(b) \,\&\, (x \in count(b) \,\&\, y = x \cup \{x\} \rightarrow y \in count(b)) \rangle$ &
$\langle \forall b,t \mid b \in t \,\&\, \langle \forall x \in t \mid x \cup \{x\} \in t \rangle \rightarrow count(b) \subseteq t \rangle$ &
$\langle \forall b \mid count(b) = \{b\} \cup \{u \cup \{u\} : u \in count(b)\} \rangle$.

DEF nats: [*vonNeumann'snaturalnumbers*] $\mathbb{N} =_{\text{Def}}$ count$(\emptyset)$.

It would be pointless to discuss here the inferential armory of Ref, because we are still in the phase of designing how to formalize the basic techniques underlying nonstandard analysis, and the expected outcome of such a formalization is best described by a plan concerning the core THEORY interfaces and by choices as to how implement some key definitions.

An important enhancement to the Zermelo-Fraenkel set theory came historically with von Neumann's introduction of an axiom,

$$\forall x \, \exists a \, \forall y \in x \; (a \in x \,\&\, y \notin a),$$

which *forbids* membership to form infinite chains $\ell_0 \ni \ell_1 \ni \ell_2 \ni \cdots$; this is tersely stated by singling out, for any given set x, a set a disjoint from x that belongs to x unless $x = \emptyset$. In Ref this principle is embodied by a construct, ***arb*** (X), such that

[13] What follows is not meant to imply that the definition of $\mathbb{N}$ shown is the ideal one.

$$\forall x\Big(\boldsymbol{arb}\,(x)\cap x=\emptyset\ \ \&\ \boldsymbol{arb}\,(x)\in x\cup\{x\}\Big)$$

(implying $\boldsymbol{arb}\,(\emptyset)=\emptyset$). The meaning of $\boldsymbol{arb}$ is competently handled by a most basic inference method of Ref.

To appreciate the usefulness of $\boldsymbol{arb}$, consider the THEORY whose interface appears on the left of Fig. 10.3. Upon receipt of a set n_0 that meets a given property P, this THEORY will return a set $\mathsf{transfInd}_{\Theta}$ still enjoying P but none of whose elements satisfies P. In its hidden internal working, transfInduction first applies the THEORY reachGlob seen in Fig. 10.2 to $g(X)=\boldsymbol{arb}\,(\{u\in X\mid P(u)\})$ and then applies the resulting glob_{Θ} to n_0 to get a set $N_0=\big\{n_0\,,\ g\,n_0\,,\ g\big(g\,n_0\big),\ldots,\emptyset\big\}$ such that $\boldsymbol{arb}\,(\{w\in N_0\mid P(w)\})$ is the sought $\mathsf{transfInd}_{\Theta}$.

In Ref the well-foundedness of membership also lies behind a definition mechanism based on $\in$-*recursion*, shown at work with the specification of img in Fig. 10.4 and which we will repeatedly use in the ongoing. A discussion about the syntax of $\in$-recursive definitions can be found in [35, pp. 216–217]); concrete illustrations of it will suffice here. A basic example is

$$\mathsf{rk}\,(X)\ =_{\text{Def}}\bigcup\Big\{\mathsf{rk}\,(y)\cup\{\mathsf{rk}\,(y)\}\ :\ y\in X\Big\}\,,$$

defining the RANK of a set X. The mechanism at stake is akin to recursion as used in computer programming; like it, it resorts to a base case to avoid circularity: in fact, $\mathsf{rk}\,(X)=\emptyset$ when $X=\emptyset$, since obviously $\{\mathsf{rk}\,(y)\cup\{\mathsf{rk}\,(y)\}\ :\ y\in\emptyset\}=\emptyset$. But $\mathsf{rk}\,(X)$ might also be an infinite set (actually, a transfinite ordinal), a situation which will occur, e.g., when X is infinite or has some infinite elements.

THEORY transfInduction$\big(\mathsf{n_0},\mathsf{P(X)}\big)$	THEORY finInduction$\big(\mathsf{n_0},\mathsf{P(X)}\big)$
$\quad\mathsf{P(n_0)}$	$\quad\mathsf{P(n_0)}\,\&\,\mathsf{Finite}\,(\mathsf{n_0})$
$\Rightarrow(\mathsf{transfInd}_{\Theta})$	$\Rightarrow(\mathsf{fInd}_{\Theta})$
$\quad\mathsf{P(transfInd_{\Theta})}\,\&\,\big\langle\forall\mathsf{k}\in\mathsf{transfInd}_{\Theta}\mid\neg\mathsf{P(k)}\big\rangle$	$\quad\mathsf{P(fInd_{\Theta})}\,\&\,\big\langle\forall\mathsf{k}\subsetneq\mathsf{fInd}_{\Theta}\mid\neg\mathsf{P(k)}\big\rangle$
END transfInduction	END finInduction

Fig. 10.3 Transfinite induction contrasted with finite induction. The former exploits the well-foundedness of $\in$ while the latter exploits the well-foundedness of $\subsetneq$ over finite sets. Other classical forms of induction, e.g., arithmetic induction or induction over ordinals, can be conveniently hooked to membership or inclusion

$\mathsf{img(I,B)}\ =_{\text{Def}}$ **if** $\mathsf{I}=\emptyset$ **then** B **else** $\boldsymbol{arb}\,\big(\big\{\mathsf{g}\big(\mathsf{img(j\,,B)}\big):\mathsf{j}\in\mathsf{I}\big\}\big)$ **fi**
$\mathsf{glob}_{\Theta}(\mathsf{B})\ =_{\text{Def}}\ \{\mathsf{img(i,B)}:\ \mathsf{i}\in\boldsymbol{\sigma}_{\infty}\}$

Fig. 10.4 A viable specification of the iterated images of g and of the output symbol glob_{Θ} inside the THEORY reachGlob of Fig. 10.2. Here $\boldsymbol{\sigma}_{\infty}$ is a Ref's built-in constant subject to the assumption that $\boldsymbol{\sigma}_{\infty}\neq\emptyset\ \&\ \langle\forall\mathsf{x}\in\boldsymbol{\sigma}_{\infty}\mid\{\mathsf{x}\}\in\boldsymbol{\sigma}_{\infty}\rangle$

Let us digress briefly. The α's satisfying the equality $\alpha = \mathsf{rk}(\alpha)$ turn out to be precisely the sets known, after von Neumann, as *ordinal numbers*;[14] and it is not hard to prove, about the indexed class of sets which satisfy the conditions

$$\begin{aligned} V_{\emptyset} &= \emptyset\,, \\ V_{\gamma\cup\{\gamma\}} &= \mathscr{P}(V_\gamma) && \text{for every ordinal number } \gamma\,, \\ V_\lambda &= \textstyle\bigcup_{\beta\in\lambda} V_\beta && \text{for every nonnull ordinal } \lambda \text{ not of the form } \gamma\cup\{\gamma\} \end{aligned}$$

—historically called the *cumulative hierarchy*—, that V_α consists, for each ordinal α, of all sets whose ranks lie below α. Now consider the property

$$\mathscr{V}(L) \leftrightarrow_{\text{Def}} L = \bigcup\{\mathscr{P}(\ell) \,:\, \ell \in L \mid \mathscr{V}(\ell)\}\,.$$

In this new instance of $\in$-recursion, the reader can recognize a streamlined definition of the stages of the cumulative hierarchy: as one readily sees, $\mathscr{V}(\emptyset)$ holds; more generally, one can show that $\mathscr{V}(L)$ is logically equivalent to the existence of an ordinal α such that $L = V_\alpha$. We do not prove this fact but do call attention to it because a similar change of perspective will motivate our formalization of superstructures in the following section.

10.7 Top-Down Recognition of Superstructure Stages

Concerning the unusual way, just hinted at, of approaching the cumulative hierarchy, one might contend that it is presumably harder—or, if anything, less transparent—to infer directly from the definition of $\mathscr{V}(L)$ a statement such as

$$\big(\mathscr{V}(L') \,\&\, \mathscr{V}(L'')\big) \;\rightarrow\; (L' \subsetneq L'' \leftrightarrow L' \in L'')$$

than to prove, for any pair α, β of ordinals, the biimplications

$$\big(V_\beta \subsetneq V_\alpha \leftrightarrow \beta \in \alpha\big) \;\&\; \big(V_\beta \in V_\alpha \leftrightarrow \beta \in \alpha\big)\,.$$

A tentative reply is that transfinite induction of the kind schematized in Fig. 10.3 (left) is often a shortcut compared to a proof pattern relying on the theory of ordinals. On

[14]A common definition of ordinals, owing to a simplification due to Raphael Robinson, is:

$$\mathsf{Ord}(U) \leftrightarrow_{\text{Def}} \forall x\,(x \in U \rightarrow x \subseteq U) \,\&\, \forall x\,\forall y\big(\{x,y\} \subseteq U \rightarrow (x \in y \vee y \in x \vee x = y)\big)\,.$$

a smaller scale, as will now be seen, we can treat superstructures without numbering their stages: with virtually no recourse to natural numbers.[15]

We can exploit recursion to describe sets L which are *stages* of a superstructure. The first of the three definitions shown below is $\in$-recursive and specifies a function seeking a set s of *individuals* (recall Sect. 10.2) such that $\mathsf{s}_m = L$ for some $m \in \mathbb{N}$; if such an s exists, it can be found by repeated extraction

$$L \ni \log L \ni \log\log L \ni \cdots \ni \mathsf{s}$$

of the 'logarithm' of L, where $\ell = \log L$ momentarily means that $L = \ell \cup \mathscr{P}(\ell)$ (needless to say, this equation has either one or no solution—in the former case, $\emptyset \in L$ and hence L cannot be regarded as a set of individuals):

$$\begin{aligned}\mathsf{basis}(L) \;=_{\text{Def}}\; & \textbf{if } \emptyset \notin L \;\&\; L \cap \bigcup L = \emptyset \textbf{ then } L\\ & \textbf{elseif } \big(\exists \ell \mid L = \ell \cup \mathscr{P}(\ell) \;\&\; \mathscr{P}(\ell) \cap \bigcup(\ell \setminus \mathscr{P}(\ell)) = \emptyset\big)\\ & \textbf{then } \boldsymbol{arb}\,(\{\mathsf{basis}(\ell) : \ell \in L \mid L = \ell \cup \mathscr{P}(\ell)\})\\ & \textbf{else } \{\emptyset\} \textbf{ fi};\end{aligned}$$

$$\mathsf{Stage}(L, S) \leftrightarrow_{\text{Def}} L = \emptyset \vee \big(\mathsf{basis}(L) = S \;\&\; S \neq \{\emptyset\}\big);$$

$$\begin{aligned}\mathsf{Ur}\,(S) \leftrightarrow_{\text{Def}} & \; \emptyset \notin S \;\&\; S \cap \bigcup S = \emptyset \;\&\\ & \big(\forall \ell \mid \mathsf{Stage}(\ell, S) \rightarrow \mathscr{P}(\ell) \cap \bigcup S = \emptyset\big).\end{aligned}$$

The chain $L = L_0$, $L_{n+1} = \log L_n$ of logarithms surely has finite length but may end with a set L_m such that either $\emptyset \in L_m$ or $L_m \cap \bigcup L_m \neq \emptyset$ holds, in which cases L_m cannot serve as a set of individuals. When this happens, $\mathsf{basis}(L)$ will flag the failure by returning $\{\emptyset\}$; but failure can be detected earlier during the descent, should $\mathscr{P}(L_n) \cap \bigcup(L_n \setminus \mathscr{P}(L_n))$ be nonnull at some point. The predicate $\mathsf{Stage}(L, S)$ indicates L as a potential stage of the superstructure—if any—generated by its 'ultimate logarithm' $S = \mathsf{basis}(L)$ when the latter is obtained without failure; but even when so, S does not qualify as a set of individuals unless one can indefinitely ascend, starting with S, through stages none of which reveals the inner structure of its elements. The property $\mathsf{Ur}\,(S)$ captures the sense of our last remark.

Under the assumption $\mathsf{Ur}\,(\mathsf{s}_0)$, we have in fact checked with the assistance of Ref that $\widehat{\mathsf{s}}_0$ behaves as desired (see Fig. 10.5), even though genuine individuals ('*urelemente*' of the nature set forth in [14, p. 11]) do not exist in the von Neumann cumulative universe of all sets.

The interface, shown in Fig. 10.5, of the THEORY superstructure may look intimidating, the cause being that it exploits the *property* $\langle \exists \ell \mid \mathsf{Stage}(\ell, \mathsf{s}_0) \;\&\; X \in \ell \rangle$

[15]Natural numbers will play an irreplaceable role in the informal arguments providing the rationale for the formal constructions that follow; within the formal treatment, their collection $\mathbb{N}$ will act as a set whose infinitude is easiest to prove (and infinite sets will be crucial in Sect. 10.8).

THEORY universe($\mathfrak{U}(X)$)

$$\mathfrak{U}(\emptyset)\ \&\ \big\langle \forall x,y \mid \mathfrak{U}(x)\,\&\,\mathfrak{U}(y) \rightarrow \mathfrak{U}(\{x,y\})\big\rangle$$
$$\big\langle \forall x,y,z \mid \mathfrak{U}(x)\,\&\,\mathfrak{U}(y)\,\&\,\{y,z\} \subseteq x \rightarrow \mathfrak{U}(z)\big\rangle$$

$\Rightarrow$

$$\big\langle \forall x,y \mid \mathfrak{U}(x)\,\&\,\mathfrak{U}(y) \rightarrow \mathfrak{U}(\langle x,y\rangle)\big\rangle$$
$$\Big\langle \forall f,x \mid \mathsf{Is\_map}(f)\,\&\,\mathfrak{U}(f)\,\&\,\big\langle \forall x \in \mathbf{dom}(f) \mid \mathfrak{U}(x)\big\rangle\ \&\ \Big(f=\emptyset \vee \big\langle \exists q \in f \mid \mathfrak{U}(q)\big\rangle\Big) \rightarrow \mathfrak{U}(f\upharpoonright x)\Big\rangle$$

END universe

THEORY superstructure(s_0)

$\mathsf{Ur}(s_0)$

$\Rightarrow$ (sstr_Θ)

$$\mathsf{Stage}(\emptyset,s_0)\,\&\,\mathsf{Stage}(s_0,s_0)\,\&\,s_0 \neq \{\emptyset\}\ \&\ \big\langle \forall \ell \mid \mathsf{Stage}(\ell,s_0)\,\&\,\ell\neq\emptyset \rightarrow s_0 \subseteq \ell\big\rangle\ \&\ s_0 \cap \bigcup(s_0\setminus\mathscr{P}(s_0))=\emptyset\,\&\,s_0\setminus\mathscr{P}(s_0)=s_0$$
$$\big\langle \forall \ell \mid \mathsf{Stage}(\ell,s_0)\,\&\,s_0=\emptyset \rightarrow \ell \subseteq \mathscr{P}(\ell)\big\rangle\ \&\ \big\langle \forall \ell \mid \mathsf{Stage}(\ell,s_0)\,\&\,s_0=\emptyset \rightarrow \mathsf{Stage}(\mathscr{P}(\ell),s_0)\big\rangle$$
$$\big\langle \forall \ell \mid \mathsf{Stage}(\ell,s_0) \rightarrow (\ell=\emptyset \vee s_0=\ell\setminus\mathscr{P}(\ell))\,\&\,\ell\cap\bigcup(\ell\setminus\mathscr{P}(\ell))=\emptyset\big\rangle$$
$$\big\langle \forall \ell \mid \mathsf{Stage}(\ell,s_0)\,\&\,(s_0=\emptyset \vee \ell\neq\emptyset) \rightarrow \mathsf{Stage}(\ell\cup\mathscr{P}(\ell),s_0)\big\rangle\ \&\ \mathsf{Stage}(s_0\cup\mathscr{P}(s_0),s_0)\,\&\,\emptyset \in s_0\cup\mathscr{P}(s_0)$$
$$\Big\langle \forall x,\ell \mid x \in \ell\setminus s_0\,\&\,\mathsf{Stage}(\ell,s_0) \rightarrow \big\langle \exists h \mid \mathsf{Stage}(h,s_0)\,\&\,\mathscr{P}(x)\subseteq\mathscr{P}(h)\big\rangle\Big\rangle$$
$$\Big\langle \forall x,\ell,y,m,z \mid x\in\ell\,\&\,y\in m\,\&\,\{y,z\}\subseteq x\,\&\,\mathsf{Stage}(\ell,s_0)\,\&\,\mathsf{Stage}(m,s_0) \rightarrow \big\langle \exists h \mid \mathsf{Stage}(h,s_0)\,\&\,z\in h\big\rangle\Big\rangle$$
$$\big\langle \forall \ell,m \mid \ell \not\subseteq m\,\&\,\mathsf{Stage}(\ell,s_0)\,\&\,\mathsf{Stage}(m,s_0) \rightarrow m\subseteq\ell\big\rangle$$
$$\Big\langle \forall x,\ell,y,m \mid x\in\ell\,\&\,y\in m\,\&\,\mathsf{Stage}(\ell,s_0)\,\&\,\mathsf{Stage}(m,s_0) \rightarrow \big\langle \exists h \mid \mathsf{Stage}(h,s_0)\,\&\,\{x,y\}\in h\big\rangle\Big\rangle$$
$$\big\langle \exists k \mid \mathsf{Stage}(k,s_0)\,\&\,\emptyset\in k\big\rangle\ \&$$
$$\Big\langle \forall x,y \mid \big\langle \exists h \mid \mathsf{Stage}(h,s_0)\,\&\,x\in h\big\rangle\ \&\ \big\langle \exists k \mid \mathsf{Stage}(k,s_0)\,\&\,y\in k\big\rangle \rightarrow \big\langle \exists k \mid \mathsf{Stage}(k,s_0)\,\&\,\{x,y\}\in k\big\rangle\Big\rangle\ \&$$
$$\Big\langle \forall x,y,z \mid \big\langle \exists h \mid \mathsf{Stage}(h,s_0)\,\&\,x\in h\big\rangle\ \&\ \big\langle \exists k \mid \mathsf{Stage}(k,s_0)\,\&\,y\in k\big\rangle\ \&\ \{y,z\}\subseteq x \rightarrow \big\langle \exists m \mid \mathsf{Stage}(m,s_0)\,\&\,z\in m\big\rangle\Big\rangle$$
$$\Big\langle \forall x,y \mid \big\langle \exists h \mid \mathsf{Stage}(h,s_0)\,\&\,x\in h\big\rangle\ \&\ \big\langle \exists k \mid \mathsf{Stage}(k,s_0)\,\&\,y\in k\big\rangle \rightarrow \big\langle \exists k \mid \mathsf{Stage}(k,s_0)\,\&\,\langle x,y\rangle\in k\big\rangle\Big\rangle\ \&$$
$$\Big\langle \forall f,x \mid \mathsf{Is\_map}(f)\,\&\,\big\langle \exists k \mid \mathsf{Stage}(k,s_0)\,\&\,f\in k\big\rangle\ \&\ \big\langle \forall u\in\mathbf{dom}(f), \exists h \mid \mathsf{Stage}(h,s_0)\,\&\,u\in h\big\rangle\ \&$$
$$\Big(f=\emptyset \vee \big\langle \exists q\in f, \exists m \mid \mathsf{Stage}(m,s_0)\,\&\,q\in m\big\rangle\Big) \rightarrow \qquad \big\langle \exists n \mid \mathsf{Stage}(n,s_0)\,\&\,f\upharpoonright x\in n\big\rangle\Big\rangle$$
$$\Big\langle \forall x \mid \big\langle \exists \ell \mid \mathsf{Stage}(\ell,s_0)\,\&\,x\in\ell\big\rangle \leftrightarrow x\in\mathsf{sstr}_\Theta\Big\rangle$$

END superstructure

Fig. 10.5 Interfaces of the THEORYS of universes and superstructures

as a temporary surrogate of the sought $\widehat{s}_0$. Only its final claim shows that the X's enjoying that property form a set, namely the output parameter sstr_Θ to be then actualized as $\widehat{s}_0$ outside the THEORY; even so we can exploit the said property as a universe to get, through the THEORY universe, derived closure properties. Observe, in fact, that the second-to-last and penultimate claim of superstructure match the assumptions of universe and its internally derived conclusions.

The moral is that our recursive characterization of the stages of a superstructure disclosed handy patterns to our formal reasoning about them; however, at one point we had to resort to a construction from below, closer in spirit to [14, Sect. 1.3]: this happened when it came to ascertaining that the union-class of all the stages is, in fact, a set. For that purpose, we applied the THEORY reachGlob (see Fig. 10.2 above) to the actual input parameter **if** $X=\emptyset\ \&\ s_0\neq\emptyset$ **then** s_0 **else** $X\cup\mathscr{P}(X)$ **fi**, thus getting a function glob whence the sought superstructure was obtained simply by taking $\mathsf{sstr}_\Theta=\bigcup\mathsf{glob}(\emptyset)$. The following triad of equations conveys the idea, in functionally equivalent terms:

$$\mathsf{nextStage}(L) = \textbf{if } L = \emptyset \,\&\, \mathsf{s}_0 \neq \emptyset \textbf{ then } \mathsf{s}_0 \textbf{ else } L \cup \mathscr{P}(L) \textbf{ fi},$$

$$\mathsf{stage}(I) = \boldsymbol{arb}\left(\{\mathsf{nextStage}(\mathsf{stage}(j)) : j \in I\}\right),$$

$$\mathsf{sstr}_\Theta = \bigcup\{\mathsf{stage}(i) : i \in \boldsymbol{\sigma}_\infty\}.$$

These equations, in fact, adjust the construction of Fig. 10.4 to the case at hand; as said under that figure, $\boldsymbol{\sigma}_\infty$ is a Ref's built-in witnessing that infinite sets exist.

10.8 Forging Companion Sets of Individuals

When undertaking the construction of a standard universe, in practice one starts with a pre-defined, infinite basis—say the set $\mathbb{R}$ of all real numbers—whose elements may have an inner structure that prevents their direct use as individuals. If so, how can we conceal their structure? We need a technique for converting a set s' whatsoever into a set s'' so that $\mathsf{Ur}\left(s''\right)$ holds and there is a one-one correspondence between s' and s''.

One plainly sees that $\mathsf{Ur}\left(s''\right)$ cannot hold if any set of finite rank belongs to s''; on the other hand, imposing that $\emptyset \notin s''$ and that all elements of elements of s'' share the same infinite rank r suffices to ensure that $\mathsf{Ur}\left(s''\right)$ holds—one shows inductively, in fact, that each stage originating from s'' is the union of a set of finite rank with a set whose elements have ranks exceeding r. This observation makes it rather easy to conceive an injection ur whose domain is the given s' and whose set of values, $s'' = \left\{\mathsf{ur}\,x : x \in s'\right\}$, can serve as basis in place of s' in the construction of the standard superstructure. Should any rationale arise for doing so, we can even tune the range of s'' by means of an auxiliary 'gauge' set c', as suggested by the interface of the THEORY $\mathsf{urification}$ in Fig. 10.6.

This THEORY receives sets s', c' such that $s' \cup c'$—and hence $\mathsf{rk}\left(s' \cup c'\right)$—is infinite; it manufactures and produces in output a function ur_Θ sending injectively each $x \in s'$ to a set $\mathsf{ur}_\Theta(x)$ all of whose elements have rank $\mathsf{rk}\left(s' \cup c'\right)^+ = \mathsf{rk}\left(s' \cup c'\right) \cup$

DEF $=_{\text{Def}}$ $\mathsf{X} \cup \{\mathsf{X}\}$

THM $\neg\mathsf{Finite}(\mathsf{R}) \,\&\, \emptyset \notin \mathsf{S} \,\&\, \langle \forall \mathsf{u} \in \bigcup \mathsf{S} \mid \mathsf{rk}(\mathsf{u}) = \mathsf{R}\rangle \rightarrow \mathsf{Ur}(\mathsf{S})$

```
THEORY urification(s', c')
    ¬Finite(s') ∨ ¬Finite(c')
⇒ (ur_Θ)
    ⟨∀x ∈ s', y ∈ s' | ur_Θ(x) = ur_Θ(y) → x = y⟩
    ⟨∀v ∈ ⋃{ur_Θ(x) : x ∈ s'} | rk(v) = rk({s' ∪ c'})⟩
    ⟨∀u ∈ {ur_Θ(x) : x ∈ s'} | rk(u) = rk({s' ∪ c'})⁺⟩
    Ur({ur_Θ(x) : x ∈ s'})
END urification
```

Fig. 10.6 Gauged transformation of a set s' whatsoever into a set of individuals

$\{\mathsf{rk}\,(s' \cup c')\}$, where $R^+ =_{\text{Def}} R \cup \{R\}$. The definition of ur_Θ—internally hidden, insofar as immaterial outside the THEORY $\mathsf{urification}$—could well be

$$\mathsf{ur}_\Theta(\mathsf{X}) =_{\text{Def}} \left\{\mathsf{s}' \setminus \{\mathsf{X}\} \cup \{\mathsf{s}' \cup \mathsf{c}'\}\right\}.$$

What really counts to us is that $\mathsf{Ur}\,(\{\mathsf{ur}_\Theta(x) : x \in s'\})$ holds, as we aimed at.

To see a more sophisticated exploitation of the THEORY at hand, suppose next that we are given a set s along with an infinite set i' that we want to use as index set for enlarging s, seen as a standard set of individuals, into a set w of nonstandard individuals. To ease the discussion, we momentarily dismiss the concurrence issue debated in Sect. 10.4; we will content ourselves with an ultrafilter none of whose elements is a finite set, over (a counterpart i'' of) i'.

First move. Convert i' into a set i'' so that all indices j in i'' have the same infinite rank r, exceeding the rank of s, and there is a one-one correspondence $\mathsf{u}(X)$ between i' and i'':

APPLY $(\mathsf{ur}_\Theta : \mathsf{u})$ $\mathsf{urification}(\,\mathsf{s}' \mapsto i',\ \mathsf{c}' \mapsto s\,) \Rightarrow \cdots$
DEF $\quad i'' =_{\text{Def}} \{\mathsf{u}(\mathsf{x}) : \mathsf{x} \in i'\}$

Second move. Observe that when $\mathscr{W}$ is a set of functions from i'' to s then each element of $\bigcup \mathscr{W}$ is an ordered pair $\langle j, x\rangle = \{\{j\}, \{j, x\}\}$, whose rank is infinite. Trivially $\emptyset \notin \mathscr{W}$ and hence $\mathsf{Ur}\,(\mathscr{W})$ holds.

Third move. Introduce an *ultrafilter* æ such that

$$\bigcup \text{æ} = i'' \quad \text{and} \quad \text{æ} \supseteq \left\{i'' \setminus \{j\} : j \in i''\right\},$$

and at this point specify $\mathscr{W}$ as follows:

$$\begin{aligned} \rho(g) &=_{\text{Def}} \boldsymbol{arb}\left(\left\{h \in s^{i''} \mid \{j \in i'' \mid h\,j = g\,j\} \in \text{æ}\right\}\right), \\ \mathscr{W} &=_{\text{Def}} \left\{\rho(g) : g \in s^{i''}\right\}. \end{aligned}$$

Now regard this $\mathscr{W}$ and its subset

$$\mathscr{S} =_{\text{Def}} \left\{h \in \mathscr{W} \mid (\exists\, y \mid \mathsf{dom}\big((i'' \times \{y\}) \cap h\big) \in \text{æ})\right\},$$

respectively, as the 'wide' and the 'small' set of all nonstandard individuals and of the standard ones: it should be clear that $\mathscr{S}$ can act as a counterpart of the original s, in view of the natural correspondence between the two.

What precedes has offered clues about how to implement the THEORY whose interface is shown in the lower part of Fig. 10.7.

$$\textsc{Def}\quad \mathsf{X}\Delta\mathsf{Y} =_{\mathrm{Def}} \mathsf{X}\cup\mathsf{Y}\setminus\mathsf{X}\cap\mathsf{Y}$$

$$\textsc{Def}\ \mathsf{Ultrafilter}(\text{Æ}) \leftrightarrow_{\mathrm{Def}} \langle\forall \mathsf{x} \mid (\mathsf{x}\cap\bigcup\text{Æ})\in\text{Æ}\vee(\bigcup\text{Æ}\setminus\mathsf{x})\in\text{Æ}\rangle\ \&$$
$$\langle\forall \mathsf{x}\in\text{Æ},\mathsf{y}\in\text{Æ} \mid \mathsf{x}\cap\mathsf{y}\in\text{Æ}\setminus\{\emptyset\}\rangle$$

$$\textsc{Thm}\ \langle\forall \mathsf{x}\in\mathsf{B},\mathsf{y}\in\mathsf{B} \mid \mathsf{x}\cap\mathsf{y}\in\mathsf{B}\setminus\{\emptyset\}\ \&\ \mathsf{x}\subseteq\ddot{\mathsf{I}}\rangle\ \&\ \mathsf{B}\neq\emptyset \rightarrow$$
$$\langle\exists \text{æ} \mid \mathsf{Ultrafilter}(\text{æ})\ \&\ \mathsf{B}\subseteq\text{æ}\ \&\ \ddot{\mathsf{I}}=\bigcup\text{æ}\rangle$$

$$\textsc{Theory}\ \mathsf{individuation}(\mathsf{s},\mathsf{i}'')$$
$$\mathsf{s}\neq\emptyset$$
$$\neg\mathsf{Finite}(\mathsf{i}'')$$
$$\langle\forall \mathsf{j}\in\mathsf{i}'' \mid \mathsf{rk}(\mathsf{s})\in\mathsf{rk}(\mathsf{j})\ \&\ \mathsf{rk}(\mathsf{j})=\mathsf{rk}(\boldsymbol{arb}(\mathsf{i}''))\rangle$$
$$\Rightarrow (\text{æ}_\Theta,\mathsf{w}_\Theta)$$
$$\mathsf{Ultrafilter}(\text{æ}_\Theta)\ \&\ \bigcup\text{æ}_\Theta=\mathsf{i}''\ \&\ \left\{\mathsf{i}''\setminus\{\mathsf{j}\} : \mathsf{j}\in\mathsf{i}''\right\}\subseteq\text{æ}_\Theta$$
$$\langle\forall \mathsf{g}\in\mathsf{w}_\Theta,\mathsf{h}\in\mathsf{w}_\Theta\setminus\{\mathsf{g}\} \mid \mathsf{Svm}(\mathsf{g})\ \&\ \mathbf{dom}(\mathsf{g})=\mathsf{i}''\ \&\ \mathbf{dom}(\mathsf{g}\Delta\mathsf{h})\in\text{æ}_\Theta\rangle$$
$$\langle\forall \mathsf{y} \mid \langle\exists \mathsf{h}\in\mathsf{w}_\Theta \mid \mathbf{dom}((\mathsf{i}''\times\{\mathsf{y}\})\cap\mathsf{h})\in\text{æ}_\Theta\rangle\leftrightarrow\mathsf{y}\in\mathsf{s}\rangle$$
$$\langle\forall \mathsf{g} \mid \mathsf{Svm}(\mathsf{g})\ \&\ \mathbf{dom}(\mathsf{g})=\mathsf{i}''\ \&\ \mathbf{dom}((\mathsf{i}''\times\mathsf{s})\cap\mathsf{g})\in\text{æ}_\Theta \rightarrow$$
$$\langle\exists \mathsf{h}\in\mathsf{w}_\Theta \mid \mathbf{dom}(\mathsf{g}\Delta\mathsf{h})\notin\text{æ}_\Theta\rangle\rangle$$
$$\neg\mathsf{Finite}(\mathsf{rk}(\boldsymbol{arb}(\mathsf{i}'')))\ \&\ \emptyset\notin\mathsf{w}_\Theta\ \&\ \langle\forall p\in\bigcup\mathsf{w}_\Theta \mid \mathsf{rk}(p)=\mathsf{rk}(\boldsymbol{arb}(\mathsf{i}''))^{++}\rangle$$
$$\mathsf{Ur}(\mathsf{w}_\Theta)$$
$$\textsc{End}\ \mathsf{individuation}$$

Fig. 10.7 Transformation of a set s into a set w_Θ of nonstandard individuals

10.9 Set-Encoding of Bounded-Quantifier Formulae

Before we can exploit Ref to state and prove propositions such as the transfer principle (not to mention Łoś's theorem, see Sect. 10.3), we must devise a set-encoding of terms and formulae that enables easy specifications of how to

(A) evaluate a term or formula under a set-assignment for its variables,
(B) determine the truth value of a sentence,
(C) replace a free variable by a constant within a term or formula,

and the like. Then we will be able to reason formally with Ref about the languages of specific universes.

The set-theoretic representation of terms and formulae can be conceived of rather liberally. By seeing each universe $\mathscr{U}$ as embedded in the class of all sets, which is Ref's domain of discourse, we will in particular

- treat the different languages $\mathscr{L}_{\mathscr{U}}$ by a single encoding instead of separately,
- specify the function val that evaluates a 'term' t under a set-assignment v for the variables occurring in it so that val(t, v) yields a result even when t does not encode a term.

$$
\begin{array}{l}
\text{THEORY termEncoding}()\\
\Rightarrow (\mathsf{cst}_\Theta, \mathsf{Pair}_\Theta, \mathsf{Appl}_\Theta, \mathsf{lft}_\Theta, \mathsf{rgt}_\Theta)\\
\quad \langle \forall c, k \mid \mathsf{cst}_\Theta(c) = \mathsf{cst}_\Theta(k) \rightarrow c = k \rangle\\
\quad \langle \forall c, i, p, q, x, y, z \mid i \in \mathbb{N} \,\&\, \mathsf{Pair}_\Theta(p) \,\&\, \mathsf{Appl}_\Theta(q) \rightarrow \{\mathsf{cst}_\Theta(c), i, p, q\} \neq \{x, y, z\} \rangle\\
\quad (\forall x, y \mid \emptyset \notin \{x, y\} \rightarrow \langle \exists p \mid \langle \forall s \mid \mathsf{Pair}_\Theta(s) \,\&\, \mathsf{lft}_\Theta(s) = x \,\&\, \mathsf{rgt}_\Theta(s) = y \leftrightarrow s = p \rangle\rangle\rangle\\
\quad \langle \forall x, y \mid \emptyset \notin \{x, y\} \rightarrow \langle \exists q \mid \langle \forall s \mid \mathsf{Appl}_\Theta(s) \,\&\, \mathsf{lft}_\Theta(s) = x \,\&\, \mathsf{rgt}_\Theta(s) = y \leftrightarrow s = q \rangle\rangle\rangle\\
\quad \langle \forall s \mid \mathsf{Pair}_\Theta(s) \vee \mathsf{Appl}_\Theta(s) \rightarrow \emptyset \notin \{\mathsf{lft}_\Theta(s), \mathsf{rgt}_\Theta(s)\} \rangle\\
\quad \langle \forall t \mid \{\mathsf{lft}_\Theta(t), \mathsf{rgt}_\Theta(t)\} \subseteq t \cup \{\emptyset\} \rangle\\
\text{END termEncoding}
\end{array}
$$

Fig. 10.8 THEORY about the set-encoding of terms

About one feature of the representation of the syntax, we see no reason for being flexible: each variable $\boldsymbol{x}_n$ will be encoded by its subscript n, a positive integer.

The THEORY interface displayed in Fig. 10.8 formulates the constraints to which we submit our encoding of terms, effected via two properties, Pair and Appl, and three functions: lft, rgt, and cst. The two properties are meant to indicate which sets encode terms of the respective forms $\langle \ell, r \rangle$ and $(\ell \upharpoonright r)$; lft(p) and rgt(p) will provide, when applied to a set p that encodes a term of the form $\langle \ell, r \rangle$, the two sets encoding the immediate subterms, ℓ and r respectively; lft(q) and rgt(q) will behave likewise when q encodes a term of the form $(\ell \upharpoonright r)$. As for cst, it will send each set c to a constant $\boldsymbol{c}$ designating it *univocally*: not only cst$(c) \neq$ cst(k) must hold whenever $c \neq k$, but we require also that ¬Pair(cst(c)), ¬Appl(cst(c)), and cst$(c) \notin \mathbb{N}$, to avoid 'collision' between cst(c) and any set encoding a non-constant term. Unambiguous readability also demands that $p \notin \mathbb{N}$, $q \notin \mathbb{N}$, and $p \neq q$ hold when Appl(p) and Pair(q) hold. This is the rationale behind the first two claims issued by the THEORY termEncoding. To understand the third, fourth, and fifth claim thereof, think of $\emptyset$ as non-encoding set *par excellence*: for every pair x, y of sets which differ from $\emptyset$, we want unique sets p, q to exist such that lft$(p) =$ lft$(q) = x$, rgt$(p) =$ rgt$(q) = y$, and Pair(p), Appl(q) hold; conversely, we want lft(s) and rgt(s) to differ from $\emptyset$ when either Pair(s) or Appl(s) holds. The last claim of termEncoding plays a technical role: since the only built-in kind of recursion in Ref is $\in$-recursion, by imposing that the immediate subterms of any compound term t (as encoded by a set) *belong* to t, this claim will ease the recursive definition of functions over all terms.

Figure 10.9 suggests one way of implementing the wanted functions and properties inside termEncoding, based on the remark that when $\emptyset \notin \{x, y\}$ the projections x, y can be retrieved from both variants $\langle x, y\rangle \cup \{x, y\}$, $\langle x, y\rangle \cup \{x, y\} \cup \{\emptyset\}$ (the former of which equals $\{x, y\}^+ \cup \{\{x\}\}$) of Kuratowski's pair $\langle x, y\rangle$.

Assuming that terms are encoded according to a quintuple such as the one produced by termEncoding, it is easy to implement their evaluation thus developing the THEORY evalTerm whose interface is shown in Fig. 10.10.

$$
\begin{array}{rll}
\mathsf{cst}(C) & =_{\text{Def}} & \{\{C\}\} \\
\mathsf{lft}(T) & =_{\text{Def}} & \boldsymbol{arb}\big(\{x:\ x\in T\,,y\in T\mid T\setminus\{\emptyset\}=\{x,y\}^{+}\cup\{\{x\}\}\}\big) \\
\mathsf{rgt}(T) & =_{\text{Def}} & \boldsymbol{arb}\big(\{y:\ x\in T\,,y\in T\mid T\setminus\{\emptyset\}=\{x,y\}^{+}\cup\{\{x\}\}\}\big) \\
\mathsf{Pair}(P) & \leftrightarrow_{\text{Def}} & \emptyset\notin P\,\&\,\emptyset\notin\{\mathsf{lft}(P)\,,\mathsf{rgt}(P)\} \\
\mathsf{Appl}(Q) & \leftrightarrow_{\text{Def}} & \emptyset\in Q\,\&\,\emptyset\notin\{\mathsf{lft}(Q)\,,\mathsf{rgt}(Q)\} \\
\hline
\mathsf{Lit}(L) & \leftrightarrow_{\text{Def}} & \big\langle\exists s,t\mid L\setminus\{\emptyset\}=\langle s,t\rangle\ \&\ t\notin\{\emptyset,s\}\ \&\, s\neq\emptyset\big\rangle \\
\mathsf{Qnt}(Q) & \leftrightarrow_{\text{Def}} & \big\langle\exists x,t\mid Q\setminus\{\emptyset\}=\langle x,t\rangle\ \&\ t\notin\{\emptyset,x\}\ \&\, x\in\mathbb{N}\big\rangle
\end{array}
$$

Fig. 10.9 A viable implementation of the quintuple needed to encode terms, followed by encodings of literals of the forms $(s\in t)$, $(s\notin t)$ and of bounded quantifiers of the forms $(\exists\, x_n\in t)$, $(\forall\, x_n\in t)$. Equality can be eliminated in terms of membership

$$
\begin{array}{l}
\text{THEORY}\ \mathsf{evalTerm}\big(\mathsf{th}(\mathsf{N},\mathsf{V}),\mathsf{cst}(\mathsf{S}),\mathsf{Pair}(\mathsf{P}),\mathsf{Appl}(\mathsf{P}),\mathsf{lft}(\mathsf{P}),\mathsf{rgt}(\mathsf{P})\big) \\
\quad \big\langle\forall \mathsf{c},\mathsf{k}\mid \mathsf{cst}(\mathsf{c})=\mathsf{cst}(\mathsf{k})\to\mathsf{c}=\mathsf{k}\big\rangle \\
\quad \big\langle\forall \mathsf{c},\mathsf{i},\mathsf{p},\mathsf{q},\mathsf{x},\mathsf{y},\mathsf{z}\mid \mathsf{i}\in\mathbb{N}\,\&\,\mathsf{Pair}(\mathsf{p})\,\&\,\mathsf{Appl}(\mathsf{q})\to\{\mathsf{cst}(\mathsf{c}),\mathsf{i},\mathsf{p},\mathsf{q}\}\neq\{\mathsf{x},\mathsf{y},\mathsf{z}\}\big\rangle \\
\quad \big\langle\forall \mathsf{x},\mathsf{y}\mid \emptyset\notin\{\mathsf{x},\mathsf{y}\}\to\big\langle\exists\mathsf{p},\forall\mathsf{s}\mid \mathsf{Pair}(\mathsf{s})\,\&\,\mathsf{lft}(\mathsf{s})=\mathsf{x}\,\&\,\mathsf{rgt}(\mathsf{s})=\mathsf{y}\leftrightarrow\mathsf{s}=\mathsf{p}\big\rangle\big\rangle \\
\quad \big\langle\forall \mathsf{x},\mathsf{y}\mid \emptyset\notin\{\mathsf{x},\mathsf{y}\}\to\big\langle\exists\mathsf{q},\forall\mathsf{s}\mid \mathsf{Appl}(\mathsf{s})\,\&\,\mathsf{lft}(\mathsf{s})=\mathsf{x}\,\&\,\mathsf{rgt}(\mathsf{s})=\mathsf{y}\leftrightarrow\mathsf{s}=\mathsf{q}\big\rangle\big\rangle \\
\quad \big\langle\forall \mathsf{s}\mid \mathsf{Pair}(\mathsf{s})\vee\mathsf{Appl}(\mathsf{s})\to\emptyset\notin\{\mathsf{lft}(\mathsf{s}),\mathsf{rgt}(\mathsf{s})\}\big\rangle \\
\quad \big\langle\forall \mathsf{t}\mid \{\mathsf{lft}(\mathsf{t}),\mathsf{rgt}(\mathsf{t})\}\subseteq\mathsf{t}\cup\{\emptyset\}\big\rangle \\
\Rightarrow(\mathsf{val}_{\Theta}) \\
\quad \big\langle\forall \mathsf{c},\mathsf{v}\mid \mathsf{val}_{\Theta}(\mathsf{cst}(\mathsf{c}),\mathsf{v})=\mathsf{c}\big\rangle \\
\quad \big\langle\forall \mathsf{p},\mathsf{v}\mid \mathsf{Pair}(\mathsf{p})\to\mathsf{val}_{\Theta}(\mathsf{p},\mathsf{v})=\langle\mathsf{val}_{\Theta}(\mathsf{lft}(\mathsf{p}),\mathsf{v}),\mathsf{val}_{\Theta}(\mathsf{rgt}(\mathsf{p}),\mathsf{v})\rangle\big\rangle \\
\quad \big\langle\forall \mathsf{q},\mathsf{v}\mid \mathsf{Appl}(\mathsf{q})\to\mathsf{val}_{\Theta}(\mathsf{q},\mathsf{v})=\mathsf{val}_{\Theta}(\mathsf{lft}(\mathsf{q}),\mathsf{v})\upharpoonright\mathsf{val}_{\Theta}(\mathsf{rgt}(\mathsf{q}),\mathsf{v})\big\rangle \\
\quad \big\langle\forall \mathsf{n},\mathsf{v}\mid \mathsf{n}\in\mathbb{N}\setminus\{\emptyset\}\to\mathsf{val}_{\Theta}(\mathsf{n},\mathsf{v})=\mathsf{th}(\mathsf{n},\mathsf{v})\big\rangle \\
\text{END}\ \mathsf{evalTerm}
\end{array}
$$

Fig. 10.10 A THEORY about the evaluation of terms

This THEORY receives, along with a quintuple of the said kind, a function $\mathsf{th}(N,\,V)$ supplying the value of the N-th variable in a set-valued assignment V; it manufactures and produces in output the evaluating function val_{Θ}. In order to represent a set-valued assignment it suffices to use a finite-length list which must, in its turn, be modeled somehow: in a manner—we propose—complying with the THEORY interface shown in Fig. 10.11.

The property Lst produced by the THEORY list is meant to indicate which sets represent lists; the dyadic function th associates with any such set ℓ the number of components of the list and the sets occupying those components.[16] Specifically, supposing that $\mathsf{Lst}(\ell)$ holds, $\mathsf{th}(0,\,\ell)$ will exceed by one the overall number of components of ℓ, and $\mathsf{th}(n,\,\ell)$ will provide the n-th component of ℓ when $0<n<\mathsf{th}(0,\,\ell)$. It should be clear from this explanation that the three claims issued by list state that:

[16] One way of implementing lists is discussed in [30, pp. 127–128].

$$
\begin{array}{l}
\textsc{Theory}\ \mathsf{list}(\,)\\
\Rightarrow (\,\mathsf{Lst}_\Theta\,,\,\mathsf{th}_\Theta\,)\\
\quad \big\langle \forall \ell \mid \mathsf{Lst}_\Theta(\ell) \rightarrow \mathsf{th}_\Theta(\emptyset,\ell) \in \mathbb{N}\setminus\{\emptyset\}\big\rangle\\
\quad \big\langle \forall \ell, \mathsf{m} \mid \mathsf{Lst}_\Theta(\ell)\,\&\,\mathsf{Lst}_\Theta(\mathsf{m})\,\&\,\big\langle \forall \mathsf{n} \in \mathsf{th}_\Theta(\emptyset,\ell) \mid \mathsf{th}_\Theta(\mathsf{n},\ell) = \mathsf{th}_\Theta(\mathsf{n},\mathsf{m})\big\rangle \rightarrow \ell = \mathsf{m}\big\rangle\\
\quad \big\langle \forall \mathsf{m},\mathsf{h},\mathsf{x} \mid \mathsf{h} \in \mathbb{N} \rightarrow \big\langle \exists \ell \mid \mathsf{Lst}_\Theta(\ell)\,\&\,\mathsf{th}_\Theta(\emptyset,\ell) = \mathsf{h}^+\,\&\\
\qquad\qquad\qquad\qquad \big\langle \forall \mathsf{n} \in \mathsf{h}\setminus\{\emptyset\} \mid \mathsf{th}_\Theta(\mathsf{n},\ell) = \mathsf{th}_\Theta(\mathsf{n},\mathsf{m})\big\rangle\,\&\\
\qquad\qquad\qquad\qquad (\mathsf{h} = \emptyset \vee \mathsf{th}_\Theta(\mathsf{h},\ell) = \mathsf{x})\big\rangle\\
\textsc{End}\ \mathsf{list}
\end{array}
$$

Fig. 10.11 A THEORY of lists

1. the length of every list is a finite ordinal;
2. the equality criterion for lists ℓ, m is that ℓ and m have the same length h and the same n-th component for $n = 1, \ldots, h$;
3. from every triple m, h, x consisting of a list m, a natural number h, and a set x, one can obtain a list ℓ of length h whose last component—if any—is x and whose n-th component is $\mathsf{th}(n, m)$ for $n = 1, \ldots, h - 1$; viz.:

$$
\ell = \begin{cases} \langle\,\rangle & \text{if } h = 0\,,\\ \langle \mathsf{th}(1,m), \ldots, \mathsf{th}(h-1,m), x\rangle & \text{otherwise.} \end{cases}
$$

For a sparing encoding of formulae, we can think of equality as a derived construct; a logical equivalence by which it can be eliminated is in fact $(s = t) \leftrightarrow (\exists \boldsymbol{x}_n \in \langle s, s\rangle)(t \in \boldsymbol{x}_n)$, where $\boldsymbol{x}_n$ does not occur in s or in t. It is also advisable to treat conjunction and disjunction as polyadic connectives, so that the only formulae which need to be encoded directly are the ones of the forms $(s \in t)$, $(s \notin t)$, $(\exists \boldsymbol{x}_n \in t)\left(\bigwedge_{i=0}^{h} \varphi_i\right)$, and $(\forall \boldsymbol{x}_n \in t)\left(\bigvee_{j=0}^{k} \psi_j\right)$, where each φ_i and each ψ_j has in its turn one of these forms. An expedient way of representing a multiple conjunction or disjunction, that owes much to Martin Davis for its dissemination in the early 1960s, is as the sets of conjuncts or disjuncts, respectively;[17] we will rely on this representation for completing our endeavor.

10.10 Related Work

> Often [$\cdots$] the nonstandard definition of a concept is simpler than the standard definition (both intuitively simpler and simpler in a technical sense, such as quantifiers over lower types or fewer alternations of quantifiers). As a result, nonstandard analysis sometimes makes it easier to find proofs.
> [4, p. 37]

[17] This way of representing formulae in conjunctive normal form is widely used today. In recent years [32] resorted to it, to give a Ref-based correctness proof for the DPLL satisfiability algorithm.

In what follows, we rely on [6] as an up-to-date comparative survey on systems which offer automated proof abilities related to real analysis. Some of the formalizations supported by such systems characterize real numbers axiomatically, as a given set with specific operations and properties; others construct real numbers either from rational Cauchy sequences or as Dedekind cuts. Nonstandard analysis is available in ACL2(r) and in Isabelle/HOL (see [26, 27], respectively): both achievements are reminiscent of [2, 3].

The semi-automated theorem prover ACL2, which ACL2(r) potentiates, offers limited support to quantifier handling (cf. [27, pp. 323–324]); in order to circumvent that difficulty, ACL2(r) focuses on the extension ${}^{*}\mathbb{R}$ of the reals. With hyperreal numbers, in fact, the quantifier alternation $\forall\,\varepsilon > 0\;\exists\,\delta > 0\ldots$ which affects the usual formulas about limits becomes unnecessary, hence the proofs benefit from a higher degree of automation. The formalism of ACL2(r) is based on an axiomatization of ${}^{*}\mathbb{R}$ as an autonomous domain.

The Isabelle/HOL-mechanization of real analysis, on the other hand, introduces the standard, along with the nonstandard, definition of each concept; thereby, 'users will have the freedom either to stick with classical (standard) techniques, use non-standard ones, or a combination of both' [26, p. 161]. 'Our first task', the author notes, 'each time we introduce a new concept from analysis, is to prove that the two definitions are equivalent' [26, p. 150]. Thus, albeit implicitly, the *transfer principle* plays a central role. It is 'neither an axiom nor a theorem, but a meta-theorem, since it applies to theorem statements' and, as such, 'it is not directly proved in Isabelle/HOL' [6]; nevertheless, since this principle informs the general pattern followed by all the equivalence proofs, the ultrapower construction of the hyperreals presupposed that a proof of Zorn's lemma and a theory of filters and ultrafilters were developed for Isabelle/HOL (cf. [26, p. 145]).

As an eventual reward of the exploration discussed in this paper, we hope to get Ref-based, nonstandard proofs of theorems of real analysis and to check by means of Ref many of the results presented in Martin Davis's chapter on hyperreal numbers [14, Sects. 2.3–2.8]. However, a formal remake of real analysis along unconventional lines is only an incidental issue here. As discussed at the beginning, we rather feel confronted with a proof-engineering issue—akin to metamathematical extensibility—which our proof assistant could tackle well because a proof of the relevant meta-theorem can be set up with relative ease in a full-fledged set theory.

After all, the guidelines for a Ref-based development of analysis which J. T. Schwartz sketched in [35, Chap. 5] stick to the tradition; the use of nonstandard methods can lead to much simpler and more elegant proofs than the classical ones, but one can contend that it calls for an extra amount of work spent on preliminary constructions, which may be out of scale with a proof of Rolle's theorem (to cite a result of analysis proper).

For a large-scale endeavor, this additional work is justified by considerations such as the following:

> Not only does nonstandard analysis provide a rigorous treatment of infinitesimals in the area of mathematics where they were originally used, it also gives elegant approaches to some ideas that developed later.
>
> [4, p. 37]

10.11 Concluding Remarks

> The well known theorem of Gödel shows that every system of logic is in a certain sense incomplete, but at the same time it indicates means whereby from a given system L of logic a more complete system L' may be obtained. By repeating the process we get a sequence L, $L_1 = L'$, $L_2 = L'_1$, $L_3 = L'_2, \ldots$ of logics each more complete than the preceding.
>
> (A. M. Turing, 1938)

The authors have at this point prepared the ground for verifying, with a proof-checker based on set theory, the propositions in the first chapter of [14].[18] A variant of the Zermelo-Fraenkel set theory, postulating global choice, regularity and infinity,[19] underlies the logical armory of the proof-checker, Ref, on which our experimental activity relies. The formally checked proofs regard, for the time being, only certain parts of our planned work: in particular, we proved the conclusions of the THEORYs about universes, superstructures, and 'urification' shown in Figs. 10.5 and 10.6, as well as the unique readability of the sets that encode terms inside the THEORY termEncoding (see Figs. 10.8 and 10.9); the proof of the ultrafilter theorem was available from the outset,[20] along with many minor but useful facts about finiteness, rank, ordinals, the set constructs $\mathscr{P}$, $\bigcup$, etc.

In the phase on which we have reported, anyway, our work has been mainly architectural: given the availability of a second-order construct, 'THEORY', supporting modularization and proof reuse in Ref, we deem it wise to invest in designing the THEORY interfaces before formalizing proofs meticulously.

---

[18] A website reporting on our experiment is at
http://www2.units.it/eomodeo/InitialSetupForNonStandardAnalysis.html,
http://aetnanova.units.it/scenarios/InitialSetupForNonStandardAnalysis/.

[19] In Ref the well-foundendess of membership and statements of the axiom of choice easily result from the availability of the construct ***arb*** discussed in Sect. 10.6, thanks to the interplay of ***arb*** with abstract set formers; infinity is embodied by Ref's built-in constant σ_∞.

[20] For a Ref-based proof of Zorn's lemma (whence the ultrafilter theorem follows easily), see [35, pp. 373–405]. This lemma was used in Ref's proof of the maximal ideal theorem for Boolean algebras as presented in [9].

We are confident that we can finish the envisaged proof-development tasks without getting entangled in unforeseen difficulties. Then, as said in the introduction, we must adopt schemes of notation and extended rules of inference that conveniently assist Ref's users in exploitations of the nonstandard methods.

Even after those enhancements, Ref's theory will be a conservative extension of the specific set theory available in Ref's initial endowment. A more challenging and intriguing view on the extensibility of proof-checkers should cope with the progressive extension of theories, in a frame of mind close to some of Alan Turing's early investigations (see [37]).

Acknowledgements Discussions with Francesco Di Cosmo helped in polishing this paper. The first author acknowledges partial support from the Polish National Science Centre research project DEC-2011/02/A/HS1/00395; and the second author from the project FRA-UniTS (2014) "*Learning specifications and robustness in signal analysis.*

References

1. Anastasio, S. (Coordinating Editor) (2015). In memory of Jacob Schwartz. *Notices of the AMS*, 473–490.
2. Ballantyne, A. M. (1991). The Metatheorist: Automatic proofs of theorems in analysis using non-standard techniques, Part II. In R. S. Boyer (Ed.), *Automated reasoning: Essays in Honor of Woody Bledsoe* (pp. 61–75). Dordrecht, The Netherlands: Kluwer Academic.
3. Ballantyne, A. M., & Bledsoe, W. W. (1977). Automatic proofs of theorems in analysis using nonstandard techniques. *Journal of the ACM*, *24*(3), 353–374.
4. Blass, A. (1978). Book reviews of *Applied nonstandard analysis*, by Martin Davis, *Introduction to the theory of infinitesimals*, by K. D. Stroyan and W. A. J. Luxemburg, and *Foundations of infinitesimal calculus*, by H. Jerome Keisler. *Bull. Amer. Math. Soc.*, 84(1):34–41, 1978.
5. Bledsoe, W. W. (1977). Non-resolution theorem proving. *Artificial Intelligence*, *9*(1), 1–35.
6. Boldo, S., Lelay, C., & Melquiond, G. (2015). Formalization of real analysis: A survey of proof assistants and libraries. *Mathematical Structures in Computer Science*, 38 pp.
7. Burstall, R., & Goguen, J. (1977). Putting theories together to make specifications. In R. Reddy (Ed.), *Proceedings of the 5th International Joint Conference on Artificial Intelligence* (pp. 1045–1058). Cambridge, MA.
8. Cantone, D., Omodeo, E. G., & Policriti, A. (2001). *Set Theory for Computing. From Decision Procedures to Declarative Programming with Sets*. Monographs in Computer Science. Springer.
9. Ceterchi, R., Omodeo, E. G., & Tomescu, A. I. (2014). The representation of Boolean algebras in the spotlight of a proof checker. In L. Giordano, V. Gliozzi, & G. L. Pozzato, (Eds.), *CILC 2014: Italian Conference on Computational Logic*, volume 1195 http://ceur-ws.org/Vol-1195/, ISSN 1613-0073, pp. 287–301. CEUR Workshop Proceedings, July 2014.
10. Chinlund, T. J., Davis, M., Hinman, P. G., & McIlroy, M. D. (1964). Theorem-proving by matching. Technical report, Bell Telephone Laboratories, Incorporated, Murray Hill, New Jersey.
11. Cohen, P. J. (1966). *Set Theory and the Continuum Hypothesis*. Mathematics Lecture Note Series. Reading, Massachusetts: W. A. Benjamin, Inc.
12. Davis, M. (1960). A program for Presburger's algorithm. *Summaries of talks presented at the Summer Institute of Symbolic Logic in 1957 at Cornell University* (vol. 2, pp. 215–223). Princeton, NJ. Communications Research Division, Institute for Defense Analyses. Reprinted as "A computer program for Presburger's algorithm" in [36, pp. 41–48].

13. Davis, M. (1963). Eliminating the irrelevant from mechanical proofs. *Proceedings of Symposia in Applied Mathematics* (vol. 15, pp. 15–30). Providence, RI: AMS. Reprinted in [36, pp. 315–330]; Russian transl. in *Kiberneticheskiy sbornik. Novaya seriya*, 7, 1970, pp. 160–179.
14. Davis, M. (1977). *Applied nonstandard analysis*. Wiley. Reprinted with corrections Dover, 2005. Russian translation, Izdatel'stvo Mir, Moscow 1980. Japanese translation 1977.
15. Davis, M. (2001). The early history of automated deduction. In J. A. Robinson & A Voronkov, (Eds.), *Handbook of Automated Reasoning* (pp. 3–15). Elsevier and MIT Press.
16. Davis, M. (2013). *Jack Schwartz meets Karl Marx*. In *[22, pp. 23–37]*.
17. Davis, M., & Fechter, R. (1991). A free variable version of the first-order predicate calculus. *Journal of Logic and Computation*, *1*(4), 431–451.
18. Davis, M., & Hersh, R. (1972). Nonstandard analysis. *Scientific American*, *226*, 78–86.
19. Davis, M., & Putnam, H. (1958). *Feasible computational methods in the propositional calculus*. Technical report, Rensselaer Polytechnic Institute, Research Division, Troy, New York.
20. Davis, M., & Putnam, H. (1960). A computing procedure for quantification theory. *Journal of the ACM*, *7*(3):201–215. Reprinted in [36, pp. 125–139].
21. Davis, M., & Schonberg, E. (2011). Jacob Theodore Schwartz 1930–2009. *Biographical Memoirs of the National Academy of Sciences*,19 pp.
22. Davis, M., & Schonberg, E. (Eds.). (2013). *From Linear Operators to Computational Biology: Essays in Memory of Jacob T. Schwartz*. Springer.
23. Davis, M. & Schwartz, J. T. (1977). Correct-program technology/Extensibility of verifiers—Two papers on Program Verification with Appendix of Edith Deak. Technical Report No. NSO-12, Courant Institute of Mathematical Sciences, New York University.
24. Davis, M. & Schwartz, J. T. (1979). Metamathematical extensibility for theorem verifiers and proof-checkers. *Computers and Mathematics with Applications*, *5*, 217–230. Also in [25, pp. 120–146].
25. Davis, M., Logemann, G., & Loveland, D. W. (1962). A machine program for theorem-proving. Communications of the Association for Computing. *Machinery*, *5*(7), 394–397.
26. Fleuriot, J. D. (2000). On the mechanization of real analysis in Isabelle/HOL. In M. Aagaard & J. Harrison, (Eds.), *Theorem Proving in Higher Order Logics, 13th International Conference, TPHOLs 2000, Portland, Oregon, USA, 14–18 August 2000, Proceedings*, volume 1869 of *Lecture Notes in Computer Science* (pp. 145–161). Springer.
27. Gamboa, R., & Kaufmann, M. (2001). Nonstandard analysis in ACL2. *Journal of Automated Reasoning*, *27*(4), 323–351.
28. Keisler, H. J. (1976). *Foundations of infinitesimal calcuus*. Boston, MA: Prindle, Weber & Schmidt, Inc.
29. Omodeo, E. G. (1982). The Linked Conjunct method for automatic deduction and related search techniques. *Computers and Mathematics with Applications*, *8*, 185–203.
30. Omodeo, E. G. (2012). The Ref proof-checker and its "common shared scenario". In M. Davis & E. Schonberg, (Eds.), *From Linear Operators to Computational Biology: Essays in Memory of Jacob T. Schwartz* (pp. 121–131). Springer.
31. Omodeo, E. G., & Schwartz, J. T. (2002). A 'Theory' mechanism for a proof-verifier based on first-order set theory. In A. Kakas & F. Sadri, (Eds.), *Computational logic: Logic programming and beyond—Essays in honour of Bob Kowalski, part II* (vol. 2408, pp. 214–230). Springer.
32. Omodeo, E. G., & Tomescu, A. I. (2008). Using ÆtnaNova to formally prove that the Davis-Putnam satisfiability test is correct. *Le Matematiche*, *63*(1), 85–105.
33. Policriti, A. (1988). Decision procedures for elementary sublanguages of set theory. IX. Unsolvability of the decision problem for a restricted class of the Δ_0-formulas in set theory. *Communications on Pure and Applied Mathematics41*(2), 221–251.
34. Robinson, J. A. (1967). Review: Martin Davis, Eliminating the irrelevant from mechanical proofs. *Journal of Symbolic Logic*, *32*(1), 118–119.
35. Schwartz, J. T., Cantone, D., & Omodeo, E. G. (2011). *Computational logic and set theory—Applying formalized logic to analysis*. Springer.
36. Siekmann, J., & Wrightson, G. (Eds.). (1983). *Automation of reasoning 1: Classical papers on computational logic 1957–1966*. Berlin, Heidelberg: Springer.

37. Turing, A. M. (1939). Systems of logic based on ordinals. *Proceedings of the London Mathematical Society*, *2*(45), 161–228.
38. Weyhrauch, R. W. (1977). A users manual for FOL. Technical Report MEMO AIM-235.1, Stanford University, Stanford, CA, USA.
39. Yarmush, D. L. (1976). The Linked Conjunct and other algorithms for mechanical theorem-proving. Technical Report IMM 412, Courant Institute of Mathematical Sciences, New York University.

Chapter 11
What Is Essential Unification?

Peter Szabo, Jörg Siekmann and Michael Hoche

Abstract A *unifier* of two terms s and t is a substitution σ such that $s\sigma = t\sigma$. For first-order terms there exists a *most general unifier* σ in the sense that any other unifier τ can be composed from σ with some substitution λ such that $\tau = \sigma \circ \lambda$. For many practical applications it turned out to be useful to generalize this notion to E-unification, where E is an equational theory, $=_E$ is equality under E and σ is an E-unifier if $s\sigma =_E t\sigma$. Depending on the equational theory E, the set of most general unifiers is always a singleton (as above) or it may have more than one unifier, either finitely or infinitely many unifiers and for some theories it may not even exist, in which case we call the theory of type nullary. The *set of most general unifiers* is denoted as $\mu\mathscr{U}\Sigma_E(\Gamma)$ for a unification problem Γ, which is a system of equations and an equational theory E. Unfortunately the set $\mu\mathscr{U}\Sigma_E(\Gamma)$ may be very large in general—even if it is finite—and for all practical purposes not really useful. For this and other reasons there is hence (i) a strong interest to compute a much smaller generating set of *minimal* unifiers and then (ii) to find efficient engineering solutions to handle these sets. *Essential unifiers*, as introduced by Hoche and Szabo, generalize the notion of a most general unifier and they have a dramatically pleasant effect: the set of essential unifiers is often much smaller than the set of most general unifiers. Essential unification may even reduce an infinitary theory to an essentially finitary theory. For example the one variable string unification problem is essentially finitary whereas it is infinitary in the usual sense. The most drastic reduction known so far is obtained for idempotent semigroups, or bands as they are called in computer science, which are of type nullary: there exist two unifiable terms s and t, but the set of most general unifiers does not exist. This is in stark contrast to essential

P. Szabo (✉)
Kurt-Schumacher-Str. 13, 75180 Pforzheim, Germany
e-mail: szabo@cs.uni-saarland.de

J. Siekmann
Saarland University/DFKI, Stuhlsatzenhausweg, 66123 Saarbrücken, Germany
e-mail: Joerg.Siekmann@dfki.de

M. Hoche
Airbus Defense and Space, Claude-Dornier-Str., 88090 Immenstaad, Germany
e-mail: michael.hoche@googlemail.com

E.G. Omodeo and A. Policriti (eds.), *Martin Davis on Computability, Computational Logic, and Mathematical Foundations*,
Outstanding Contributions to Logic 10, DOI 10.1007/978-3-319-41842-1_11

unification: the set of essential unifiers for bands always exists and is finite. The key idea for essential unification is to base the notion of generality not on the standard subsumption order for terms with the associated subsumption order for substitutions, but on the *encompassment order* for terms and substitutions. Hence we propose the encompassment order as a more natural order relation for minimal and complete sets of E-unifiers and call these sets *essential unifiers,* denoted as $e\mathcal{U}\Sigma_E(\Gamma)$. This paper introduces *essential unification,* provides a definitional framework based on order relations and surveys what is presently known. We conclude with a list of some of the more important open problems, including the main open problem, namely how to build essential unification into an automated reasoning system.

Keywords E-Unification · Order relations for unification · Most general unifiers · Essential unifiers · Unification theory

11.1 Introduction

Unification is a well established concept in artificial intelligence and automated theorem proving, in computational linguistics and universal algebra as well as in theoretical and applied computer science like for example in the semantics of programming languages, for the semantic web and in many other areas (see [50, 65, 76] for several application areas). Surveys of unification theory can be found in [7, 8, 31, 50, 76] and there is more recent work on unification in description logics [4, 10, 11], modal logics [5], nominal unification [81], disunification [9] and other application areas. The current state of the art is represented at the UNIF workshop series.[1] A survey of the related topic of rewriting systems is presented in [22] and in the "emerging" textbook [48]; a list of open problems can be found in [68]. A standard textbook is by Franz Baader and Tobias Nipkow, *Term Rewriting and All That* [6]. An interesting collection of open and solved unification problems for several common algebraic structures like groups, vector spaces, commutative rings, Boolean algebras and others is collected by Stanley Burris[2] and a recent survey on higher order unification is presented in [41].

Unification algorithms were first invented for automated theorem proving systems and the historically first computer generated mathematical proof for a theorem was found by a program from **Martin Davis** in 1954. It postulated the remarkable fact that the product of two even numbers is again even, formulated in a decidable fragment of first order logic, called Pressburger Arithmetic. The first complete unification algorithm for first order logic is due to Prawitz [64], but a complete algorithm as

[1]First workshop in Val d'Ajol in 1987 and since then annually. Since 1997, there is a website UNIF1997, UNIF1998, UNIF1999 up to UNIF2005 in Japan and UNIF2006 at the FLOC conference in Seattle, UNIF2007 and UNIF2008 at the Schloss Hagenberg, Linz, Austria. The current UNIF's can be found at UNIF2013, UNIF2014 and UNIF2015.

[2]http://www.math.uwaterloo.ca/~snburris/htdocs/WWW/PDF/e_unif.pdf.

we know it today was working already in 1963 in Martin Davis's "Davis-Putnam Procedure" [20], where improvements were implemented by D. McIlroy and Peter Hinman. All of this, including Martin's linked conjunct method, was later subsumed by Alan J. Robinson's resolution principle [67], where the unification algorithm is the corner stone of this method. Resolution dominated the field of automated reasoning then for many decades to come and is still the most wellknown inference rule in artificial intelligence.

Unification as we see it today is a general mechanism to solve equational problems. For practical applications it is often crucial to have a minimal representation of the solutions, from which all other solutions (unifiers) can be derived. This is an essential feature of any of todays resolution, matrix, rewrite or tableaux based automated reasoning systems, where the most general unifier represents the infinitely many elements from the Herbrand Universe that had to be enumerated and instantiated into the universally quantified variables of earlier automated deduction systems. All of these early sytems implemented the key idea of Herbrands work [84], that a first order formula can be proven by instantiating in a systematic fashion the quantified universal variables by ground terms, (now often called the Herbrand Universe)[3] and then prove it by some decision procedure for propositional logic. These early systems like Gilmore [33], Wang [83], Kangar [45], Davis [20], Veenker [82] and others differed in their struggle to find the "right" instances out of the infinite set of potential ground terms and it took a little more than a decade until the notion of a complete first order unification algorithm became standard. Martin Davis's article "The prehistory and early history of automated deduction" [17] gives a lively historical account of these early developments and the two volumes [78] collect the most important contributions in these early days.

Martin also worked with Julia Robinson on Hilbert's Tenth Problem [18], a topic we shall pick up again in paragraph 4.2 where we discuss its relationship to string unification. For all these well known and influential contributions and many more personal reasons this article is dedicated to Martin Davis.

For unification problems in the free algebra of terms (also known as syntactic unification), there exists always a unique unifier for solvable unification problems from which all other unifiers can be derived by instantiation. This unique (up to renaming) unifier is called the *most general unifier*, but for equational algebras the situation is completely different: the minimal complete set of unifiers is not always finite and it may not even exist, which was conjectured by Gordon Plotkin [63] in his seminal paper in 1972. This paper introduced the idea to take some troublesome axioms like associativity or commutativity out of the data base and to build them directly into the deduction machinery. Since then unification problems and equational theories have been classified with respect to the cardinality of their minimal complete set of unifiers. These results led to the development of general approaches and algorithms, which can be applied to a whole class of theories. This is the topic of *universal unification*, see e.g. [75, 80].

---

[3] Actually several other logicians of the time had this idea and it is not known who came first.

More specifically, a unification problem $s =^? t$ for two given terms s and t is the problem to find a unifier σ such that $s\sigma = t\sigma$. A substitution σ is more general than a substitution τ if there is a substitution λ such that $\tau = \sigma \circ \lambda$. We will also say σ *subsumes* τ. The unifier σ is a *most general unifier,* if for any other unifier τ of s and t we can find a substitution λ such that $\tau = \sigma \circ \lambda$. We often have the need to limit the equality on substitutions to a set of variables and write $\sigma =^V \tau$ if $x\sigma = x\tau$ for all variables $x \in V$. Generalizing this notion to E-unification, where E is an equational theory, $=_E$ is equality under E and σ is an E-unifier for s and t with $s\sigma =_E^V t\sigma$, we may have more than one most general unifier. A minimal and complete set of E-unifiers, denoted $\mu\mathscr{U}\Sigma_E$ for s and t, is a set such that for every $\sigma \in \mu\mathscr{U}\Sigma_E$ we have $s\sigma =_E^V t\sigma$. The set is *complete* if for any E-unifier τ there exists some σ in $\mu\mathscr{U}\Sigma_E$ such that $\tau =_E^V \sigma \circ \lambda$. The set $\mu\mathscr{U}\Sigma_E$ is *minimal* in the sense that for every two unifiers σ, τ in $\mu\mathscr{U}\Sigma_E$ there is no λ with $\sigma =_E^V \tau \circ \lambda$, that is all unifiers in $\mu\mathscr{U}\Sigma_E$ are independent. We say that a unification problem is *unitary* if $\mu\mathscr{U}\Sigma_E$ is a singleton, it is *finitary* if $\mu\mathscr{U}\Sigma_E$ is finite for every s and t and it is *infinitary* if there are terms s and t such that $\mu\mathscr{U}\Sigma_E$ is infinite. Unfortunately there are theories such that two terms are unifiable, but the set $\mu\mathscr{U}\Sigma_E$ is not recursively enumerable. In this case we call the problem *nullary* or of *type zero*. This classification according to the type naturally leads to a hierarchy of equational theories called the *unification hierarchy*.

It turned out that this well established view of unification theory changes drastically, if we redefine the notion of a most general unifier. Recall that a unifier σ *subsumes* another unifier τ if:

$$\tau =_E^V \sigma \circ \lambda$$

Hence standard unification theory is based on the *subsumption relation*. We generalize this notion and define *an encompassment relation* on substitutions: a substitution σ is encompassed by a substitution τ, if there exist substitutions λ_1 and λ_2 such that

$$\tau =_E^V \lambda_1 \circ \sigma \circ \lambda_2$$

where λ_1 has to have certain properties to be defined in the next paragraph below. The idea is that λ_2 is used to establish the known subsumption relation between τ and σ as in standard unification theory and is composed as usual "from the right" in the tripartition $\lambda_1 \circ \sigma \circ \lambda_2$. The substitution λ_1 allows us also to compose "from the left" and this can drastically reduce the cardinality of the set of *minimal* E-unifiers, which we now call *essential E-unifiers*: an E-unifier σ is an essential E-unifier if for any other unifier τ there exist substitutions λ_1 and λ_2 such that $\tau =_E^V \lambda_1 \circ \sigma \circ \lambda_2$. We say τ *encompasses* σ and the *set of essential E-unifiers*, denoted as $e\mathscr{U}\Sigma_E$, is the set of E-unifiers such that for any unifier τ there is some $\sigma \in e\mathscr{U}\Sigma_E$, such that $\tau =_E^V \lambda_1 \circ \sigma \circ \lambda_2$.

We say a unification problem is *e-unitary* (is *e-finitary*) if the set of essential unifiers is always a singleton (is always finite). A unification problem is *e-infinitary* (*e-nullary*) if there are two terms such that the set of essential unifiers is infinite (does not exist).

11.2 Notions and Notation

Notation and basic definitions in unification theory are well known and have found their way into many and diverse academic fields and most monographs and textbooks on automated reasoning contain sections on unification.

In the following we unify the various presentations of the necessary concepts for unification towards a concise notation which serves our purpose and we show how the additional concepts for essential unification can be built upon these definitions. The notion of an algebra given below embraces algebraic structures and the original notions in computational logic, recursive function theory, theory of automata or automated theorem proving are compatible and natural applications.

11.2.1 Signatures, Terms and Term Algebras

A *signature* is a finite set F of function symbols with a nonnegative integer n, called *arity*, that is assigned to each member f of F and f is an n-ary function symbol. The subset of n-ary function symbols in F is denoted by F_n. An *algebra* of type F is an ordered pair $\mathscr{A} = \langle A, F\rangle$ where A is a nonempty set and F is a family of finitary operations on A indexed by the signature F such that corresponding to each n-ary function symbol f in F_n there is an n-ary operation f^A on A. The set A is called the carrier of the algebra $\mathscr{A} = \langle A, F\rangle$.

Let X be a set of (distinct) variables. Let F be a signature. The set $T(F, X)$ of (syntactic) *terms* of F over X is the smallest set

(i) comprising X and F_0 and

(ii) if $t_1, \ldots, t_n$ in $T(F, X)$ and f in F_n then $f(t_1, \ldots, t_n)$ in $T(F, X)$

The set of variable-free terms are called *ground terms*. The set of variables occurring in a term t is denoted by $\mathbf{Var}(t)$. The set of *sub terms* of a term $f(t_1, \ldots, t_n)$ contains the term itself and is closed recursively by containing $t_1, \ldots, t_n$. It is denoted by $\mathbf{Sub}(t)$.

Given F and X, then the *term algebra* of type F over X, denoted by $\langle T(F, X), F\rangle$, has as its universe the set of terms $T(F, X)$ and the fundamental operations satisfying

$$f^{\langle T(F,X),F\rangle}(t_1 \ldots, t_n) = f(t_1 \ldots, t_n)$$

for f in F_n and terms $t_1, \ldots, t_n$ in $T(F, X)$. Term algebras give an algebraic structure to (syntactic) terms and focus the attention on Herbrand interpretations, where the set of terms itself is the carrier.

11.2.2 Substitutions

A *substitution* is a (unique) homomorphism in the term algebra generated by a mapping $\sigma : X \longrightarrow T(F, X)$ from a finite set of variables to terms. Substitutions are generally denoted by small Greek letters $\alpha, \beta, \gamma, \sigma$ etc. and they are represented explicitly as a function by a set of variable bindings $\sigma = \{x_1 \mapsto s_1, \ldots, x_m \mapsto s_m\}$. $\mathcal{S}_{F,X}$ denotes the set of all substitutions. The application of the substitution σ to a term t, denoted $t\sigma$, is defined by induction on the structure of terms

$$t\sigma = \begin{cases} s_i & \text{if } t = x_i \\ f(t_1\sigma, \ldots, t_n\sigma) & \text{if } t = f(t_1, \ldots, t_n) \\ t & \text{otherwise} \end{cases}$$

The substitution $\varepsilon = \{\}$ with $t\varepsilon = t$ for all terms t in $\mathcal{T}_{F,X}$ is called the *identity*. A substitution $\sigma = \{x_1 \mapsto s_1, \ldots, x_m \mapsto s_m\}$ has the finite *domain:*

$$\mathbf{Dom}(\sigma) := \{x | x\sigma \neq x\} = \{x_1, \ldots, x_m\};$$

The *range* is the set of terms

$$\mathbf{Ran}(\sigma) := \bigcup_{x \in \mathbf{Dom}(\sigma)} \{x\sigma\} = \{s_1, \ldots, s_{m'}\}, m' \leqslant m$$

The set of variables occurring in the range is $\mathbf{VRan}(\sigma) := \mathbf{Var}(\mathbf{Ran}(\sigma))$ and $\mathbf{Var}(\sigma) = \mathbf{Dom}(\sigma) \cup \mathbf{VRan}(\sigma)$. The *restriction* of a substitution σ to a set of variables $Y \subseteq X$, denoted by $\sigma |_Y$, is the substitution which is equal to the identity everywhere except over $Y \cap \mathbf{Dom}(\sigma)$, where it is equal to σ. The *composition* of two substitutions σ and θ is written $\sigma \circ \theta$ (to emphasize the composition) or just as $\sigma\theta$ and its application is defined by $t\sigma\theta = (t\sigma)\theta$. This is fine if $\sigma\theta$ has no contradictory variable bindings, otherwise if $x\sigma \neq x\theta$ for some variable x, this binding in θ is applied to σ and eliminated in $\sigma\theta$, (see [8] p. 451, for details). A substitution σ is *idempotent* if $\sigma\sigma = \sigma$ and this is true iff $Dom(\sigma) \cap VRan(\sigma) = \varnothing$. The application of a substitution to a term can be tricky, if it is not idempotent, for example if it contains infinite cycles or contradictory bindings, and there are several solutions proposed for this problem in the literature. In the area of automated reasoning there is the convention that the variables in s_i are always renamed into new variables and contradictory bindings are removed. If σ is not idempotent, then the set representation of a substitution is inadequate, as the application order of the individual bindings matters. In that case $\sigma = \{x_1 \mapsto s_1,\ x_2 \mapsto s_2, \ldots,\ x \mapsto s_m\}$, is often rewritten into "triangle form" [8]:

$$\{x_1 \mapsto s_1\}\{x_2 \mapsto s_2\} \ldots \{x_m \mapsto s_m\}$$

and then applied sequentially and component wise, sometimes called dag solvent form.

Relations such as $=, \geqslant, \ldots$ between substitutions sometimes hold only if they are restricted to a certain set of variables V. A relation R which is restricted to V is denoted as R^V, and defined as $\sigma R^V \tau \iff x\sigma R x\tau$ for all x in V. Two substitutions σ and θ are *equal*, denoted $\sigma = \theta$ iff $x\sigma = x\theta$ for every variable x, they are *equal restricted to V*, $x\sigma =^V x\theta$, iff $x\sigma = x\theta$ for all variables x in V.

11.2.3 Congruences and Equation

An equivalence relation Θ on the underlying set (the carrier) of an algebra $\mathscr{A}$ of type F is a *congruence*, if for each n-ary function symbol f in F and elements a_i, b_i of A, for all i in $1 \leqslant i \leqslant n$ we have

$$a_i \Theta b_i \Rightarrow f^A(a_1, \ldots, a_n) \Theta f^A(b_1, \ldots, b_n)$$

The quotient algebra $\mathscr{A}_\Theta$ is the algebra whose carrier are the equivalence classes A/Θ and whose operations satisfy

$$f^{A/\Theta}(a_1/\Theta, \ldots, a_n/\Theta) = f^A(a_1, \ldots, a_n)/\Theta$$

We are interested in quotient algebras, where the congruence is defined by a set of equations E, which is denoted as $=_E$. For a term t in $T(F, X)$ and the congruence E the equivalence class of t is denoted as $[t]_E$.

11.2.4 Orders

A term t is (syntactically) an *instance* of a term s, denoted $s \leqslant t$, if $s\sigma = t$ for some substitution σ:

$$s \leqslant t \iff \exists \sigma : s\sigma = t$$

We say s *subsumes* t and this relation is a quasiorder (or preorder as it is sometimes called. That is, it is reflexive and transitive). We call it the *subsumption order* on terms.

A term t (syntactically) *encompasses* a term s, denoted $s \sqsubseteq t$, if a sub term of t is an instance of s:

$$s \sqsubseteq t \iff \exists \sigma : s\sigma \in \mathbf{Sub}(t)$$

Syntactically encompassment conveys the notion that s appears in t with a context "above" and a substitution instance "below". We say *t encompasses s* or *s is encompassed* by t and $\sqsubseteq$ is the *encompassment order*. In particular we define $s \sqsubset t$, called *strict encompassment*, if $s\sigma$ is a proper sub term of t.[4]
So in summary we have the following *orders on terms:*

Definition 11.1 (*syntactic*)

1. A term s is a *sub term* of t if $s \in \mathbf{Sub}(t)$ and we denote this order by $s \trianglelefteq t$.
2. A term s *subsumes* t, denoted $s \leqslant t$, if there exists a substitution σ with $s\sigma = t$
3. A term s is *encompassed* by t, denoted $s \sqsubseteq t$, if there exists a substitution σ such that $s\sigma \in \mathbf{Sub}(t)$.

These standard order relations are now extended to equality modulo E for the congruences induced by the equations in E.

Definition 11.2 (*modulo E*)

1. A term s is a *sub term of t modulo E, denoted* $s \trianglelefteq_E t$, if there is a term $t' =_E t$ and $s \trianglelefteq t'$.
2. A term s subsumes t *modulo E*, $s \leqslant_E t$, if there exists a substitution σ with $s\sigma =_E t$
3. A term s is *encompassed by t modulo E*, $s \sqsubseteq_E t$ if there is a substitution σ such that $s\sigma \trianglelefteq_E t$.

We will now lift these order relations on terms component-wise to order relations on substitutions: for all variables in the domain of the substitution we require that the images fulfill the corresponding relation.

Definition 11.3 (*syntactic*)
Let V be some set of variables.

1. A substitution σ is a *sub-substitution* of τ, denoted as $\sigma \trianglelefteq \tau$, if $\mathbf{Dom}(\sigma) = \mathbf{Dom}(\tau)$ and for all x in this domain $x\sigma$ is a sub term of $x\tau$, that is, $x\sigma \in \mathbf{Sub}(x\tau)$. $\mathbf{SUB}(\tau)$ denotes the set of sub-substitutions of τ.
2. A substitution σ *subsumes* a substitution τ or τ is an instance of σ, denoted $\sigma \leqslant^V \tau$, if there exists a substitution λ such that $\tau =^V \sigma\lambda$. The relation $\leqslant$ is a quasiorder, called the *subsumption order for substitutions.*
3. A substitution σ is *encompassed* by τ, denoted by $\sigma \sqsubseteq^V \tau$, if there exists λ, such that $(\sigma\lambda) \mid_V$ is a sub-substitution of τ. That is $(\sigma\lambda) \mid_V \in \mathbf{SUB}(\tau)$.

The corresponding *order on substitutions modulo E*, $\sigma \leqslant_E \tau$ and $\sigma \sqsubseteq_E \tau$ is defined as:

Definition 11.4 (*substitution ordering modulo E restricted to a set of variables*)
Let V be some set of variables.

---

[4]Signs and notation are still not uniform in all related fields, in particular our notation is used more often in the literature on automated theorem proving and unification theory [6], whereas term rewriting systems usually prefer notational conventions such as $<$and $>$; see [22, 23].

1. A substitution σ is a *sub-substitution* of τ *modulo E*, denoted as $\sigma \trianglelefteq_E^V \tau$, if $\mathbf{Dom}(\sigma) = \mathbf{Dom}(\tau)$ and for all x in this domain, $x\sigma$ is a sub term of $x\tau$ modulo E.
2. A substitution σ *subsumes* a substitution τ modulo E restricted to V, denoted as $\sigma \leqslant_E^V \tau$, if there exists a substitution λ such that $\tau =_E^V \sigma\lambda$. The relation $\leqslant_E^V$ is called the *subsumption order for substitutions modulo E* restricted to V. We denote *subsumption equivalence* as $\sigma \sim_E^V \tau$, if $\sigma \leqslant_E^V \tau$ *and* $\tau \leqslant_E^V \sigma$.
3. A substitution σ is *encompassed* by τ modulo E restricted to V, denoted $\sigma \sqsubseteq_E^V \tau$, if there exists λ, such that $(\sigma\lambda)\mid_V$ is a sub-substitutionof τ modulo E restricted to V. We denote *encompassment equivalence* as $\sigma \approx_E^V \tau$, it holds if $\sigma \sqsubseteq_E^V \tau$ *and* $\tau \sqsubseteq_E^V \sigma$.

An example for the syntactic sub-substitutions of $\tau = \{x \mapsto f(a, z)\}$ is:

$$\begin{aligned}\mathbf{SUB}(\tau) = \{&\{x \mapsto a\},\\ &\{x \mapsto z\},\\ &\{x \mapsto f(a, z)\}\ \}\end{aligned}$$

because $x\{x \mapsto a\} = a \in \mathbf{Sub}(x\tau) = \mathbf{Sub}(f(a, z)) = \{f(a, z), a, z\}$; and similarly for the other elements of $\mathbf{SUB}(\tau)$.
To demonstrate the analogy between the better known encompassment definition for terms with the new encompassment order on substitutions, consider the terms s and t and substitutions σ and τ:

$$s = f(x, y),\ t = f(x, g(a, b)),\ \text{and}\ \lambda = \{y \mapsto g(a, b)\}$$

then $s \notin \mathbf{Sub}(t)$, but $s\lambda \in \mathbf{Sub}(t)$, i.e. $s \sqsubseteq t$.
Now consider the substitutions

$$\begin{aligned}\sigma &= \{x \mapsto a,\ y \mapsto g(a, z)\}\\ \tau &= \{x \mapsto f(a, b),\ y \mapsto f(a, g(a, b))\}\\ \lambda_2 &= \{z \mapsto b\}\end{aligned}$$

where $\sigma \notin \mathbf{SUB}(\tau)$ but $\sigma\lambda_2 \in \mathbf{SUB}(\tau)$, that is $\sigma \sqsubseteq \tau$.
With

$$\lambda_1 = \left\{x \mapsto f(x, b),\ y \mapsto f(a, y)\right\}$$

we can brake τ apart into a tripartition $\lambda_1\sigma\lambda_2$, with $V = \{x, y\}$:

$$
\begin{aligned}
\tau &=^V \lambda_1\sigma\lambda_2 \\
&=^V \left\{x \mapsto f(x,b),\ y \mapsto f(a,y)\right\}\{x \mapsto a,\ y \to g(a,z)\}\{z \mapsto b\} \\
&=^V \left\{x \mapsto f(x,b),\ y \mapsto f(a,y)\right\}\{x \mapsto a,\ y \mapsto g(a,b)\} \\
&=^V \{x \mapsto f(a,b),\ y \mapsto f(a,g(a,b))\}
\end{aligned}
$$

and hence σ encompasses τ, $\sigma \sqsubseteq \tau$.

Our claim is that the encompassment order is better suited for a general framework for E-unification than the standard order $\leqslant_E$. To get a feeling for the advantage let us look at the following example.

Substitutions form a semigroup with respect to their composition and this fact was used to define the subsumption order on unifiers, namely

$$\sigma \leqslant_E \tau \iff \exists\lambda_2 : \tau =_E^V \sigma \circ \lambda_2,$$

which led to the notion of a most general unifier.

Now consider the equational theory of associativity $A = \{x(yz) = (xy)z\}$, that is the free semigroup, and the unification problem $\Gamma = \{ax =_A^? xa\}$. This has an infinite set of most general unifiers $\sigma_n = \{\{x \mapsto a^n\} |\ n \geqslant 1\}$. However, the essential unifier in this set seems to be intuitively $\sigma_0 = \{x \mapsto a\}$, because every most general unifier contains this unifier in a certain sense, namely with

$$\lambda_n = \{x \mapsto a^{n-1}x\}$$

we have

$$\sigma_n = \lambda_n \circ \sigma_0,\ n > 1.$$

Since substitutions form a semigroup, the dual of the subsumption order, namely left-composition instead of right-composition would induce a semigroup as well with the advantage that the above infinitary problem would become a unitary one. So if we use the order $\leqslant_A$ with left composition $\exists\lambda : \sigma =_A \lambda\tau$ we have a *unitary* problem. But this is not compatible with the original notion of generality and of course it would not quite work in general as long as the λ can be any substitution. Our solution is based on *lifting* the encompassment order on terms to the *encompassment order on substitutions modulo E*. In particular the encompassment order on substitutions allows us to represent τ as a tripartition with left *and* right composition:

$$\sigma \sqsubseteq_E^V \tau \Longrightarrow \exists\,\lambda_1 \exists\lambda_2 : \tau =_E^V \lambda_1\sigma\lambda_2$$

If λ_1 is empty this is the usual subsumption relation and if λ_2 is empty then τ is not an instance and σ is a proper sub-substitution and so it is encompassed by τ.

11.2.5 E-Unification

Let E be an equational theory and let Σ be the signature of the term algebra. An *E-unification problem* is a finite set of equations

$$\Gamma = \{s_1 =^?_E t_1, \ldots, s_n =^?_E t_n\}$$

An *E-unifier* of Γ is a substitution σ, such that

$$s_1\sigma =_E t_1\sigma, \ldots, s_n\sigma =_E t_n\sigma$$

The set of all E-unifiers of Γ is denoted $\mathcal{U}\Sigma_E(\Gamma)$. A *complete* set of E-unifiers $c\mathcal{U}\Sigma_E(\Gamma)$ for Γ is a set of E-unifiers, such that for every E-unifier τ there exists $\sigma \in c\mathcal{U}\Sigma_E(\Gamma)$ with $\sigma \leqslant_E \tau$. The set $\mu\mathcal{U}\Sigma_E(\Gamma)$ is called a *minimal complete set* of E-unifiers for Γ, if it is complete and for all distinct elements σ and σ' in $\mu\mathcal{U}\Sigma_E(\Gamma)$ $\sigma \leqslant_E \sigma'$ implies $\sigma =_E \sigma'$.

When a minimal complete set of E-unifiers of a unification problem Γ exists, it is unique up to subsumption equivalence $\sim_E$. Minimal complete sets of E-unifiers need not always exist, and if they do, they might be singular, finite, or infinite. Since minimal complete sets of E-unifiers are isomorphic whenever they exist, they can be used to classify theories with respect to their corresponding unification problem as well. This leads naturally to the concept of a *unification hierarchy* which was first introduced in Siekmann's doctoral thesis in 1976 [73] and further refined and extended by himself and his later students as well as by many subsequent workers in the field of unification theory, see [7, 8, 31, 50, 76] for the standard surveys on this aspect.

A unification problem Γ is:

- *nullary*, if Γ is unifiable, but the minimal complete set of E-unifiers does not exist.
- *unitary*, if it is not nullary and the minimal complete set of E-unifiers for Γ is of cardinality less or equal to 1.
- *finitary*, if it is not nullary and the minimal complete set of E-unifiers is always finite.
- *infinitary*, if it is not nullary and the minimal complete set of E-unifiers is infinite.

An equational theory E is:

- *unitary*, if all unification problems for E are unitary
- *finitary*, if all unification problems are finitary.
- *infinitary*, if there is at least one infinitary unification problem and all unification problems have minimal complete sets of E-unifiers.
- If there exists a solvable unification problem Γ not having a minimal complete set of E-unifiers, then the equational theory E is *nullary* or of *type zero*.

The subsumption order and the encompassment order for E-unifiers are inherited from the above order on substitutions, that is an E-unifier σ is *more general* than an E-unifier τ, denoted as $\sigma \leqslant_E \tau$, if there exists a substitution λ such that $\tau =^V_E \sigma\lambda$.

11.2.6 Essential E-Unification

In Sect. 2.4 we showed the steps for the extension of the subsumption quasiorder for terms to substitutions and then extended it to equational theories $=_E$ in order to get the concept of a most general E-unifier.

However for all practical purposes, we do not only have the unpleasant situation that there are nullary E-unification theories [2, 30, 69], but even more importantly there is such an enormous plenitude of most general unifiers that it defies its original purpose: there are very simple theories such as associativity that are infinite and even the finite theories may have sets of most general unifiers that are beyond practical usefulness. For example the theory of associativity and commutativity AC has exponentially many unifiers, in fact for a base B there are B with a tower of exponentials many unifiers [12, 25, 46, 47]. So the question is: can we find a more general concept for the generating set, than the set of most general unifiers. As a step in this direction we propose the notion of an essential unifier, whose definition is not based on the generality order of the past, but on the encompassment order lifted to substitutions and extended to equational theories. So the concept of an essential E-unifier is as follows:

Definition 11.5 (*essential E-unifier*)

1. An E-unifier σ for a unification problem Γ modulo the equational theory E and the variables $V = \mathbf{Var}(\Gamma)$, is *encompassed* by an E-unifier τ for Γ, denoted as above by $\sigma \sqsubseteq_E^V \tau$, if there exists a substitution λ, such that $(\sigma\lambda)\mid_V$ is a subsubstitution of τ.
2. An E-unifier σ for a unification problem Γ modulo the equational theory E that does not encompass any other E-unifier for Γ is called an *essential E-unifier*. We denote the set of essential E-unifiers as $e\mathcal{U}\Sigma_E(\Gamma)$. Two unifiers σ and τ are *encompassment equivalent modulo E*, denoted $\approx_E^V$, if $\sigma \sqsubseteq_E^V \tau$ and $\tau \sqsubseteq_E^V \sigma$.
3. A *complete* set of *essential* E-unifiers for Γ is a set of E-unifiers, such that for each E-unifier τ there exists σ in the set with $\sigma \sqsubseteq_E^V \tau$.
4. The set $e\mathcal{U}\Sigma_E(\Gamma)$ is called a *minimal complete set* of *essential* E-unifiers for Γ, or simply *the set of essential E-unifiers for* Γ, if it is a complete set and for all σ and σ' in $e\mathcal{U}\Sigma_E(\Gamma)$ σ and σ' are encompassment equivalent.

Proposition 11.1 *The encompassment order on substitutions is a quasiorder, that is, it is reflexive and transitive.*

Proof reflexivity: $\sigma \sqsubseteq_E \sigma$ means that there are substitutions $\lambda_1, \lambda_2 : \sigma =_E \lambda_1\sigma\lambda_2$, setting λ_1 and λ_2 to the substitution identity ε we have $\sigma =_E \varepsilon\sigma\varepsilon = \sigma$.

transitivity: $\sigma \sqsubseteq_E^V \tau$ and $\tau \sqsubseteq_E^V \psi$ implies $\sigma \sqsubseteq_E^V \psi$, where by definition we have $\mathbf{Dom}(\sigma) = \mathbf{Dom}(\tau) = \mathbf{Dom}(\psi) =: V$, so

$$\tau =_E^V \lambda_{1,1}\sigma\lambda_{2,1}$$
$$\psi =_E^V \lambda_{1,2}\tau\lambda_{2,2}$$

which implies

$$\psi =_E^V (\lambda_{1,2}\lambda_{1,1}\sigma\lambda_{2,1}\lambda_{2,2}) \Rightarrow \sigma \sqsubseteq_E^V \psi$$

□

A well known property of unification in the free term algebra is that the most general unifier is unique (up to renaming of variables), if the unification problem is solvable. This is a property we would like to have for sets of essential unifiers as well.

One way to generate the set of all E-unifiers from the minimal set of most general unifiers, $\mu\mathscr{U}\Sigma_E(\Gamma)$, is via a closure operator. The notion of a *closure operator* is well established and plays a central role in many areas of algebra, computational logic and mathematics in general (see for example [29]). For a set M a mapping $\mathscr{C} : \mathscr{P}(M) \rightarrow \mathscr{P}(M)$ from the power set of M to itself, is a *closure operator* if it is extensive, monotone increasing and idempotent. It can be shown that the set $e\mathscr{U}\Sigma_E(\Gamma)$ of essential unifiers can be closed and generates all E-unifiers for Γ as well.

Proposition 11.2 *The set of essential unifiers $e\mathscr{U}\Sigma_E(\Gamma)$ is unique up to part equivalence $\approx_E$.*

Proof Suppose it is not unique, then there would be two complete sets of essential unifiers $e\mathscr{U}\Sigma_E^1$ and $e\mathscr{U}\Sigma_E^2$. Let σ_2 be in $e\mathscr{U}\Sigma_E^2 \backslash e\mathscr{U}\Sigma_E^1$, now because $e\mathscr{U}\Sigma_E^1$ is complete, there exists some τ_1 in $e\mathscr{U}\Sigma_E^1$ which encompasses σ_2: $\tau_1 \sqsubseteq_E \sigma_2$. On the other hand, because $e\mathscr{U}\Sigma_E^2$ is a set of essentials, i.e. in particular it is also complete, there exists σ_3 in $e\mathscr{U}\Sigma_E^2$ with $\sigma_3 \sqsubseteq_E \tau_1$. But then $\sigma_3 \sqsubseteq_E \tau_1 \sqsubseteq_E \sigma_2$ and by transitivity we have $\sigma_3 \sqsubseteq_E \sigma_2$, contradicting the assumption that $e\mathscr{U}\Sigma_E^2$ is minimal. □

Lemma 11.1 *The set of essential unifiers of a non nullary unification problem Γ is a subset of the set of most general unifiers: $e\mathscr{U}\Sigma_E(\Gamma) \subseteq \mu\mathscr{U}\Sigma_E(\Gamma)$.*

Proof This follows easily from the fact that the subsumption order is a special case of encompassment, where λ_1 is the empty substitution ε. More explicitly: if σ is an essential E-unifier, it is not encompassed by any other unifier, hence it is not subsumed by any other unifier either. □

The important observation is that the set of essential E-unifiers can be *lovely*, that is, it can be extremely small in comparison to its superset of most general unifiers. The results currently known in this small subfield of unification theory—that is still at its infancy and under development—are summarized in Sect. 11.4.

11.3 Essentially Nullary Theories

Gordon Plotkin [63] conjectured in his seminal paper in 1972, that there may be sets of most general unifiers modulo E, that are not recursively enumerable, that is, subsumption modulo E is not a "well quasiorder".

Definition 11.6 A *quasiorder* $\geqslant$ *on a set S is a well quasiorder*, if for any infinite descending chain $s_0 \geqslant s_1 \geqslant s_2 \geqslant \cdots$ in S there exists some $i < j$ such that $s_i \leqslant s_j$ and there are no infinite anti chains, where an anti chain consists of incomparable elements with respect to $\geqslant$.

In the early 1980 s the first equational theories of unification type nullary were discovered [30] and in particular idempotent semigroups (with the axioms of associativity and idempotency, also called bands [2, 69]), became well known. So a natural question is: are there essentially nullary theories as well? Franz Baader gave in [3] an example of an equational theory F and a nullary matching problem for F, which is illuminating for our demonstration here as well. We show first, that this particular problem, which is nullary in the traditional sense, is *essentially unitary, that is it is e-unitary* in this new sense. This and other examples presented in 4.1 and 4.2 below gave rise to the early hope that there are no e-nullary theories and that many infinitary problems may collapse to e-finitary or even e-unitary problems. However a slight modification of F and of the matching problem shows that there are unfortunately still essentially nullary (*e-nullary*) problems as well.
Let F be the following equational theory with constant symbols a and b and the function symbols f, g, h and q

$$\begin{aligned}
F1 : g(x, f(a, z), f(a, y)) &= g(x, z, y),\\
F2 : h(x, y, f(a, z)) &= h(x, y, z),\\
F3 : h(x, y, b) &= b,\\
F4 : q(g(x, y, z)) &= h(x, y, z)
\end{aligned}$$

Let Γ_1 be the equational unification problem $q(x) =^?_F b$, with $V = \mathbf{Var}(\Gamma_1) = \{x\}$ and let $\varphi_n(x)$ be defined as $\varphi_0(x) = x$ and $\varphi_{i+1}(x) = f(a, \varphi_i(x))$ for $i \geqslant 0$.
Franz Baader shows in [3], that the following complete set of F-unifiers of Γ_1, $c\mathscr{U}\Sigma_F(\Gamma) = \{\theta_0, \theta_1, \theta_2, \ldots\}$ with $\theta_i = \{x \mapsto g(x', y', \varphi_i(b))\}$, has an infinite decreasing chain with respect to $\geqslant_F$, namely $\theta_0 \geqslant_F \theta_1 \geqslant_F \theta_2 \geqslant_F \ldots$
Clearly $\theta_i \geqslant_F \theta_{i+1}$, because with $\lambda = \{y' \mapsto f(a, y')\}$ and with the axioms in F we have:

$$\begin{aligned}
\theta_{i+1}\lambda &= \{x \mapsto g(x', y', \varphi_{i+1}(b))\}\{y' \mapsto f(a, y')\}\\
&= \{x \mapsto g(x', f(a, y'), \varphi_{i+1}(b)), y' \mapsto f(a, y')\}\\
&=^V \{x \mapsto g(x', f(a, y'), \varphi_{i+1}(b))\}\\
&= \{x \mapsto g(x', f(a, y'), f(a, \varphi_i(b)))\}\\
&=_{F1} \{x \mapsto g(x', y', \varphi_i(b))\}\\
&= \theta_i
\end{aligned}$$

The chain $\theta_0 >_F \theta_1 >_F \theta_2 >_F \cdots$ has no lower bound, hence Γ_1 is indeed nullary. Now, let us look at the encompassment ordering modulo F for the same problem.

As the following reasoning shows, the theory F is no longer *nullary* in the ordinary sense, but in fact it collapses to an *e-unitary* theory.
As before θ_0 encompasses θ_1, i.e. $\theta_0 \sqsupseteq_F \theta_1$, because with $\lambda = \{y' \mapsto f(a, y')\}$:

$$\begin{aligned}
\theta_1\lambda &= \{x \mapsto g(x', y', \varphi_1(b))\}\{y' \mapsto f(a, y')\} \\
&=^V \{x \mapsto g(x', f(a, y'), \varphi_1(b))\} \\
&= \{x \mapsto g(x', f(a, y'), f(a, b))\} \text{ by definition of } \varphi_i(\mathrm{x}) \\
&=_{F1} \{x \mapsto g(x', y', b)\} \\
&= \theta_0
\end{aligned}$$

But in this case θ_1 also encompasses θ_0 modulo F, since there are substitutions $\lambda_{1,0} = \{x \mapsto g(x', y', f(a, q(x)))\}$ and $\lambda_{2,0} = \varepsilon$ such that $\theta_1 =_F \lambda_{1,0}\,\theta_0\,\lambda_{2,0}$, because:

$$\begin{aligned}
\lambda_{1,0}\,\theta_0\,\lambda_{2,0} &= \{x \mapsto g(x', y', f(a, q(x)))\}\{x \mapsto g(x', y', b)\} \\
&= \{x \mapsto g(x', y', f(a, q(g(x', y', b))))\} \\
&=_{F4} \{x \mapsto g(x', y', f(a, h(x', y', b))\} \\
&=_{F3} \{x \mapsto g(x', y', f(a, b)\} \\
&= \theta_1
\end{aligned}$$

Therefore θ_i encompasses θ_0 for $\mathrm{i} \geqq 1$, because $\theta_i = \{x \mapsto g(x', y', \varphi_i(b)\} =_F \{x \mapsto g(x', y', \varphi_i(q(x))\}\theta_0$. Since $\sqsupseteq_F$ is transitive, θ_0 encompasses all θ_i modulo F, because with $\lambda_i = \{y' \mapsto \varphi_i(y')\}$ we have $\theta_0 =_F \theta_i\lambda_i = \{x \mapsto g(x', \varphi_i(y'), \varphi_i(b))\}$ $=_{F1} \{x \mapsto g(x', y', b)\}$. Consequently they all are encompassment equivalent. Taking θ_0 as the representative for this equivalent class, θ_0 is the only essential F-unifier for Γ_1, which means, that Γ_1 is now an *e-unitary problem*.

But unfortunately not all is well: this collapse does not hold in general and in fact a slight modification of F and Γ_1 turns this back into an e-nullary problem. Consider the equational theory H defined by the following axioms, which are almost identical to the theory F, except in axiom H3:

$$\begin{aligned}
&H1\, g(x, f(a, z), f(a, y)) = g(x, z, y), \\
&H2\, h(x, y, f(a, z)) = h(x, y, z), \\
&H3\, h(x, y, b) = h(b, b, b), \\
&H4\, q(g(x, y, z)) = h(x, y, z)
\end{aligned}$$

Using $\varphi_i(x)$ as defined before take the following H-unification problem:
$\Gamma_2 = \{q(x) =^?_H h(b, b, b)\}$
Let the rewrite $\rightarrow_{Hi}$ denote the rule from left to wright of the axiom Hi, i=1,2,3,4. For example the rule for H2 is $h(x, y, f(a, z)) \rightarrow_{H2} h(x, y, z)$. Analogously $\leftarrow_{Hi}$ denotes the rule from right to left.

We now want to show that the complete set of unifiers for Γ_1, i.e. $c\mathcal{U}\Sigma_F(\Gamma_1)$ is the same (up to renaming) as the set $c\mathcal{U}\Sigma_H(\Gamma_2)$. For this we need a few observations.[5]

Proposition 11.3 *Every unifier for Γ_1 or Γ_2 is of the form $\sigma = \{x \mapsto g(s_1, s_2, s_3)\}$, where the s_i are some terms.*

This is easy to see as $q(x)\sigma =^?_F b$ (or $q(x)\sigma =^?_H h(b, b, b)$) must first apply axiom F_4 (or H_4) respectively.

Proposition 11.4 *For every unifier $\sigma = \{x \mapsto g(s_1, s_2, s_3)\}$ for Γ_2, the correctness of $q(x)\sigma =^?_H h(b, b, b)$ can be shown by a minimal sequence of rewrites of the form $q(g(s_1, s_2, s_3)) \rightarrow_{H4} h(s_1, s_2, s_3) \rightarrow^n_{H2} h(s_1, s_2, b) \rightarrow_{H3} h(b, b, b)$, $n \geqq 0$, where n>0 is the smallest number of steps.*

Proof The first step with $\rightarrow_{H4}$ is obvious, as no other axiom applies and as immediate is the last step with $\rightarrow_{H3}$.

For the intermediate sequence we show by induction:
$h(s_1, s_2, s_3) \overset{n}{\rightarrow}_{H2} h(s_1, s_2, b)$ if and only if $s_3 = \varphi_n(b)$, where n is the smallest number of steps and $\varphi_n(b) = f(a, \varphi_{n\text{-}1})$ is defined as above.
Using induction over the minimal number of rewrite steps:
"$\Longleftarrow$"

$$n = 1 : s_3 = \varphi_1(b) = f(a, b) \Longrightarrow h(s_1, s_2, s_3) = h(s_1, s_2, f(a, b)) \rightarrow_{H2} h(s_1, s_2, b)$$

$n \rightarrow n + 1 : s_3 = \varphi_{n+1}(b) = f(a, \varphi_n(b))) \Longrightarrow$
$h(s_1, s_2, f(a, \varphi_n(b))) \rightarrow_{H2} h(s_1, s_2, \varphi_n(b))$ and by induction hypothesis $h(s_1, s_2, \varphi_n(b)) \rightarrow^n_{H2} h(s_1, s_2, b)$.
"$\Longrightarrow$"

$n = 1 : h(s_1, s_2, s_3) \rightarrow_{H2} h(s_1, s_2, b) \Longrightarrow s_3 = f(a, b)$ i.e. $s_3 = \varphi_1(b)$.
$n \rightarrow n + 1 : h(s_1, s_2, s_3) \rightarrow_{H2} h(s_1, s_2, s'_3) \rightarrow^n_{H2} h(s_1, s_2, b) \Longrightarrow s'_3 = \varphi_n(b)$ using the induction hypothesis. Therefore $h(s_1, s_2, s_3) \rightarrow_{H2} h(s_1, s_2, \varphi_n(b))$.
But this is only possible if $f(a, \varphi_n(b)) = \varphi_{n+1}(b)$.
Hence $h(s_1, s_2, s_3) \overset{n+1}{\rightarrow}_{H2} h(s_1, s_2, b)$ and $s_3 = \varphi_{n+1}(b)$. □

Lemma 11.2 *The unification problems $\Gamma_1 : q(x) =^?_F b$ and $\Gamma_2 : q(x) =^?_H h(b, b, b)$ have the same complete set of unifiers (up to renaming):*
$c\mathcal{U}\Sigma_F(\Gamma_1) = c\mathcal{U}\Sigma_H(\Gamma_2)$.

Proof "$\Longrightarrow$"
Every unifier in $c\mathcal{U}\Sigma_F(\Gamma_1)$ is in $c\mathcal{U}\Sigma_H(\Gamma_2)$:
By Proposition 11.3 every unifier is of the form $\sigma = \{x \mapsto g(s_1, s_2, s_3)\}$. So let $\sigma_F = \{x \mapsto g(s_1, s_2, s_3)\}$ be in $c\mathcal{U}\Sigma_F(\Gamma_1)$ then there exists a chain of equational steps:

[5]Unfortunately we do not know if the axioms H1 to H4 can be directed into a canonical rewrite system as the axioms F1 to F4 in ([2]). So we make a little detour and look at the actual derivation, instead of the more elegant proof by Franz Baader for F1 to F4, based on the canonical rewrite system for F, in ([2]). Thanks to Franz Baader for this hint.

$q(x)\sigma_F = q(g(s_1, s_2, s_3)) \rightarrow_{F4} h(s_1, s_2, s_3) \rightarrow^n_{F2} h(s_1, s_2, b) \rightarrow_{F3} b, n \geqq 0,$
where each F-step involves just one axiom from F.

W.l.o.g. we may also assume that the axiom $F_3 = h(x, y, b)$ is only used once as the last step. But then

$q(x)\sigma_H = q(g(s_1, s_2, s_3)) \rightarrow_{H4} h(s_1, s_2, s_3) \rightarrow^n_{H2} h(s_1, s_2, b) \rightarrow_{H3} h(b, b, b)$, $n \geqq 0$, is also a valid sequence, since H and F have the same axioms except for $H3$ and $F3$.

Hence σ_F is also a unifier for Γ_2.

"$\Longleftarrow$"

Now for the other direction, let σ_H be a unifier in $c\mathcal{U}\Sigma_H(\Gamma_2)$. By Proposition 11.3 it is of the form $\sigma_H = \{x \mapsto g(s_1, s_2, s_3)\}$ and by Proposition 11.4 there exists a chain

$q(x)\sigma_H = q(g(s_1, s_2, s_3)) \rightarrow_{H4} h(s_1, s_2, s_3) \rightarrow^n_{H2} h(s_1, s_2, b) \rightarrow_{H3} h(b, b, b)$,
with $n \geqq 0$, which has no application of $H3$ except for the last step. But then

$q(x)\sigma_F = q(g(s_1, s_2, s_3)) \rightarrow_{F4} h(s_1, s_2, s_3) \rightarrow^n_{F2} h(s_1, s_2, b) \rightarrow_{F3} b$ is a valid sequence in H as well and hence σ_H is a unifier for Γ_1 and so it is in $c\mathcal{U}\Sigma_F(\Gamma_1)$ (up to renaming). □

Theorem 11.1 *There are essentially nullary and essentially infinitary theories.*

Proof The infinite chain $\theta_0 \sqsupset_H \theta_1 \sqsupset_H \theta_2 \sqsupset_H \dots$ *has no lower bound.* Otherwise, there would be indices i, j with $1 \leqq i < j$, such that $\theta_j \sqsupseteq_H \theta_i$. That is θ_i is encompassed by θ_j, hence $x\theta_i\lambda_i \trianglelefteq_H x\theta_j$, with $\lambda_i = \{y' \mapsto \varphi_{j-i}(y')\}$. But this is impossible, because $x\theta_j = g(x', \varphi_j(y'), \varphi_j(b))$ and there is no H-equivalent term, $x\theta_j =_H p_j$ with a sub term of p_j which contains a term beginning with the symbol g. Only the axiom H1 contains g. This axiom introduces, or deletes only f symbols, so a sub term beginning with g never comes up. This means, that there is no tripartition of any $\theta_i =^V_H \theta_{i+1}\lambda$. But then the encompassment relation $\sqsupseteq_H$ specializes to the instantiation relation $\geqslant_H$ and the above chain is identical to the $\geqslant_F$ chain, so the minimality condition cannot be fulfilled (see [3]). Hence H must be an e-nullary theory. □

Intuitively the point is that in the theory F it was possible to derive a term with $g(x', y', b)$ as a sub term. For example:

$$\begin{aligned} x\theta_2 &= g(x', y', f(a, f(a, \mathbf{b}))) \\ &=_F g(x', y', f(a, f(a, \mathbf{h}(\mathbf{x}', \mathbf{y}', \mathbf{b})))) \\ &=_F g(x', y', \mathbf{q}(\mathbf{g}(\mathbf{x}', \mathbf{y}', \mathbf{b}))) \end{aligned}$$

But this chain of equations is not valid in the term algebra modulo H.

Currently we are experimenting with a stronger relation than encompassment so that there are no nullary theories.

11.4 What Is Currently Known

We shall now summarize and reference what is known so far (by the end of 2015).

11.4.1 Associativity and Idempotency

The equational theory of associativity and idempotency (AI) for one dyadic function symbol f, called *idempotent semigroups* as defined by

$$AI = \{f(x, f(y, z)) = f(f(x, y), z) \text{ and } f(x, x) = x\}$$

demonstrates an interesting case for essential unifiers. It has been studied very early in the history of unification theory as it is a standard data structure in computer science and artificial intelligence called *bands* [40]. It was the first case to prove Gordon Plotkin's conjecture that there are theories where the minimal set of unifiers $\mu\mathscr{U}\Sigma_E(\Gamma)$ does not always exist, see [2, 30, 69].
However, with respect to the encompassment order $\sqsubseteq_E$ this well-known situation changes completely as this theory is in fact e-finitary. Associativity and idempotency constitute the algebra of idempotent strings and for technical convenience these two axioms can be reformulated into the equivalent theory for strings

$$AI = \{xx = x,\ xyz = xz \ \ if\ \mathbf{Symb}(y) \subseteq \mathbf{Symb}(x) = \mathbf{Symb}(z)\}$$

where $\mathbf{Symb}(s)$ denotes the symbols occurring in s. This encoding is due to [40] and in [77] we showed, that it can be directed into a canonical (i.e. confluent and terminating) conditional rewrite system. Based on this result Hoche and Szabo showed in [39] that:

Proposition 11.5 *The theory of AI is not nullary with respect to essential unifiers.*

Looking at the proof of this theorem one may suspect that this theory is even unitary, however there are AI-unification problems with more than one essential unifiers:

Proposition 11.6 *AI is not unitary with respect to essential unifiers.*

So finally we have the most striking result:

Theorem 11.2 *The theory AI is finitary with respect to essential unifiers.*

11.4.2 Associativity

The unification problem in free semigroups, where

$$A = \{f(x, f(y, z)) = f(f(x, y), z)\}$$

and the set of terms are built up as usual over constants, variables, but only one function symbol f with **arity**(f) = 2 is called *string unification*, since we can just drop the fs and brackets and write strings (or words as they are more commonly called in the mathematical literature [56, 57]) over the alphabet of constants and variables. In addition we will simply write $=$ for the equality of strings instead of $=_A$. This is probably the most famous unification problem, respectively called the solvability of *word equations* and the question is: can similar results as in 4.1. above be obtained as well for strings.

In the 1950s A. A. Markov was interested in Hilbert's 10th problem, which is one of the 23 famous problems Hilbert proposed in 1900 during his seminal talk in Paris. It is the following problem: Does there exist an algorithm (a decision procedure) to compute whether a Diophantine equation has a solution in rational integers. Martin Davis and Julia Robinson, working first separately and then, joined by Hilary Putnam, in collaboration, proved that it would follow that there is no such algorithm, if a single polynomial equation were found with a particular exponential growth property [21]. Finally, the young mathematician Yuri Matiyasevich [60] solved the problem by producing such an equation, something the three had been unable to accomplish despite a decade of trying [19].[6]

A.A. Markov tried to reduce it to the solvability of word equations in free semigroups: he noted that every word equation over a two constant alphabet can be translated into a set of diophantine equations [59]. Using this translation he hoped to find a proof for the unsolvability of Hilbert's 10th problem by showing that the solvability of word equations is undecidable [61]. This put the problem firmly on the map and others joined in: Lentin and Schützenberger [54], J.I. Hmelevskij [34–36], V.K. Bulitko [13], A. Lentin [53], V.G. Durnev [28] and many others, see [1] for a survey as well as the volumes edited by several mathematicians under the pseudonym of M. Lothaire on *Algebraic Combinatorics on Words* [56, 57].

The problem was finally solved in the affirmative in the seminal work by G.S. Makanin [58]. An exposition of Makanin's algorithm with several improvements is presented inter alia by Klaus Schulz [70, 71] and by Volker Diekert [24]. Algorithms for the computation of a minimal set of unifiers are given in [26, 43] and there is a history of improved algorithms and their complexity bounds, a standard reference

[6] See also http://www.springer.com/article/10.1007%2FBF03024472#page-1. There is actually a nice film about the three and how John McCarthy informed Martin about the result, see http://www.zalafilms.com/films/jrbackground4.html.

is [62]. Some more recent articles are for example [14, 27, 51] and since then the amount of works and results for this and related problems has exploded even more.[7]

The most interesting observation is probably that the problem is decidable, whereas H10 is undecidable—hence Markov's idea would not have worked anyway.

The decision procedure for word equations due to G. Makanin is one of the most complex known algorithms and it marks in an interesting way the borderline between decidable and undecidable problems.

Apart from its theoretical and mathematical interest, the problem became more widely known, because of its relevance in computer science, artificial intelligence and automated reasoning. Examples are equations over lists with concatenation, the data structure string in pattern invoked procedures in AI and finally building associativity into a resolution style theorem prover. Gordon Plotkin [63], Jörg Siekmann [55, 72, 73] and André Lentin [53] independently found an algorithm to enumerate the set of most general unifiers for strings, which is infinite in general.

As opposed to the above cited works on decidability, which just enumerate all solutions and make the decidability or the existence of a solution their primary focus, in our community we are more interested in the latter works, inspired by automated theorem proving, where the set $\mu\mathscr{U}\Sigma$ of the *most general* unifiers is the focus of attention.

The most common and simple example to show that string unification in free semigroups is infinitary is the following already mentioned case:

$$xa = ax$$

with the set of most general unifiers

$$\mu\mathscr{U}\Sigma = \{\{x \mapsto a\}, \{x \mapsto aa\}, \{x \mapsto aaa\}, \ldots\}.$$

It is easy to see that indeed this is a solution set and it is not as immediate, but still not too hard to show that there does not exist any other more general set of unifiers $\mu\mathscr{U}\Sigma$ for this problem. Finally $\mu\mathscr{U}\Sigma$ is minimal, which again is obvious, as a^n is ground and thus the unifiers do not yield to instantiation. Hence in general

string unification is infinitary.

As we have said, this is a well known fact since the mid seventies and it is probably the most often quoted example in any lecture or monograph on unification theory.

[7] Google scholar finds 62,600,000 entries in 0.21 s for word equations this year (not all of which is relevant for our topic of course, but narrowing it down to "word equations" still leads to 1500 entries in 0.16 s) and several 100,000 more entries if one is patient enough to continue the search and to filter gold from garbage. In the year 2008, at the unification workshop, where we published a preliminary result, we asked Dr. Google and "he" found 70,300 entries for "word equations" in 0.13 s—so what are we to make of this fact?

A similar example

$$xa = bx$$

is usually chosen to demonstrate that the naive string unification algorithms as for example in [55, 63, 72, 73] are not decision procedures: although it is obvious that the above example is not unifiable, the naive algorithms would run forever. However J. Jaffar's algorithm [43], which is built upon Makanin's work, would recognize this situation and halt.

In contrast to string unification as it has been understood up to now, the first problem has a finite set (in fact an even e-unitary set) of essential unifiers

$$e\mathcal{U}\Sigma = \{\{x \mapsto a\}\} = \{\sigma_1\}$$

and any other most general unifier can be obtained with $\lambda_n = \{x \mapsto a^{n-1}x\}$, $n > 1$. In other words, for any unifier $\sigma_n = \{x \mapsto a^n\}, n > 1$:

$$\begin{aligned}\sigma_n &= \lambda_n \sigma_1 \\ &= \{x \mapsto a^{n-1}x\} \circ \sigma_1 \\ &= \{x \mapsto a^{n-1}x\} \circ \{x \mapsto a\} \\ &= \{x \mapsto a^n\}\end{aligned}$$

where λ_n obeys the structural property, as defined in Sect. 2.6.

Once this observation had been made many years ago, there was an intense struggle to find the correct definitional framework in order to generalize this observation to the whole string unification problem and to prove the conjecture

string unification is e-finitary.

We have shown in [37, 38] that this conjecture is false in general, albeit it holds for subclasses of strings, for example the one variable strings.
Let us summarize the main results and denote the set of string equations as $\Gamma = \{u_1 = v_1, \ldots, u_n = v_n\}$ where the u_i and the v_i are strings. **Var**(Γ) is the set of free variable symbols occurring in u_i and v_i. Let $V = \mathbf{Var}(\Gamma)$, then a (string-) unifier $\sigma : V \mapsto \Sigma^*$ is a solution for Γ if $u_i\sigma = v_i\sigma$, $1 \leqslant i \leqslant n$. The set of all unifiers is denoted as $\mathcal{U}\Sigma_A(\Gamma)$ and we may now drop the A.

Let us look first at a few motivating examples, which show that indeed an infinite set of most general unifiers $\mu\mathcal{U}\Sigma$ collapses to a finite set of *essential unifiers* $e\mathcal{U}\Sigma$. Our first example is the well known string unification problem mentioned in the introduction:

$$ax =^? xa \text{ with } \sigma_n = \{x \to a^n\}, n > 0$$

has infinitely many most general unifiers σ_n, but there is just *one* e-unifier $\sigma_0 = \{x \to a\}$ because of

$$\sigma_n = \{x \to a^{n-1}x\} \circ \sigma_0.$$

The next example taken from the Burris-Problem-List[8] has two variables:

$$xy =^? yx$$

and it has infinitely many most general unifiers $\sigma_{i,j} = \{x \to z^i, y \to z^j\}, i, j > 0$, where i and j are relative prime, i.e. $gcd(i, j) = 1$. The condition for i and j to be relatively prime ensures that we get only most general unifiers, not just any unifier (see example 15 in the Burris-Problem-List).

But it has only one e-unifier $\sigma_0 = \{x \to z, y \to z\}$ because of

$$\sigma_{i,j} = \{x \to z^{i-1}x, y \to z^{j-1}y\} \circ \sigma_0$$

Our next example is taken from J. Karhumäki in *Combinatorics of Words* [15] see also [56, 57]. The system

$$\begin{Bmatrix} xaba =^? baby \\ abax =^? ybab \end{Bmatrix}$$

has infinitely many most general unifiers

$$\sigma_n = \{x \to b(ab)^n, y \to (ab)^n a\}, n \geqslant 0$$

But it has only one e–unifier $\sigma_0 = \{x \to b, y \to a\}$ because of

$$\sigma_n = \{x \to x(ab)^n, y \to (ab)^n y\} \circ \sigma_0.$$

Exploiting the analogy between the first and the second example above, we can easily construct more examples in this spirit.

Our fourth example is taken from J. Karhumäki as well:

$$axxby =^? xaybx$$

It has infinitely many most general unifiers

$$\sigma_{i,j} = \{x \to a^i, y \to (a^i b)^j a^i\}, i \geqslant 1, j \geqslant 0$$

but it has only one e–unifier $\sigma_{1,0} = \{x \to a, y \to a\}$ which is essential because of

$$\sigma_{i,j} = \{x \to ya^{i-1}, y \to (a^i b)^j xa^{i-1}\} \circ \sigma_{1,0}$$

that is $\sigma_{1,0} \sqsubset \sigma_{i,j}$ and $\sigma_{1,0}$ does not encompass any other unifier. The final example is a bit more elaborate but still in the same spirit:

$$zaxzbzy =^? yyzbzaz$$

[8]See http://www.math.uwaterloo.ca/~snburris/htdocs/WWW/PDF/e_unif.pdf, example 15.

has infinitely many most general unifiers

$$\sigma_n = \{x \to b^{2n}a, y \to b^n ab^n, z \to b^n\}, n > 0$$

but it has only one e–unifier, namely $\sigma_1 = \{x \to bba, y \to bab, z \to b\}$ because of

$$\sigma_n = \{x \to b^{2n-2}x, y \to b^{n-1}yb^{n-1}, z \to b^{n-1}z\} \circ \sigma_1$$

String unification with at most one variable is e-finitary

Let us assume our unification problem

$$\Gamma = \{u_1 =^? v_1, \ldots, u_n =^? v_n\}$$

over the signature Σ has at most one variable, but arbitrarily many constants. Without loss of generality, each set of string equations can be encoded into a single string equation preserving the solutions, which is well known (for example see J.I. Hmeleyskij [36]). Volker Diekert in [24] used the following construction

$$\{u_1 a \ldots u_n a u_1 b \ldots u_n b =^? v_1 a \ldots v_n a v_1 b \ldots v_n b\}$$

where a and b are distinct constants. It can be shown, that the two equational problems have the same solutions.
The following facts are known, see [16] and also [44], and we shall use them for our main result as well. Note, a word is primitive if it is not a power of any other word.

Theorem 11.3 *A string unification problem Γ in one variable has either no solution or $\mu\mathcal{U}\Sigma_A(\Gamma) = F \cup \{x \longmapsto (pq)^{i+1}p, i \geqslant 0\}$ for some p, q in Σ, pq is primitive and F is a finite set of unifiers, which is bounded by $\mathcal{O}(log \mid \Gamma \mid)$.*

Proof see Theorem 3 and Lemma 1 in [16] □

It is also known that

Proposition 11.7 *Let $\Gamma = \{u_0 x u_1 \ldots x u_n = x v_1 \ldots x v_n\}$ be a solvable string equation with at most* one *variable x. Then all unifiers are ground substitutions:*
$\forall \sigma \in \mathcal{U}\Sigma_A(\Gamma) : x\sigma \in \Sigma^*$

Proof Suppose by contradiction with an arbitrary unifier $\{x \mapsto w\} \in \mathcal{U}\Sigma_A(\Gamma)$: $w = w_1 z w_2$ where z is a new variable $z \neq x$ such that w_1 is the ground prefix of w. Applying the unifier $x \to w_1 z w_2$ yields

$$u_0 w u_1 \ldots = w v_1 \ldots = u_0 w_1 z w_2 u_1 \ldots = w_1 z w_2 v_1 \ldots$$

Consider the prefixes $u_0 w_1 \ldots = w_1 z \ldots$ Since $|u_0 w_1| \geqslant |w_1 z|$ and u_0 is nonempty, z must be a symbol in $u_0 w_1$, which is impossible, since u_0 and w_1 are ground by assumption.

Hence $\mathbf{Var}(w)$ is empty and the set of solutions is even minimal:
$\mathscr{U}\Sigma_A(\Gamma) = \mu\mathscr{U}\Sigma_A(\Gamma)$. □

So finally we have

Theorem 11.4 *String unification with one variable is e-finitary and the number of unifiers is bounded by* $\mathscr{O}(log \mid \Gamma \mid)$.

Proof In [38] □

String unification in general is e-infinitary

String unification with at most one variable in the signature Σ is e-finitary as we have seen above and surely there are many more special cases of signature restrictions, where the set of e-unifiers is always finite or even unitary.

However the general result for essential string unification is:

Theorem 11.5 *String unification with more than one variable is e-infinitary*

Proof see [38] □

A general A-theorem

Let E be a set of equational axioms containing the associativity axiom of a binary operator $*$, i.e. $A = \{x * (y * z) = (x * y) * z\}$ and $E = A \cup R$, where R is some set of equations. We call the theory modulo E *A-separate*, if any equation in R can not be applied to a pure string $s_1 * s_2 * \cdots * s_n$ (the brackets are suppressed).

For instance consider distributivity (which is an infinitary unification theory), see [6, 80]

$$D = \{x * (y + z) = (x * y) + (x * z), (x + y) * z = (x * z) + (y * z)\},$$

then the theory of $E = A \cup D$ is A-separate. To see this, note that no equation of D can be applied to a string of $x_1 * x_2 * \cdots * x_n$, simply because there are no sums involving the plus sign $+$, but each equation in D has the sum symbol $+$ on its left and on its right hand side.

Formally, $E = A \cup R$ is A-separate, if for all elements u of the A-theory $u =_R v$ implies $u = v$.

Theorem 11.6 *All not e-nullary A-separate E-theories are e-infinitary*

Proof see [38] □

As noted above the not e-nullary theory $A \cup D$ is A-separate and hence:

Theorem 11.7 *The theory $A \cup D$ is e-infinitary.*

Note that the theorem does not imply that D alone is e-infinitary: D is infinitary [79, 80], but the essential case for D is not yet known.

11.4.3 Commutativity

The equational theory C consisting of the single axiom $C = \{f(x, y) = f(y, x)\}$ is also one of the first axioms that has been investigated alone and in combination with other axioms. It is well-known that this theory is finitary [74] and since the set of essential unifiers is a subset of the set of most general unifiers, we have:

Theorem 11.8 *The set of essential unifiers for commutativity is e-finitary*

Unfortunately however it does not collapse into an e-unitary theory within our current definitional framework:

Theorem 11.9 *The theory of commutativity is not e-unitary*

Proof Consider the problem $\Gamma = \{f(x, y) =^?_C f(a, b)\}$ which has two unifiers: $\sigma_1 = \{x \mapsto a, y \mapsto b\}$ and $\sigma_2 = \{x \mapsto b, y \mapsto a\}$. Both unifiers are ground with a single constant symbol, so obviously they do not encompass any other unifier. Hence $e\mathcal{U}\Sigma_C(\Gamma) = \{\sigma_1, \sigma_2\}$. □

11.4.4 Idempotency

The theory $I = \{f(x, x) = x\}$ is also well-known as a finitary theory and similar to C it is also not e-unitary:

Theorem 11.10 *The theory of idempotency is not e-unitary*

Consider the problem $\Gamma = \{f(f(a, x), f(y, b)) =^?_I f(a, b)\}$ which has two unifiers: $\sigma_1 = \{x \mapsto a, y \mapsto b\}$ and $\sigma_2 = \{x \mapsto b, y \mapsto a\}$ and since both unifiers are ground and they do not encompass any other unifier, they are both in $e\mathcal{U}\Sigma_I(\Gamma)$, hence the theory is e-finitary.

Finally we have

Theorem 11.11 *The theory of idempotency and commutativity is e-finitary*

Proof It is known that IC is finitary [66], but could it be e-unitary? Unfortunately not, because consider the problem $\Gamma = \{f(f(a, x), y) =^?_{CI} f(a, b)\}$. Since $f(a, b) =_C f(b, a)$ and $f(f(a, b), f(a, b)) =_I f(a, b)$ we have two unifiers: $\sigma_1 = \{x \mapsto a, y \mapsto b\}$ and $\sigma_2 = \{x \mapsto b, y \mapsto f(a, b)\}$. Both unifiers are ground and not encompassment equivalent, hence the result. □

11.5 Conclusion

At this stage of development there are more problems open than solved:

(i) The most obvious thing to do is to investigate more equational theories to compare $e\mathscr{U}\Sigma$ with the known results for the corresponding set of most general unifiers $\mu\mathscr{U}\Sigma$. A particularly interesting case is the theory of associativity and commutativity (AC) because of its huge proliferation of most general unifiers.

(ii) The most pressing open problem is to find a practically useful integration of e-unifiers into the deductive machinery of a reasoning system. The standard lifting lemma for resolution does not work, so we are currently looking at other proof techniques as for example the abstract lifting lemma due to Pat Hayes. Another promising alley may be to look at constraint resolution, where set(s) of e-unifiers are carried along in the constraints.

(iii) As we collect more essential knowledge about the landscape of equational theories, we may find that the encompassment order is still not the best choice: the reduction of the size of the set of unifiers is not always worth the effort with respect to the overall computation in space and time. The problem is to find the right balance between the effort to compute the set $e\mathscr{U}\Sigma$ and the effort to compute the specific unifier at each resolution step (from the set $e\mathscr{U}\Sigma$). At one extreme we would spend no computational time on the first task and just enumerate the Herbrand universe to instantiate the universally quantified variables as we go along and then spend all the time within the deductive search, as in early deduction systems. On the other hand we could have a very general finite representation (which by the way could be the unification problem itself) and then spend all computational effort on computing the unifiers: for example $\{ax =^?_E xa\}$ is obviously a finite representation for the infinitely many unifiers $\{x \mapsto a^n \mid n \geqslant 1\}$. So the open problem is to find feasible ways out of these two unreasonable extremes. Instance based automated theorem proving as in the early systems by P. C. Gilmore, Martin Davis, Hao Wang and others found a revival and for several decades now there is a new interest in this work. Since the work on hyperlinking by Lee and Plaisted [52] there are numerous publications as surveyed in Harald Ganzinger and Konstantin Korovin [32, 49] and Swen Jacobs and Uwe Waldmann [42].

(iv) The essential idea for our new representation of a generator set $e\mathscr{U}\Sigma_E(\Gamma)$ is that we instantiate not only "from the right" as in $\delta = \sigma\lambda$, where σ is an mgu, but also from "the left" as in $\delta = \lambda_1\sigma\lambda_2$ where σ is an essential unifier. Now this opens up possibilities to search separately for λ_1 and λ_2. Depending on the search space and the structure of the term algebra modulo E under scrutiny, different heuristics may apply in automated reasoning, planning, computer vision, the semantic web or wherever the actual unification problem may arise.

References

1. Adian, S. I., & Durnev, V. G. (2000). Decision problems for groups and semigroups. *Russian Mathematical Surveys, 55*(2), 207.
2. Baader, F. (1986). Unification in idempotent semigroups is of type zero. *Journal of Automated Reasoning, 2*(3), 283–286.
3. Baader, F. (1988). A note on unification type zero. *Information Processing Letters, 27*, 91–93.
4. Baader, F., Binh, N. Th., Borgwardt, S., & Morawska, B. (2015). Deciding unifiability and computing local unifiers in the description logic $\mathscr{E}\mathscr{L}$ without top constructor. *Notre Dame Journal of Formal Logic*.
5. Baader, F., & Ghilardi, S. (2011). Unification in modal and description logics. *Logic Journal of the IGPL, 19*(6), 705–730. http://jigpal.oxfordjournals.org/content/19/6/705.abstract.
6. Baader, F., & Nipkow, T. (1998). *Term rewriting and all that*. Cambridge University Press.
7. Baader, F., & Siekmann, J. (1994). General unification theory. In D. Gabbay, C. Hogger & J. Robinson (Eds.), *Handbook of logic in artificial intelligence and logic programming* (pp. 41–126). Oxford University Press.
8. Baader, F., & Snyder, W. (2001). Unification theory. In A. Robinson & A. Voronkov (Eds.), *Handbook of automated reasoning* (Vol. 1). Elsevier Science Publishers.
9. Baader, F., Borgwardt, S., & Morawska, B. (2015). Dismatching and local disunification in $\mathscr{E}\mathscr{L}$. In M. Fernández (Ed.), *Proceedings of the 26th International Conference on Rewriting Techniques and Applications (RTA'15)*. Volume 36 of *Leibniz International Proceedings in Informatics*, Warsaw, Poland. Dagstuhl Publishing.
10. Baader, F., Gil, O. F., & Morawska, B. (2013). Hybrid unification in the description logic $\mathscr{E}\mathscr{L}$. In B. Morawska & K. Korovin (Eds.), *Proceedings of the 27th International Workshop on Unification (UNIF'13)*. The Netherlands: Eindhoven.
11. Baader, F., & Morawska, B. (2014). Matching with respect to general concept inclusions in the description logic $\mathscr{E}\mathscr{L}$. In C. Lutz & M. Thielscher (Eds.), *Proceedings of the 37th German Conference on Artificial Intelligence (KI'14)*. Volume 8736 of *Lecture Notes in Artificial Intelligence* (pp. 135–146). Springer.
12. Buerckert, H. J., Herold, A., Kapur, D., Siekmann, J., Stickel, M., Tepp, M., et al. (1988). Opening the AC-unification race. *Journal of Automated Reasoning, 4*(4), 465–474.
13. Bulitko, V. K. (1970). Equations and inequalities in a free group and semigroup. *Geometr. i Algebra Tul. Gos. Ped. Inst. Ucen. Zap. Mat. Kafedr., 2*, 242–252.
14. Charatonik, W., & Pacholski, L. (1993). Word equations with two variables. *Word Equations and Related Topics, IWWERT* (pp. 43–56). Berlin: Springer.
15. Choffrut, Ch., & Karhumaeki, J. (1997). Combinatorics of words. In *Handbook of formal languages* (pp. 329–438). Berlin: Springer.
16. Dabrowski, R., & Plandowski, W. (2002, 2011). On word equations in one variable. *Lecture Notes in Computer Science, 2420*, 212–220 and In *Algorithmica, 60*(4), 819–828.
17. Davis, M. (1983). The prehistory and early history of automated deduction. In J. Siekmann & G. Wrightson (Eds.), *Automation of reasoning 1: Classical papers on computational logic 1957–1966* (pp. 1–28). Berlin: Springer.
18. Davis, M. (1973). Hilbert's tenth problem is unsolvable. *American Mathematical Monthly* 233–269.
19. Davis, M., & Hersh, R. (1984). Hilbert's 10th problem. *Mathematics: People, Problems, Results, 2*, 136–148.
20. Davis, M., & Putnam, H. (1960). A computing procedure for quantification theory. *Journal of the ACM (JACM), 7*(3), 201–215.
21. Davis, M., Putnam, H., & Robinson, J. (1961). The decision problem for exponential diophantine equations. *Annals of Mathematics, 74*, 425–436.
22. Dershowitz, N., & Jouannaud, J.-P. (1990). Rewrite systems. In J. van Leeuwen (Ed.), *Handbook of theoretical computer science* (pp. 244–320). North-Holland: Elsevier Science Publishers.
23. Dershowitz, N., & Jouannaud, J.-P. (1991). Notations for rewriting. *Bulletin of the EATCS, 43*, 162–174.

24. Diekert, V. (2002). Makanin's algorithm. In M. Lothaire (Ed.), *Algebraic combinatorics on words* (Chap. 12, pp. 387–442). Cambridge University Press.
25. Domenjoud, E. (1992). A technical note on AC-unification: The number of minimal unifiers of the AC equation. *Journal of Automated Reasoning*, *8*(1), 39–44.
26. Dougherty, D. J., & Johann, P. (1994). An improved general E-unification method. *Journal of Symbolic Computation*, *11*, 1–19.
27. Durnev, V. (1997). Studying algorithmic problems for free semi-groups and groups. In *Logical foundations of computer science* (pp. 88–101). Berlin: Springer.
28. Durnev, V. G. (1974). On equations in free semigroups and groups. *Mathematical Notes of the Academy of Sciences of the USSR*, *16*(5), 1024–1028.
29. Erne, M. (2009). Closure. *Journal of the American Mathematical Society*. In F. Mynard & E. Pearl (Ed.), *Beyond topology, contemporary mathematics* (pp. 163–283). American Mathematical Society.
30. Fages, F., & Huet, G. (1986). Complete sets of unifiers and matchers in equational theories. *Theoretical Computer Science*, *43*(1), 189–200.
31. Gallier, J. H. (1991). Unification procedures in automated deduction methods based on matings: A survey. Technical Report CIS-436, University of Pensylvania, Department of Computer and Information Science.
32. Ganzinger, H., & Korovin, K. (2003). New directions in instantiation-based theorem proving. In *Proceedings. 18th Annual IEEE Symposium on Logic in Computer Science, 2003* (pp. 55–64). IEEE.
33. Gilmore, P. C. (1960). A proof method for quantification theory: Its justification and realization. *IBM Journal of research and development*, *4*(1), 28–35.
34. Hmelevskij, J. I. (1964). The solution of certain systems of word equations. *Dokl. Akad. Nauk SSSR Soviet Math.*, *5*, 724.
35. Hmelevskij, J. I. (1966). Word equations without coefficients. *Soviet Math. Dokl.*, *7*, 1611–1613.
36. Hmelevskij, J. I. (1967). Solution of word equations in three unknowns. *Dokl. Akad. Nauk SSSR Soviet Math.*, *5*, 177.
37. Hoche, M., Siekmann, J., & Szabo, P. (2008). String unification is essentially infinitary. In M. Marin (Ed.), *The 22nd International Workshop on Unification (UNIF'08)* (pp. 82–102). Hagenberg, Austria.
38. Hoche, M., Siekmann, J., & Szabo, P. (2016). String unification is essentially infinitary. *IFCoLog Journal of Logics and their Applications*.
39. Hoche, M., & Szabo, P. (2006). Essential unifiers. *Journal of Applied Logic*, *4*(1), 1–25.
40. Howie, J. M. (1976). *An introduction to semigroup theory*. Academic Press.
41. Huet, G. (2002). Higher order unification 30 years later. In *Theorem proving in higher order logics* (Vol. 2410, pp. 3–12). Springer LNCS.
42. Jacobs, S., & Waldmann, U. (2007). Comparing instance generation methods for automated reasoning. *Journal of Automated Reasoning*, *38*(1–3), 57–78.
43. Jaffar, J. (1990). Minimal and complete word unification. *JACM*, *27*(1), 47–85.
44. Jez, A. (2013). One-variable word equation in linear time. *ICALP, Lecture Notes in Computer Science*, *7966*, 324–335.
45. Kanger, S. (1963). A simplified proof method for elementary logic. *Studies in Logic and the Foundations of Mathematics*, *35*, 87–94.
46. Kapur, D., & Narendran, P. (1992). Complexity of unification problems with associative-commutative operators. *Journal of Automated Reasoning*, *9*(2), 261–288.
47. Kapur, D., & Narendran, P. (1992). Double-exponential complexity of computing a complete set of AC-unifiers. In *Proceedings of the Seventh Annual IEEE Symposium on Logic in Computer Science, LICS92*.
48. Kirchner, C., & Kirchner, H. (2006). Rewriting solving proving. http://www.loria.fr/~ckirchne/=rsp/rsp.pdf.
49. Korovin, K. (2009). An invitation to instantion-based reasoning: From theory to practice. In *Proceedings of CADE 22, Springer Lecture Notes on AI* (Vol. 5663, pp. 163–166). Springer.

50. Knight, K. (1989). Unification: A multidisciplinary survey. *ACM Computing Surveys (CSUR), 21*(1), 93–124.
51. Laine, M., & Plandowski, W. (2011). Word equations with one unknown. *International Journal of Foundations of Computer Science, 122*(2), 345–375.
52. Lee, S.-J., & Plaisted, D. A. (1992). Eliminating duplication with the hyper-linking strategy. *Journal of Automated Reasoning, 9*(1), 25–42.
53. Lentin, A. (1972). Equations in free monoids. In M. Nivat (Ed.), *Automata, languages and programming*. North Holland.
54. Lentin, A., & Schuetzenberger, M. P. (1967). A combinatorial problem in the theory of free monoids. In R. C. Bose & T. E. Dowling (Eds.), *Combinatorial mathematics* (pp. 112–144). University of North Carolina Press.
55. Livesey, M., & Siekmann, J. (1975). Termination and decidability results for string unification. Report Memo CSM-12, Computer Centre, Essex University, England.
56. Lothaire, M. (1997). *Combinatorics on words*. Volume 17 of *Encyclopedia of Mathematics*. Addison-Wesley. (Reprinted from Cambridge University Press, Cambridge Mathematical Library (1983)).
57. Lothaire, M. (2002). *Algebraic combinatorics on words*. Cambridge University Press.
58. Makanin, G. S. (1977). The problem of solvability of equations in a free semigroup. *Original in Russian: Mathematicheskii Sbornik, 103*(2), 147–236 (*Math. USSR Sbornik, 32*, 129–198).
59. Markov, A. A. (1954). Theory of algorithms. *Trudy Mathematicheskogo Instituta Imeni VA Steklova, Izdat.Akad. Nauk SSSR, 17*, 1038.
60. Matijasevich, Ju. V. (1970). Enumerable sets are diophantine. *Dokl. Akad. Nauk SSSR 191=Soviet Math. Dokl., 11*, 354-358, 279–282.
61. Matiyasevich, Y. (1970). The connection between Hilbert's 10th problem and systems of equations between words and lengths. *Seminars in Mathematics, 8*, 61–67.
62. Plandowski, W. (2004). Satisfiability of word equations with constants is in Pspace. *JACM, 51*(3), 483–496.
63. Plotkin, G. (1972). Building-in equational theories. In B. Meltzer & D. Michie (Eds.), *Machine Intelligence* (Vol. 7, pp. 73–90). Edinburgh University Press.
64. Prawitz, D. (1960). An improved proof procedure. *Theoria, 26*, 102–139.
65. Raulefs, P., Siekmann, J., Szabo, P., & Unvericht, E. (1979). A short survey on the state of the art in matching and unification problems. *ACM Sigsam Bulletin, 13*(2), 14–20.
66. Raulefs, P., & Siekmann, J. H. (1978). Unification of idempotent functions. Technical report, Fachbereich Informatik, Universität Kaiserslautern.
67. Robinson, J. A. (1965). A machine-oriented logic based on the resolution principle. *Journal of the ACM, 12*(1), 23–41.
68. RTA list of open problems (2008). http://rtaloop.pps.jussieu.fr.
69. Schmidt-Schauss, M. (1986). Unification under associativity and idempotence is of type nullary. *Journal of Automated Reasoning, 2*(3), 277–281.
70. Schulz, K. (1993). Word unification and transformations of generalized equations. *Journal of Automated Reasoning, 11*, 149–184.
71. Schulz, K. U. (1992). Makanin's algorithm for word equations: Two improvements and a generalization. In *Proceedings of the First International Workshop on Word Equations and Related Topics (IWWERT90)* (Vol. 572, pp. 85–150). Springer LNCS.
72. Siekmann, J. (1975). String unification. Report CSM-7, Computer Centre, University of Essex.
73. Siekmann, J. (1976). *Unification and matching problems*. PhD thesis, Essex University, Computer Science.
74. Siekmann, J. (1979). Unification of commutative terms. In *Notes in Computer Science*. Volume 72 of *Proceedings of the International Symposium on Symbolic and Algebraic Manipulation* (pp. 531–545). EUROSAM'79, Springer.
75. Siekmann, J. (1984). Universal unification. In *Proceedings of the 7th International Conference on Automated Deduction* (pp. 1–42). London: Springer.
76. Siekmann, J. (1989). Unification theory. *Journal of Symbolic Computation, 7*(3 & 4), 207–274.

77. Siekmann, J., & Szabo, P. (1982). A noetherian and confluent rewrite system for idempotent semigroups. *Semigroup Forum*, *25*, 83–100.
78. Siekmann, J., & Wrightson, G. (1983). *Automation of reasoning: Classical papers on computational logic* (Vols. 1 & 2). Berlin: Springer.
79. Szabo, P., & Unvericht, E. (1982). *D-unification has infinitely many mgus*. Technical report, University of Karlsruhe, Inst. f. Informatik I.
80. Szabo, P. (1982). *Unifikationstheorie erster Ordnung*. PhD thesis, University Karlsruhe.
81. Urban, C., Pitts, A. M., & Gabbay, M. J. (2004). Nominal unification. *Theoretical Computer Science*, *323*(1), 473–497.
82. Veenker, G. (1967). Beweisalgorithmen für die prädikatenlogik. *Computing*, *2*(3), 263–283.
83. Wang, H. (1960). Toward mechanical mathematics. *IBM Journal of Research and Development*, *4*(1), 2–22.
84. Wirth, C.-P., Siekmann, J., Benzmüller, C., & Autexier, S. (2009). Jacques herbrand: Life, logic, and automated deduction. In D. Gabbay & J. Woods (Eds.), *Handbook of the History of Logic, Volume 5—Logic from Russell to Church*. Elsevier.

Chapter 12
DPLL: The Core of Modern Satisfiability Solvers

Donald Loveland, Ashish Sabharwal and Bart Selman

Abstract Propositional theorem provers have become a core technology in a range of areas, such as in hardware and software verification, AI planning, and mathematical discovery. The theorem provers rely on fast Boolean satisfiabilty (SAT) solving procedures, whose roots can be traced back to the work by Martin Davis and colleagues in the late 1950s. We review the history of this work with recent advances and applications.

Keywords Boolean satisfiability · SAT solvers · Davis-Putnam procedure · DPLL · Theorem proving

12.1 Introduction

For practical reasons there is considerable interest in Boolean satisfiability (SAT) solvers. This interest extends to complete SAT solvers, many based on the Davis-Putnam-Logemann-Loveland (DPLL) procedure, a variant of the Davis-Putnam (DP) procedure. Complete solvers determine both satisfiability and unsatisfiability. The DPLL procedure focuses on one-literal clauses, branches on truth assignments, and performs a backtrack search in the space of partial truth assignments. Since its introduction in 1962, the major improvements to DPLL have been smart branch selection

---

This paper is dedicated to Martin Davis, in recognition of his foundational contributions to the area of automated reasoning.

D. Loveland (✉)
Duke University, Durham, USA
e-mail: dwl@cs.duke.edu

A. Sabharwal
Allen Institute for AI, Seattle, USA
e-mail: AshishS@allenai.org

B. Selman
Cornell University, Ithaca, USA
e-mail: selman@cs.cornell.edu

E.G. Omodeo and A. Policriti (eds.), *Martin Davis on Computability, Computational Logic, and Mathematical Foundations*,
Outstanding Contributions to Logic 10, DOI 10.1007/978-3-319-41842-1_12

heuristics, extensions like clause learning and randomized restarts, and well-crafted data structures. The DP and DPLL procedures, and the major improvements that followed, will be reviewed in this paper.[1]

The DPLL procedure, even half a century after its introduction, remains a foundational component of modern day SAT solvers. Through SAT solvers [4], as well as through satisfiability modulo theory (SMT) [40] and answer set programming (ASP) [19] solvers that build upon SAT techniques, the DPLL procedure has had a tremendous practical impact on the field with applications in a variety of areas such as formal verification, AI planning, and mathematical discovery.

The DP and DPLL programs were among the earliest propositional provers. It happens that all these earliest algorithms or programs were formulated in 1958 (excepting DPLL), in independent efforts.[2] It is instructive to briefly consider these earliest programs for they collectively contain all the major features of the DPLL procedure except the most basic feature, introduced with the DP procedure.

12.2 The DP and DPLL Procedures with Historical Context

It was no accident that the first propositional provers were developed simultaneously. For one thing, computers of substantial speed and memory size were just becoming available outside of special locations. Moreover, IBM was looking to support projects that demonstrated that computers could do tasks other than numerical work. (All of the projects we review were executed at IBM except for the DP and DPLL projects.) Also, logicians became aware of work done by Newell and Simon on automating human problem solving and reasoning. We discuss this below.

Actually, the earliest program that addressed a task in logic was the implementation of the Presburger algorithm by Martin Davis in 1954. Davis programmed the IAS computer to execute the Presburger algorithm, which decides queries in elementary number theory involving only addition [8].[3] The IAS computer was a very early computer built at the Institute for Advanced Study directly under John von Neumann's direction and its specifications were followed by a number of the early computers. That included the storage: 1024 40-bit words.

Two years later, in 1956, Newell and Simon completed the Logic Theorist, a program that explored how humans might reason when tackling some theorems from Whitehead and Russell's *Principia Mathematica* [39]. A few logicians were also aware of the ongoing Geometry-Theorem Proving Machine project, an automated deduction program that proved theorems in high school geometry, making use of the diagrams presented to students [17, 18].[4] The appearance of the Newell-Simon

---

[1]The papers introducing DP [12] and DPLL [11] each have over 3000 citations listed in Google Scholar (May 2015), which is indicative of their tremendous influence over the years.

[2]The Dunham, Fridshal, and Sward project might have begun in 1957.

[3]Almost all of the papers cited in the introduction can be found in [46].

[4]This project was also executed at IBM.

paper lead some logicians to observe that algorithms using tools of logic certainly could well outperform the Logic Theorist. This is not to state that these logicians were unsympathetic to studying computer simulation of human reasoning, but they felt that any study of the power of computers in deduction warranted understanding what the best procedures can yield.

There were three major projects that were developed simultaneously with the DP procedure. We start with the method by P.C. Gilmore that is explicitly examined in the DP paper [12]. Shortly after joining IBM, in 1958 Paul Gilmore undertook programming a theorem prover to teach himself elementary programming. He also wished to attend the IFIP conference in Paris (1959), so he thought he could learn some programming and have a paper for submission to the conference at the same time. He knew of the Logic Theorist, and was sure that he could do better using some results from logic [20]. His proof method was motivated by the semantic tableaux of Beth and Hintikka, but Gilmore acknowledges that the result owes more to Herbrand and Gentzen.

Like most of the other provers we consider, Gilmore's prover tested for validity of predicate logic formulas rather than unsatisfiability of propositional formulas [21, 22]. However, his program centered on a propositional prover for unsatisfiability, and we focus on this propositional prover.

Gilmore's propositional prover effectively involved a succession of conjunctive normal form (CNF) formulas (a conjunction of disjunctions or "clauses") being folded into an existing disjunctive normal form (DNF) formula, a disjunction of conjunctions. This involves "multiplying out" the new CNF formula over the DNF formula using the familiar distributive law $((a \vee b) \wedge c) \leftrightarrow ((a \wedge c) \vee (b \wedge c))$. Using arithmetical notation (+ for $\vee$), $(a + b + c + d)(ef + gh + ij)$ yields 12 partial products, and one sees the explosion that quickly overwhelms any computer. Gilmore, of course, saw this and used clever programming encodings to compress and speed the computation. One device, the *truth list*, allowed Gilmore to save variable names, important to Gilmore's coding design. It also introduced a one-literal deletion rule, and propagation of this rule, that later was a key component of the DP algorithm.

Gilmore did not explicitly note the possible propagation effect, and could never observe it because any computer run was almost immediately terminated with memory overflow.

To understand the truth list, consider the CNF formula $(a + b)\overline{c}$ multiplied by the existing DNF formula $cd + cf + ef + eg + be$.[5] After removal of contradictory clauses and redundant literals, we have $a\overline{c}ef + a\overline{c}eg + a\overline{c}be + b\overline{c}ef + b\overline{c}eg + b\overline{c}e$. The literal $\overline{c}$ occurs in each clause, as could be anticipated because we were "multiplying" a one-literal clause $\overline{c}$ into the existing DNF formula. When a literal appears in each clause it is added to the truth list and eliminated from each clause. Thereafter, any new occurrence of this literal or its complementary literal is ignored.[6] Also, after the contradictory clauses containing c were removed during simplification, the variable e now occurs in each clause, as seen above, and is treated as a one-literal clause;

[5] $\overline{c}$ denotes *not* c here.

[6] A complementary literal has the same variable name but opposite negation status.

it is removed and entered on the truth list. In this sense the one-literal clause property propagates. To justify the truth list action, note that were the literal retained, later reentering that literal in any clause would cause merging with the existing literal; adding the complementary literal would cause that clause to be contradictory.

Dunham, Fridshal and Sward (DFS) developed a method strictly for testing validity in the propositional calculus [13, 14]. The DFS method shares with DPLL the branching property, and seems to be the first method to use this property. By splitting the given formula recursively, the reduced formula allows simplification at each level, which for many formulas allows shallow trees and fast processing. Formulas were placed in negative normal form (NNF) which requires that all negation symbols be next to variables; the formula is not processed further. (This is not a true normal form.)

A literal is in a *partial state* if there are no complementary literal occurrences in the formula and no literal occurrence in the scope of an $\leftrightarrow$ symbol. A literal not in a partial state is in a *full state*. At each level literals in a partial state are removed by setting their truth value to 0 (false) and using the simplification rules given below. (This is a generalization of the affirmative-negative rule of the DP procedure.) Since no partial state literal is in the scope of a negation or an $\leftrightarrow$ symbol, assigning false to any such literal is consistent with any falsifying assignment, if one exists.

The method initiates the validity test by simplifying the given formula and then branching on a full state variable, if one exists. *Simplifying* a formula involves use of the following simplification rules along with the associative and commutative properties of $\wedge$, $\vee$, $\leftrightarrow$, and returns the reduced formula to NNF if needed.

$$
\begin{array}{lll}
\Phi \vee 1 \Leftrightarrow 1 & \Phi \vee 0 \Leftrightarrow \Phi & \\
\Phi \wedge 1 \Leftrightarrow \Phi & \Phi \wedge 0 \Leftrightarrow 0 & \\
\Phi \leftrightarrow 1 \Leftrightarrow \Phi & \Phi \leftrightarrow 0 \Leftrightarrow \neg\Phi & \\
L \leftrightarrow L \Leftrightarrow 1 & L \vee L \Leftrightarrow L & L \wedge L \Leftrightarrow L \\
L \leftrightarrow \overline{L} \Leftrightarrow 0 & L \vee \overline{L} \Leftrightarrow 1 & L \wedge \overline{L} \Leftrightarrow 0
\end{array}
$$

Here L denotes a literal, and Φ denotes a formula. Note that, upon simplification, one always has a nonempty formula or a 1 or 0.

We give the program as a recursive algorithm although the implementation was an iterative program. This presentation invites comparison with the DPLL presentation of this paper; the similarity is notable, but is primarily a consequence of the use of branching. We note that this paper appeared two years before the DPLL paper. (However, the DPLL method design was not influenced by this paper. See the summary of the DPLL creation process given later.) It should be mentioned that Dunham, Fridshal and Sward explicitly promote the branching device, with the split formulas conducive to further simplification.

In the DFS- RECURSIVE algorithm, $F|_\ell$ denotes the simplified formula after all occurrences of literal ℓ have been replaced by 0.

Hao Wang, looking far ahead, envisioned proof methods that could tackle problems such as proving that $\sqrt{2}$ is irrational and well-known analysis theorems (e.g., the Heine-Borel theorem). He argued, however, that it would be wrong to attempt to skip over provers for the predicate calculus in the pursuit of mechanical mathematics [49].

Algorithm 1: `DFS-recursive`(F, ρ)

Input : A NNF formula F and an initially empty partial assignment ρ
Output: {the top-level **return** value} VALID or NON- VALID
begin
  $(F, \rho) \leftarrow$ `PartialVariable` (F, ρ)
  if $F = 1$ **then return** VALID
  if $F = 0$ **then return** NON- VALID
  $\ell \leftarrow$ a variable not assigned by ρ `// the branching step`
  if `DFS-recursive` $(F|_{\ell}, \rho \cup \{\ell\})$ = NON- VALID **then return** NON- VALID
  return `DFS-recursive` $(F|_{\neg\ell}, \rho \cup \{\neg\ell\})$

sub `PartialVariable` (F, ρ)
begin
  while *there exists a literal* ℓ *in partial state* **do**
    $F \leftarrow F|_{\ell}$
    $\rho \leftarrow \rho \cup \{\ell\}$
  return (F, ρ)

Wang produced programs for the predicate calculus that tackled several decision problems, subdomains of the predicate calculus, and along the way proved all 350 theorems in *Principia Mathematica* that fell within the realm of the predicate calculus with equality. Surprisingly, this could be done with a program that could handle propositional formulas, a restricted part of the AE predicate calculus and a scattering of other formulas [49–51].[7]

Wang developed a complete validity-testing propositional prover for use within his first-order logic prover, and did not dedicate himself to optimal speed in this prover. He, in fact, acknowledged that the Dunham, Fredshal, and Sward and the DP provers were superior on complex formulas. (Wang knew of the DP provers only in preparation of his paper for publication, whereas he likely knew of the other programs through the IBM connection.)

In spite of Wang's lack of focus on the speed of his propositional prover, we do comment on this prover as it was developed coincident to the other early propositional provers. Also, this allows us to note the major accomplishments of Wang's work, executed in this same time window. The prover was based on a Herbrand-Gentzen cut-free logical framework. He employed ten connective introduction rules formulated in terms of sequents. Axioms are of the form $\lambda \rightarrow \pi$, where λ and π are nonempty strings of variables (atomic formulas), and share a common variable.

[7] An AE formula is a formula with all quantifiers leftmost with no existential quantifier to the left of a universal quantifier.

There are two connective introduction rules for each of the five connectives $\neg$, $\vee$, $\wedge$, $\supset$, and $\leftrightarrow$. The proof search is done in reverse order, from theorem statement to axioms, forming a tree as induced by the rules that have two premise sequents. We give four of these rules followed by a simple example to illustrate the proof technique. Here $\eta, \zeta, \lambda, \rho, \pi$ are formula strings (perhaps empty or a single formula) and φ and ψ are formulas. (The rules for negation are correct; the introduction of negation is applied to formulas in extreme positions.) The proof search reduces the leftmost symbol. The proof search system is propositionally complete with the full rule set.

(1a) Rule $\rightarrow \neg$: If $\varphi, \zeta \rightarrow \lambda, \rho$, then $\zeta \rightarrow \lambda, \neg\varphi, \rho$.

(1b) Rule $\neg \rightarrow$: If $\lambda, \rho \rightarrow \pi, \varphi$, then $\lambda, \neg\varphi, \rho \rightarrow \pi$.

(2a) Rule $\rightarrow \vee$: If $\zeta \rightarrow \lambda, \varphi, \psi, \rho$, then $\zeta \rightarrow \lambda, \varphi \vee \psi, \rho$.

(2b) Rule $\vee \rightarrow$: If $\lambda, \varphi, \rho \rightarrow \pi$, and $\lambda, \psi, \rho \rightarrow \pi$ then $\lambda, \varphi \vee \psi, \rho \rightarrow \pi$.

We give a proof of a valid sequent in the manner of proof search Wang employed, but with a different line notation than adopted by Wang. On the left is the line number. To the right we indicate the line for which this line is the antecedent by the rule named. The task is to reduce the purported theorem to axioms by applying the rules in reverse. Rule (2b) causes a split, resulting in a simple search tree. Wang explored search trees depth first.

(1)	$\neg(P \vee Q), \neg Q \vee \neg R \rightarrow \neg P \vee \neg R$	Given
(2)	$\neg Q \vee \neg R \rightarrow \neg P \vee \neg R,\ P \vee Q$	(1)[1b]
(3)	$\neg Q \rightarrow \neg P \vee \neg R,\ P \vee Q$	(2)[2b], branch 1
(4)	$\rightarrow \neg P \vee \neg R, P \vee Q, Q$	(3)[1b]
(5)	$\rightarrow \neg P, \neg R, P \vee Q, Q$	(4)[2a]
(6)	$P \rightarrow \neg R, P \vee Q, Q$	(5)[1a]
(7)	$R, P \rightarrow P \vee Q, Q$	(6)[1a]
(8)	$R, P \rightarrow P, Q, Q$	(7)[2a]
	Axiom	
(9)	$\neg R \rightarrow \neg P \vee \neg R,\ P \vee Q$	(2)[2b], branch 2
(10)	$\rightarrow \neg P \vee \neg R, P \vee Q, R$	(9)[1b]
(11)	$\rightarrow \neg P, \neg R, P \vee Q, R$	(10)[2a]
(12)	$P \rightarrow \neg R, P \vee Q, R$	(11)[1a]
(13)	$R, P \rightarrow P \vee Q, R$	(12)[1a]
(14)	$R, P \rightarrow P, Q, R$	(13)[2a]
	Axiom	

The given formula is a valid sequent, and a valid formula if the sequent arrow is replaced by the implication connective.

The fourth method undertaken in 1958 is the Davis-Putnam procedure. (Although the labels *method* and *procedure* usefully define schema and their complete specification, respectfully, we have followed tradition and refer to the DP and DPLL procedures. The label "procedure" is correct if one views unspecified decision points as selection of the first qualified item. The refinements to the DPLL "procedure" discussed here certainly dramatize that DP and DPLL are methods.) Like Gilmore's method, the DP procedure is a complete prover for validity in the predicate calculus using a propositional prover that tests for unsatisfiability. The propositional procedure was conceived in the summer of 1958, and expanded to the quantification theory procedure in 1959. The reason work on the procedure spread over two summers is that their focus was elsewhere, on Hilbert's 10th problem. Although their 1960 paper reads as if they were motivated by the propositional inefficiencies of the Gilmore and Wang methods, Davis and Putnam had no knowledge of the parallel efforts when they undertook this project [9]. The setting that spawned the DP procedure is described in [10].

Many of the readers of this paper know the DP rules, but it is important to note that not only did Davis and Putnam introduce significant rules, they also introduced the entire structure now used by most of the automated deduction provers. If one recalls the three methods previously discussed, one notes *none used CNF pursuing an unsatisfiability test.* One advantage of the DP format is in formula preparation. A conjectured theorem is negated by simply negating the theorem assertion and placing each axiom individually in CNF. This is very significant if the axiom set is large. Also, seeking *satisfying* assignments is more intuitive than seeking *falsifying* assignments. Certainly the pursuit of the SAT problem is much more appealing than pursuing the falsifiability problem. At the first-order level they introduced the use of Skolem functions.

We now consider the DP procedure. The goal of the procedure is to report a model (i.e., satisfying truth assignment) for a formula, if the formula is satisfiable, or otherwise, report "unsatisfiable."

We present the rules of the DP procedure not as one would implement them, but as a concept. The focus of these rules is on variable elimination. At each stage a variable p is chosen and the formula is written in the form $(A \vee p) \wedge (B \vee \neg p) \wedge R$, where A, B and R are formulas in CNF free of the variable p (or its negation). This is done using the distributive laws, if necessary. CNF formula A is the conjunction of all clauses that contained p, but with the p removed; similarly for B.

I *The one-literal rule.* If A and B both contain the empty clause, then the algorithm terminates with a declaration of unsatisfiability. The empty clause is always false for any truth assignment, so both conjunctions A and B are always false and can be discarded. This leaves p and $\neg p$ as one-literal clauses, which is contradictory. Otherwise, suppose (only) A contains the empty clause. Then p must be true if the given formula is satisfiable, so one can delete A, p and $\neg p$. That is, the formula $(A \vee p) \wedge (B \vee \neg p) \wedge R$ is unsatisfiable iff $B \wedge R$ is. The working clause set becomes $B \wedge R$ and we can add p to the partial model. We have the symmetric case if B contains the empty clause.

II *The affirmative-negative rule.* If p is absent in the working clause set then $\neg p$ may be declared true and added to the partial model. Since $(A) \wedge (B \vee \neg p) \wedge R$ is unsatisfiable iff $A \wedge R$ is, the working clause set becomes $A \wedge R$. Again, we have the symmetric case for $\neg p$.

III *Rule for eliminating atomic formulas.* $(A \vee p) \wedge (B \vee \neg p) \wedge R$ is unsatisfiable iff $(A \vee B) \wedge R$ is unsatisfiable. The working clause set becomes the CNF formula for $(A \vee B) \wedge R$. Note that each clause of the CNF of $(A \vee B)$ is the disjunction of a clause from A and a clause from B. In any interpretation, one of p and $\neg p$ is false, so one of A or B has a true literal in each clause. Thus, Rule III preserves models of the formula upon which it operates. Any tautological clause is removed.[8]

If the working clause set is the empty set of clauses, then the original formula is satisfiable. To find a model for the given formula one has a start with the partial model, but must do further work with the other variables to find a satisfying truth assignment. Although each rule eliminates a variable, we stick with the original label for rule III which declares that explicitly. It is now referred to as the *resolution rule*.

The implementation is, of course, somewhat different from the above definition. One selects one literal clauses in turn and removes clauses containing that literal and removes the complementary literal from all clauses. One then seeks to remove clauses that have a literal with no complementary literal elsewhere (Rule II). After Rules I and II are exhausted, the algorithm calls for Rule III to be applied. Here the original algorithm called for selecting a literal from the first clause of minimal length and eliminating that literal. One then returns to Rule I. This whole program is repeated until either the empty clause is obtained (unsatisfiability), or there are no more clauses to treat (satisfiability).

We might note that the affirmative-negative rule is a global one, whereas the other rules are not. However, if all clauses are in fast memory, and one builds literal lists that point to all occurrences of that literal, then one can quickly apply the affirmative-negative rule to any literal where the complement literal list is empty.

Davis [10] mentions that the splitting rule was considered as an option. It would have been surprising if Davis and Putnam had not considered any such reasonable alternative. The choice of the resolution-style Rule III was an intelligent one on paper. Rule III is certainly the more elegant one, and better suited their focus of immediately eliminating one variable per operation, for certainly one of the split formulas would be considerably delayed in its treatment. We state the splitting rule as a lead into the discussion of the DPLL variant.

III* *The Splitting Rule.* Let the given formula F be put in the form $(A \vee p) \wedge (B \vee \neg p) \wedge R$ where A, B and R are free of p. Then F is unsatisfiable if and only if $A \wedge R$ and $B \wedge R$ are both unsatisfiable.

Although the DP procedure was conceived in the summer of 1958, the birth only occurred in the summer of 1960, and it was a still-birth. The rule III proved to be too space-consuming; it was the distributive law application that killed the Gilmore procedure, in miniature. One took the product of all the clauses containing literal p with those containing $\neg p$, minus those two literals. The concept of eliminating a

[8] A tautological clause contains $p \vee \neg p$ for some variable p.

variable, so elegant in theory, proved too costly in implementation. So, the Davis-Putnam-Logemann-Loveland variant, discussed below, was born of necessity.

In the spring of 1960, Martin Davis recruited George Logemann and Donald Loveland to undertake the DP implementation that summer. The two students, early in their Ph.D. pursuit, were glad to have a summer job, and an exciting one at that. George Logemann came to the Courant Institute of Mathematical Sciences (CIMS) at NYU to study applied math at then perhaps the top applied math program. (In 2014 it is so ranked.[9]) He was a skilled programmer, acquiring some of that experience at CalTech where he did his undergraduate work. (After receiving his Ph.D. Logemann worked as a computer scientist for a few years, then, an accomplished cellist, he pursued a musical career, which included computer music. He passed away in 2012.) Donald Loveland considered himself a probability/statistics student, as there was no logic opportunity at the CIMS, Davis having moved to Yeshiva University. (See [10] for Davis's view of this period.) Loveland's interest was in artificial intelligence (AI), and both logic and probability theory were good mathematics background for that. He already could be said to be in the field of Automated Deduction, had that field and name then existed, having been a programmer on the Geometry-Theorem Proving Machine [18].

The task of implementing the DP procedure was split. Logemann undertook the parsing of the CNF formula, entered in Polish notation, the formula preparation having been done by hand. Logemann also handled the structuring of the set of clauses, while Loveland took on the testing of the clause set for consistency. The program was written in SAP, the Symbolic Assembler Program, the assembly language for the IBM 704. After the first runs, which quickly saturated the 32,768 36-bit words of available storage, George Logemann suggested that Rule III be replaced with a splitting rule. He noted that it was easy to save the current environment on tape, and retrieve it on backtracking. As for the other systems with splitting rules, this led to depth-first search. Thus, with the new program, instead of clause addition there was clause removal. Interspersed with applications of Rules I and II, the program recursed on Rule III*, saving the environment on tape for backtracking. This solved the space issue, at least until the input clause set overwhelmed the memory. Now very important, but not emphasized at the time, was the easy definition of a satisfying assignment should there be one. It is unclear why "turning off" and then "turning on" various clauses and literals on backtracking was not pursued, as writing to tape is a slow operation. Using the appropriate list structures, top level routines existed that quickly deleted all occurrences of a specified literal and eliminated the appropriate related clauses. These could just black out these entities instead. But the chosen implementation worked well; considerable pains had been taken in data structure design and in coding to minimize run times. There was awareness of the importance of choosing the right one-literal clause for Rule I, although no experimentation was done in this regard.

[9]U.S. News and World Report: Best Grad Schools 2014.

We now briefly consider an actual implementation of the DPLL procedure. Algorithm 2, `DPLL-recursive`(F, ρ), sketches the basic DPLL procedure on CNF formulas. The idea is to repeatedly select an unassigned literal ℓ in the input formula F and recursively search for a satisfying assignment for $F|_\ell$ and $F|_{\neg\ell}$. (The notation $F|_\ell$ denotes the formula F with ℓ set to TRUE followed by formula simplification, done by the removal of true clauses and false literals.) The step where such an ℓ is chosen is commonly referred to as the *branching* step. Setting ℓ to TRUE or FALSE when making a recursive call is called a *decision*, and is associated with a *decision level* which equals the recursion depth at that stage. The end of each recursive call, which takes F back to fewer assigned variables, is called the *backtracking* step.

Algorithm 2: `DPLL-recursive`(F, ρ)

```
Input  : A CNF formula F and an initially empty partial assignment ρ
Output: UNSAT, or an assignment satisfying F
begin
    (F, ρ) ← UnitPropagate(F, ρ)
    if F contains the empty clause then return UNSAT
    if F has no clauses left then
        Output ρ
        return SAT
    ℓ ← a literal not assigned by ρ              // the branching step
    if DPLL-recursive(F|ℓ, ρ ∪ {ℓ}) = SAT then return SAT
    return DPLL-recursive(F|¬ℓ, ρ ∪ {¬ℓ})

sub UnitPropagate(F, ρ)
begin
    while F contains no empty clause but has a unit clause x do
        F ← F|x
        ρ ← ρ ∪ {x}
    return (F, ρ)
```

A partial assignment ρ is maintained during the search and output if the formula turns out to be satisfiable. If $F|_\rho$ contains the empty clause, the corresponding clause of F from which it came is said to be *violated* by ρ. To increase efficiency, unit clauses are immediately set to TRUE as outlined in Algorithm 2; this process is termed *unit propagation*. *Pure literals* (those whose complement does not appear) are also set to TRUE as a preprocessing step and, in some implementations, during the simplification process after every branch. Modern solvers can set millions of variables per second in the unit propagation process. The number of backtracks per second is in the order of several hundred thousands. Unit propagation is generally the most time consuming component of the SAT solving process. Therefore, the more efficient the unit propagation process is implemented, the better the SAT solver.

Variants of this algorithm form the most widely used family of complete algorithms for formula satisfiability. They are frequently implemented in an iterative rather than recursive manner, resulting in significantly reduced memory usage. The key difference in the iterative version is the extra step of *unassigning* variables when one backtracks. In this step, generally large numbers of variables (hundreds to tens of thousands) assigned via unit propagation need to be unassigned. The naive way of unassigning variables in a CNF formula is computationally expensive, requiring one to examine every clause in which the unassigned variable appears. However, a clever data structure approach involving so-called *watched literals* gets around these inefficiencies.

12.3 The Long Wait

The DP and DPLL procedures were developed in the two year period from 1958 to 1960. Somewhat incredibly, there were no further developments for propositional reasoning or SAT solving for another thirty years. Only in the early 1990s, did we start to see new developments and research in this area. However, the progress since then has been dramatic. Around 1990, using the basic DPLL procedure, one could solve CNF formulas with a few hundred variables and a few hundred clauses in a reasonable amount of time (up to several hours of CPU time), but practical formulas with more than around 500 variables and clauses were out of reach. This kept propositional reasoning largely of academic interest because most real-world reasoning tasks, e.g., in verification or AI planning, require tens of thousands or hundreds of thousands of variables and clauses when encoded as CNFs. Fortunately, we have since seen dramatic progress. In the last two decades, new SAT solvers have been developed that can now handle formulas with up to 10 million variables and over 100 million clauses. This has opened up a whole range of practical applications of SAT solver technology, ranging from program verification, to program synthesis, to AI planning, and mathematical discovery. It is safe to say that no-one could have foreseen such dramatic advances. Much of the progress has been made by building on the ideas behind DPLL. We will summarize the extensions to DPLL below.

Before we proceed, let us briefly reflect on why there was a thirty year "standstill" in the area of propositional reasoning or SAT solving. Firstly, in the early days of theorem proving, it was noted that more expressive formalisms were needed to obtain interesting results e.g. for proving theorems in mathematics and other applications. This led to an extensive research effort on first-order theorem proving based on predicate calculus. Robinson's resolution based solvers provided a major step forward. However, the search space for first-order theorem proving is substantially more difficult to explore than in the propositional domain. The difficulty with such provers led to a shift towards proof assistance and reasoning support systems instead of fully automated solvers. Another difficulty for a propositional approach is that even if translations to SAT were possible, they tended to be much too large for main memory at the time.

The 1970s saw rapid developments in computational complexity theory, introducing the problem class NP, with SAT as the first problem demonstrated to be complete for the class NP [7, 33]. Under the widely believed assumption that P≠NP, this means that no polytime algorithm for SAT exists. Given the worst-case exponential scaling of SAT procedures, the general sense was that encoding reasoning tasks and other problems into SAT was not a promising avenue to pursue. Of course, much of this point of view is predicated upon the idea that worst-case complexity provides a good measure for the scaling behavior of algorithms and search methods on practical problem instances. Our recent experience with SAT solvers has shown that this perspective on problem complexity needs to be reconsidered. More specifically, although the basic DPLL backtrack search procedure for SAT shows exponential scaling behavior on almost all problem domains, the recently added extensions of the procedure brings the scaling down to a very slow growing exponential or even a polynomial in a number of practical problem domains. The work on backdoor variables discussed in Sect. 12.4 provides a formal analysis of this phenomenon [52]. To understand this phenomena at an intuitive level, it is useful to think of modern SAT solvers as combining a basic backtrack search with clever polynomial time algorithmic strategies that can exploit hidden problem structure in the underlying domains. These algorithmic strategies are invoked repeatedly throughout the backtrack search process, and are often so effective that hardly any backtracking is required in order to find a solution or prove that the formula is unsatisfiable. The most basic, but already powerful, algorithmic strategy is unit propagation, which occurs after each branching point in the search. Unit propagation was already part of DPLL but the propagation process has been implemented with incredible efficiency in modern solvers. Unit propagation is combined with other clever algorithmic ideas such as clause learning, branching variable selection, non-chronological backtracking, and random restarts (see Sect. 12.4). Modern SAT solvers have shown that developing new algorithmic strategies in the context of NP-complete problems is in fact a useful pursuit with real practical impact. It would be incorrect to think of these improvements as just clever "heuristics." The term "heuristics" is generally reserved for algorithmic ideas that are somewhat adhoc and often cannot be formally analyzed. However, the ideas implemented in modern SAT solvers combined with an understanding of general domain properties do allow us to derive formal guarantees on the overall behavior of the solver, at least in principle.

In the early nineties, there was renewed interest in the field of knowledge representation and reasoning (KR&R), a subfield of AI, about how to deal with worst-case intractable reasoning tasks. During the 1980s, researchers had developed a series of knowledge representation languages that allowed for worst-case tractable reasoning but the formalisms were generally considered too restrictive for real-world application. This raised the question as to whether one could allow some representation language constructs that would lead to intractable inference in the worst-case but perhaps such worst-case behavior might not be observed in practice. These questions were first explored in a propositional setting by considering SAT encodings of reasoning problems and the behavior of SAT solvers on such encodings. It was found that on certain random SAT problem distributions, DPLL scaling could go from polynomial to exponential depending on the choice of problem distribution [36]. The work also led to new types of SAT solvers based on local search style stochastic algorithms, such as GSAT and WalkSAT [44, 45]. These solvers are incomplete in that they can be quite effective in finding satisfying assignment but cannot show unsatisfiability. The WalkSAT solver could solve problem instances beyond what the basic DPLL could handle. This led to promising practical results on SAT encodings of AI planning problems [32]. Interestingly, propositional AI planning is actually a PSPACE-complete problem, because certain plans can be exponential in the size of the problem instances. To bring the problem down to NP-complete, one simply considers the bounded plan length version of AI planning, where one only considers plans up to a predefined length of k steps (k is polynomial in the original problem size). The work on translating bounded length AI planning into SAT provided the impetus to explore a similar approach for hardware and software verification, by considering the so-called bounded model-checking problem and translating it into SAT. (In the bounded model-checking problem, one explores whether there exists an execution of up to a fixed number of steps that leads to an undesirable state. Such a "bug trace" is very similar to a plan that reaches a certain goal state from a given initial state in an AI planning formulation.) SAT-based planning and verification approaches have become very successful over the last decade or so. The success of stochastic incomplete solvers led researchers to search for extensions of the DPLL strategy to improve the complete solvers. This led to the dramatic increase of performance of such solvers. In most applications, the complete solvers are now more effective than the stochastic solvers but stochastic solvers are also still being improved upon. Moreover, by moving to distributed solvers running on cloud based platforms, a scale up to 10 million variable and 100 million clause problems would appear to be within reach. This would again open up a range of new applications, for example in program synthesis and mathematical discovery.

12.4 Modern Complete SAT Solvers

The efficiency of state-of-the-art complete SAT solvers, extending the basic DPLL approach, relies heavily on various features that have been developed, analyzed, and tested over the last two decades. These include fast unit propagation using watched literals, learning mechanisms, deterministic and randomized restart strategies, non-chronological backtracking, effective constraint database management (clause deletion mechanisms), and smart static and dynamic branching heuristics. We give a flavor of some of these later in this section, referring to [4, 24] for more detailed descriptions.

Algorithm 3: `DPLL-WithClauseLearning`

```
Input  : A CNF formula
Output: UNSAT, or SAT along with a satisfying assignment
begin
    decision level ← 0
    status ← UnitPropagate
    if status = CONFLICT then return UNSAT
    while TRUE do
        if no more free variables then
            return (SAT, current variable assignment)
        DecideNextBranch
        while TRUE do
            status ← UnitPropagate
            if status = CONFLICT then
                if decision level = 0 then return UNSAT
                blevel ← AnalyzeConflictAndLearn
                Backtrack(blevel)
            else break
```

We begin with a variation of Algorithm 2 that mirrors more closely the high level flow of modern day SAT solvers. While fundamentally similar to the recursive formulation of DPLL, this iterative version, outlined in Algorithm 3, makes explicit certain key aspects such as unit propagation, conflict detection and analysis, backtracking to a dynamically computed search tree level, and stopping search on a satisfiable formula when all variables have been assigned a value. One important feature that is not included in this pseudocode for simplicity is restarting the solver every so often. The details of various sub-procedures are beyond the scope of this article, but we briefly mention some highlights.

The algorithm maintains a decision level, which starts at zero, is incremented with each branching decision, and decremented (by one or more) upon backtracking. `UnitPropagate`, as discussed earlier, simplifies the formula by effectively removing true clauses and false literals, setting any resulting unit clauses (i.e., those containing only one active variable) to true, and repeating until no more simplification is possible. This process has a unique fixed point, irrespective of the order in which propagation steps are performed. Algorithm 3 returns UNSAT if it encounters an empty, and hence unsatisfiable, clause during unit propagation at decision level zero. On the other hand, if all variables have been successfully assigned a value without generating the empty clause (a "conflict"), the algorithm returns SAT. Note that this check for satisfiability is different from asking whether the current partial assignment yields a true literal in every clause; this latter test is inefficient when using lazy data structures that will be discussed shortly.

`DecideNextBranch` heuristically identifies a free variable, assigns it a value, and increments the decision level by one. Again, this is somewhat different from explicitly branching "both ways" on a variable. Instead, clause learning makes the solver implicitly flip the value of the assigned variable upon backtracking. The decision level to which the algorithm backtracks upon encountering a conflict is determined dynamically using a graph-based analysis of the unit propagation steps leading to the conflict. This analysis, performed in `AnalyzeConflictAndLearn`, also makes the algorithm learn (i.e., add to its set of clauses) a new clause that explains the conflict being analyzed and prevents unnecessary future branching on variables that would lead to a conflict for a similar reason.

With this overall structure in mind, we are ready to explore some of the most prominent features of modern SAT solvers that build upon the DPLL framework. Many of these features directly involve, are guided by, or are necessitated by clause learning.

12.4.1 Key Features

Variable (and value) selection heuristic is a feature that often varies from one SAT solver to another. Also referred to as the *decision strategy*, it can have a significant impact on the efficiency of the solver (see e.g. [34] for a survey). The commonly employed strategies vary from randomly fixing literals to maximizing a moderately complex function of the current variable- and clause-state, such as the MOMS (Maximum Occurrence in clauses of Minimum Size) heuristic [30] or the BOHM heuristic [6]. One could select and fix the literal occurring most frequently in the yet unsatisfied clauses (the DLIS or Dynamic Largest Individual Sum heuristic [35]), or choose a literal based on its weight which periodically decays but is boosted if a clause in which it appears is used in deriving a conflict, like in the VSIDS (Variable State Independent Decaying Sum) heuristic [37]. Solvers like `BerkMin` [23], `Jerusat` [38], `MiniSat` [15], and `RSat` [42] employ further variations on this theme.

Clause learning has played a critical role in the success of modern complete SAT solvers. The idea here is to cache "causes of conflict" in a succinct manner (as learned clauses) and utilize this information to prune the search in a different part of the search space encountered later. A DPLL style procedure backtracks when it reaches an inconsistency, i.e., the settings of the variables selected on the current branch of the backtrack search tree combined with the implied unit propagations violates one or more of the original problem clauses. The procedure will then backtrack to explore another truth value setting of at least one of the variables on the branch. In clause learning, the procedure explores what variables lie at the core of reaching the inconsistency. This process is called conflict analysis, and may identify a relatively small subset of the variables set on the current branch. For example, on the current branch, variables x_1 and x_5 set to False and True, respectively. In the remaining search, one would not want to revisit this particular setting of these two variables when exploring other branches. To prevent this from happening, one can add the clause $(x_1 \vee \neg x_5)$ to the problem instance. This is called a "learned clause." Setting x_1 to False on a branch will now immediately force x_5 to be set to False via unit propagation, avoiding another exploration of x_5 to True (with x_1 to False). The clause is implied (since the negation added to the formula leads to an inconsistency), so the satisfiability of the problem instance is not affected. However, the clause helps to prune parts of the remaining search space. It can be shown formally that adding such clauses can in fact exponentially reduce the search space of the basic DPLL procedure.

There is extensive research on how to analyze conflicts quickly, what learned clauses are most effective for pruning, and how many learned clauses should be stored. Storing too many learned clauses will start to slow down the solver. To highlight the importance of the learned clause strategies in modern SAT solvers, these solvers are sometimes referred to as Conflict Driven Clause Learning (CDCL) based solvers, even though DPLL still provides for the core framework. For further details, see [4, 24].

The watched literals scheme of Moskewiez et al. [37], introduced in their solver `zChaff`, is now a standard method used by most SAT solvers for efficient constraint propagation. This technique falls in the category of lazy data structures introduced earlier by Zhang [53] in the solver `Sato`. The key idea behind the watched literals scheme, as the name suggests, is to maintain and "watch" two special literals for each active (i.e., not yet satisfied) clause that are not FALSE under the current partial assignment; these literals could either be set to TRUE or be as yet unassigned. Recall that empty clauses halt the DPLL process and unit clauses are immediately satisfied. Hence, one can always find such watched literals in all active clauses. Further, as long as a clause has two such literals, it cannot be involved in unit propagation. These literals are maintained as follows. Suppose a literal ℓ is set to FALSE. We perform two maintenance operations. First, for every clause C that had ℓ as a watched literal, we examine C and find, if possible, another literal to watch (one which is TRUE or still unassigned). Second, for every previously active clause C' that has now become satisfied because of this assignment of ℓ to FALSE, we make $\neg\ell$ a watched literal for C'. By performing this second step, positive literals are given priority over unassigned literals for being the watched literals.

With this setup, one can test a clause for satisfiability by simply checking whether at least one of its two watched literals is TRUE. Moreover, the relatively small amount of extra book-keeping involved in maintaining watched literals is well paid off when one unassigns a literal ℓ by backtracking—in fact, one needs to do absolutely nothing! The invariant about watched literals is maintained as such, saving a substantial amount of computation that would have been done otherwise. This technique has played a critical role in the success of SAT solvers, in particular those involving clause learning. Even when large numbers of very long learned clauses are constantly added to the clause database, this technique allows propagation to be very efficient—the long added clauses are not even looked at unless one assigns a value to one of the literals being watched and potentially causes unit propagation.

Randomized restarts, introduced by Gomes et al. [26], allow clause learning algorithms to arbitrarily stop the search and restart their branching process from decision level zero. Restarts are motivated by the observation that the runtime of DPLL-style backtrack search methods can vary dramatically depending on the variable and value selection heuristics. The inherent exponential nature of the backtrack search process appears to magnify the unpredictability of search procedures, making it common to observe a solver "hang" on a given instance, whereas a different heuristic, or even just another randomized run, solves the instance quickly. Indeed, the runtime of such solvers depicts *heavy tailed* behavior [16, 25, 29], which can be avoided by periodically restarting the solver. For DPLL solvers employing clause learning, all clauses learned so far are retained and now treated as additional initial clauses. Most of the current SAT solvers, starting with `zChaff` [37], employ aggressive restart strategies, sometimes restarting after as few as 20 backtracks. This has been shown to help immensely in reducing the solution time. Theoretically, if allowed sufficiently many restarts, SAT solvers are known to be able to realize the full power of the underlying resolution proof system [2, 43].

Related to this is the notion of **backdoor variables** introduced by Williams et al. [52], which explains the surprisingly short runs also observed in backtrack search solvers exhibiting heavy-tailed behavior. Intuitively, a backdoor is a subset of variables such that once the solver assigns values to them, the rest of the formula becomes very easy to solve with a polynomial time sub-solver. For example, the residual formula may take the 2-SAT or Horn form, or be solvable simply by unit propagation. Even though computing minimum backdoor sets is worst-case intractable [48], extremely small backdoors have been shown to exist in many interesting real-world instances. For example, a logistics planning formula with nearly a thousand variables and several thousand clauses can have a backdoor of size only about a dozen variables. This highlights an implicit structure present in many real-world instances that DPLL-based SAT solvers are able to exploit.

Conflict-directed backjumping, originally introduced by Stallman and Sussman [47], allows a solver identifying a conflict when branching on variable x at decision level d to backtrack directly to a decision level $d' < d$ (often $d' < d - 1$) if all variables other than x involved in the conflicts at both branches on x have decision levels at most d'. This simulates what would have happened had the solver chosen to branch on x at level $d' + 1$ to begin with, namely, that the conflicts currently observed at level d would have been observed at level $d' + 1$. Skipping the unnecessary intermediate branching levels thus maintains completeness of the search while often significantly enhancing efficiency.

In the context of Algorithm 3, a related technique, sometimes referred to as **fast backjumping**, is employed. It is relevant mostly to the now-popular *1-UIP* learning scheme used in SAT solvers `Grasp` [35] and `zChaff` [37]. The idea is to let the solver jump directly to a decision level $d' < d$ when even only one branch at level d leads to a conflict involving variables at levels at most d' (in addition to the variable x at level d). One then simply selects a new variable and value for level $d' + 1$, and continues search with a new learned clause added to the database as well as a potentially a new implied literal (which may or may not be a literal of x). With clause learning, this still maintains completeness of search and is experimentally observed to often increase efficiency. Note, however, that while conflict-directed backjumping is always beneficial (in that it only discards redundant branches), fast backjumping may not be so; the latter discards intermediate decisions at levels $d' + 1$ through $d - 1$, which may, in the worst case, be made again unchanged after fast backjumping.

Assignment stack shrinking based on conflict clauses is a relatively new technique introduced by Nadel [38] in the solver `Jerusat`, and is now used in other solvers as well. When a conflict occurs because a clause C' is violated and the resulting conflict clause C to be learned exceeds a certain threshold length, the solver backtracks to almost the highest decision level of the literals in C. It then starts assigning to FALSE the unassigned literals of the violated clause C' until a new conflict is encountered, which is expected to result in a smaller and more pertinent conflict clause to be learned.

Conflict clause minimization was introduced by Eén and Sörensson [15] in their solver `MiniSat`. The idea is to try to reduce the size of a learned conflict clause C by repeatedly identifying and removing any literals of C that are implied to be FALSE when the rest of the literals in C are set to FALSE. This is achieved using the subsumption resolution rule, which lets one derive a clause A from $(x \vee A)$ and $(\neg x \vee B)$ where $B \subseteq A$ (the derived clause A subsumes the antecedent $(x \vee A)$). This rule can be generalized, at the expense of extra computational cost that usually pays off, to a sequence of subsumption resolution derivations such that the final derived clause subsumes the first antecedent clause.

Glue clauses were introduced by Audemard and Simon [1] in their solver `Glucose`. The idea emerged from the study of practical ways to determine which of the millions of learned clauses derived by a solver were important and which weren't—a critical piece of the puzzle when deciding which learned clauses to periodically discard in order to keep the overhead induced by rapid clause learning to a minimum. While shorter clauses are generally more powerful than longer ones, the idea behind glue clauses is to look instead at literals that "propagate together" in the current search context, as characterized by the decision level associated with them, and treat them as a single unit. The so-called LBD (Literal Block Distance) level of a clause then is the number of different such units in it, and a clause with LBD level 2 is termed a glue clause. A common heuristic is to never discard glue clauses, as they, despite generally being longer, behave essentially like powerful 2-clauses (i.e., those with 2 literals) in terms of propagation strength and formula simplification.

Parallel and distributed SAT solvers are increasingly gaining attention as multi-core and cloud-based systems become popular. The most common approach is a portfolio style one, where one exploits the variability across solver parameters and formulas by running multiple, differently parameterized, independent solvers in parallel, each on the entire problem instance [27]. This approach can be extended to include sharing of clauses across multiple compute cores on a single machine [3] and across several connected machines [5]. With the advent of very fast graphic processing units (GPUs), attention has also shifted to utilizing such special-purpose hardware for suitable aspects of SAT solver computation such as unit propagation, supporting parallel propagation on different parts of the data while the main computation remains sequential [41]. While effectively exploiting parallel computing hardware in a SAT solver beyond a few dozen compute cores remains a challenge [28, 31], it is also a highly promising direction that may help reduce computation time on real-world instances from several days to a few hours.

12.5 Conclusions

We presented work on automated propositional reasoning using Boolean satisfiability (SAT) solvers as initiated by Martin Davis and colleagues in the late 1950s and early 1960s. They introduced two key procedures, which are now widely known by the names of their developers, DP, the Davis-Putnam procedure, and DPLL, the Davis-Putnam-Logemann-Loveland7 procedure. Complete modern SAT solvers, as well as SMT and ASP solvers that build upon these solvers, derive directly from this early work. We saw how, after a thirty year gap, work on SAT solvers dramatically accelerated in the early 1990s. Even though SAT is an NP-complete problem, and therefore believed to be worst-case intractable, modern SAT solvers can handle real-world problem encodings with millions of variables and tens of millions of clauses. This development has changed our view on "intractable" combinatorial problems, and has opened up a wide range of applications tackled via SAT-based encodings, such as software and hardware verification, program synthesis, AI planning, and mathematical discovery, resulting in a rich and vibrant research community centered around propositional reasoning. On one hand, Davis's DPLL work shows the power of algorithmic approaches through a clear practical impact of modern SAT, SMT, and ASP solvers in a range of domains, while, on the other hand, his work on Hilbert's tenth problem represents a truly foundational advance providing a concrete example of the fundamental limits of algorithmic approaches.

References

1. Audemard, G., & Simon, L. (2009, July). Predicting learnt clauses quality in modern SAT solvers. In *21st IJCAI* (pp. 399–404), Pasadena, CA.
2. Beame, P., Kautz, H., & Sabharwal, A. (2004). Understanding and harnessing the potential of clause learning. *JAIR, 22*, 319–351.
3. Biere, A. (2012). Plingeling: Solver description. In *SAT Challenge 2012*.
4. Biere, A., Heule, M., van Maaren, H., & Walsh, T. (Eds.) (2009). *Handbook of satisfiability*. IOS Press.
5. Bloom, B., Grove, D., Herta, B., Sabharwal, A., Samulowitz, H., & Saraswat, V. A. (2012). SatX10: A scalable plug&play parallel sat framework—(tool presentation). In *15th SAT* (pp. 463–468).
6. Böhm, M., & Speckenmeyer, E. (1996). A fast parallel SAT-solver—efficient workload balancing. *Annals of Mathematics and Artificial Intelligence, 17*(3–4), 381–400.
7. Cook, S. A. (1971, May). The complexity of theorem proving procedures. In *Conference Record of 3rd STOC* (pp. 151–158), Shaker Heights, OH.
8. Davis, M. (1983). A computer program for Presburger's algorithm. In J. Siekmann and G. Wrightson (Eds.) *Automated reasoning: Classical papers on computational logic* (pp. 41–48). Springer.
9. Davis, M. (2014). *Private communication*.
10. Davis, M. (2015). *My life as a logician*. In this volume. Springer.
11. Davis, M., Logemann, G., & Loveland, D. (1962) A machine program for theorem-proving. *Communication of ACM, 5*, 394–397. ISSN: 0001-0782.
12. Davis, M., & Putnam, H. (1960). A computing procedure for quantification theory. *Journal of the Association for Computing Machinery, 7*, 201–215. ISSN: 0004-5411.

13. Dunham, B., Fridshal, R., & Sward, G. (1959). A non-heuristic program for proving elementary logical theorems. In *IFIP Congress* (pp. 282–284).
14. Dunham, B., Fridshal, R., & Sward, G. (1959). A non-heuristic program for proving elementary logical theorems (abstract). *CACM*, *2*, 19–20.
15. Eén, N., & Sörensson, N. (2005, June). MiniSat: A SAT solver with conflict-clause minimization. In *8th SAT*, St. Andrews, U.K.
16. Frost, D., Rish, I., & Vila, L. (1997). Summarizing CSP hardness with continuous probability distributions. In *Proceedings of the Fourteenth National Conference on Artificial Intelligence (AAAI-97)* (pp. 327–334). New Providence, RI: AAAI Press.
17. Gelernter, H. (1960). Realization of a geometry theorem proving machine. In *Information processing* (pp. 273–282). UNESCO, Paris: R. Oldenbourg, Munich: Butterworths, London.
18. Gelernter, H., Hansen, J. R., & Loveland, D. W. (1960). Empirical explorations of the geometry theorem machine. In *Papers presented at the May 3–5, 1960, Western Joint IRE-AIEE-ACM computer conference* (pp. 143–149). ACM.
19. Gelfond, M. (2008). Answer sets. In F. van Harmelen, V. Lifschitz and B. Porter (Eds.), *Handbook of knowledge representation. Foundations of artificial intelligence*, Chapter 7 (Vol. 3, pp. 285–316). Elsevier.
20. Gilmore, P. (1958). *Private communication.*
21. Gilmore, P. C. (1959). A program for the production of proofs of theorems derivable within the first order predicate calculus. In *CACM* (Vol. 2, pp. 19–19). ACM.
22. Gilmore, P. C. (1960). A proof method for quantification theory: Its justification and realization. *IBM Journal of Research and Development*, *4*, 28–35. ISSN: 0018-8646.
23. Goldberg, E., & Novikov, Y. (2007). BerkMin: A fast and robust SAT-solver. *Discrete Applied Mathematics*, *155*(12), 1549–1561.
24. Gomes, C., Kautz, H., Sabharwal, A., & Selman, B. (2008). Satisfiability solvers. In F. Van Harmelen, V. Lifschitz and B. Porter (Eds.), *Handbook of knowledge representation* (pp. 89–134). Elsevier.
25. Gomes, C., Selman, B., & Crato, N. (1997). Heavy-tailed distributions in combinatorial search. In *3rd CP* (pp. 121–135).
26. Gomes, C. P., Selman, B., & Kautz, H. (1998, July). Boosting combinatorial search through randomization. In *15th AAAI* (pp. 431–437), Madison, WI.
27. Hamadi, Y., Jabbour, S., & Sais, L. (2009). Manysat: A parallel SAT solver. *Journal on Satisfiability*, *6*(4), 245–262.
28. Hamadi, Y., & Wintersteiger, C. M. (2012). Seven challenges in parallel SAT solving. In *26th AAAI.*
29. Hogg, T., & Williams, C. (1994). Expected gains from parallelizing constraint solving for hard problems. In *Proceedings of the Twelfth National Conference on Artificial Intelligence (AAAI-94)* (pp. 1310–1315). Seattle, WA: AAAI Press.
30. Jeroslow, R. G., & Wang, J. (1990). Solving propositional satisfiability problems. *Annals of Mathematics and Artificial Intelligence*, *1*(1-4), 167–187.
31. Katsirelos, G., Sabharwal, A., Samulowitz, H., & Simon, L. (2013). Resolution and parallelizability: Barriers to the efficient parallelization of SAT solvers. In *27th AAAI.*
32. Kautz, H., & Selman, B. (1996). Pushing the envelope: Planning, propositional logic, and stochastic search. In *Proceedings of AAAI-96*, Portland, OR.
33. Levin, L. A. (1973). Universal sequential search problems. *Problems of Information Transmission*, *9*(3), 265–266. Originally in Russian.
34. Marques-Silva, J. P. (1999, September). The impact of branching heuristics in propositional satisfiability algorithms. In *9th Portuguese Conference on AI.* LNCS (Vol. 1695, pp. 62–74).
35. Marques-Silva, J. P., & Sakallah, K. A. (1996, November). GRASP—a new search algorithm for satisfiability. In *ICCAD* (pp. 220–227), San Jose, CA.
36. Mitchell, D., & Levesque, H. (1996). Some pitfalls for experimenters with random SAT. *Artificial Intelligence*, *81*(1–2), 111–125.
37. Moskewicz, M. W., Madigan, C. F., Zhao, Y., Zhang, L., & Malik, S. (2001, June). Chaff: Engineering an efficient SAT solver. In *38th DAC* (pp. 530–535), Las Vegas, NV.

38. Nadel, A. (2002). Backtrack search algorithms for propositional logic satisfiability: Review and innovations. Master's thesis, Hebrew University of Jerusalem.
39. Newell, A., Shaw, J. C., & Simon, H. (1957). Empirical explorations with the logic theory machine: A case study in heuristics. *Western Joint Comp. Conf.*, *1*, 49–73.
40. Nieuwenhuis, R., Oliveras, A., & Tinelli, C. (2006). Solving SAT and SAT modulo theories: From an abstract Davis-Putnam-Logemann-Loveland procedure to DPLL(T). *Journal of Association for Computing Machinery*, *53*(6), 937–977.
41. Palù, A. D., Dovier, A., Formisano, A., & Pontelli, E. (2015). CUD@SAT: SAT solving on GPUs. *Journal of Experimental and Theoretical Artificial Intelligence*, *27*(3), 293–316.
42. Pipatsrisawat, K., & Darwiche, A. (2006). RSat 1.03: SAT solver description. Technical Report D-152, Automated Reasoning Group, Computer Science Department, UCLA.
43. Pipatsrisawat, K., & Darwiche, A. (2011). On the power of clause-learning SAT solvers as resolution engines. *AI Journal*, *175*(2), 512–525.
44. Selman, B., Kautz, H., & Cohen, B. (1996). Local search strategies for satisfiability testing. In D. S. Johnson and M. A. Trick (Eds.), *Cliques, coloring, and satisfiability: The second DIMACS implementation challenge*. DIMACS Series in Discrete Mathematics and Theoretical Computer Science (Vol. 26, pp. 521–532). American Mathematical Society.
45. Selman, B., Levesque, H. J., & Mitchell, D. G. (1992, July). A new method for solving hard satisfiability problems. In *10th AAAI* (pp. 440–446), San Jose, CA.
46. Siekmann, J., & Wrightson, G. (Eds.) (1983). *Automation of Reasoning1: Classical Papers on Computational Logic 1957–1966*. Springer Publishing Company Incorporated.
47. Stallman, R. M., & Sussman, G. J. (1977). Forward reasoning and dependency-directed backtracking in a system for computer-aided circuit analysis. *AI Journal*, *9*, 135–196.
48. Szeider, S. (2005). Backdoor sets for DLL subsolvers. *Journal of Automated Reasoning*, *35*(1–3).
49. Wang, H. (1960, April). Proving theorems by pattern recognition—I. *Communications of the ACM*, *3*(4), 220–234. ISSN: 0001-0782.
50. Wang, H. (1960). Toward mechanical mathematics. *IBM Journal of Research and Development*, *4*, 2–22. ISSN: 0018-8646.
51. Wang, H. (1961). Proving theorems by pattern recognition—II. *Bell System Technical Journal*, *40*(1), 1–41. ISSN: 1538-7305.
52. Williams, R., Gomes, C., & Selman, B. (2003). Backdoors to typical case complexity. In *18th IJCAI*.
53. Zhang, H. (1997, July). SATO: An efficient propositional prover. In *14th CADE*. LNCS (Vol. 1249, pp. 272–275), Townsville, Australia.

Chapter 13
On Davis's "Pragmatic Platonism"

Hilary Putnam

Abstract (added by editor). A comparison is made between Martin Davis's realism in mathematics and the forms of mathematical realism defended by Hilary Putnam. Both reject the idea that mathematics should be interpreted as referring to immaterial objects belonging to a "second plane of reality" and put emphasis on the use of quasi-empirical arguments in mathematics. The author defends Hellman's use of the formalism of modal logic to explicate his own modal realism.

Keywords Mathematical truth · Objectivity · Modal realism · Indispensability argument · Consilience

When Martin Davis learned that I was going to write about his fascinating essay,[1] he emailed me as follows, "Reading your old 'What is Mathematical Truth?'[2] (which was new to me) it seems to me that the position expressed there is pretty close to what I was suggesting." That (1975) essay did, in fact, say some things that jibe with what Davis was to write in "Pragmatic Platonism", and I still believe those things. Does that mean I agree with "Pragmatic Platonism"? It does. Not only does it formulate Davis's (and my) view that mathematics includes (but, of course, does not solely consist in) what I called "quasi-empirical" (and Davis calls "inductive") arguments in a remarkably clear way, it argues persuasively that, while this is something Gödel

[1] Martin Davis published "Pragmatic Platonism" online:
http://foundationaladventures.files.wordpress.com/2012/01/platonic.pdf; shortly after I completed the present essay, Davis sent me an expanded (forthcoming) version, "Pragmatic Realism; Mathematics and the Infinite", in Roy T. Cook and Geoffrey Hellman (eds.), *Putnam on Mathematics and Logic* (Cham, Switzerland: Springer International Publishing, forthcoming). The online version was read at a conference celebrating Harvey Friedman's 60th birthday. All the passages from "Pragmatic Platonism" I quote here are retained verbatim in the expanded version.

[2] "What is Mathematical Truth?" *Historia Mathematica* 2 (1975): 529–543. Collected in my *Mathematics, Matter and Method* (Cambridge: Cambridge University Press, 1975), 60–78. The expanded version of Davis's "Pragmatic Platonism" referred to in the previous note contains a fine discussion of "What is Mathematical Truth", for which I am grateful.

H. Putnam (✉)
University of Harvard, Cambridge, USA
e-mail: hputnam@fas.harvard.edu

E.G. Omodeo and A. Policriti (eds.), *Martin Davis on Computability, Computational Logic, and Mathematical Foundations*,
Outstanding Contributions to Logic 10, DOI 10.1007/978-3-319-41842-1_13

too believed, and while it is a support for what one might call *realism* with respect to mathematics, one does not have to be a Gödelian Platonist to agree with this. 'Pragmatic Platonism' is realism enough. This claim is what I shall write about here.

The structure of this essay is as follows: I first describe the form of realism that Martin defends (being prepared, of course, to learn that I misunderstand it), and say something about the sort of realism ("modal realism") that I defended in a "Mathematics Without Foundation"[3] as well as in "What is Mathematical Truth", and have subsequently gone on to elaborate and defend[4] (together with Geoffrey Hellman who has brilliantly worked out the details[5]), and discuss the relation between Davis's view in "Pragmatic Platonism" and the views I defend. I will close by describing an argument I have given in the past for realism with respect to mathematics, an argument that has been *misdescribed* as the "Quine-Putnam indispensability argument",[6] and close by discussing a question that occurred to me on reading "Pragmatic Platonism", the question as to whether my indispensability argument is needed, or whether the considerations Davis offers in favor of regarding mathematical truth as objective are actually sufficient.

13.1 Martin Davis's Realism in Mathematics

In "Pragmatic Platonism", Davis points out that an ancestor of the integral calculus, "the method of indivisibles"[7] was used by Torricelli in the seventeenth century to obtain results—one of the most surprising at the time being the existence of a solid (the "Torricelli trumpet") with infinite surface area and finite volume. These results could be checked (but not discovered) by other methods, and the method of indivisibles (and other methods that lacked rigorous justification until the nineteenth century, including the use of complex numbers in calculating the real roots of an equation) became part of the mathematician's repertoire. Nor does the story stop in the nineteenth century; the Axiom of choice was introduced by Zermelo in 1904, but cannot be justified from the other axioms of Zermelo-Frankel set theory."[8] Yet, as Davis remarks (op. cit. p.9), "The obligation to always point out a use of the axiom

---

[3]"Mathematics without Foundations," *Journal of Philosophy* 64.1 (19 January 1967): 5–22. Collected in *Mathematics, Matter and Method*, 43–59. Repr. In Paul Benacerraf and Hilary Putnam (eds.). *Philosophy of Mathematics: Selected Readings*, 2nd ed. (Cambridge: Cambridge University Press, 1983), 295–313.

[4]In "Set Theory, Replacement, and Modality", collected in *Philosophy in an Age of Science* (Cambridge, MA: Harvard University Press, 2012), and "Reply to Steven Wagner", forthcoming in *The Philosophy of Hilary Putnam* (Chicago: Open Court, 2015).

[5]Geoffrey Hellman, *Mathematics without Numbers* (Oxford: Oxford University Press, 1989).

[6]For a description of the argument and its misunderstandings see my "Indispensability Arguments in the Philosophy of Mathematics", in *Philosophy in an Age of Science*, 181–201.

[7]The method of indivisibles was invented by Bonaventura Cavalieri in 1637.

[8]For a detailed account, see Kanamori, Akihiro (2004), "Zermelo and set theory", *The Bulletin of Symbolic Logic* **10** (4): 487–553, doi:10.2178/bsl/1102083759, ISSN 1079-8986, MR 2136635.

of choice is a thing of the past." And he adds, "I haven't heard of anyone calling the proof of Fermat's Last Theorem into question because of the large infinities implicit in Grothendieck universes." (In "What is Mathematical Truth", I similarly pointed out that since Descartes the isomorphism of the geometrical line with the continuum of real numbers has become fundamental to virtually all of analysis without a "proof" from other axioms.) Davis boldly concludes, "What can we say about Torricelli's methodology? He was certainly not seeking to obtain results by 'cogent proofs from the definitions' or 'in ontological terms, from the essences of things'. He was experimenting with a mathematical technique that he had learned, and was attempting to see whether it would work in an uncharted realm. In the process, something new about the infinite was discovered. I insist that this was induction from a body of mathematical experience."

Although the remarks I have just quoted are primarily epistemological, both the use of the term "discovered" and the title "Pragmatic *Platonism*" (emphasis added) indicate that Davis believes that mathematical knowledge is *objective*, and, in fact, he goes on to say so explicitly. (I shall quote the place in a moment.) But Davis (like myself) cannot go along with Gödel's view that mathematical objects exist in a Platonic realm that (parts of which) the mind is somehow capable of perceiving. I now quote two paragraphs from Davis's essay that seem to me to capture the essence of what I am calling his "realism".

> If the objects of mathematics are not in nature and not in a "second plane of reality," then where are they? Perhaps we can learn something from the physicists. Consider for example, the discussion of the "Anthropic Principle" [1]. The advocates of this principle note that the values of certain critical constants are finely tuned to our very existence. Given even minor deviations, the consequence would be: no human race. It is not relevant here whether this principle is regarded as profound or merely tautological. What I find interesting in this discussion of alternate universes whose properties exclude the existence of us, is that no one worries about their ontology. There is simply a blithe confidence that the same reasoning faculty that serves physicists so well in studying the world that we actually do inhabit, will work just as well in deducing the properties of a somewhat different hypothetical world. A more mundane example is the ubiquitous use of idealization. When Newton calculated the motions of the planets assuming that each of the heavenly bodies is a perfect sphere of uniform density or even a mass particle, no one complained that the ontology of his idealized worlds was obscure. The evidence that our minds are up to the challenge of discovering the properties of alternative worlds is simply that we have successfully done so. Induction indeed! This reassurance is not at all absolute. Like all empirical knowledge it comes without a guarantee that it is certain.
>
> My claim is that what mathematicians do is very much the same. We explore simple austere worlds that differ from the one we inhabit both by their stark simplicity and by their openness to the infinite. It is simply an empirical fact that we are able to obtain apparently reliable and objective information about such worlds. And, because of this, any illusion that this knowledge is certain must be abandoned.

The key notions in these paragraphs are "hypothetical worlds", "idealized worlds", and "objective information". For Davis, mathematics is not about worlds that actually exist in some hyper-cosmology (unlike David Lewis's "possible worlds"[9]), but

[9]David Lewis, *On the Plurality of Worlds* (Oxford: Blackwell, 1986).

about what *would be the case* if certain idealized worlds existed, worlds that *would* contain infinities (in some cases "large infinities") *if* they really *did* exist.[10] If a reader were to ask me whether the difference between Davis's hypothetical worlds, which I find it reasonable to talk about, and Lewis's possible worlds, which I don't, isn't "rather thin", I would reply that it is like the difference between saying that a solid gold mountain actually exists somewhere, only not in this world, which was Lewis's view ("real" is just what we call our world; all possible worlds are equally real to their inhabitants) and saying that physically possibly a solid gold mountain *could* exist, which is the case according to present day physics. I interpret Davis's term "hypothetical" to mean that he, like me, conceives of mathematical structures as ones that, in some sense of "could", *could* exist. If I have him right, they exist *hypothetically*, but not *actually*, not even in a Platonic heaven. I don't find the difference between saying that certain worlds or structures are *possible* and saying that they *exist* "thin" at all. We reason about such hypothetical worlds by using our human abilities to imagine and to idealize and to deduce from given assumptions. Because it is obtained in this way, mathematical knowledge is fallible (*pace* Gödel, there is nothing "perceptual" about it), but the consilience of the results[11] justifies our taking the results to be *objective information*. There is a fact of the matter about *what would be the case if* those "hypothetical worlds" were real.

13.2 The Sort of Realism I Defend

I recently (Dec. 2014) described my philosophy of mathematics[12] in three posts on my blog (putnamphil.blogspot.com). In brief, the main points were:

(1) An interpretation of mathematics must be compatible with scientific realism. It is not enough that the *theorems* of pure mathematics used in physics come out true under one's interpretation of mathematics—even some antirealist interpretations arguably meet that constraint—the *content* of the "mixed statements" of science (empirical statements that contain some mathematical terms and some empirical terms) also needs to be interpretable in a realist way. For example, if a theory talks about electrons, according to me it is talking about things we cannot see with the naked eye, and not simply about what measuring instruments would do under certain circumstances, as operationalists and logical positivists maintained. I believe many proposed interpretations fail that test.[13]

---

[10]Here I am going by Davis's reference to the use of Grothendieck's infinity topoi by Wiles and Taylor in their proof of Fermat's "Last Theorem".

[11]By "consilience" I mean that the results are not only consistent, but that they extend one another, often in unexpected directions.

[12]The relevant publications are, in addition to the already mentioned "What is Mathematical Truth" and "Mathematics without Foundations", are "Set Theory, Replacement, and Modality", collected in *Philosophy in an Age of Science* (Cambridge, MA: Harvard University Press, 2012), and "Reply to Steven Wagner", forthcoming in *The Philosophy of Hilary Putnam* (Chicago: Open Court, 2015).

[13]Brouwer's Intuitionism was my example of an interpretation that is incompatible with scientific realism in "What is Mathematical Truth", 75.

(2) Both objectualist interpretations (interpretations under which mathematics presupposes the mind-independent existence of sets as "intangible objects"[14] and potentialist/structuralist interpretations (interpretations under which mathematics only presupposes the *possible* existence of *structures* that exemplify the structural *relations* ascribed to sets), may meet the foregoing constraint. For example, under both Gödel's (or Quine's) Platonist interpretations and Hellman's and my modal logical interpretation the logical connectives are interpreted classically. In contrast to this, under Brouwer's interpretation, the logical connectives (including "or" and "not") are interpreted in terms of (Brouwer's version of) *provability*. For example, in Intuitionism, "P or Q" means "There is a proof that either there is a proof of P or there is a proof of Q". But according to scientific realists, the statement that a physical system either has a property P or has a property Q, does not entail that either disjunct can be proved, or even empirically verified. A statement can be true without being verifiable at all.[15] But if statements of pure mathematics are interpreted intuitionistically, mustn't statements of physics also be interpreted in terms of the same non-classical understanding of the logical connectives?
(3) But, while positing the actual existence of sets as "intangible objects" may justify the use of classical logic, it suffers not only from familiar epistemological problems (not to mention conflicting with naturalism, which is the reason Davis gives for rejecting it), but from a generalization of a problem first pointed out by Paul Benacerraf,[16] a generalization I call "Benacerraf's Paradox", namely that too many identities (or proposed identities) between different categories of mathematical "objects" seem undefined on the objectualist picture—e.g. *are sets a kind of function or are functions a sort of set? Are the natural numbers sets, and if so which sets are they? etc.* For me, the objectualist's lack of an answer that isn't completely arbitrary tips the scales decisively in favor or potentialism/structuralism.
(4) Rejecting objectualism (as Martin and I both do) does not requires one to say that sets, functions, numbers, etc., are *fictions*. (I hope Martin agrees.)

In "Mathematics without Foundations", where I first proposed the modal logical interpretation), I claimed that objectualism and potentialism are "equivalent descriptions", which was a mistake. I now defend the view that potentialism is *a rational reconstruction* of our talk of "existence" in mathematics, rather than an "equivalent" way of talking. Rational reconstruction does not "deny the existence" of sets (or, to change the example), of "a square root of minus one"; it provides a construal of such

---

[14]Gödel's Platonism is a prototypical "objectualist" interpretation, but the term "intangible objects" was used by Quine in *Theories and Things*, (Cambridge, MA: Harvard University Press, 1981), 149.

[15]For a fine defense of the claim that a statement can be true but unverifiable, see Tim Maudlin "Confessions of a Hard-Core, Unsophisticated Metaphysical Realist", forthcoming in *The Philosophy of Hilary Putnam*. Maudlin rightly criticizes me for giving it up in my "internal realist" period (1976–1990); after I returned to realism *sans phrase* in 1990 I defended the same claim in a number of places, e.g. "When 'Evidence Transcendence' Is Not Malign: A Reply to Crispin Wright," *Journal of Philosophy* 98.11 (November 2001), 594–600.

[16]Paul Benacerraf (1965), "What Numbers Could Not Be" *Philosophical Review* Vol. 74, pp. 47–73.

talk that avoids the paradoxes. In Davis's language, the mathematician is talking about, for example, entities that *play the role* of a square root of minus one in certain hypothetical worlds, but unlike Gödel she does not suppose that such entities exist in some Platonic realm. (Gödel claimed we can *perceive* them with the aid of a special mental faculty.)

13.3 Relations Between Davis's and My Forms of Mathematical Realism

Strange as it may seem, the largest difference between Davis's "Pragmatic Platonism" and my "Mathematics without Foundations" and its "modal logical interpretation" is not metaphysical but mathematical. It is not metaphysical, because both essays, my 1967 essay and Davis's recent on-line essay, reject the idea that mathematics must be interpreted as referring to immaterial objects, *à la* either Gödel or Quine.[17] Both essays argue that mathematics can and does discover objective truths about what would be the case if certain abstract structures were real, and that the success of mathematics and the consilience of results obtained by different mathematical methods, including ones whose justification is "quasi-empirical" (my term) or "inductive" (Davis's term), justifies the belief that this is so. This is common ground between us, and it is substantial.

However, my "Mathematics without Foundations" sketched a program for "translating" assertions that quantify over set into explicitly modal statements, a program carried out and then extended in new directions by Hellman; a program that suggests new ways of motivating key axioms of set theory and some of its large cardinal extensions, while Davis's brief essay basically leaves set theory as it is.

This difference exists because already in "Mathematics without Foundations" I was concerned to be consistent with the idea, that I believe to be correct, that there is no such thing as "the totality of all sets"—not even in a "hypothetical world". *Any hypothetical world (to use Davis's language) of sets is only an initial segment of another possible world of sets. Possible models of set theory are inherently extendable.* (This is the idea that led to Hellman's current efforts to deploy extendability principles to motivate possible existence of large cardinals in standard models of set theory.) The key idea of my "Mathematics without Foundations" was to reformulate statements of set theory that are "unbounded", in the sense of quantifying over sets of all ranks, without assuming the existence of even a *possible* totality of such sets. As Hellman describes[18]:

---

[17]Quine is often described as a "reluctant" Platonist because of statements like this one: "I have felt that if I must come to terms with Platonism, the least I can do is keep it extensional", *Theories and Things* (Cambridge, MA: Harvard University Press, 1990), 100.

[18]Hellman, ibid, second page (the page proofs I have seen do not indicate the forthcoming page numbers).

> Putnam took initial steps in illustrating how modal translation would proceed without falling back on set-theoretic language normally associated with "models." To this end, he introduced models of Zermelo set theory as "concrete graphs" consisting of "points" and "arrows" indicating the membership relation, so that, except for the modal operators, the modal logical translation required only nominalistically acceptable language. Finally (in this brief summary), Putnam proposed an intriguing translation pattern[19] for set-theoretic sentences of unbounded rank (standardly understood as quantifying over arbitrary sets of the whole cumulative hierarchy or set-theoretic universe) in which all quantifiers are restricted to items of a (concrete, standard[20]) model but the effect of "unbounded rank" is got by modally quantifying over arbitrary possible extensions of models.

But it is time to return to philosophy.

Both the differences and the similarities between Davis's views and mine, in the essays I have been discussing stem from the fact that Davis considers whole hypothetical worlds without discussing relations (such as one world's being an extension of another) between these "worlds".

On the side of "similarities". First, there is the already-emphasized fact that both of us believe that it is right to recognize the objectivity, the "there-being-a-fact-of-the-matter", of mathematical statements, and that this is realism enough. Coupled with rejection of the claim that all mathematical knowledge is apriori,[21] and our (independent) emphasis on the use of quasi-empirical argument in mathematics, this is a large measure of agreement indeed.

It may seem to be a difference that I worry about Benacerraf's problem (or paradox) and Davis does not, but in fact once one gives up the idea that talk of numbers and sets is talk of real objects in favor of a conception of them as elements of a *possible* (or "hypothetical") model, there is no call to worry about which otherwise-specified object the number two is (or the square root of minus one is, although that one did worry British algebraists for a hundred years[22]). Such problems simply disappear.

What may be a difference is that Davis only worries about *pure* mathematics, and from the beginning I am concerned with finding an interpretation of mathematical truth that is consistent with a scientific realist interpretation of empirical statements, of what I called "mixed statements" above. How to interpret mixed statements in a modal-logical framework is something that Hellman and I discussed over the years,

---

[19] An example of my translation method (from "Mathematics without Foundations") is this: If the statement has the form $(x)(Ey)(z)Mxyz$, where M is quantifier-free, then the translation is: Necessarily: If G is any graph that is a standard model for Zermelo set theory and if x is any point in G, then it is possible that there is a graph G' that extends G and is a standard concrete model for Zermelo set theory and a point y in G' such that $\Box$ (if G'' is any standard concrete model for Zermelo set theory that extends G' and z is any point in G'', then $Mxyz$ holds in G'').

[20] A model of Zermelo (or Zermelo-Fraenkel) set theory is standard just in case (1) it is well-founded (no infinite descending membership chains), and (2) power sets are maximal.

[21] Actually, I believe that all so-called "a priori" truths presuppose a background conceptual system, and that no conceptual system is guaranteed to never need revision. For this reason, I prefer to speak of truths being conceptually necessary relative to a conceptual background. I would not be surprised if Martin Davis agreed with this.

[22] Menahem Fisch, "The Emergency Which has Arrived: The Problematic History of 19th Century British Algebra—A Programmatic Outline", *The British Journal for the History of Science*, 27: 247–276, 1994.

and some of the most ingenious work in *Mathematics without Numbers* is devoted to it. For example, in applied mathematics one needs to talk (to put it heuristically[23]) about possible worlds *in which the physical objects are as they actually are*, not about possible worlds simpliciter, and formalizing this is non-trivial. And I repeat, *extension relations* among possible worlds (or "hypothetical worlds", or "idealized worlds", or whatever you want to call them), and between possible worlds and the actual world, need to be considered if potentialist approaches are to be spelled out in a rigorous way.

13.4 Indispensability Arguments—What Mine Was, and Are They Necessary

If one consults the *Stanford Encyclopedia of Philosophy* on the topic "Indispensability Arguments in the Philosophy of Mathematics,"[24] one finds (as part of a moderately lengthy entry written by Mark Colyvan) the following statements:

> From the rather remarkable but seemingly uncontroversial fact that mathematics is indispensable to science, some philosophers have drawn serious metaphysical conclusions. In particular, Quine...[25] and Putnam...[26] have argued that the indispensability of mathematics to empirical science gives us good reason to believe in the existence of mathematical entities....This argument is known as the Quine-Putnam indispensability argument for mathematical realism.

From my point of view, Colyvan's description of my argument(s) is far from right. In "What is Mathematical Truth" what I argued was that *the internal success and*

---

[23]Officially, Hellmann and I avoid literal quantification over possible worlds or *possibilia*, relying entirely on modal operators that officially we avoid literal quantification over possible worlds or *possibilia*, relying entirely on modal operators.

[24]Mark Colyvan, "Indispensability Arguments in the Philosophy of Mathematics," in E.N. Zalta, ed., *The Stanford Encyclopedia of Philosophy* (Fall 2004 Edition), http://Plato.stanford.edu/archives/fall2004/entries/mathphil-indis/. Colyvan is also the author of *The Indispensability of Mathematics* (Oxford: Oxford University Press, 2001).

[25]The author of this entry, Mark Colyvan, is referring to W.V. Quine, "Carnap and Logical Truth," reprinted in *The Ways of Paradox and Other Essays*, revised edition (Cambridge, Mass.: Harvard University Press, 1976), 107–132 and in Paul Benacerraf and Hilary Putnam, eds., *Philosophy of Mathematics, Selected Readings* (Cambridge: Cambridge University Press,1983), 355–376; W.V. Quine, "On What There Is," *Review of Metaphysics*, 2 (1948): 21–38; reprinted in *From a Logical Point of View* (Cambridge, Mass.: Harvard University Press, 19802), 1–19; W.V. Quine, "Two Dogmas of Empiricism," *Philosophical Review*, 60, 1 (January 1951): 20–43; reprinted in his *From a Logical Point of View* (Cambridge, Mass.: Harvard University Press, 1961), 20–46; W. V. Quine, "Things and Their Place in Theories," in *Theories and Things* (Cambridge, Mass.: Harvard University Press, 1981), 1–23; W.V. Quine, "Success and Limits of Mathematization," in *Theories and Things* (Cambridge, Mass.: Harvard University Press, 1981), 148–155.

[26]Colyvan is referring to "What is Mathematical Truth" and Hilary Putnam, *Philosophy of Logic* (New York: Harper and Row, 1971), reprinted in *Mathematics, Matter and Method: Philosophical Papers Vol. 1*, 2nd edition, (Cambridge: Cambridge University Press, 1979), 323–357.

coherence of mathematics is evidence that it is true under some interpretation, and that its *indispensability for physics* is evidence that it is true under a realist interpretation—the antirealist interpretation I considered there was Intuitionism. This is a distinction that Quine nowhere draws. It is true that in *Philosophy of Logic* I argued that at least some set theory is indispensable in physics *as well as logic* (Quine had a very different view on the relations of set theory and logic, by the way), but both "What Is Mathematical Truth?" and "Mathematics without Foundations" were published in *Mathematics, Matter and Method* together with "Philosophy of Logic," and in both of *those* essays I said that set theory did not have to be interpreted Platonistically. In fact, in "What Is Mathematical Truth?"[27] I said, "*the main burden of this essay is that one does not have to 'buy' Platonist epistemology to be a realist in the philosophy of mathematics. The modal logical picture shows that one doesn't have to 'buy' Platonist ontology either.*" Obviously, a careful reader of *Mathematics, Matter and Method* would have had to know that I was in no way giving an argument for realism about sets as *opposed* to realism about truth values on a modal interpretation.

Unlike my argument in "What is Mathematical Truth", Davis's argument against Gödel's version of "Platonism" does not mention "indispensability for physics", and this raised for me the question I mentioned at the beginning of this essay, the question as to whether my "indispensability argument is needed, or whether the considerations Davis offers in favor of regarding mathematical truth as objective are actually sufficient." To discuss this question, we have to return to the notion of objectivity.

Assuming that Davis and I are on the same wavelength with respect to that notion, and recalling that antirealist philosophies of mathematics all identify truth with provability, in one sense or another of "provability", this reduces to the question as to whether his arguments really rule out the possibility that mathematical truth is the same thing as provability. Let us begin with two clarifications.

First, the question isn't whether mathematical truth = provability in some one fixed formal system that "we can see to be correct". Even before the Gödel Incompleteness Theorems were proved, Brouwer's Intuitionism did not depend on assuming—in fact, Brouwer didn't believe—that constructive provability could be captured by any one formal system. And after the Gödel theorems were proved, Turing thought that we can see from Gödel's argument that reflection[28] on any formal system that is strong enough for arithmetic and that we can intuitively see to be correct will enable us to find a more powerful system, in fact a constructive transfinite sequence of stronger and stronger systems, such that a proof in any one of them would still intuitively count

---

[27] "What is Mathematical Truth?", 72.

[28] "Reflection" here denotes producing a stronger system by adding a consistency statement for a given system. If the systems are indexed by *notations* for constructive ordinals—that is, elements of a recursive well-ordering—and the ordering is already *proved* to be a well-ordering, one can continue "reflection" into the transfinite, when one comes to a "limit notation" by adding a suitably formalized statement to the effect that the union of the systems with indexes below the limit notation is a consistent system.

as an acceptable "mathematical proof".[29] (However, it had better not be possible to "see from below" just how far up the hierarchy of constructive ordinals such metamathematical reflection can take us, if we are not to run into contradiction.) So there may be a sense of "proof" in which what is "provable" outruns what is *formally* provable in one system that we can see to be correct.
And the Gödel results are not enough to exclude identification of mathematical truth with provability is such a sense. Do Davis's arguments exclude such an identification?

I assume they are meant to. In philosopher's jargon, views according to which mathematical truth is just provability (from axioms human mathematicians can see to be correct) count as *antirealist* and I have been assuming that, like me, Davis is a realist. It would be a disappointment to find out he isn't, and I have been laboring under a serious misconception!

One reason for supposing that Davis is not an antirealist is that he clearly thinks that our means of mathematical discovery are often "inductive", that is, they are not just deductions from self-evident axioms. Indeed, that is the main point of his essay. So I am not *seriously* worried that I have misunderstood "Pragmatic Platonism". But the question now arises in another form: if "Pragmatic Realism" is an argument against identifying truth with provability *is the argument good enough*? In "What is Mathematical Truth" I had written that "the consistency and fertility of classical mathematics is evidence that it—or most of it—*is true under some interpretation.* But the interpretation might not be a realist interpretation."[30] And I went on to rule out this possibility with the aid of two arguments I have already mentioned: the indispensability argument and the argument that an antirealist interpretation of pure mathematics does not fit together with a scientific realist interpretation of physics. Was this last step actually unnecessary?

What Davis's arguments show is that mathematicians do not proceed by proof alone. They also use "induction", that is, quasi-empirical methods. But does the fact that mathematical *discovery* is first made—in many cases—without formal proof show that *correctness of the result* isn't simply provability? Arguably, Torricelli's results were correct *in the sense that, and only in the sense that*, they were provable from acceptable axioms, even if Torricelli himself didn't have either the proof or the axioms.

Well, both Davis and I (in "What is Mathematical Truth") mention that new axioms sometimes get accepted in mathematics. But (1) this does not happen very often; and (2) when it does happen, it happens because the new axioms are appealing for *mathematical* reasons. The indispensability argument considers the need for an interpretation of the mathematical concepts *when they function in empirical science*, and argues that antirealism has no satisfactory account of this. Davis' argument considers only what goes on in pure mathematics. It certainly confirms a claim I made in "What is Mathematical Truth", the claim that mathematics does not only use deduction, but is full of quasi-inductive elements. But is that something an antirealist

[29]Turing, A.M. (1939), 'Systems of Logic Based on Ordinals', *Proceedings of the London Mathematical Society*, Ser. 2 45, pp. 161–228.
[30]"What is Mathematical Truth", 73.

need be impressed by? Couldn't an antirealist say that Davis's essay has to do with the context of discovery, and, perhaps, has also to do with an often-unrecognized context of "quasi-empirical justification", but not with the question of realism. If mathematics yields, as Davis says, "objective information", is that "objective information" about more than what we humans count as proof? We both think it is about more than that, but I still think that the success of *applied* mathematics needs to be brought into the picture in order to make the best case. But I look forward happily and affectionately to Davis's response.

Chapter 14
Pragmatic Platonism

Martin Davis

Abstract It is argued that to a greater or less extent, all mathematical knowledge is empirical.

Although I have never thought of myself as a philosopher, Harvey Friedman has told me that I am "an extreme Platonist". Well, extremism in defense of truth may be no vice, but I do feel the need to defend myself from that description.

Gödel's Platonism

When one thinks of Platonism in mathematics, one naturally thinks of Gödel. In a letter to Gotthard Günther in 1954, he wrote:

> When I say that one can ...develop a theory of classes as objectively existing entities, I do indeed mean by that existence in the sense of ontological metaphysics, by which, however, I do not want to say that abstract entities are present in nature. They seem rather to form a second plane of reality, which confronts us just as objectively and independently of our thinking as nature.[1]

If indeed that's extreme Platonism, it's not what I believe. I don't find myself confronted by such a "second plane of reality".

In his Gibbs lecture of 1951, Gödel made it clear that he rejected any mechanistic account of mind, claiming (with no citations) that

[1]See [5], vol IV, pp. 502–505.

M. Davis (✉)
Department of Mathematics, University of California, Berkeley, CA, USA
e-mail: martin@eipye.com

M. Davis (✉)
Courant Institute of Mathematical Sciences, New York University, New York, NY, USA

E.G. Omodeo and A. Policriti (eds.), *Martin Davis on Computability, Computational Logic, and Mathematical Foundations*,
Outstanding Contributions to Logic 10, DOI 10.1007/978-3-319-41842-1_14

> …some of the leading men in brain and nerve physiology …very decidedly deny the possibility of a purely mechanistic explanation of psychical and nervous processes.[2]

In a 1974 letter evidently meant to help comfort Abraham Robinson who was dying of cancer, he was even more emphatic:

> The assertion that our ego consists of protein molecules seems to me one of the most ridiculous ever made.[3]

Alas, I'm stuck with precisely this ridiculous belief. Although I wouldn't mind at all having the transcendental mind Gödel suggests, I'm aware of no evidence that our mental activity is anything but the work of our physical brains.

In his Gibbs lecture Gödel suggests another possibility:

> If mathematics describes an objective world just like physics, there is no reason why inductive methods should not be applied in mathematics just the same as in physics. The fact is that in mathematics we still have the same attitude today that in former times one had toward all science, namely we try to derive everything by cogent proofs from the definitions (that is, in ontological terminology, from the essences of things). Perhaps this method, if it claims monopoly, is as wrong in mathematics as it was in physics.[4]

I will claim that mathematicians have been using inductive methods, appropriately understood, all along. There is a simplistic view that induction simply means the acceptance of a general proposition on the basis of its having been verified in a large number of cases, so that for example we should regard the Riemann Hypothesis as having been established on the basis of the numerical evidence that has been obtained. But this is unacceptable: no matter how much computation has been carried out, it will have verified only an infinitesimal portion of the infinitude of the cases that need to be considered. But inductive methods (even those used in physics) need to be understood in a much more comprehensive sense.

Gödel Incompleteness and the Metaphysics of Arithmetic

Gödel has claimed that it was his philosophical stance that made his revolutionary discoveries possible and that his Platonism had begun in his youth. However, an examination of the record shows something quite different, namely a gradual and initially reluctant embrace of Platonism as Gödel considered the philosophical implications of his mathematical work [3]. It is at least as true that Gödel's philosophy was the result of his mathematics as that the latter derived from the former.

In 1887, in an article surveying transfinite numbers from mathematical, philosophical, and theological viewpoints, Georg Cantor made a point of attacking a little pamphlet on counting and measuring written by the great scientist Hermann von Helmholtz. Cantor complained that the pamphlet expressed an "extreme empirical-psychological point of view with a dogmatism one would not have thought possible …" He continued:

[2]See [5] vol. III, p. 312.

[3]See [5] vol. V, p. 204.

[4]See [5], vol. III, p. 313.

> Thus, in today's Germany we see, as a reaction against the overblown Kant-Fichte-Hegel-Schelling Idealism, an *academic-positivistic skepticism* that powerfully dominates the scene. This skepticism has inevitably extended its reach even to *arithmetic*, in which domain it has led to its most fateful conclusions. Ultimately, this may turn out most damaging to this positivistic skepticism itself.

In reviewing a collection of Cantor's papers dealing with the transfinite, Frege chose to emphasize the remark just quoted, writing [6]:

> Yes indeed! This is the very reef on which this doctrine will founder. For ultimately, the role of the infinite in arithmetic is not to be denied; yet, on the other hand, there is no way it can coexist with this epistemological tendency. Thus we can foresee that this issue will provide the setting for a momentous and decisive battle.

In a 1933 lecture, Gödel, considering the consequences of his incompleteness theorems, and perhaps not having entirely shaken off the positivism of the Vienna Circle, showed that the "battle" Frege had predicted was taking place in his own mind:

> The result of our previous discussion is that our axioms, if interpreted as meaningful statements, necessarily presuppose a kind of Platonism, which cannot satisfy any critical mind and which does not even produce the conviction that they are consistent.[5]

The axioms to which Gödel referred were an unending sequence produced by permitting variables for ever higher "types" (in contemporary terminology, sets of ever higher rank) and including axioms appropriate to each level. He pointed out that to each of these levels there corresponds an assertion of a particularly simple arithmetic form, what we now would call a Π^0_1 sentence, which is not provable from the axioms of that level, but which becomes provable at the next level. In the light of later work,[6] a Π^0_1 sentence can be seen as simply asserting that some particular equation

$$p(x_1, x_2, \ldots, x_n) = 0,$$

where p is a polynomial with integer coefficients, has no solutions in natural numbers. To say that such a proposition is *true* is just to say that for each choice of natural number values $a_1, a_2, \ldots, a_n$ for the unknowns,

$$p(a_1, a_2, \ldots, a_n) \neq 0.$$

Moreover a proof for each such special case consists of nothing more than the sequence of additions and multiplications needed to compute the value of the polynomial together with the observation that that value is not 0. So in the situation to which Gödel is calling attention, at a given level there is no single proof that subsumes this infinite collection of special cases, while at the next level there is such a proof.

This powerful way of expressing Gödel incompleteness is not available to one who holds to a purely formalist foundation for mathematics. For a formalist, there is no "truth" above and beyond provability in a particular formal system. Post had reacted

[5] See [5] vol. III, p. 50.

[6] See [4] pp. 331–339.

to this situation by insisting that Gödel's work requires "at least a partial reversal of the entire axiomatic trend of the late nineteenth and early twentieth centuries, with a return to meaning and truth as being of the essence of mathematics".[7] Frege's reference to the "role of the infinite in arithmetic" is very much to the point here. It is the infinitude of the natural numbers, the infinitude of the sequence of formal systems, and finally, the infinitude of the special cases implied by a Π_1^0 proposition that point to some form of Platonism.

Infinity in the Seventeenth Century

Hilbert saw the problem of the infinite as central to resolving foundational issues. Perhaps succumbing a bit to hyperbole, he said:

> The infinite has always stirred the emotions of mankind more deeply than any other question; the infinite has stimulated and fertilized reason as few other ideas have; but also the infinite, more than any other notion is in need of clarification.[8]

People have pronounced and speculated about what is and isn't true about infinity since they began thinking abstractly. Aristotle's views on the subject in particular had a great influence. A discovery made by the Italian mathematician Torricelli in 1641 provides a very revealing example.[9] He found that the volume of a certain solid of infinite extent is finite. The solid in question is obtained by rotating about an axis a certain plane figure with infinite area. Specifically, in modern terminology, it is the figure bounded by the hyperbola whose equation is $y = 1/x$, the line $x = 1$ and the horizontal asymptote of the hyperbola, namely the X-axis. Torricelli's solid is formed by rotating this figure about the X-axis. Although showing that this solid of revolution has a finite volume is a routine "homework" problem in a beginning calculus course,

$$\pi \int_1^\infty \frac{1}{x^2} dx = \pi,$$

at the time it created a sensation because it contradicted prevalent views about the infinite. Torricelli himself remarked "...if one proposes to consider a solid, or a plane figure, infinitely extended, everybody immediately thinks that such a figure must be of infinite size." In 1649, Petri Gassendi wrote,

> Mathematicians ...weave those famous demonstrations, some so extraordinary that they even exceed credibility, like what ...Torricelli showed of a certain...solid infinitely long which nevertheless is equal to a finite cylinder.

Writing in 1666, Isaac Barrow found Torricelli's result contradicting what Aristotle had taught. He referred to Aristotle's dictum, "there is no proportion between the finite and the infinite":

> The truth of which statement, a very usual and well known axiom, has been in part broken by ...modern geometricians [who] demonstrate ...equality of ...solids protracted to infinity with other finite ...solids which prodigy ...Torricelli exhibited first.

---

[7]See [9] p. 295.

[8]See [10] p. 371.

[9]This discussion, including the quotations, is based on Paolo Mancosu's wonderful monograph [7].

Much can be learned from this example about the way in which mathematicians expand the applicability of existing methods to new problems and with how they deal with the philosophical problems that may arise. Torricelli used a technique called the method of *indivisibles*, a method pioneered by Cavalieri that provided a short-cut for solving area and volume problems. Torricelli used this technique to prove that his infinite body had the same volume as a certain finite cylinder. The method conceived of each of the two bodies being compared as constituted of a continuum of plane figures. Although there was no rigorous foundation for this, Cavalieri and later Torricelli showed how effective it could be in easily obtaining interesting results. They were well aware of the Eudoxes-Archimedes method of exhaustion (which they called "the method of the ancients"), and used it to confirm their results and/or to convince skeptics.[10] But, Torricelli insisted on the validity of the new method.

What can we say about Torricelli's methodology? He was certainly not seeking to obtain results by "cogent proofs from the definitions" or "in ontological terms, from the essences of things". He was *experimenting* with a mathematical technique that he had learned, and was attempting to see whether it would work in an uncharted realm. In the process, something new about the infinite was discovered. I insist that this was induction from a body of mathematical experience.

Robustness of Formalism

An interesting example is provided by the development of complex numbers. The fact that the square of any non-zero real number is positive had been generally accepted as implying that there could be no number whose square is negative. Sixteenth century algebra brought this into question. The quadratic formula, essentially known since antiquity, did seem to lead to solutions which did involve square roots of negative quantities. But those were simply regarded as impossible. But the analogous formula for cubic equations, discovered by Tartaglia and published in Cardano's book of 1545, forced a rethinking of the matter. In the case of a cubic equation with real coefficients and three real roots, the formula led to square roots of negative numbers as intermediary steps in the computation. Bombelli discussed this in his book of 1572. In particular, he noted that although the equation $x^3 - 15x - 4 = 0$ had the three roots $4, -2 + \sqrt{3}, -2 - \sqrt{3}$, the Tartaglia formula forced one to consider $\sqrt{-109}$. Soon mathematicians were working freely with complex numbers without questioning whether they really exist in some "second plane of reality". What this experience illustrates is the robustness of mathematical formalisms. These formalisms often point the way to expansions of the subject matter of mathematics before any kind of convincing justification can be supplied. This is again a case of induction in mathematical practice.

Leibniz referred to this very experience when asked to justify the use of infinitesimals. As Mancosu explains

[10]The method of exhaustion typically required one to have the answer at hand, whereas with indivisibles the answer could be computed.

> ...the problem for Leibniz was not, Do infinitely small articles exist? but, Is the use of infinitely small quantities in calculus reliable?[11]

In justifying his use of infinitesimals in calculus, Leibniz compared this with the use of complex numbers which had become generally accepted although at the time, there was no rigorous justification.

In another example, the rules of algebra, including the manipulation of infinite series was applied to operators with scant justification. This can be seen in Boole's [2] massive tract on differential equations in which marvelous manipulative dexterity is deployed with not a theorem in sight.

The Ontology of Mathematics

If the objects of mathematics are not in nature and not in a "second plane of reality," then where are they? Perhaps we can learn something from the physicists. Consider for example, the discussion of the "Anthropic Principle" [1]. The advocates of this principle note that the values of certain critical constants are finely tuned to our very existence. Given even minor deviations, the consequence would be: no human race. It is not relevant here whether this principle is regarded as profound or merely tautological. What I find interesting in this discussion of alternate universes whose properties exclude the existence of us, is that no one worries about their ontology. There is simply a blithe confidence that the same reasoning faculty that serves physicists so well in studying the world that we actually do inhabit, will work just as well in deducing the properties of a somewhat different hypothetical world. A more mundane example is the ubiquitous use of idealization. When Newton calculated the motions of the planets assuming that each of the heavenly bodies is a perfect sphere of uniform density or even a mass particle, no one complained that the ontology of his idealized worlds was obscure. The evidence that our minds are up to the challenge of discovering the properties of alternative worlds is simply that we have successfully done so. Induction indeed! This reassurance is not at all absolute. Like all empirical knowledge it comes without a guarantee that it is certain.

My claim is that what mathematicians do is very much the same. We explore simple austere worlds that differ from the one we inhabit both by their stark simplicity and by their openness to the infinite. It is simply an empirical fact that we are able to obtain apparently reliable and objective information about such worlds. And, because of this, any illusion that this knowledge is certain must be abandoned. If, on a neo-Humean morning, I were to awaken to the skies splitting open, hearing a loud voice bellowing, "This ends Phase 1; Phase 2 now begins," I would of course be astonished. But I will not say that I *know* that this will not happen. If presented with a proof that PA is inconsistent or even that some huge natural number is not the sum of four squares, I would be very very skeptical. But I will not say that I *know* that such a proof must be wrong.

[11] See [7] p. 172.

Infinity Today

Mathematical practice obtains information about what it would be like if there were infinitely many things. It is not at all evident a priori that we can do that. But mathematicians have shown us that we can. Our steps are tentative, but as confidence is acquired we move forward. Our theorems are proved in many different ways, and the results are always the same. Our formalisms are robust and yield information beyond the original intent. To doubt the significance of the concrete evidence for the objectivity of mathematical knowledge is like anti-evolutionists doubting the evidence of paleontology by suggesting that those fossils were part of creation. As was discussed above, Gödel's work has left us with a transfinite sequence of formal systems involving larger and larger sets. Models of these systems can be obtained from initial segments of the famous hierarchy obtained by iterating transfinitely the power set operation $\mathscr{P}$:

$$V_0 = \emptyset; \quad V_{\alpha+1} = \mathscr{P} V_\alpha; \quad V_\lambda = \bigcup_{\alpha<\lambda} V_\alpha, \ \lambda \text{ a limit ordinal}$$

Thus, $V_{\omega 2}$ is a model of the original Zermelo axioms. To obtain a model of the more comprehensive Zermelo-Fraenkel (ZF) axioms, no ordinal whose existence is provable in ZF will do.[12] To continue the transfinite sequence of formal systems, it is necessary to enter the realm of large cardinals in which there has been intensive research. Workers in this realm are pioneers on dangerous ground: although we know that no proof of the consistency with ZF of the existence of these enormous sets is possible, it is always conceivable that a proof in ZF of the inconsistency of one of them will emerge thereby destroying a huge body of work. But the empirical evidence is encouraging. Although the defining characteristics of the various large cardinal types that have been studied seeming quite disparate, they line themselves up neatly in order of increasing consistency strength. Moreover, they have shown themselves to be the correct tool for resolving open questions in descriptive set theory.

So far Gödel incompleteness has had only a negligible effect on mathematical practice. Cantor's continuum hypothesis remains a challenge: although the Gödel-Cohen results prove its undecidability from ZF, if the iterative hierarchy is taken seriously, it does have a truth value whether we can ever find it or not. In the realm of arithmetic many important unsolved problems, including the Riemann Hypothesis and the Goldbach Conjecture, are equivalent to Π^0_1 sentences. However, so far no undecidable Π^0_1 sentences have been found that are provably equivalent to questions previously posed (as has been done for uncomputability). However, Harvey Friedman has produced a remarkable collection of Π^0_1 and Π^0_2 arithmetic sentences with clear combinatorial content that can only be resolved in the context of large cardinals.

---

[12]Because otherwise the consistency of ZF would be provable in ZF contradicting Gödel's second incompleteness theorem. For that matter the set $V_{\omega 2}$ cannot be proved to exist from the Zermelo axioms alone; in ZF its existence follows using Replacement.

The Chimerical Effort to Seek Certainty

Mark Twain suggested the lovely notion of a "Sunday truth": something fervently believed in church on Sunday but having no effect on behavior in the rest of the week. Many mathematicians will profess a belief in formalism when foundational matters are discussed. But in their day-to-day work as mathematicians, they remain thoroughgoing Platonists. The "crisis" in foundations from the turn of the 20th century to the 1920s has quietly dissipated. Set theory as a foundation is evident in the initial chapter of many graduate-level textbooks. The obligation to always point out a use of the axiom of choice is a thing of the past. I haven't heard of anyone calling the proof of Fermat's Last Theorem into question because of the large infinities implicit in Grothendiek universes.[13] But there are those who wish to draw a line between safe and unsafe proof methods. The line is drawn by some who insist on some variety of constructivity. Others demand predicativity. Contemporary foundational research makes such notions precise and obtains theorems on the relative strengths of different methods. But there is no pointless attempt to restrict mathematicians. History suggests that they will use whatever methods work including the higher realms of the infinite.

References

1. Barrow, J. D., & Tipler, F. J. (1986). *The anthropic cosmological principle*. Oxford: Oxford University Press.
2. Boole, G. (1865). *A treatise on differential equations*. London: Macmillan and Co.
3. Davis, M. (2005). What did Gödel believe and when did he believe it? *Bulletin of Symbolic Logic*, *11*, 194–206.
4. Davis, M., Matiyasevich, Yu., & Robinson, J. (1976). Hilbert's tenth problem. Diophantine equations: Positive aspects of a negative solution. In *Proceedings of Symposia in Pure Mathematics: Positive Aspects of a Negative Solution* (Vol. XXVIII, pp. 323–378).
5. Feferman, S., et al. (1986–2003). *Kurt Gödel Collected Works* (Vols. I–V). Oxford: Oxford University Press.
6. Frege, G. (1892). Rezension von: Georg Cantor. Zum Lehre vom Transfiniten. *Zeitschrift für Philosophie und philosophische Kritik, new series*, *100*, 269–272.
7. Mancosu, P. (1996). *Philosophy of mathematics & mathematical practice in the seventeenth century*. Oxford: Oxford University Press.
8. McLarty, C. (2010). What does it take to prove Fermat's last theorem? Grothendiek and the logic of number theory. *Bulletin of Symbolic Logic*, *16*, 359–377.
9. Post, E. L. (1944). Recursively enumerable sets of positive integers and their decision problems. *Bulletin of the American Mathematical Society*, *50*, 284–316. Reprinted: M. Davis (Ed.), The undecidable Raven Press, New York 1965; Dover, New York 2004. Reprinted:: M. Davis (Ed.), *Solvability, provability, definability: The collected works of Emil L. Post*, Birkhäuser 1994.
10. van Heijenoort, J. (Ed.) (1967). *From Frege to Gödel: A source book in mathematical logic, 1879–1931*. Cambridge: Harvard University Press.

[13]Number theorists regard the use of Grothendiek universes as a mere convenience. See [8] for a careful discussion.

Chapter 15
Concluding Comments by Martin

Martin Davis

Abstract After a very brief comment on Yuri Matiyasevich's contribution, I discuss at greater length proposals to use modal logic to clarify foundational issues in set theory. Finally, I very sadly bid farewell to my friend and collaborator Hilary Putnam.

15.1 Comments on Yuri Matiyasevich's Essay

First I want to express my thanks to my good friend Yuri for his generous account of my contributions to the solution of Hilbert's Tenth Problem. The theorem that every listable set is Diophantine that, as Yuri explains, I had conjectured in my doctoral dissertation, is often referred to as Matiyasevich's Theorem because he supplied the crucial final step. He kindly suggests calling it DPRM to emphasize the role each of us played in its eventual proof. It is also sometimes referred to as MRDP.

I would like to comment briefly on a few of the matters he discusses. Although, as Yuri emphasizes, my conjecture was widely disbelieved because of its counter-intuitive consequences, I want to mention one argument in its favor that impressed me, perhaps unduly. Namely it was easy to prove (non-constructively) that there is a Diophantine set whose complement is not Diophantine. Namely, because the class of Diophantine subsets of N^n is closed under existential quantification (i.e., projection), if it were also closed under complementation, it would be closed under universal quantification as well. Therefore it would include all arithmetic sets. But this is impossible because all Diophantine sets are listable and there are arithmetic sets that are not listable. Thus I knew that the class of Diophantine sets shares with the class of listable sets the properties of being closed under union, intersection and existential quantification, but not under complementation.

M. Davis (✉)
Department of Mathematics, University of California, Berkeley, CA, USA
e-mail: martin@eipye.com

M. Davis
Courant Institute of Mathematical Sciences, New York University,
New York, NY, USA

E.G. Omodeo and A. Policriti (eds.), *Martin Davis on Computability, Computational Logic, and Mathematical Foundations*,
Outstanding Contributions to Logic 10, DOI 10.1007/978-3-319-41842-1_15

In connection with the arithmetic representation of listable sets involving only a single bounded universal quantifier (what Raphael Robinson called Davis Normal Form), I'd like to point out that while from one point of view it is a simple reduction of Gödel's representation with several universal quantifiers, given what Hilary Putnam, Julia Robinson, and I knew in 1959, Davis Normal Form was crucial for our proof of the DPR theorem. This was because without assuming JR, the variable exponents introduced by the elimination of the innermost universal quantifier from Gödel's representation, could not be eliminated to permit iterating the process. Of course after Yuri proved JR, this was no longer an issue, and contemporary proofs of DPR no longer need mention Davis Normal Form.

Yuri mentions his own crucial contribution in a single modest sentence. His wonderful proof of JR, showing that the relation between a number n and the $2n$th Fibonacci number is Diophantine by an explicit construction, accomplished something that we others had been trying unsuccessfully to do for twenty years.

Although Yuri mentions my recent conjecture in connection with Bjorn Poonen's work on rings of rational numbers, referring to me as a "guru", I am not very optimistic about the usefulness of that conjecture. However, he does not mention my most successful conjecture of all. During the period when DPR had been proved so that it was known that the truth of my conjecture and thus the unsolvability of Hilbert's Tenth Problem would follow if JR were proved, I gave a number of talks in which I emphasized the consequences of either the truth or the falsity of JR, noting that, in either case, some of those consequences were rather implausible. Asked during the question period for my own opinion as to the truth of JR, I would reply, half in jest: Oh, I think that JR is true and will be proved by a clever young Russian.

Martin Davis, June 13, 2015

15.2 Comments on Hilary Putnam's Remarks on "Pragmatic Platonism"

I was delighted to learn that Hilary and I agree about so much concerning the nature of mathematical knowledge. Here I will concern myself with a few aspects where his remarks suggest that our views may differ, as well as to see say a little more about certain topics than I did in my original essay.

Mathematics and Natural Science

To a mathematician it is certainly gratifying that our field is so richly applicable in science, if only for the economic advantages that accrue even to those of us whose work is remote from applications. And of course there are important and difficult philosophical problems in understanding this relationship. As was indicated in a famous essay by Eugene Wigner, it all seems almost too good to be true. But I don't see that this relationship sheds any light on the question with which my essay deals: how is that we can obtain objective knowledge about infinite entities.

Hilary suggests that this connection helps to show that intuitionism is unsatisfactory as a foundation of mathematics. I am more persuaded by a semi-facetious remark that Hilary himself made to me in conversation many years ago: "Do the intuitionists intend to put people who use non-intuitionisitic methods in jail?" Mathematicians will just use whatever methods seem to work and when faced with methodological difficulties will not long retreat to "safe methods" but learn how best to work around the difficulties. The evolution of the ideas of the great mathematician Hermann Weyl illustrates this well. Convinced that full-blooded mathematical analysis was methodologically unsustainable, even remarking that it was a "house built on sand", he became a disciple of Brouwer, writing (much to the chagrin of his teacher Hilbert) "Brouwer, Das ist die Revoultion!" Many years later, writing Hilbert's obituary, still admiring Brouwer, Weyl wrote that trying to develop mathematics in an intuitionistic setting leads to "an almost unbearable awkwardness" [4]. Weyl was well acquainted with mathematical physics, especially with relativity, but never referred to that as a reason to abandon intuitionism.

The Benacerraf Problem

Benacerraf deems inappropriate, properties that set-theoretic objects intended to serve a specific mathematical function possess that are not relevant to that function. Thus in von Neumann's explication:

$$1 = \{\emptyset\};\ \ 3 = \{\emptyset, \{\emptyset\}, \{\emptyset, \{\emptyset\}, \{\emptyset, \{\emptyset\}\}\}\}$$

we get the "inappropriate" $1 \in 3$. My tendency when faced with a philosophical problem regarding mathematics is to look to mathematical history and practice for help. Beginning in the 19th century, mathematicians were faced with a number of important equivalence relations and the need to see the equivalence as a kind of equality. Perhaps the first was Gauss's use of the congruence relation, where $a \equiv b \bmod r$ is defined to mean that the natural number r, called the modulus, is a divisor of the integer $b - a$. The equivalence classes form a ring, and in the case that r is prime, a field. It has become customary to designate each such residue class by its least non-negative member. So for example, with the modulus 5, we have a finite field whose elements one writes as $\{0, 1, 2, 3, 4\}$. And we write such equations as $3 + 4 = 2$ and $2 \cdot 3 = 1$. In effect the equivalence classes are each represented by one of its members. But the Benacerraf "problem" applies here. The property $2 < 4$ is "inappropriate" in just the same way as in the example above. It wouldn't occur to a mathematician to be concerned with the question of whether these irrelevant properties show that e.g., the number 2 is not truly the class of numbers congruent to 2 modulo 5. Is there really a philosophical puzzle here?

The Frege-Russell attempt to define the cardinal numbers as, in effect, the equivalence classes corresponding to the relation between a pair of sets of the existence of a one-one correspondence between them. Of course the attempt failed because the classes were too large. What von Neumann did was to choose a member of each equivalence class to designate the class according to the elegant recursion:

$$0 = \emptyset;\ \ n + 1 = \{0, 1, \ldots, n\}$$

so that the number n is designated by a set that does have exactly n members. This should trouble only philosophers who truly seek to know what a number "really" is.

New Axioms

Since Hilary does mention new axioms, I'll take the opportunity, to say a little about the topic, although I suspect that Hilary will not disagree with much of what I say. Zermelo's axiom of choice is the obvious example of a proposed axiom that has gained general acceptance although it excited considerable controversy in the early years of the twentieth century. There was a prominent group of French mathematicians who went to considerable lengths to avoid the axiom, so that, for example, they didn't permit themselves to say without qualification: *The union of a countable set of countable sets is countable.* By the second half of the century, it had become an indispensable tool in various branches of mathematics.

The branch of mathematics called *descriptive set theory,* pioneered in Eastern Europe during the first half of the twentieth century, provides an interesting example of the power of a new axiom. We write $\mathscr{R}$ for the set of real numbers and for a set $B \in \mathscr{R}^{n+1}$, we write $\mathbf{Proj}(B)$ for the set

$$A = \{\langle x_1, \ldots, x_n\rangle \in \mathscr{R}^n \mid \exists y \in \mathscr{R}[\langle x_1, \ldots, x_n, y\rangle \in B]\}.$$

The hierarchy of *projective sets* is defined simultaneously in all $\mathscr{R}^n$ as follows:

$$\begin{aligned} \Sigma^1_0 &= \text{the set of Borel sets} \\ \Pi^1_m &= \{A \subseteq \mathscr{R}^n \mid \mathscr{R}^n - A \in \Sigma^1_m\} \\ \Sigma^1_{m+1} &= \{A \subseteq \mathscr{R}^n \mid \exists B \in \Pi^1_m[A = \mathbf{Proj}(B)]\} \end{aligned}$$

Lusin proved the key hierarchy theorem: For non-negative integers m, $\Sigma^1_m \subset \Sigma^1_{m+1}$ and $\Pi^1_m \subset \Pi^1_{m+1}$. Also, for $m > 0$, $\Sigma^1_m - \Pi^1_m \neq \emptyset$.

Souslin proved that the Borel sets are exactly those that are in both Σ^1_1 and Π^1_1, and in 1917 Lusin showed that every set in Σ^1_1 is Lebesgue measurable. It seemed plausible to researchers, noting that known proofs of the existence of non-measurable sets used the axiom of choice, that sets that had explicit definitions should be measurable. This leads to the conjecture that projective sets, evidently being explicitly definable, should all be Lebesgue measurable. But efforts to prove this failed. When Cohen developed his forcing method, it became clear why success was so elusive. The proposition that all projective sets are Lebesgue measurable turned out to be undecidable in ZFC.

A new axiom seemed to be called for to settle the question, and it turned out that the concept of *determinacy* provided the key: Associated with a set A of real numbers is an infinite game defined as follows: Players I and II alternately move by each specifying a binary digit 0 or 1. They thus specify the binary expansion of a real number x in the unit interval. If x is the fractional part of a member of A, then I wins; otherwise II wins. The set A is *determined* if either I or II has a winning strategy. The axiom of *projective determinacy* (PD), states that every projective set of real numbers is determined. And PD does yield the desired result that every projective set is Lebesgue measurable. In fact a number of other open questions about projective sets can be settled when PD is assumed. Tony Martin and John Steel were able to derive PD from a suitable large cardinal axiom, thus providing a satisfying conclusion to those who view large cardinals with equanimity.[1]

On the other hand, axioms asserting the existence of "large" cardinals are certainly not being widely accepted. Harvey Friedman has found a considerable number of propositions in combinatorial mathematics, some of them rather attractive, that can only be proved by assuming such axioms.

Modal Logic and Mathematical Existence

I agree with Hilary in general terms that when set theorists talk about *all* sets, it needs to be understood in a relative manner. Relative to what? To the ordinals available to serve as ranks. It is the large cardinals that extend this range. So uncovering a new large cardinal concept augments the universe of sets. As Hellman [2] points out, this conception can already be seen, at least in embryo, in Zermelo's account in his [5].

But Hilary suggests more: a program to use the formal apparatus of modal logic develop the idea that the class of sets and proper classes generally have only a possible existence.[2] And his student Geoffrey Hellman has made a bravura effort to carry out this program in his [2]. However, to paraphrase a trenchant comment of Poincaré:

> It is difficult to see that the word *possibly* acquires when written $\diamondsuit$, a virtue it did not possess when written possibly.[3]

Of course Frege, Russell, and Hilbert had answers for Poincaré: Frege and Russell used a conceptual apparatus using $\supset$ and other symbols to demonstrate that mathematics could be formalized in a formal system, and Hilbert proposed to use that knowledge to overcome the doubts about set-theoretic mathematics. And what came after would have been a surprise to them as well as to Poincaré.

Hellman uses second order S5 to prove that a "modalist" need not give up anything that's available to the mathematical platonist, that, as it were, for mathematical purposes possible existence is an adequate substitute for actual existence. But his formalism is syntactic: to provide meaning to his formalism would land him back in the platonic soup. Moreover his second order formalism includes full comprehension axioms, and as Quine pointed out long ago, this is to admit at least a modicum of set theory. So while I admire Hellman's heroic effort, I wonder whom it is for. Will it convince mathematical constructivists or predicativists to give up their doubts about set theory? Not the ones I know! Will set theorists who, while making free use of proper classes in their technical work, have qualms about about the concept, be reassured? Again I doubt it. Certainly no one will propose a full modal formalism

---

[1]The cardinal in question is in fact quite large: a countable infinity of Woodin cardinals with a measurable cardinal above them.

[2]Recent work by Joel Friedman on modalism [1] should also be mentioned.

[3]"It is difficult to see that the word *if* acquires when written $\supset$, a virtue it did not possess when written if." [3], p. 156.

for the purpose of computer proof verification. So I am skeptical, but I will be very pleased if I am proved wrong, if this remarkable project turns out to yield fruitful results.

Martin Davis, June 13, 2015

Farewell to Hilary Putnam (1926–2016)

I have been very fortunate in having Hilary Putnam in my life as a close friend and a collaborator. Our families lived together in a house in Ithaca, New York in the summer of 1957 where Hilary and I were attending a five week Institute for Logic at Cornell University. We spent the following three summers working together, 1958 and 1959 in Eastern Connecticut, where I was on the faculty of the Connecticut branch of Rensselaer Polytechnic Institute, and 1960 at the University of Colorado Boulder where we attended a conference on physics for mathematicians.

In our time together there was hardly a topic in the full range of human intellectual inquiry into which our conversations did not range. This was in addition to our technical work which certainly includes contributions of which I'm very proud. Also our educations had been sufficiently complementary that we were really able to learn from each other; this included matters remote from our technical work. When I showed Hilary a copy of my first book that had just arrived from the publisher smelling of printer's ink, he offered to find an error on any page. When I offered the reverse side of the title page, certain that the few lines of text on that page would be free of error, Hilary noticed that the word "permission" was missing its first "i"!

Hilary's sharp mind, wit, and humane attitude toward life made his company a pleasure and our work together always fun. I miss him very much.

Martin Davis, June 13, 2015

References

1. Friedman, J. (2005). Modal platonism: An easy way to avoid ontological commitment to abstract entities. *Journal of Philosophical Logic*, *34*, 227–273.
2. Hellman, G. (1989). *Mathematics without numbers: Toward a modal-structural interpretation*. Oxford.
3. Poincaré, H. (2012). *Science and method,* translated from French by Francis Maitland, Thomas Nelson and Sons, London 1914. Facsimile Reprint: Forgotten Books. http://www.forgottenbooks.org.
4. Weyl, H. (1950). David Hilbert and his mathematical work. *Bulletin of the American Mathematical Society*, *50*, 612–654.
5. Zermelo, E. (1996). Über Grenzzahlen und Mengenbereiche: neue Untersuchungen über die Grundlagen der Mengenlehre, *Fundamenta Mathematicae*, vol. 16, pp. 29–47. English Translation: On Boundary numbers and domains of sets: New investigations in the foundations of set theory. In W. B. Ewald (Ed.), *From Kant to Hilbert: A source book in the foundations of mathematics* (pp. 1219–1233). Oxford University Press.

Chapter 16
Martin Davis's Bibliography 1950–2015

Eugenio G. Omodeo

Abstract This appendix offers a comprehensive list of articles and books which Martin Davis has published till the present day, cross-referenced with a list of bibliographic entries regarding conference proceedings, paper collections, and books, to which he has contributed. Our list does not include the many reviews written by Martin Davis, in particular the ones which have appeared on *The Journal of Symbolic Logic*.

This appendix offers a comprehensive list of articles and books which Martin Davis has published till the present day, cross-referenced with a list of bibliographic entries regarding conference proceedings, paper collections, and books, to which he has contributed.

Our list does not include the many reviews written by Martin Davis, in particular the ones which have appeared on *The Journal of Symbolic Logic*.

(1) M. Davis. Arithmetical problems and recursively enumerable predicates (abstract). *J. Symbolic Logic*, 15(1):77–78, 1950.
(2) M. Davis. *On the theory of recursive unsolvability*. PhD thesis, Princeton University, May 1950.
(3) M. Davis. Relatively recursive functions and the extended Kleene hierarchy. In *Proceedings of the International Congress of Mathematicians* (Harvard University, Cambridge, MA, August 30–September 6, 1950), volume 1, page 723. AMS, Providence, RI, 1952.
(4) M. Davis. Arithmetical problems and recursively enumerable predicates. *J. Symbolic Logic*, 18(1):33–41, 1953. Russian transl. in [1, pp. 15–22].

E.G. Omodeo (✉)
University of Trieste, Trieste, Italy
e-mail: eomodeo@units.it

E.G. Omodeo and A. Policriti (eds.), *Martin Davis on Computability, Computational Logic, and Mathematical Foundations*,
Outstanding Contributions to Logic 10, DOI 10.1007/978-3-319-41842-1_16

(5) M. Davis. A note on universal Turing machines. In Claude E. Shannon and John McCarthy, editors, *Automata Studies*, volume 34 of *Ann. of Math. Stud.*, pages 167–175. Princeton University Press, 1956.

(6) M. Davis. The definition of universal Turing machine. *Proc. Amer. Math. Soc.*, 8(6):1125–1126, December 1957.

(7) M. Davis. *Computability and Unsolvability*. McGraw-Hill, New York, 1958. Reprinted with an additional appendix, Dover 1983. Japanese translation 1966. Italian translation: (8).

(8) M. Davis. *Computabilità e Insolubilità—Introduzione alla teoria della computabilità e alla teoria delle funzioni ricorsive*. Collana di Epistemologia diretta da Evandro Agazzi. Edizioni Abete, Roma, 1975. Italian translation of (7); foreword by Mariano Bianca (ed.).

(9) M. Davis and Hilary Putnam. Reductions of Hilbert's tenth problem. *J. Symbolic Logic*, 23(2):183–187, 1958. Russian transl. in [1, pp. 49–53].

(10) M. Davis and Hilary Putnam. Feasible computational methods in the propositional calculus. Technical report, Rensselaer Polytechnic Institute, Research Division, Troy, New York, October 1958. (Reprinted in this book, pp. 371–405.)

(11) M. Davis and Hilary Putnam. On Hilbert's tenth problem. *Notices Amer. Math. Soc.*, 6(5):544, 1959.

(12) M. Davis and Hilary Putnam. A computational proof procedure; Axioms for number theory; Research on Hilbert's Tenth Problem. Technical Report AFOSR TR59-124, U.S. Air Force, October 1959. (Partly reprinted in this book, pp. 407–426.)

(13) M. Davis. A program for Presburger's algorithm. In *Summaries of talks presented at the Summer Institute of Symbolic Logic in 1957 at Cornell University*, volume 2, pages 215–223, Princeton, NJ, 1960. Communications Research Division, Institute for Defense Analyses. Reprinted as "A computer program for Presburger's algorithm" in [5, pp. 41–48].

(14) M. Davis. Computable functional of arbitrary finite type. In *Summaries of talks presented at the Summer Institute of Symbolic Logic in 1957 at Cornell University*, pages 242–246. Institute for Defense Analyses, 1960.

(15) M. Davis and Hilary Putnam. A computing procedure for quantification theory. *J. ACM*, 7(3):201–215, 1960. Preprinted as (12, Part I); reprinted in [5, pp. 125–139].

(16) M. Davis, Hilary Putnam, and Julia Robinson. The decision problem for exponential Diophantine equations. *Ann. of Math. (2)*, 74(3):425–436, 1961. Reprinted in [14, pp. 77–88]. Russian transl. in [1, pp. 69–79].

(17) M. Davis. Aspects of mechanical theorem-proving. In *Proceedings of Third International Congress on Cybernetics (Namur, Belgium, 11–15 September 1961)*, Association Internationale de Cybernétique, pages 415–418. Gauthier-Villars, Paris, 1965.

(18) M. Davis, George Logemann, and Donald Loveland. A machine program for theorem-proving. Technical Report AFOSR 819, IMM-NYU 288, New York University, Institute of Mathematical Sciences, June 1961.

(19) M. Davis, George Logemann, and Donald W. Loveland. A machine program for theorem-proving. *Commun. ACM*, 5(7):394–397, 1962. Preprinted as (18); reprinted in [5, pp. 267–270].

(20) M. Davis. Applications of recursive function theory to number theory. In J. C. E. Dekker, editor, *Recursive Function Theory*, volume 5 of *Proc. Sympos. Pure Math.*, pages 135–138. American Mathematical Society, Providence, RI, 1962. (Third printing, with corrections, 1979).

(21) M. Davis. Unsolvable problems: A review. In Jerome Fox and Rose Meyerson, editors, *Mathematical Theory of Automata: Proc. Symp. Math. Th. Aut., New York, N.Y., April 24,25,26, 1962*, volume 12 of *Microwave Research Institute symposia*, pages 15–22. Polytechnic Press, 1963.

(22) M. Davis and Hilary Putnam. Diophantine sets over polynomial rings. *Illinois J. Math.*, 7(2):251–256, 1963. Russian transl. in [1, pp. 85–90].

(23) M. Davis. Extensions and corollaries of recent work on Hilbert's tenth problem. *Illinois J. Math.*, 7(2):246–250, 1963. Russian transl. in [1, pp. 80–84].

(24) M. Davis. Eliminating the irrelevant from mechanical proofs. In *Proc. Symp. Appl. Math.*, volume 15, pages 15–30, Providence, RI, 1963. AMS. Reprinted in [5, pp. 315–330]; Russian transl. in [2, pp. 160–179].

(25) Thomas J. Chinlund, M. Davis, Peter G. Hinman, and Malcolm Douglas McIlroy. Theorem-proving by matching. Technical report, Bell Telephone Laboratories, Incorporated, Murray Hill, New Jersey, 1964.

(26) M. Davis, editor. *The Undecidable—Basic Papers in Undecidable Propositions, Unsolvable Problems and Computable Functions*. Raven Press, New York, 1965. Corrected reprinted edition, Dover Publications, Mineola, NY, 2004.

(27) M. Davis. Meeting of the Association for Symbolic Logic. *J. Symbolic Logic*, 31(4):697–706, 1966.

(28) M. Davis. Diophantine equations and recursively enumerable sets. In Eduardo R. Caianiello, editor, *Automata Theory*, pages 146–152. Academic Press, 1966.

(29) M. Davis. Recursive functions—An introduction. In Eduardo R. Caianiello, editor, *Automata Theory*, pages 153–163. Academic Press, 1966.

(30) M. Davis. Computability. In *Proceedings of the Symposium on System Theory*, pages 127–131, Brooklyn, N.Y., 1966.

(31) M. Davis. *Lectures on Modern Mathematics*. Gordon and Breach, 1967. 196 pp.

(32) M. Davis. *A First Course in Functional Analysis*. Gordon and Breach, 1967. 110 pp., Reprinted Dover 2013.

(33) M. Davis. Recursive function theory. In Paul Edwards, editor, *Encyclopedia of Philosophy*, volume 7, pages 89–98. Macmillan and Free Press, New York, 1967.

(34) M. Davis. One equation to rule them all. Technical Report RM-5494-PR, The RAND Corporation, Santa Monica, CA, February 1968.

(35) M. Davis. One equation to rule them all. *Transactions of the New York Academy of Sciences*. Series II, 30(6):766–773, 1968.

(36) M. Davis. An explicit Diophantine definition of the exponential function. *Comm. Pure Appl. Math.*, XXIV(2):137–145, 1971.

(37) M. Davis. On the number of solutions of Diophantine equations. *Proc. Amer. Math. Soc.*, 35(2):552–554, 1972.

(38) M. Davis and Reuben Hersh. Nonstandard analysis. *Scientific American*, 226:78–86, 1972.

(39) M. Davis and R. Hersh. Hilbert's 10th problem. *Scientific American*, 229:84–91, 1973. Reprinted in [4, pp. 555–571] and in [6, pp. 136–148]; Italian translation [3, pp. 138–146].

(40) M. Davis. Hilbert's tenth problem is unsolvable. *Amer. Math. Monthly*, 80(3):233–269, 1973. Reprinted with corrections as Appendix 2 of the Dover edition of *Computability and Unsolvability* (7, pp. 199–235).

(41) M. Davis. Speed-up theorems and Diophantine equations. In Randall Rustin, editor, *Courant Computer Science Symposium 7: Computational Complexity*, pages 87–95. Algorithmics Press, Inc., New York, NY, 1973.

(42) M. Davis. Computability, 1973–1974. Notes by Barry Jacobs, Courant Institute of Mathematical Sciences, New York University, 1974. Based on the course "Computers and computability" as given at the Courant Institute of Mathematical Sciences, in the Fall 1973 semester.

(43) M. Davis, Yuri Matijasevič, and Julia Robinson. Hilbert's tenth problem. Diophantine equations: positive aspects of a negative solution. In *Mathematical Developments Arising From Hilbert Problems*, volume 28 of *Proc. Sympos. Pure Math.*, pages 323–378, Providence, RI, 1976. AMS. Reprinted in [14, pp. 269–378].

(44) M. Davis. *Applied nonstandard analysis*. John Wiley & Sons, Inc., 1977. Reprinted with corrections Dover, 2005. Russian translation, Izdatel'stvo Mir, Moscow 1980. Japanese translation 1977.

(45) M. Davis. Unsolvable problems. In Jon Barwise, editor, *Handbook of Mathematical Logic*, pages 567–594. North-Holland, Amsterdam, 1977.

(46) M. Davis. A relativity principle in quantum mechanics. *Internat. J. Theoret. Phys*, 16(11):867–874, 1977.

(47) M. Davis and Jacob T. Schwartz. Correct-program technology / Extensibility of verifiers—Two papers on Program Verification with Appendix of Edith Deak. Technical Report No. NSO-12, Courant Institute of Mathematical Sciences, New York University, September 1977.

(48) M. Davis. What is a computation? In Lynn Arthur Steen, editor, *Mathematics Today—Twelve Informal Essays*, pages 241–267. Springer-Verlag, 1978.

(49) M. Davis and Jacob T. Schwartz. Metamathematical extensibility for theorem verifiers and proof-checkers. *Comput. Math. Appl.*, 5:217–230, 1979. Also in (47, pp. 120–146).

(50) M. Davis. The mathematics of non-monotonic reasoning. *Artif. Intell.*, 13(1–2):73–80, 1980.

(51) M. Davis. Obvious logical inferences. In *Proceedings of the 7th IJCAI* Volume 1, pages 530–531, San Francisco, CA, USA, 1981. Morgan Kaufmann Publishers Inc.

(52) M. D. Davis and Elaine J. Weyuker. Pseudo-oracles for non-testable programs. In *Proceedings of the ACM'81 Conference*, pages 254–257, New York, NY, USA, 1981. ACM.
(53) M. Davis, Carl Smith, and Paul Young. Introduction. *Information and Control*, 52(1):1, 1982.
(54) M. Davis, Carl Smith, and Paul Young. Introduction. *Information and Control*, 54(1/2):1, 1982.
(55) M. Davis. Why Gödel didn't have Church's thesis. *Information and Control*, 54(1/2):3–24, 1982.
(56) M. Davis. Lectures at "Atlanta State". *Annals of the History of Computing*, 4(4):370–371, 1982.
(57) M. Davis. The prehistory and early history of Automated Deduction. In Siekmann and Wrightson [5], pages 1–28.
(58) M. D. Davis and Elaine J. Weyuker. *Computability, Complexity, and Languages —Fundamentals of Theoretical Computer Science*. Academic Press, 1983.
(59) M. D. Davis and Elaine J. Weyuker. A formal notion of program-based test data adequacy. *Information and Control*, 56(1/2):52–71, 1983.
(60) M. Davis, E. G. K. López-Escobar, and Wilfried Sieg. Meeting of the Association for Symbolic Logic: Washington, D.C., 1985. *J. Symbolic Logic*, 51(4):1085–1092, 1986.
(61) M. Davis. *Church's Thesis*. In [7, vol. 1, pp. 99–100]. John Wiley & Sons, 1987.
(62) M. Davis. Mathematical logic and the origin of modern computers. In [8, pp. 137–165], 1987. Reprinted in [9, pp. 149–174].
(63) M. Davis. Influences of mathematical logic on computer science. In [9, pp. 315–326]. Spanish translation: "Influencias de la Lógica Matemática en las Ciencias de la Computación" by Facundo García Valverde available at http://www.econ.uba.ar/www/departamentos/humanidades/plan97/logica/legris/apuntes/davis.pdf.
(64) M. Davis and Rohit Parikh. Meeting of the Association for Symbolic Logic: New York City, May 1987. *J. Symbolic Logic*, 53(4):1270–1274, 1988.
(65) M. D. Davis and Elaine J. Weyuker. Metric space-based test-data adequacy criteria. *The Computer Journal*, 31(1):17–24, 1988.
(66) M. Davis. Trends in Logic: Relations with Computer Science. In [10, pp. 357–359], 1989.
(67) M. Davis. Teaching the incompleteness theorem. In [10, pp. 385–392], 1989.
(68) M. Davis. Emil Post's contributions to computer science. In [11, pp. 134–136], 1989.
(69) M. Davis. Is mathematical insight algorithmic? *Behavioral and Brain Sciences*, 13(4):659–660, 1990.
(70) M. Davis. Review: Kurt Gödel, Collected Works, vol. 1. *J. Symbolic Logic*, 55(1):340–348, 1990.
(71) M. Davis. Review: Richard L. Epstein and Walter A. Carnielli, Computability; (computable functions, logic, and the foundations of mathematics). *Bull. Amer. Math. Soc. (N.S.)*, 25(1):106–111, 1991.

(72) M. Davis and Ronald Fechter. A free variable version of the first-order predicate calculus. *J. Logic Comput.*, 1(4):431–451, 1991.
(73) M. Davis. How subtle is Gödel's theorem? More on Roger Penrose. *Behavioral and Brain Sciences*, 16:611–612, 9 1993.
(74) M. Davis. First order logic. In [12, pp. 31–65], 1993.
(75) M. Davis. Foreword to [13], pages xiii–xvii. 1993.
(76) M. Davis. Lecture notes in Logic, Courant Institute of Mathematical Sciences, New York University, 1993.
(77) M. D. Davis, Ron Sigal, and Elaine J. Weyuker. *Computability, Complexity, and Languages —Fundamentals of Theoretical Computer Science*. Academic Press, 2nd edition, 1994.
(78) Emil Leon Post and M. Davis. *Solvability, Provability, Definability: The collected works of Emil L. Post.* Contemporary mathematicians. Birkhäuser, Boston, 1994. Including M. Davis's article "Emil L. Post: His life and work".
(79) M. Davis. American logic in the 1920s. *Bull. Symbolic Logic*, 1(3):273–278, 1995.
(80) M. Davis. The collaboration in the United States. In [15, pp. 91–97], 1996.
(81) M. Davis. From logic to computer science and back. In [16, pp. 53–85], 1999.
(82) M. Davis. *The Universal Computer: The Road from Leibniz to Turing*. W.W. Norton, 2000. Turing Centenary Edition, CRC Press, Taylor & Francis 2012.
(83) M. Davis. *Engines of Logic: Mathematicians and the Origin of the Computer*. W.W. Norton, 2001. Paperpack edition of (82).
(84) M. Davis. The early history of Automated Deduction. In [17, pp. 3–15], 2001.
(85) M. Davis. The myth of hypercomputation. In Christof Teuscher, editor, *Alan Turing: Life and Legacy of a Great Thinker*, pages 195–211. Springer Berlin Heidelberg, 2004.
(86) M. Davis. An appreciation of Bob Paige. *Higher-Order and Symbolic Computation*, 18(1–2):13–13, 2005.
(87) M. Davis. What did Gödel believe and when did he believe it? *Bull. Symbolic Logic*, 11(2):194–206, 2005.
(88) M. Davis. The Church-Turing thesis: Consensus and opposition. In [18, pp. 125–132], 2006.
(89) M. Davis. Why there is no such discipline as hypercomputation. *Appl. Math. Comput.*, 178(1):4–7, 2006. (Special issue on Hypercomputation, edited by F. A. Doria and J. F. Costa).
(90) M. Davis. SAT: Past and future. In [19, pp. 1–2], 2007.
(91) M. Davis. Inexhaustibility: A Non-Exhaustive Treatment by Torkel Franzén; Gödel's Theorem: An Incomplete Guide to Its Use and Misuse by Torkel Franzén. *Amer. Math. Monthly*, 115(3):270–275, 2008.
(92) M. Davis. I fondamenti dell'aritmetica. In Claudio Bartocci and Piergiorgio Odifreddi, editors, *La matematica —Problemi e teoremi,* Volume II, Grandi Opere, pages 33–60. Einaudi, 2008.
(93) George Paul Csicsery. *Julia Robinson and Hilbert's Tenth Problem.* Zala Films, Oakland, CA, 2008, a documentary film with M. Davis, Yu. Matiya-

sevich, H. Putnam. See http://www.ams.org/ams/julia.html and http://www.zalafilms.com/films/juliarobinson.html.

(94) Allyn Jackson. Interview with Martin Davis. *Notices Amer. Math. Soc.*, 55(5):560–571, 2008.

(95) M. Davis. Sex and the mathematician: The high school prom theorem. *Games and Economic Behavior*, 66(2):600, 2009.

(96) M. Davis. Diophantine equations and computation. In [20, pp. 4–5], 2009.

(97) M. Davis. Representation theorems for r.e. sets and a conjecture related to Poonen's larges subring of $\mathbb{Q}$. *Zapiski Nauchnykh Seminarov Peterburgskogo Otdeleniya Matematicheskogo Instituta im. V.A. Steklova RAN (POMI)*, 377:50–54, 2010. Reproduced as (98).

(98) M. Davis. Representation theorems for recursively enumerable sets and a conjecture related to Poonen's large subring of $\mathbb{Q}$. *J. Math. Sci. (N. Y.)*, 171(6):728–730, 2010. Reproduction of (97).

(99) M. Davis. Il decimo problema di Hilbert: equazioni e computabilità. In Claudio Bartocci and Piergiorgio Odifreddi, editors, *La matematica—Pensare il mondo,* Volume IV, Grandi Opere, pages 135–160. Einaudi, 2010.

(100) M. Davis. Foreword to [21], pages vii–viii. 2011.

(101) M. Davis. Pragmatic Platonism. https://foundationaladventures.files.wordpress.com/2012/01/platonic.pdf, 2012. (A slightly revised version inside this book, pp. 349–356.)

(102) M. Davis. Three proofs of the unsolvability of the Entscheidungsproblem. In Cooper & Leeuwen, editor, *Alan Turing: His Work and Impact*, pages 49–52. Elsevier, 2012.

(103) M. Davis. Foreword to [22], 2012.

(104) M. Davis. Computability and Arithmetic. In [23, pp. 35–54], 2013.

(105) M. Davis. Logic and the development of the computer. In [24, pp. 31–38], 2014.

(106) M. Davis. Preface to [25], pp. v–viii, 2016.

(107) M. Davis and Wilfried Sieg. Conceptual Confluence in 1936: Post & Turing. In Sommaruga and Strahm [25, pp. 3–27], 2016.

(108) M. Davis. Algorithms, equations, and logic. In [26, pp. 4–19], 2016.

(109) M. Davis. Universality is ubiquitous. In [27].

References

1. *Matematika*, 8(5), 1964. (MR 24 #A3061).
2. *Kiberneticheskiy sbornik. Novaya seriya*, 7, 1970.
3. *Le Scienze Italian edition of Scientific American*, 66, 1974.
4. Abbott, J. C. (Ed.). (1978). *The Chauvenet papers*, vol. 2. Mathematical Association of America.
5. Siekmann, J., & Wrightson, G. (Eds.). (1983). *Automation of Reasoning 1: Classical Papers on Computational Logic 1957–1966*. Berlin, Heidelberg: Springer.

6. Campbell, D. M., & Higgins, J. C. (Eds.). (1984). *Mathematics: People, Problems, Results* (Vol. 2). Belmont, CA: Wadsworth International.
7. Shapiro, S. C., & Eckroth, D. (Eds.). (1987). *Encyclopedia of Artificial Intelligence*. John Wiley & Sons.
8. Phillips, E. R. (Ed.). (1987). *Studies in the History of Mathematics*, (Vol. 26). Mathematical Association of America.
9. Herken, R. (Eds.). (1988). *The Universal Turing Machine—A half-century survey*. Verlag Kammerer & Unverzagt, Hamburg, Berlin 1988, Oxford University Press. (Springer 2nd edition, 1995).
10. Ferro, R., Bonotto, C., Valentini, S., & Zanardo, A. (Eds.). (1989). *Logic Colloquium 1988: Proceedings*. (Elsevier North Holland Publishing Co., Amsterdam).
11. *Proceedings of the Fourth Annual Symposium on Logic in Computer Science (LICS '89), Pacific Grove, California, USA, June 5-8, 1989*. IEEE Computer Society Press, Washington, D.C., 1989.
12. Gabbay, D. M., Hogger, C. J., & Robinson, J. A. (Eds.). (1993). *Handbook of Logic in Artificial Intelligence and Logic Programming. Vol. 1: logical foundations*. Clarendon Press, Oxford.
13. Matiyasevich, Yu. V. (1993). *Hilbert's tenth problem*. The MIT Press, Cambridge (MA) and London, pp. xxii+264.
14. J. Robinson. (1996). *The collected works of Julia Robinson*, volume 6 of *Collected Works*. AMS, Providence, RI, pp. xliv+338 . ISBN 0-8218-0575-4. (With an introduction by Constance Reid. Edited and with a foreword by Solomon Feferman).
15. Reid. C. (1996). Julia: A life in mathematics. (With contributions by I. L. Gaal, M. Davis, and Yu. V. Matijasevich). MAA Spectrum. The Mathematical Association of America, Washington, DC.
16. Calude, C. S. (Ed.). (1999). *People & ideas in theoretical computer science*. Discrete Mathematics and Theoretical Computer Science. Singapore; New York: Springer.
17. Robinson, J. A., & Voronkov A. (Eds.). (2001). *Handbook of Automated Reasoning (in 2 volumes)*. Elsevier and MIT Press.
18. Beckmann, A., Berger, U., Löwe, B., & Tucker, J. V. (Eds.). (2006). Logical Approaches to Computational Barriers. In *Second Conference on Computability in Europe, CiE 2006*, Swansea, UK, June 30–July 5, 2006, Proceedings, volume 3988 of *Lecture Notes in Computer Science*. Springer.
19. Marques-Silva J., & Sakallah, K. A. (Eds.). (2007). *Theory and Applications of Satisfiability Testing—SAT 2007, 10th International Conference*, Lisbon, Portugal, May 28–31, 2007, Proceedings, volume 4501 of *Lecture Notes in Computer Science*. Springer.
20. Calude, C. S., Costa, J. F., Dershowitz, N., Freire, E., & Rozenberg, G. (Eds.). (2009). *Unconventional Computation, 8th International Conference, UC 2009*, Ponta Delgada, Azores, Portugal, September 7–11, 2009. Proceedings, volume 5715 of *Lecture Notes in Computer Science*. Springer.
21. Schwartz, J. T., Cantone, D., & Omodeo, E. G. (2011). *Computational Logic and Set Theory—Applying Formalized Logic to Analysis*. Springer.
22. Turing, S. (2012). *Alan M. Turing: Centenary Edition*. New York, NY, USA: Cambridge University Press.
23. Jack Copeland, B., Posy, C. J., & Shagrir, O. (Eds.). (2013). *Computability: Turing, Gödel, Church, and beyond*. MIT Press.
24. Siekmann, J. H. (Ed.). (2014). *Computational logic*, volume 9 of Handbook of the History of Logic. Elsevier.
25. Sommaruga, G., & Strahm, T. (Eds.). (2016). *Turing's Revolution—The Impact of His Ideas about Computability*. Basel: Birkhäuser.
26. Barry Cooper, S., & Hodges, A. (Ed.). (2016). *The Once and Future Turing—Computing the World*. Cambridge University Press.
27. Bokulich, A., & Floyd, J. (Ed.). (2012). *Philosophical explorations of the legacy of Alan Turing—Turing 100 (An anthology based on the Turing 100 meeting in November 2012 at Boston University)*. Boston Studies in the Philosophy and History of Science. Springer Verlag, forthcoming.

Appendix A
"Feasible Computational Methods in the Propositional Calculus", the Seminal Report by M. Davis and H. Putnam

> "Our report for the NSA, entitled *Feasible Computational Methods in the Propositional Calculus* is dated October 1958. It emphasizes the use of conjunctive normal form for satisfiability testing (or, equivalently, the dual disjunctive normal form for tautology testing). The specific reduction methods whose use together have been linked to the names Davis-Putnam are all present in this report." (M. Davis, this volume p. 15)

> "The DPLL procedure, even half a century after its introduction, remains a foundational component of modern day SAT solvers. Through SAT solvers $\cdots$, as well as through satisfiability modulo theory $\cdots$ and answer set programming $\cdots$ solvers that build upon SAT techniques, the DPLL procedure has had a tremendous practical impact on the field with applications in a variety of areas such as formal verification, AI planning, and mathematical discovery." (D. Loveland et al., this volume p. 316)

A research report which Martin Davis and Hilary Putnam jointly wrote in 1958 is faithfully reproduced in this appendix, where it gets published for the first time; three papers which promoted its wide influence on later research in the field of automated deduction are:

[1] *A Computing procedure for Quantification Theory* by Davis and Putnam, 1960 (a typescript of which appears inside a research report of 1959);
[2] *A Machine Program for Theorem-Proving* by Davis, George Logemann, and Donald W. Loveland, 1962 (preprinted as a research report in 1961);
[3] *Eliminating the Irrelevant from Mechanical Proofs* by Davis alone, 1963 (translated into Russian in 1970).

The 1958 report tackles propositional calculus from a broader angle than the subsequent papers just cited. Its first part discusses the advantage of putting formulas into some normal form (such as the Gégalkine polynomial form); it notes that not all normal forms have the same properties and argues that *conjunctive* normal form is usually preferable to the disjunctive one for treatment by satisfiability testing methods; moreover, it adds, it is convenient to have 'more than one such method available $\cdots$, since in case one method fails one can then always attempt the others'.

E.G. Omodeo and A. Policriti (eds.), *Martin Davis on Computability, Computational Logic, and Mathematical Foundations*,
Outstanding Contributions to Logic 10, DOI 10.1007/978-3-319-41842-1

In [1], Davis and Putnam will propose a satisfiability decision algorithm for propositional formulas in conjunctive normal form. This will consist of three rules drawn or adapted from the kit of rules described in the 1958 report. Two of these, named *elimination of one-literal clauses* and *affirmative-negative rule*, will be retained also in [2, 3]. The third, named *rule for eliminating atomic formulas*, will be replaced in [2, 3] by another one called, ever since, the *splitting rule*. The new rule is theoretically equivalent to the superseded one but preferable, for practical reasons duly explained in [2].

The revised procedure has today acquired great renown under the name DPLL (from the initials of its inventors). The two interchanged rules were specified in [1, 2], respectively, in the following terms:

Rule for Eliminating Atomic Formulas. Let the formula F [given in conjunctive normal form] be put into the form $(A \vee p)$ & $(B \vee \bar{p})$ & R, where A, B, and R are free of p. (This can be done simply by grouping together the clauses containing p and "factoring out" occurrences of p to obtain A, grouping the clauses containing $\bar{p}$ and "factoring out" $\bar{p}$ to obtain B, and grouping the remaining clauses to obtain R.) Then F is inconsistent if and only if $(A \vee B)$ & R is inconsistent.

Splitting Rule. Let the given formula F be put in the form $(A \vee p)$ & $(B \vee \bar{p})$ & R, where A, B, and R are free of p. Then F is inconsistent if and only if A & R and B & R are both inconsistent.

In the 1958 report as reproduced below, these rules are specified in dual form (that is, they refer to tautology testing of an F which is given in disjunctive normal form) and they bear the respective names '*rule for eliminating variables*' and '*rule of case analysis*'; the latter, in particular, reads:

> Let F' and F'' be obtained from F by substituting 0 and 1 respectively for all occurrences of p in F and making the obvious simplifications. Then F is a tautology if and only if F' and F'' both are.

The second part of the 1958 Davis-Putnam report presents a Gentzen-type system for a version of propositional calculus embodying the exclusive disjunction connective. It shows that such a system is complete and that it remains such if the cut inference rule is withdrawn; through this system, another decision procedure alternative to truth-table methods is obtained.

Davis and Putnam were showing so much interest in *feasible computational methods* for propositional calculus ten years before the notion of **NP**-completeness began to emerge and Stephen Cook brought the propositional satisfiability problem under the spotlight of the very challenging question as to whether $\mathbf{P} = \mathbf{NP}$.

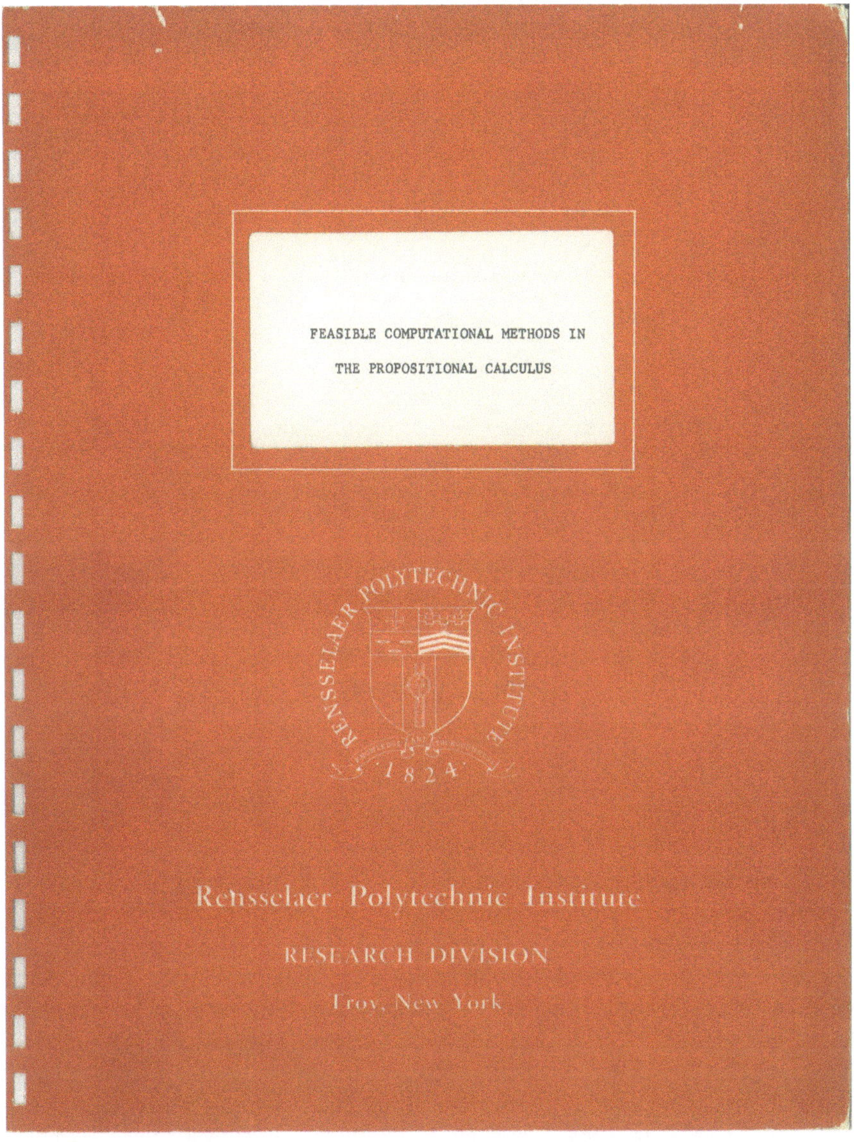
FEASIBLE COMPUTATIONAL METHODS IN

THE PROPOSITIONAL CALCULUS

RENSSELAER POLYTECHNIC INSTITUTE

1824

Rensselaer Polytechnic Institute

RESEARCH DIVISION

Troy, New York

Feasible Computational Methods in the Propositional Calculus

Submitted by

Martin Davis
Associated Professor of Mathematics
Rensselaer Polytechnic Institute
Hartford Graduate Division

and

Hilary Putnam
Assistant Professor of Philosophy
Princeton University

October 1958

It is a consequence of a well-known result of Church and Turing that there is no effective calculating procedure for determining whether or not a given schema of elementary quantification theory is or is not valid. But if we restrict ourselves to schemata of the propositional calculus, then, as has long been known (since Post's thesis of 1921, cf. Post [1]), an effective calculating procedure does exist, namely the method of truth-tables. Unfortunately, this solution of the decision problem for the propositional calculus is mainly of theoretical interest. Since the truth-table for a formula containing n variables must contain 2^n lines, truth-table methods are quite unfeasible for use in practical computation once the number of variables becomes at all large.

The present report summarizes our investigations of alternative computational methods which could be employed where truth table methods are inapplicable. Part I of this report presents several techniques, none of which is superior to truth-table methods in *all* cases, but which are nonetheless vastly better than truth-tables in the sorts of cases which can be expected to arise in practice. In Part II, a modified version of the Gentzen-type system of Kleene [1] for propositional calculus (modified to allow exclusive disjunction as a primitive connective) is discussed. Its completeness and a version of the Gentzen Hauptsatz are derived. This is shown to lead to a decision procedure for the propositional calculus, which, however, is not particularly superior to truth-table methods. Nevertheless, we feel that this represents a promising line of investigation. For, if a complete system could be constructed which has all of the present desirable properties and which enjoys one additional property to be described below, such a system could be used to provide a decision procedure which would probably be quite useful.

Part I: Case Methods, Boolean Methods, and a Combinatorial Method for Propositional Calculus

1. *Validity and Disjunctive Normal Form*

(i) *An important special case of the decision problem for propositional calculus.* No known method of testing formulas of the propositional calculus for consistency exists which is *uniformly* superior to the truth-table method. The search for feasible methods of computation in the propositional calculus when truth-table developments are unfeasible naturally concentrates, therefore, on important special cases. The most important of these special cases will now be described, since the methods to be explained below are all presented with this case in mind.

We shall use $\sim$ for negative, $\cdot$ for conjunction, $\vee$ for inclusive disjunction, + for exclusive disjunction, $\supset$ for material implication, and $\equiv$ for material equivalence. Parentheses will be omitted wherever possible without confusion (in Part I), and where possible because of associative operations. Also, for single letters we write e.g. $\bar{p}$ for $\sim p$. Also we often omit the dot of conjunction, writing e.g. $pq\bar{r}$ for $p \cdot q \cdot \sim r$.

The principal case we consider is that of a formula or system of formulas to be tested for consistency. Let us suppose that the formulas in the system have been actually written down as a formalization of some practical problem. What specialization does this introduce?

In the first place, we can *not* assume that the number of variables is small. One formula can easily contain over 20 variables without being very long, e.g.[1] $(pqrs \vee tu) \cdot (u\bar{v} \vee w) \cdot (pabc \vee de \vee gh) \cdot (ijkl \vee mno)$. Thus, a system of, say, 10 formulas could easily contain over a hundred distinct propositional variables. But we *can* assume that the formulas are relatively short. Indeed, this is already assumed when we suppose that the system is capable of being written down by a human being. A formula or system of formulas with 100 distinct propositional variables could be as long as, say, 2^{50} symbols without having, so far as is known, any shorter normal equivalent. However, any system of formulas actually written down by human beings will presumably have far fewer than 2^{50} symbols.

We shall, therefore, concentrate on the following case: The case of a system of formulas each one of which is short enough to be "manageable" (i.e., short enough for a clerk to perform algebraic manipulations on the formula, such as putting the formula in normal form), although the number of formulas in the system may be

[1] In the present part formulas may be thought of as names of themselves.

too large for the system as a whole[2] to be "manageable" in the same sense. E.g. a conjunction of 30 formulas, each of which has 10 clauses when put in "conjunctive normal form" (explained below), may have 10^{30} clauses in its "conjunctive normal form". Thus such a system is certainly not "manageable" taken as a whole, although its individual component formulas are.

Emphasis on consistency, rather than validity, in the case of a conjunction of formulas is likewise natural. For a conjunction is valid if and only if each of its constituents is valid. Thus one can test a conjunction for validity piece by piece: but one cannot proceed in this way to test for consistency, since a conjunction of formulas may be inconsistent notwithstanding the fact that each of the formulas is consistent taken by itself. It is thus no accident that the cases actually arising of systems too complex to be dealt with by truth-tables are of *conjunctions* to be tested for *consistency* (or, dually, of *disjunctive* systems to be tested for *validity*.

(ii) *Normal forms.* The advantage of putting formulas into normal form is that we thereby impose a simple and relatively transparent structure. Unfortunately (for our purposes) not all normal forms have the same properties. Thus the "polynomial normal form" (described below) tells us whether a formula is (a) consistent, and (b) a tautology.[3] The "disjunctive normal form" tells us whether a formula is consistent: but there is no known uniform method (except the truth-table method, and methods which are in general of at least equal complexity) for determining whether or not a formula in disjunctive normal form is valid; and, finally, the "conjunctive normal form" tells us whether or not a formula is valid, but not whether or not it is consistent.

We proceed to define these normal forms and certain related terms which will be much used in the sequel.

Definition 1.1. A *literal* is a propositional variable or a single negated propositional variable.

Examples: $p, q, r, \bar{p}, \bar{q}, \bar{r}$, etc.

Definition 1.2. A *clause* is any conjunction[4] of literals appearing as a disjunctive component of a larger formula, or any disjunction of literals appearing as a conjunctive component of a larger formula.

Examples: $pq\bar{r}$ and st are clauses of $pq\bar{r} \vee st$, and $(p \vee q \vee \bar{r})$ and $(s \vee t)$ are clauses of $(p \vee q \vee \bar{r})(s \vee t)$.

Definition 1.3. A disjunction of clauses is said to be *a formula in disjunctive normal form*, provided that no propositional variable occurs both negated and not negated in any clause.

Example: $pq\bar{r} \vee st$ is a formula in disjunctive normal form.

Definition 1.4. If A is in disjunctive normal form and A is logically equivalent to B, A is called a *disjunctive normal form* of B.

---

[2]A system of formulas will, when convenient, be regarded as a single formula,—namely as the conjunction of its component formulas.

[3]We follow customary usage in referring to valid formulas of the propositional calculus as "tautologies".

[4]In Part I, disjunctions and conjunctions are permitted to have any positive integral number of constituents. E.g. we write $p \vee q \vee r$ not restricting ourselves to $(p \vee q) \vee r$ and $p \vee (q \vee r)$. A single literal may also be considered as a disjunction or conjunction (with one member).

Example: $pq \vee \bar{p}\bar{q}$ is a disjunctive normal form of $p \equiv q$.

Definition 1.5. A conjunction of clauses is said to be *a formula in conjunctive normal form*, provided that no propositional variable occurs both negated and not negated in any clause.

Example: $(p \vee q \vee \bar{r})(s \vee \bar{t})$ is a formula in conjunctive normal form.

Definition 1.6. If A is in conjunctive normal form and A is logically equivalent to B, then A is called a *conjunctive normal form of* B.

Example: $(p \vee \bar{q})(q \vee \bar{p})$ is a conjunctive normal form of $p \equiv q$.

For further discussion of conjunctive and disjunctive normal forms the reader may consult Hilbert and Ackermann [1], which also establishes the following properties: (i) a formula *has* a disjunctive normal form if and only if it is consistent; (ii) a formula *has* a conjunctive normal form as defined here if and only if it is *not* valid.

We define:

Definition 1.7. Let P be any polynomial with zero and one as coefficients, and with no exponents >1. P may be regarded as a formula of the propositional calculus by simply interpreting the variables as propositional variables. Such a formula P will be called a *formula in polynomial normal form.*

Definition 1.8. If A is in polynomial normal form and A is logically equivalent to B, A is called *the polynomial normal form of* B.

Example[5]: $1 + p + pq$ is the polynomial normal form of $p \supset q$.

Every formula of the propositional calculus has *unique* polynomial normal form. The polynomial normal form of a contradiction is 0; that of a tautology is 1. (These results were first obtained by Gegalkine (1927). The polynomial normal form was used by M.H. Stone (1936) to prove an important representation theorem. (Cf. Church [1], pp. 103–104.)

Procedures for putting a formula into disjunctive and conjunctive normal form may be found in Hilbert Ackermann [1]. A formula may be put into polynomial normal form by first writing it in terms of +, ·, and 1, and then making the usual algebraic simplifications. The following rules are used to eliminate exponents and coefficients >1:

$$p \cdot p \equiv p$$

$$p + p \equiv 0.$$

(That is, coefficients may be reduced modulo 2, and all non-zero exponents may be replaced by 1.)

(iii) *Consistency and conjunctive normal form.* If the given system of equations is short enough to be put into normal form, then our problem is solved. Namely, we try to put the system into (disjunctive) normal form. If the entire system "cancels" (i.e., every clause contains a propositional (variable) and its negation), then the system is inconsistent. Otherwise, a normal form of the system will be obtained, and the system is then consistent. This method does dispose of one important special case: namely, the case in which the number of formulas is small (<10). Thus, suppose we

[5]In this report, 0 and 1 denote the truth-values "falsity" and "truth", respectively.

are given a system of six formulas, each of which has six clauses in its disjunctive normal form. Suppose the system has 30 distinct propositional variables. Then a truth-table would have 2^{30} lines, which would make the truth-table method unfeasible; but, the disjunctive normal form of the whole system has only 6^6 clauses, which would be quite manageable for a modern digital computer. We shall return to this remark later. For the moment we note: *the complexity of a normal form development depends upon different factors than the complexity of a truth-table development.* The complexity of a truth-table development is a function of the number of variables; that of a disjunctive normal form development is a function of the number of formulas, and of the number of clauses in the normal form of each formula.

Our present problem, however, is how to deal with cases in which the number of formulas is too large to make it feasible to put the whole system into disjunctive normal form (or, for the same reason, into polynomial normal form). In such cases there is still one normal form we can use: namely, the *conjunctive normal form.*

That the conjunctive normal form can be employed follows from the remark that to put a whole system into conjunctive normal form we have only to put the individual formulas into conjunctive normal form. Thus, even if the system has hundreds or thousands of formulas, it can be put into the conjunctive normal form "piece by piece", without any "multiplying out." This is a feasible (if laborious) task even for *hand* computation: thus no specialization is introduced here beyond the one we have already made in supposing that the individual formulas in the system are "manageable" and that the whole system can be written down by a human being.

Henceforth, therefore, we shall be concerned with the following problem: the problem of providing methods to determine whether or not conjunctive normal form is consistent.

Also, throughout this paper we shall make free use of the following Principle of Duality:

A formula is inconsistent if and only if its dual[6] is valid.

In view of the duality principle, our problem is equivalent to the following: to find a method for determining whether or not a formula in disjunctive normal form is valid.

Since conjunctive normal form does not reveal consistency, the reader may wonder why we take the trouble to put the given system into conjunctive normal form. The answer is simply that we thereby vastly reduce the structural complexity that we have to deal with. For instance, every formula F in conjunctive normal form has the structure ABR where A is the conjunction of the clauses containing a given propositional variable (say, p), B is the conjunction of the clauses containing the negation of that variable (say, $\bar{p}$), and R is the conjunction of the remaining clauses. Moreover, it can be shown that F is inconsistent if and only if $A'R$ and $B'R$ are both inconsistent, where A' is obtained form A by deleting occurrences of p, and B' is

[6]The *dual* of a truth-table is obtained by interchanging 0 and 1 throughout the table. Connectives are *dual* if their truth-tables are dual (e.g. $\vee$ and $\cdot$ are dual; negation is self-dual). The *dual* of a formula is obtained by replacing each connective by its dual.

obtained from B by deleting occurrences of $\bar{p}$. Such regularities are hardly to be hoped for in the case of arbitrary formulas of the propositional calculus.

In summary: we shall be concerned with the problem of testing formulas in conjunctive normal form for consistency; or, dually, with the problem of testing formulas in disjunctive normal form for validity. We have seen that a solution to this problem would yield a complete solution to the special case of the decision problem that concerns us. Although a general solution to this problem has not yet been obtained, in the sequel we shall present methods that solve this problem in important special cases.

To ask whether a formula in disjunctive normal form is valid is equivalent to asking whether or not it has $p \vee \bar{p}$ as a normal form,—and, in fact, as a *shortest* normal form. Thus our problem is a special case of the problem which has been investigated by Quine cf. Quine [1], and others: *viz*, given a formula in disjunctive normal form, to find a shortest normal equivalent. The methods produced by Quine all deal with the case where truth-tables are feasible: thus they are of no help to us here. However it seems likely that further progress can be made, both in connection with the problem of "shortest normal equivalents", and in connection with the extremely interesting special case that concerns us: telling whether or not a formula in disjunctive normal form is a tautology.

(iv) *Taking advantage of chance*. The methods to be described in Part I all have the common feature that they try to "make advantage of chance." That is, there is no proof that they will be *uniformly* superior to truth-tables, but they are constructed so that in fact, in typical cases they will be vastly superior to truth-tables. In employing them one is therefore taking a "gamble": if the problem proves tractable, one will obtain a decision; but there is always the risk that the formulas obtained will prove unmanageable in length or in number, and in such a case one will have to record a failure. There is, therefore, an advantage to having more than one such method available (provided the methods are essentially different), since in case one method fails one can then always attempt the others.

2. *A Modified Case Method*

(i) *Characteristics of the method to be presented*. The method to be described in the present section is one that will frequently work in the case of even moderately large formulas (e.g. a disjunction of 100 clauses). Moreover, certain "earmarks" can be listed by means of which one can recognize a formula on which the method is likely to succeed. Namely, the method is most likely to succeed on a formula that has short ($<$10 literals) clauses, and in which some of the variables occur in many clauses. In particular if striking out, say, 10 of the variables produces a formula with some one-literal clauses (i.e. clauses consisting of a single literal), and if striking out the variable which occurs in the one-literal clause (in a manner to be described below) produces further one-literal clauses, etc., then the method is guaranteed to work.

(ii) *Elimination of one-literal clauses*. The key idea of the method is a way of eliminating one-literal clauses from a formula. Of course the given formula may not contain any one-literal clauses to begin with. Then we apply a rule of case-analysis (explained below) to eliminate, say, 10 of the variables. The resulting formulas may then contain one-literal clauses. If so, we can start applying the rule to be described. The result may contain further one-literal clauses. If so we can continue the process until we strike out the whole formula, or reduce the formula to 1, or obtain a formula with no one-literal clauses. In the latter case we can try further use of the rule of case analysis, and so on until either a decision is reached or the number of formulas to be tested becomes unmanageably large.

The following is the *rule for the elimination of one-literal clauses*:

(a) If a formula F in disjunctive normal form contains a variable p as a one-literal clause and also contains $\bar{p}$ as a one-literal clause then F may be replaced by 1. (I.e., F is a tautology).

(b) If case (a) does not apply, and if a propositional variable p appears as a clause in a formula F in disjunctive normal form, then one may modify F by striking out all clauses that contain p affirmatively,[7] and deleting all occurrences of $\bar{p}$ from the remaining clauses, thus obtaining a formula F' which is a tautology if any only if F is.

(c) If case (a) does not apply and $\bar{p}$ appears as a clause in a formula F in disjunctive normal form, then one may modify F by striking out all clauses that contain $\bar{p}$, and deleting all occurrences of p from the remaining clauses, again obtaining a formula F' which is a tautology if and if F is.

Justification. The justification of case (a) of the rule is obvious. For case (b), let the formula F be $p \vee \mathrm{A}$. Then F is clearly true when $p = 1$: Hence F is a tautology, provided F is true when $p = 0$. Substituting 0 for p in F and simplifying has the following effect: All clauses that contain p affirmatively reduce to 0 and may be deleted. All clauses that contain p negatively reduce to 1 (in case the whole clause was $\bar{p}$) or to $1B$, where B is the remainder of the clause. But there cannot be any clauses which consist of *just* $\bar{p}$ (otherwise case a) would apply); and $1B = B$. Hence the effect of substituting 0 for p in F and simplifying is just to strike out all clauses that contain p affirmatively, and delete all occurrences of $\bar{p}$ from the remaining clauses. Thus F' is a tautology if and if only if F is true whenever $p = 0$ if and only if F is a tautology. Case (c) is symmetrical to case (b).

Examples: (1) Let us put the self-distributive law of implication (i.e., the formula $(p \supset (q \supset r)) \supset ((p \supset q) \supset (p \supset r)))$ into normal form. The result is:

$$pq\bar{r} \vee p\bar{q} \vee \bar{p} \vee r$$

There are two one-literal clauses. Elimination of these leads immediately to $q \vee \bar{q} = 1$.

(2) Let us consider the formula $pq \vee \bar{p} \vee \bar{p}q\bar{r}$. Elimination of the one-literal clause yields $p \vee \bar{p}\bar{r}$, which in turn yields $\bar{r}$. Hence this formula is not valid.

(iii) *A further rule*. In addition to the elimination of one-literal clauses, there are further simplifications that can sometimes be made on formulas in disjunctive normal form. In particular, it is desirable to include as part of our method the following rule:

---

[7] Any occurrence of p which is not in the scope of negation is called an *affirmative* occurrence (in a normal formula). Any occurrence of $\bar{p}$ is called a *negative* occurrence of p.

The *affirmative-negative* rule: If a variable p occurs in a formula F in disjunctive normal form only affirmatively, or if p occurs only negatively, then all clauses containing p may be deleted. The resulting formula F' is a tautology if and only if F is. (If F' is empty, then F is not tautology).

Justification. Let p occur in F only affirmatively, and let F be $A \vee R$, where A is the disjunction of all the clauses that contain p. Then if F is a tautology, F is true when $p = 0$. But when $p = 0$ we have $A = 0$, and therefore $A \vee R \equiv R$ when $p = 0$. Hence, if F is a tautology, so is R. But, since $R \supset A \vee R$, if R is a tautology so is F. (If R is empty, $F = 0$ when $p = 0$, and therefore F is not a tautology.) The argument is similar when p occurs only negatively, using $p = 1$ instead of $p = 0$.

The affirmative-negative rule justifies us in restricting our attention to normal formulas in which *every variable occurs both affirmatively and negatively*: for any formula without this property can be immediately simplified by the rule.

Example: The formula $p\bar{q} \vee \bar{p}q \vee q\bar{r} \vee \bar{q}\bar{r}$ contains r only negatively. By the affirmative-negative rule it is a tautology only if $p\bar{q} \vee \bar{p}q$ is.

(iv) *The rule of case analysis*. To complete the method we add the following rule:

The rule of case analysis: Let F' and F'' be obtained from F by substituting 0 and 1 respectively for all occurrences of p in F and making the obvious simplifications. Then F is a tautology if and only if F' and F'' both are.

Justification: Obvious.

It will be observed that the rule of case analysis by itself is only an instruction for constructing a truth-table. But by using it in conjunction with the other rules presented above, one obtains decisions much more rapidly than by using truth-tables. E.g., consider the following formula:

$$pq \vee \bar{p}\bar{q} \vee pr \vee \bar{p}\bar{r} \vee qr \vee \bar{q}\bar{r}\ .$$

We have constructed this formula so that neither the rules for eliminating one-literal clauses, nor the affirmative-negative rule can be immediately applied. We begin, therefore, by applying the rule of the case analysis:

$$pq \vee \bar{p}\bar{q} \vee pr \vee \bar{p}\bar{r} \vee qr \vee \bar{q}\bar{r}$$

$(p = 0)$	$(p = 1)$
$\bar{q} \vee \bar{r} \vee qr \vee \bar{q}\bar{r}$	$q \vee r \vee qr \vee \bar{q}\bar{r}$
$\bar{r} \vee r$	$r \vee \bar{r}$
1	1

whereas a full truth-table would have had 8 rows, one application of the rule of case-analysis yields two formulas, each of which reduces to $r \vee \bar{r}$ (or $\bar{r} \vee r$) upon a single application of the rule for the elimination of one-literal clauses.

3. *Boolean Methods for the Propositional Calculus*

(1) *Boolean methods and truth-table methods*. It appears to be widely assumed that truth-table methods and Boolean methods are completely equivalent (with respect to

complexity) in the propositional calculus. In fact this is not so: case methods (truth-tables) depend in a simple exponential way on the number of propositional variables; whereas, as remarked above, Boolean methods depend on the number of calculus to be "multiplied". In the case of formulas in disjunctive normal form, this dependence assumes a simple form, given by the following:

Theorem 3.1. A formula in disjunctive normal form is valid if and only if every conjunctions formed by selecting one literal from each clause is contradictory.

Proof: Let C be a conjunction formed by choosing one literal from each clause of a formula F in disjunctive normal form. If C does not contain any propositional variable *both* affirmatively and negatively (i.e., if C is consistent) then we may falsify F simply by assigning 0 as a value to every variable that occurs affirmatively in C and 1 to each variable that occurs negatively in C. Hence F is not a tautology. Conversely, if F is not a tautology, then some assignment γ falsifies F. Then γ must make (at least) one literal in each clause take on the value 0. Thus, we obtain a C of the kind required by the theorem by simply choosing one such literal from each clause. Q.e.d.

Let m be the maximum number of literals in any clause. Suppose there are k clauses in F. Then the number of ways of choosing one literal from each clause $\leqq m^k$. Once again we observe that the exponent k is not the number of propositional variables, but rather the number of clauses.

Thus there appear to be two fundamentally different ways of ascertaining the *character* (*valid or not valid*) of formulas in disjunctive normal form: (a) *ways depending on the number of variables* (case methods); (b) *ways depending on the number of clauses*. (Boolean methods). In the preceding section, we described a modified case method, designed to "take advantage of chance" (or of the presence of short clauses.) In the present section we shall present a similarity modified Boolean method.

(ii) *Characteristics of the method to be presented.* The method to be presented is a method for eliminating variables once at a time from the given formula. The method may be continued to obtain a decision until and unless the complexity of the algebra exceeds the capacity of the computer being employed. In practical cases the complexity will frequently remain manageable for two reasons:

(1) There are cancellations which almost invariably appear at each step, and making these keeps the actual complexity far below the theoretical upper limit.

(2) The complexity depends on the number of clauses containing a given variable. In particular, if the formula has the property that any *one* variable occurs in only a few clauses, then the complexity will be low.

(iii) *Statement of the method.* The entire method is given by the following rule:

Rule for eliminating variables: Let the given formula F be put into the form $Ap \vee B\bar{p} \vee R$ where A, B, and R are free of p. (This can be done simply by grouping together the clauses containing p and deleting occurrences of p to obtain A, grouping the clauses containing $\bar{p}$ and deleting occurrences of $\bar{p}$ to obtain B, and grouping together the remaining clauses to form R.) Then F is a tautology if and only if $AB \vee R$ is a tautology.

Justification. By the rule of case-analysis, F is a tautology if and only if $A \vee R$ and $B \vee R$ are both tautologies. Hence F is a tautology if and only if $(A \vee R)(B \vee R)$ is, and $(A \vee R)(B \vee R) \equiv AB \vee R$.

In applying the rule it is necessary to "multiply out" AB in order to get the formula $AB \vee R$ back into disjunctive normal form. We see, therefore, why it is that the complexity will be small if p occurs affirmatively or negatively in only a few clauses.

It is clear in a great many cases of interest, the formula under consideration will have short (<10 literal) clauses, although the number of these may be large. In these cases, the methods described in this and the preceding section are obviously complementary: if every letter appears in a great many clauses, then by eliminating a few (<10) letters by the method of case analysis, there is a good chance that we will obtain a formula with many one-literal clauses. Hence it is advisable to use the method described in Sect. 2. If, on the other hand, any given variable occurs in only a few clauses, the method of the present section is indicated.

Example:

$$pr \vee p\bar{s} \vee \bar{p}s \vee \bar{p}\bar{r} \vee s\bar{r} \vee \bar{s}r$$

$$(r \vee \bar{s})\, p \vee (\bar{r} \vee s)\, \bar{p} \vee s\bar{r} \vee \bar{s}r$$

$$sr \vee \bar{s}\bar{r} \vee s\bar{r} \vee \bar{s}r \quad (p \text{ eliminated})$$

$$(s \vee \bar{s})\, r \vee (s \vee \bar{s})\, \bar{r}$$

$$s \vee \bar{s} \quad (\bar{r} \text{ eliminated}).$$

Note that the number of steps is $2(n-1)$, where n is the number of variables.

4. *Complementary Elimination*

(i) *Characteristics of the method to be presented.* In the present section we study a novel approach to propositional calculus: the rule of *complementary elimination* proposed by Symonds and Chisholm [1]. Although we have not been able to construct a feasible decision procedure based on this approach, we believe that it is very much worthy of study. In the first place, it is a completely new combinatorial approach to the propositional calculus. Just because it stands so completely apart from the usual methods, it appears likely that further studies into this and related methods may yield real insights. The results presented here cannot claim to more than scratch the surface in this direction. Even so, we have been able to extend the method so as to obtain a method which is demonstrably *complete*. In other words, if a formula *is* a tautology, then one can always *prove* that it is by our extended method. Moreover, the proof is usually quite short. In this sense, the method to be presented is feasible as a *proof* procedure if not as a decision procedure.

(ii) *The rule of complementary elimination.* In the present section we shall consider formulas in *conjunctive normal form*. The problem that will concern us is that of determining the *consequences* of such formulas. (Testing for *consistency* is obviously a special case of this problem, since a formula is inconsistent if and only if it implies both p and $\bar{p}$. This is the dual[8] of the previously considered problem of testing formulas in disjunctive normal form for validity.)

Before starting the rule of complementary elimination, it is useful to introduce one following definition:

Definition 4.1. (i) Each of $A_1, \ldots, A_n$ will be called a *disjunct* of the formula $A_1 \vee A_2 \vee \cdots \vee A_n$ (ii) A disjunct of a disjunct of a formula F is a disjunct of F. (iii) No formula is a disjunct of a formula F unless its being so follows from (i) and (ii).

The following is the rule of complementary elimination:

Rule of complementary elimination: Let $A_1, \ldots, A_n$ be given formulas of the propositional calculus. Form the disjunction $A_1 \vee A_2 \vee \cdots \vee A_n$. From this disjunction form a formula F by deleting up to $n - 1$ *complementary* pairs of disjuncts, each pair consisting of a formula and its negation. (One also deletes parentheses and occurrences of $\vee$ if necessary, so that F will be well formed.) Then F is implied by the conjunction of the premises $A_1, A_2, \ldots, A_n$.

Example: Consider the premises:

(a) $p \supset q$

(b) $q \supset r$

Writing these in terms of $\vee$ and $\sim$ we have:

$\bar{p} \vee q$

$\bar{q} \vee r$

Applying the rule of complementary elimination (c.e.) we get:

(i) $\bar{p} \vee q \vee \bar{q} \vee r$

(ii) $\bar{p} \vee r$

Thus $\bar{p} \vee r$ (or $p \supset r$) is implied by (a) and (b). This is the transitive law of material implication.

Note that (ii) is not implied by (i). In fact (i) is a tautology. Nevertheless, the rule correctly tells us that (ii) *is* implied by the conjunction of (a) and (b).

Justification: Assume the given premises $A_1, \ldots, A_n$ are true (under some assignment γ of truth-values). Then $A_1 \vee A_2 \vee \cdots \vee A_n$ contains (at least) n true disjuncts under this assignment. Deleting $n - 1$ complementary pairs removes at most $n - 1$ true disjuncts (and an equal number of false ones). Hence some true disjunct must remain. Q.e.d.

Corollary. If the elimination of $n-1$ complementary pairs reduces the disjunction $A_1 \vee A_2 \vee \cdots \vee A_n$ to the empty formula, then the premise set $A_1, A_2, \ldots, A_n$ is inconsistent.

(iii) *Extension of the method*; *Conjunction of results*. Let F be a formula in conjunctive normal form, say $F = A_1 A_2 A_3 \cdots A_n$. Then the rule of c.e. allows us to be delete up to $n - 1$ pairs consisting of a single variable and its negation from F.

[8]The *dual* of a truth-table is obtained by interchanging 0 and 1 throughout the table. Connectives are *dual* if their truth-tables are dual (e.g. $\vee$ and $\cdot$ are dual; negation is self-dual). The *dual* of a formula is obtained by replacing each connective by its dual.

However, this rule is still not strong enough to reveal all cases of implication, even between formulas in conjunctive normal form. For instance, consider the formula:

$$(p \vee q)(\bar{p} \vee \bar{q})(p \vee \bar{q})(\bar{p} \vee q)$$

This is inconsistent: Hence the disjunction $p \vee q \vee \bar{p} \vee \bar{q} \vee p \vee \bar{q} \vee \bar{p} \vee q$ should reduce to the empty formula. But the best one can get by eliminating three complementary pairs is $q \vee \bar{q}$, hardly a significant consequence.

We therefore extend the method as follows:

Rule of conjunction: (i) If $A_{i_1}, A_{i_2}, \ldots, A_{i_k}$ and $A_{j_1}, A_{j_2}, \ldots, A_{j_e}$ are subsets, not necessarily disjoint, of $A_1, \ldots, A_n$, and if F_1 is implied by $A_{i_1}A_{i_2}\cdots A_{i_k}$ while F_2 is implied by $A_{j_1}A_{j_2}\cdots A_{j_e}$, then F_1, F_2, and F_1F_2 are implied by $A_1A_2\cdots A_n$. (ii) If B is implied by A and C is implied by B, then C is implied by A.

The second part of this rule formalizes the transitive property of logical implication, while the first part of the rule allows us to conjoin formulas that are implied by some or all of the given premises, and to assert that the conjunction is implied by the given premises. Returning now to our example:

$$(p \vee q)(\bar{p} \vee \bar{q})(p \vee \bar{q})(\bar{p} \vee q)$$
$$(p \vee q)(p \vee \bar{q}) \text{ yields } p \vee p \quad (\equiv p)$$
$$(\bar{p} \vee q)(\bar{p} \vee \bar{q}) \text{ yields } \bar{p} \vee \bar{p} \quad (\equiv \bar{p})$$
$$p\bar{p} \text{ yields } \phi, \text{ the empty formula.}$$

In the first line we have used the rule of c.e. to infer $p \vee p$ from two of the given clauses, namely, from $(p \vee q)(p \vee \bar{q})$. In line ii) we have similarly inferred $\bar{p} \vee \bar{p}$ from $(\bar{p} \vee \bar{q})(\bar{p} \vee q)$. The conjunction of the results (after simplification) is $p\bar{p}$, which is immediately contradictory (this is shown in the proof by the fact that $p\bar{p}$ reduces to the empty formula by the rule of c.e.). In order to justify the simplifications made in lines (i) and (ii) we must however state explicitly the following rule:

Rule of simplification: $p \vee p$ may be replaced by p in any formula.

For a further example, let us consider:

$$(p \vee q)(\bar{p} \vee \bar{q})(p \vee r)(\bar{p} \vee \bar{r})(q \vee r)(\bar{q} \vee \bar{r}) .$$

We get:

(i) $(p \vee q)(p \vee r)(\bar{q} \vee \bar{r})$ yields $p \vee p \quad (\equiv p)$

(ii) $(\bar{p} \vee \bar{q})(\bar{p} \vee \bar{r})(q \vee r)$ yields $\bar{p} \vee \bar{p} \quad (\equiv \bar{p})$

(iii) $p\bar{p}$ yields ϕ

(iv) *Proof of completeness*. We now prove:

Theorem 4.1. If F is a contradiction in conjunctive normal form, then F can be reduced to the empty formula by using the rules of c.e., conjunction, and simplification.

Proof: By induction on the number of clauses in F.

If F consists of 1 clause, F is consistent.[9]

Assume the theorem is true whenever F has fewer than n clauses. Suppose F has n clauses. Since F is a contradiction, some variable must occur both affirmatively and negatively if F. (Otherwise we could make F true by assigning the value 1 to every variable occurring affirmatively and the value 0 to every variable occurring negatively.) Let p be such a variable. Then $F = ABR$, where A is the conjunction of the clauses containing p, B is the conjunction of the clauses containing $\bar{p}$, and R is the conjunction of the remaining clauses (R may be empty). F is thus equivalent to the formula $(A' \vee p)(B' \vee \bar{p})R$, where A' is obtained by factoring out[10] all occurrences of p from A and B' is obtained by factoring out all occurrences of $\bar{p}$ from B.

Case (a) $A' \neq 0$ and $B' \neq 0$. $A'R$ and $B'R$ must have at least one clause fewer than F. By the induction hypothesis, $A'R$ can be reduced to the empty formula by using rules of c.e., conjunction and simplification. Making the same series of applications, but starting with AR, yields a formula which must be of the form $p \vee p \vee \cdots \vee p$ (or a conjunction of such formulas[11]) and which therefore reduces to p by simplification. Similarly, making the series of applications that reduces $B'R$ to ϕ, but starting with BR, yields a formula which must be of the form $\bar{p} \vee \bar{p} \vee \cdots \vee \bar{p}$ (or a conjunction of such formulas[10]) and which therefore reduces to $\bar{p}$ by simplification. Conjoining the "p" obtained from the subset $\{A, R\}$ with the "$\bar{p}$" obtained from the subset $\{B, R\}$ yields $p\bar{p}$, which reduces to ϕ by c.e.

Case (b) $A' = 0$ and $B' = 0$. Then A must be of the form $A_1 p A_2$, where either of A_1 and A_2 may be empty. Since p follows from $\{p\}$ by c.e. (eliminating zero pairs), p follows from $\{A, R\}$ by the rule of conjunction. $\bar{p}$ follows from $\{B, R\}$ by the argument of the preceding case, and hence $p\bar{p}$ follows from $ABR = F$.

Case (c) $A' = 0$ and $B' = 0$. Symmetrical to the preceding case.

Case (d) $A' = 0$ and $B' = 0$. Then A must be of the form $A_1 p A_2$ and B must be of the form $B_1 \bar{p} B_2$, where any of A_1, A_2, B_1, and B_2 may be empty. Hence p follows from $\{A, R\}$ (as in case b) and $\bar{p}$ follows from $\{B, R\}$, so $p\bar{p}$ follows from $ABR = F$.

Q.e.d.

5. *A Complicated Example*

(i) *Using case analysis*. In the preceding sections we have given only such examples as were necessary to make clear the nature of the various methods being presented. However, we have tried these methods on a number of more complicated examples, and, in the case of every example we have constructed, extremely satisfactory results have been obtained. In the present section we shall present one such example and apply in turn each of the preceding methods.

The following is the example in question:

---

[9]Since F is in *conjunctive* normal form, and a *disjunction* of literals cannot be inconsistent.

[10]I.e., one uses the fact that $\vee$ is distributive with respect to·

[11]If the last line is a conjunction of formulas of the form $p \vee p \vee \cdots \vee p$, then it suffices to infer any one of these formulas from itself by the rule of c.e. (eliminating zero pairs); and hence (by the rule of conjunction) one can obtain a single formula of the form $p \vee p \vee \cdots \vee p$ from the whole conjunction.

$$cd \vee \bar{c}dab \vee c\bar{d}abuw \vee \bar{c}\bar{d}abuwpqrs \vee a\bar{b}uwpqrs \vee a\bar{b}uwpqrs \vee$$
$$\bar{a}\bar{b}uwpqrs \vee u\bar{w}pqrs \vee \bar{u}wpqrs \vee \bar{u}\bar{w}pqrs \vee pqr\bar{s} \vee pq\bar{r}s \vee$$
$$pq\bar{r}\bar{s} \vee p\bar{q} \vee \bar{p}q \vee \bar{p}\bar{q}$$

Of course this formula is still relatively quite short; but it should be realized that we were considering only examples which could be worked by pencil and paper (i.e., without a large-scale digital computer). Moreover, this example is already quite sufficient to illustrate the superiority of the present methods over truth-table methods. In fact a truth-table for the above formula would have over 1000 rows!

Let us begin by employing our method of modified case-analysis.

(i) (a) $p = 0$ then the last two clauses reduce to $q \vee \bar{q}$. Hence the entire formula reduces to 1 (by the rule for eliminating one-literal clauses.)

(b) $p = 1$. Then the third from the last clause reduces to $\bar{q}$, so the one-literal clause rule applies, yielding:

$$cd \vee \bar{c}dab \vee c\bar{d}abuw \vee \bar{c}\bar{d}abuwrs \vee a\bar{b}uwrs \vee \bar{a}buwrs \vee \bar{a}\bar{b}uwrs \vee$$
$$u\bar{w}rs \vee \bar{u}wrs \vee \bar{u}\bar{w}rs \vee r\bar{s} \vee \bar{r}s \vee \bar{r}\bar{s}$$

(ii) (a) $r = 0$. Then the last two clauses reduce to $s \vee \bar{s}$.

(b) $r = 1$. Then the third from the last clause reduces to $\bar{s}$, so the one-literal clause rule applies, yielding:

$$cd \vee \bar{c}dab \vee c\bar{d}abuw \vee \bar{c}\bar{d}abuw \vee a\bar{b}uw \vee \bar{a}buw \vee \bar{a}\bar{b}uw \vee$$
$$u\bar{w} \vee \bar{u}w \vee \bar{u}\bar{w}$$

(iii) (a) $u = 0$. Last two clauses become $w \vee \bar{w}$.

(b) $u = 1$. Third from last becomes $\bar{w}$. As before, we get:

$$cd \vee \bar{c}dab \vee c\bar{d}ab \vee \bar{c}\bar{d}ab \vee a\bar{b} \vee \bar{a}b \vee \bar{a}\bar{b}$$

(iv) (a) $a = 0$. Last two clauses become $b \vee \bar{b}$.

(b) $a = 1$. Third from last becomes $\bar{b}$. As before, we get:

$$cd \vee \bar{c}d \vee c\bar{d} \vee \bar{c}\bar{d}$$

(v)

$c = 0$	$c = 1$
$d \vee \bar{d}$	$d \vee \bar{d}$
1	1

Hence, our formula is valid.

(ii) *Using the Boolean method for eliminating variables*. Here is the same example worked by our second method:

(i) $c(d \vee \bar{d}abuw) \vee \bar{c}(\bar{d}ab \vee \bar{d}abuwpqrs) \vee$ (remaining clauses not containing c).
(ii) $dab \vee \bar{d}abuwpqrs \vee a\bar{b}uwpqrs \vee \bar{a}buwpqrs \vee \bar{a}\bar{b}uwpqrs \vee u\bar{w}pqrs \vee \bar{u}wpqrs \vee \bar{u}\bar{w}pqrs \vee pqr\bar{s} \vee pq\bar{r}s \vee pq\bar{r}\bar{s} \vee p\bar{q} \vee \bar{p}q \vee \bar{p}\bar{q}$. ($c$ eliminated).
(iii) $d(ab) \vee \bar{d}(abuwpqrs) \vee$ (remaining clauses not containing d).
(iv) $abuwpqrs \vee a\bar{b}uwpqrs \vee \bar{a}buwpqrs \vee \bar{a}\bar{b}uwpqrs \vee u\bar{w}pqrs \vee \bar{u}wpqrs \vee \bar{u}\bar{w}pqrs \vee pqr\bar{s} \vee pq\bar{r}s \vee pq\bar{r}\bar{s} \vee p\bar{q} \vee \bar{p}q \vee \bar{p}\bar{q}$. ($d$ eliminated).
(v) $a(buwpqrs \vee \bar{b}uwpqrs) \vee \bar{a}(buwpqrs \vee \bar{b}uwpqrs) \vee$ (remaining clauses not containing a).
(vi) $buwpqrs \vee \bar{b}uwpqrs \vee u\bar{w}pqrs \vee \bar{u}wpqrs \vee \bar{u}\bar{w}pqrs \vee pqr\bar{s} \vee pq\bar{r}s \vee pq\bar{r}\bar{s} \vee p\bar{q} \vee \bar{p}q \vee \bar{p}\bar{q}$. ($a$ eliminated).
(vii) $b(uwpqrs) \vee \bar{b}(uwpqrs) \vee$ (remaining clauses not containing b)
(viii) $uwpqrs \vee u\bar{w}pqrs \vee \bar{u}wpqrs \vee \bar{u}\bar{w}pqrs \vee pqr\bar{s} \vee pq\bar{r}s \vee pq\bar{r}\bar{s} \vee p\bar{q} \vee \bar{p}q \vee \bar{p}\bar{q}$ (b eliminated)
(ix) $u(wpqrs \vee \bar{w}pqrs) \vee \bar{u}(wpqrs \vee \bar{w}pqrs) \vee$(remaining clauses not containing u).
(x) $wpqrs \vee \bar{w}pqrs \vee pqr\bar{s} \vee pq\bar{r}s \vee pq\bar{r}\bar{s} \vee p\bar{q} \vee \bar{p}q \vee \bar{p}\bar{q}$ (u eliminated).
(xi) $w(pqrs) \vee \bar{w}(pqrs) \vee$ (remaining clauses not containing w).
(xii) $pqrs \vee pqr\bar{s} \vee pq\bar{r}s \vee pq\bar{r}\bar{s} \vee p\bar{q} \vee \bar{p}q \vee \bar{p}\bar{q}$ (w eliminated).
(xiii) $p(qrs \vee qr\bar{s} \vee q\bar{r}s \vee q\bar{r}\bar{s} \vee \bar{q}) \vee \bar{p}(q \vee \bar{q})$
(xiv) $qrs \vee qr\bar{s} \vee q\bar{r}s \vee q\bar{r}\bar{s} \vee \bar{q}$ (p eliminated).
(xv) $q(rs \vee r\bar{s} \vee \bar{r}s \vee \bar{r}\bar{s}) \vee \bar{q}$
(xvi) $rs \vee r\bar{s} \vee \bar{r}s \vee \bar{r}\bar{s}$ (q eliminated).
(xvii) $r(s \vee \bar{s}) \vee \bar{r}(s \vee \bar{s})$
(xviii) $s \vee \bar{s}$ (r eliminated).

The number of steps is $18 = 2(n-1)$ where n is the number of variables. The important thing is not the number of steps, but rather the fact that the algebra did not in any case lead to a more complicated formula than one started with. Since this method does not involve any "branching" it seems extremely attractive as a method to try when the number of variables becomes large.

Of course, in this presentation the method is purely mechanical. Actually, for example, it is quite clear that line xvi is a tautology. And, anyone working the problem would be strongly tempted to factor $pqrs$ in line viii:

$$pqrs(uw \vee u\bar{w} \vee \bar{u}w \vee \bar{u}\bar{w}) \vee pqr\bar{s} \vee pq\bar{r}s \vee pq\bar{r}\bar{s} \vee p\bar{q} \vee \bar{p}q \vee \bar{p}\bar{q} \, .$$

But since $uw \vee u\bar{w} \vee \bar{u}w \vee \bar{u}\bar{w}$ is obviously valid, this becomes

$$\begin{aligned} & pqrs \vee pqr\bar{s} \vee pq\bar{r}s \vee pq\bar{r}\bar{s} \vee p\bar{q} \vee \bar{p}q \vee \bar{p}\bar{q} \\ \equiv\ & pq(rs \vee r\bar{s} \vee \bar{r}s \vee \bar{r}\bar{s}) \vee p\bar{q} \vee \bar{p}q \vee \bar{p}\bar{q} \\ \equiv\ & pq \vee p\bar{q} \vee \bar{p}q \vee \bar{p}\bar{q} \\ \equiv\ & 1. \end{aligned}$$

This may serve to emphasize that fact that in both methods presented, any simplifying techniques may be used at any stage.

(iii) *Using complementary elimination.* To work the same example using complementary elimination, we must first write down the dual of our formula:

$$(c \vee d)(\bar{c} \vee d \vee a \vee b)(c \vee \bar{d} \vee a \vee b \vee u \vee w)$$
$$(\bar{c} \vee \bar{d} \vee a \vee b \vee u \vee w \vee p \vee q \vee r \vee s)$$
$$(a \vee \bar{b} \vee u \vee w \vee p \vee q \vee r \vee s)(\bar{a} \vee b \vee u \vee w \vee p \vee q \vee r \vee s)$$
$$(\bar{a} \vee \bar{b} \vee u \vee w \vee p \vee q \vee r \vee s)(u \vee \bar{w} \vee p \vee q \vee r \vee s)$$
$$(\bar{u} \vee w \vee p \vee q \vee r \vee s)(\bar{u} \vee \bar{w} \vee p \vee q \vee r \vee s)(p \vee q \vee r \vee \bar{s})$$
$$(p \vee q \vee \bar{r} \vee s)(p \vee q \vee \bar{r} \vee \bar{s})(p \vee \bar{q})(\bar{p} \vee q)(\bar{p} \vee \bar{q})$$

Then to prove our formula *valid* we have to show that its dual is *inconsistent*. Here is the proof using complementary elimination.

(i) $(p \vee q)(\bar{p} \vee \bar{q})$ yields $\bar{p}$
(ii) $(p \vee \bar{q})\bar{p}$ yields $\bar{q}$
(iii) $(p \vee q \vee \bar{r} \vee \bar{s})\bar{p}\bar{q}$ yields $\bar{r} \vee \bar{s}$
(iv) $(p \vee q \vee \bar{r} \vee s)\bar{p}\bar{q}$ yields $\bar{r} \vee s$
(v) $(\bar{r} \vee \bar{s})(\bar{r} \vee s)$ yields $\bar{r}$
(vi) $(p \vee q \vee r \vee \bar{s})\bar{p}\bar{q}\bar{r}$ yields $\bar{s}$
(vii) $(\bar{u} \vee \bar{w} \vee p \vee q \vee r \vee s)\bar{p}\bar{q}\bar{r}\bar{s}$ yields $\bar{u} \vee \bar{w}$
(viii) $(\bar{u} \vee w \vee p \vee q \vee r \vee s)\bar{p}\bar{q}\bar{r}\bar{s}$ yields $\bar{u} \vee w$
(ix) $(\bar{u} \vee \bar{w})(\bar{u} \vee w)$ yields $\bar{u}$
(x) $(u \vee \bar{w} \vee p \vee q \vee r \vee s)\bar{u}\bar{p}\bar{q}\bar{r}\bar{s}$ yields $\bar{w}$
(xi) $(\bar{a} \vee b \vee u \vee w \vee p \vee q \vee r \vee s)\bar{u}\bar{w}\bar{p}\bar{q}\bar{r}\bar{s}$ yields $\bar{a} \vee b$
(xii) $(\bar{a} \vee \bar{b} \vee u \vee w \vee p \vee q \vee r \vee s)\bar{u}\bar{w}\bar{p}\bar{q}\bar{r}\bar{s}$ yields $\bar{a} \vee \bar{b}$
(xiii) $(\bar{a} \vee b)(\bar{a} \vee \bar{b})$ yields $\bar{a}$
(xiv) $(a \vee \bar{b} \vee u \vee w \vee p \vee q \vee r \vee s)\bar{a}\bar{u}\bar{w}\bar{p}\bar{q}\bar{r}\bar{s}$ yields $\bar{b}$
(xv) $(\bar{c} \vee \bar{d} \vee a \vee b \vee u \vee w \vee p \vee q \vee r \vee s)\bar{a}\bar{b}\bar{u}\bar{w}\bar{p}\bar{q}\bar{r}\bar{s}$ yields $\bar{c} \vee \bar{d}$
(xvi) $(c \vee \bar{d} \vee a \vee b \vee u \vee w)\bar{a}\bar{b}\bar{u}\bar{w}$ yields $c \vee \bar{d}$
(xvii) $(\bar{c} \vee \bar{d})(c \vee \bar{d})$ yields $\bar{d}$
(xviii) $(\bar{c} \vee d \vee a \vee b)\bar{d}\bar{a}\bar{b}$ yields $\bar{c}$
(xix) $(c \vee d)\bar{c}\bar{d}$ yields ϕ. Q.e.d

Comments:

(1) In the above proof we have used the rules of conjunction in a very convenient form: At the left of any line we allowed ourselves to write the conjunction of any formulas previously obtained and/or any of the given clauses; next, we wrote the word "yields", and then, finally, whatever we obtained using the rule of c.e.

(2) It was not necessary to use simplification in the above proof.

(3) The above proof is quite short. Although at present c.e. is a *proof* procedure rather than a decision procedure, further investigation seems desirable to determine whether the above sort of proof-technique might not be mechanized.

Part II: The Gentzen Hauptsatz and Computations in the Propositional Calculus

1. *A Gentzen-Type Formal System*

We introduce a formal system as follows:

Alphabet:

p q r s p′ q′ r′ s′ p″ etc.	(propositional variables)
$\sim$ ⊃ · ∨ +	(propositional connectives)
[ ] ,	(punctuation)
→	(arrow)

Well-formed formulas (w.f.f's):
A formula (i.e. a finite sequence of the above symbols) is called *well-formed* (w.f.) if there is a finite sequence of formulas, of which it is the last, each of which is either a propositional variable, or has the form $\sim$ A where A is a preceding formula in the sequence, or has one of the forms $[A \supset B]$, $[A \cdot B]$, $[A \vee B]$, $[A + B]$ where A and B are preceding formulas in the sequence.

Formula sequence: The formula $A_1, A_2, \ldots, A_n$ is called a *formula sequence* if the A_i's are well-formed formulas. *It is not excluded that* $n = 0$. The A_i's are called the formulas of the formula sequence.

We shall use capital Greek letters to denote formula sequences.

Sequents: The formula $\Gamma \rightarrow \Theta$ is called a *sequent* if Γ and Θ are formula sequences. Here Γ is called the *antecedent*, and Θ the *succedent* of the sequent. A w.f.f. is called a *formula* of $\Gamma \rightarrow \Theta$ if it is a formula of Γ or of Θ.

E.g. the following formulas are all sequents:

$$p, \sim p \rightarrow p$$
$$[p \supset q] \rightarrow$$
$$\rightarrow q, [\sim p \vee q]$$
$$\rightarrow \quad .$$

p is a formula of the first sequent but not of the second or third.

With each w.f.f. A we associate a truth-function A° obtained by regarding the propositional variables as genuine variables having the truth values 0, 1 as their range, and interpreting the symbols as in Part I.

If S is the sequent

$$A_1, A_2, \ldots, A_n \rightarrow B_1, B_2, \ldots, B_m \,,$$

then S°

is the truth-function given by the expression:

$$[A_1^\circ \cdot A_2^\circ \cdot \ldots \cdot A_n^\circ] \supset [B_1^\circ \vee B_2^\circ \vee \ldots \vee B_m^\circ]\,.$$

Here the vacuous conjunction is taken to be the constant 1, and the vacuous disjunction the constant 0.

A w.f.f. A (or a sequent S) is called a tautology if A° (or S°) is a tautology.

Axioms. If A is any w.f.f., the sequent $A \to A$ is an axiom, and, there are no others.

Rules of Inference. The rules of inference are divided naturally into three groups as follows;

Group 1. Rule for introducing connectives.

	Introducing in succedent	in antecedent
$\sim$	$\dfrac{A, \Gamma \to \Theta}{\Gamma \to \Theta, \sim A}$	$\dfrac{\Gamma \to \Theta,\ A}{\sim A, \Gamma \to \Theta}$
$\supset$	$\dfrac{A, \Gamma \to \Theta,\ B}{\Gamma \to \Theta, [A \supset B]}$	$\dfrac{\Gamma \to \Theta, A \quad B, \Gamma \to \Theta}{[A \supset B], \Gamma \to \Theta}$
$\cdot$	$\dfrac{\Gamma \to \Theta,\ A \quad \Gamma \to \Theta, B}{\Gamma \to \Theta, [A \cdot B]}$	$\dfrac{A, \Gamma \to \Theta}{[A \cdot B], \Gamma \to \Theta}$ $\dfrac{B, \Gamma \to \Theta}{[A \cdot B], \Gamma \to \Theta}$
$\vee$	$\dfrac{\Gamma \to \Theta,\ A}{\Gamma \to \Theta, [A \vee B]}$ $\dfrac{\Gamma \to \Theta,\ B}{\Gamma \to \Theta, [A \vee B]}$	$\dfrac{A, \Gamma \to \Theta \quad B, \Gamma \to \Theta}{[A \vee B], \Gamma \to \Theta}$
$+$	$\dfrac{\Gamma \to \Theta, A \quad B, \Gamma \to \Theta}{\Gamma \to \Theta, [A + B]}$ $\dfrac{A, \Gamma \to \Theta \quad \Gamma \to \Theta,\ B}{\Gamma \to \Theta, [A + B]}$	$\dfrac{A, \Gamma \to \Theta \quad B, \Gamma \to \Theta}{[A + B], \Gamma \to \Theta}$ $\dfrac{\Gamma \to \Theta,\ A \quad \Gamma \to \Theta,\ B}{[A + B], \Gamma \to \Theta}$

Group 2. Structural Rules.

	in succedent	in antecedent
Thinning	$\dfrac{\Gamma \rightarrow \Theta}{\Gamma \rightarrow \Theta, \mathrm{C}}$	$\dfrac{\Gamma \rightarrow \Theta}{\mathrm{C}, \Gamma \rightarrow \Theta}$
Contraction	$\dfrac{\Gamma \rightarrow \Theta, \mathrm{C}, \mathrm{C}}{\Gamma \rightarrow \Theta, \mathrm{C}}$	$\dfrac{\mathrm{C}, \mathrm{C}, \Gamma \rightarrow \Theta}{\mathrm{C}, \Gamma \rightarrow \Theta}$
Interchange	$\dfrac{\Gamma \rightarrow \Lambda, \mathrm{C}, \mathrm{D}, \Theta}{\Gamma \rightarrow \Lambda, \mathrm{D}, \mathrm{C}, \Theta}$	$\dfrac{\Delta, \mathrm{D}, \mathrm{C}, \Gamma \rightarrow \Theta}{\Delta, \mathrm{C}, \mathrm{D}, \Gamma \rightarrow \Theta}$

Group 3. Cut.

$$\frac{\Pi \rightarrow \Phi, \mathrm{M} \qquad \mathrm{M}, \Sigma \rightarrow \Omega}{\Pi, \Sigma \quad \rightarrow \quad \Phi, \Omega\,.}$$

An array of sequents consisting of a finite number of rows, each containing a finite number of sequents is called a *proof* of the sequent $\Gamma \rightarrow \Theta$ if:

(1) the final row consists of the sequent $\Gamma \rightarrow \Theta$, and

(2) each sequent which occurs in a row of the proof is either an axiom or follows from formulas of the *preceding row* by one of the rules of inference.

We shall write $\vdash \Gamma \rightarrow \Theta$ to mean that there is a proof of $\Gamma \rightarrow \Theta$. $\Gamma \rightarrow \Theta$ is then called a theorem. We write $\vdash$ A (where A is a w.f.f.) to mean $\vdash \rightarrow$ A.

We shall indicate the rule for introducing, e.g. $\supset$, in the succedent, by $\rightarrow\supset$, proceeding analogously for the other connectives; likewise, e.g. $\cdot \rightarrow$ represents the rule for introducing $\cdot$ in the antecedent. T, C, I, will abbreviate thinning, contraction, and interchange, respectively. A double underline will indicate one or more uses of T, C, and/or I.

We now derive sequents which, in effect, express the truth-tables for the various connectives.

$$\cfrac{\cfrac{\mathrm{B} \rightarrow \mathrm{B}}{\mathrm{A}, \mathrm{B} \rightarrow \mathrm{B}}\ (\text{double line})}{B \rightarrow [A \supset B]}\ \rightarrow\supset$$

$$\text{Therefore: } \vdash \mathrm{B} \rightarrow [\mathrm{A} \supset \mathrm{B}] \tag{A.1}$$

$$\dfrac{\dfrac{\dfrac{\dfrac{\mathrm{A} \rightarrow \mathrm{A}}{\mathrm{A} \rightarrow \mathrm{B}, \mathrm{A}} \qquad \dfrac{\mathrm{B} \rightarrow \mathrm{B}}{\mathrm{B}, \mathrm{A} \rightarrow \mathrm{A}}}{[\mathrm{A} \supset \mathrm{B}], \mathrm{A} \rightarrow \mathrm{B}}\;\supset\rightarrow}{\mathrm{A} \rightarrow \mathrm{B}, \sim [\mathrm{A} \supset \mathrm{B}]}\;\rightarrow\sim}{\mathrm{A}, \sim \mathrm{B} \rightarrow \sim [\mathrm{A} \supset \mathrm{B}]}\;\sim\rightarrow$$

Therefore: $\mathrm{A}, \sim \mathrm{B} \rightarrow \sim [\mathrm{A} \supset \mathrm{B}]$. (A.2)

$$\dfrac{\dfrac{\dfrac{\mathrm{A} \rightarrow \mathrm{A}}{\mathrm{A} \rightarrow \mathrm{A}, \mathrm{B}}}{\rightarrow \mathrm{A}, [\mathrm{A} \supset \mathrm{B}]}\;\rightarrow\supset}{\sim \mathrm{A} \rightarrow [\mathrm{A} \supset \mathrm{B}]}\;\sim\rightarrow$$

Therefore: $\vdash \sim \mathrm{A} \rightarrow [\mathrm{A} \supset \mathrm{B}]$. (A.3)

$$\dfrac{\dfrac{\mathrm{A} \rightarrow \mathrm{A}}{\mathrm{A}, \mathrm{B} \rightarrow \mathrm{A}} \qquad \dfrac{\mathrm{B} \rightarrow \mathrm{B}}{\mathrm{A}, \mathrm{B} \rightarrow \mathrm{B}}}{\mathrm{A}, \mathrm{B} \rightarrow [\mathrm{A} \cdot \mathrm{B}]}\;\rightarrow\cdot$$

Therefore: $\vdash \mathrm{A}, \mathrm{B} \rightarrow [\mathrm{A} \cdot \mathrm{B}]$ (A.4)

$$\dfrac{\dfrac{\dfrac{\mathrm{B} \rightarrow \mathrm{B}}{[\mathrm{A} \cdot \mathrm{B}] \rightarrow \mathrm{B}}\;\cdot\rightarrow}{\rightarrow \mathrm{B}, \sim [\mathrm{A} \cdot \mathrm{B}]}}{\sim \mathrm{B} \rightarrow \sim [\mathrm{A} \cdot \mathrm{B}]}\;-$$

Therefore: $\vdash \sim \mathrm{B} \rightarrow \sim [\mathrm{A} \cdot \mathrm{B}]$ (A.5)

$$\dfrac{\dfrac{\dfrac{\mathrm{A} \rightarrow \mathrm{A}}{[\mathrm{A} \cdot \mathrm{B}] \rightarrow \mathrm{A}}\;\cdot\rightarrow}{\rightarrow \mathrm{A}, \sim [\mathrm{A} \cdot \mathrm{B}]}\;\rightarrow\sim}{\sim \mathrm{A} \rightarrow \sim [\mathrm{A} \cdot \mathrm{B}]}\;\sim\rightarrow$$

Therefore: $\vdash \sim \mathrm{A} \rightarrow \sim [\mathrm{A} \cdot \mathrm{B}]$ (A.6)

$$\dfrac{\mathrm{A} \rightarrow \mathrm{A}}{\mathrm{A} \rightarrow [\mathrm{A} \vee \mathrm{B}]}\;\rightarrow\vee$$

Therefore: $\vdash \mathrm{A} \rightarrow \mathrm{A} \vee \mathrm{B}]$. (A.7)

$$\frac{B \to B}{B \to [A \vee B]} \to \vee$$

$$\text{Therefore: } \vdash B \to A \vee B]. \tag{A.8}$$

$$\cfrac{\cfrac{\cfrac{\cfrac{\cfrac{A \to A}{A \to A,B} \qquad \cfrac{B \to B}{B \to A,B}}{[A \vee B] \to A, B} \vee\to}{\sim B, [A \vee B] \to A} \sim\to}{\sim A, \sim B, [A \vee B] \to} \sim\to}{\sim A, \sim B \to\sim [A \vee B]} \to\sim$$

$$\text{Therefore: } \vdash\sim A, \sim B \to\sim [A \vee B]. \tag{A.9}$$

$$\cfrac{\cfrac{\cfrac{A \to A}{A,B \to A} \qquad \cfrac{B \to B}{A,B \to B}}{[A + B], A, B \to} + \to}{A, B \to\sim [A + B].} \to\sim$$

$$\text{Therefore: } \vdash A, B \to\sim [A + B]. \tag{A.10}$$

$$\cfrac{\cfrac{\cfrac{A \to A}{A \to B, A} \qquad \cfrac{B \to B}{B, A \to B}}{A \to B, [A + B]} \to +}{A, \sim B \to [A + B]} \sim\to$$

$$\text{Therefore: } \vdash A, \sim B \to [A + B]. \tag{A.11}$$

A symmetrical proof gives:

$$\vdash\sim A, B \to [A + B]. \tag{A.12}$$

$$\cfrac{\cfrac{\cfrac{\cfrac{\cfrac{A \to A}{A \to A,B} \qquad \cfrac{B \to B}{B \to A,B}}{[A + B] \to A, B} + \to}{\sim B, [A + B] \to A} \sim\to}{\sim A, \sim B, [A + B] \to} \sim\to}{\sim A, \sim B \to\sim [A + B]} \to\sim$$

$$\text{Therefore: } \vdash\sim A, \sim B \to\sim [A + B]. \tag{A.13}$$

2. *A Proof of Completeness*

By a truth-value assignment γ for a w.f.f. A we understand a mapping which associates one of the truth-values 0, 1 with each of a certain finite set of propositional variables including all those occurring in A. Then, by $A^\circ(\gamma)$ we mean the value of the truth-function A° for the assignment γ. Furthermore, by A^γ we mean just A if $A^\circ(\gamma) = 1$ and $\sim$ A if $A^\circ(\gamma) = 0$. Then we have:

Theorem 2.1. (Kalmár's lemma). Let $b_1, b_2, \ldots, b_m$ be a list of propositional variables *including* all of those which occur in A, and let γ be a truth-value assignment for A. Then,

$$\vdash b_1^\gamma, b_2^\gamma, \ldots, b_m^\gamma \rightarrow A^\gamma.$$

Proof: By induction on the number n of propositional connectives occurring in A. For n = 0, A must be one of the b_i's, and the result follows at once using thinning. Let n = k + 1 where the result is assumed known for k connectives and let Γ abbreviate $b_1^\gamma, b_2^\gamma, \ldots, b_m^\gamma$.

Case 1a. A is $\sim$ B, $B^\circ(\gamma) = 1$. Then, B^γ is B, A^γ is $\sim$ A, i.e. $\sim\sim$ B and, by induction hypothesis:

$$\vdash \Gamma \rightarrow B.$$

The proof of this can be continued as follows giving the desired conclusion:

$$\dfrac{\Gamma \rightarrow B \qquad \dfrac{\dfrac{B \rightarrow B}{\sim B, B \rightarrow}\sim\rightarrow}{B \rightarrow \sim\sim B}\rightarrow\sim}{\Gamma \rightarrow \sim\sim B}\text{Cut}$$

Case 1b. A is $\sim$ B, $B^\circ(\gamma) = 0$. Then, B^γ is $\sim$ B, A^γ is A, i.e. $\sim$ B, and the induction hypothesis is precisely what needs to be shown.

Case 2a. A is [B $\supset$ C], $C^\circ(\gamma) = 1$. Then, C^γ is C, A^γ is A, i.e. [B $\supset$ C], and the induction hypothesis yields:

$$\vdash \Gamma \rightarrow C.$$

The proof of this can be continued as follows using (1) above:

$$\dfrac{\Gamma \rightarrow C \qquad C \rightarrow [B \supset C]}{\Gamma \rightarrow [B \supset C]}\text{Cut}$$

Case 2b. A is [B $\supset$ C], $B^\circ(\gamma) = 0$. Then, B^γ is $\sim$ B, A^γ is A, i.e., [B $\supset$ C], and the induction hypothesis yields:

$$\vdash \Gamma \rightarrow \sim B.$$

Continuing, using (3),

$$\frac{\Gamma \rightarrow \sim B \qquad \sim B \rightarrow [B \supset C]}{\Gamma \rightarrow [B \supset C]} \text{ Cut}$$

Case 2c. A is [B ⊃ C], $B^\circ(\gamma) = 1$, $C^\circ(\gamma) = 0$. Then, B^γ is B, C^γ is ∼ C, A^γ is ∼ A, i.e. ∼ [B ⊃ C], and the induction hypothesis yields:

$$\vdash \Gamma \rightarrow \text{ B and } \vdash \Gamma \rightarrow \sim C$$

Continuing, using (2),

$$\cfrac{\cfrac{\cfrac{\Gamma \rightarrow B \quad B, \sim C \rightarrow \sim [B \supset C]}{\Gamma, \sim C \rightarrow \sim [B \supset C]} \text{ Cut}}{\Gamma \rightarrow \sim C \qquad \sim C, \Gamma \rightarrow \sim [B \supset C]}}{\cfrac{\Gamma, \Gamma \rightarrow \sim [B \supset C]}{\Gamma, \rightarrow \sim [B > C]}} \text{ Cut}$$

Case 3a. A is [B · C], $B^\circ(\gamma) = 1$, $C^\circ(\gamma) = 1$. Then, B° is B, C° is C, A° is A, namely [B · C], and the induction hypothesis yields:

$$\vdash \Gamma \rightarrow \text{ B and } \vdash \Gamma \rightarrow C$$

Continuing, using (4),

$$\cfrac{\cfrac{\cfrac{\Gamma \rightarrow B \quad B, C \rightarrow [B \cdot C]}{\Gamma, C \rightarrow [B \cdot C]} \text{ Cut}}{\Gamma \rightarrow C \quad C, \Gamma \rightarrow [B \cdot C]}}{\cfrac{\Gamma, \Gamma \rightarrow [B \cdot C]}{\Gamma, \rightarrow [B \cdot C]}} \text{ Cut}$$

Case 3b. A is [B · C], $B^\circ(\gamma) = 0$. Then, B^γ is ∼ B, A^γ is ∼ A, namely ∼ [B · C], and the induction hypothesis yields:

$$\vdash \Gamma \rightarrow \sim B.$$

Continuing, using (6),

$$\frac{\Gamma \rightarrow \sim B \qquad \sim B \rightarrow \sim [B \cdot C]}{\Gamma \rightarrow \sim [B \cdot C]} \text{ Cut}$$

Case 3c. A is [B · C], $C^\circ(\gamma) = 0$. Handled symmetrically to case 3b, using (5).

Case 4a. A is $[\mathrm{B} \vee \mathrm{C}]$, $\mathrm{B}^\circ(\gamma) = 1$. Then, B^γ is B, A^γ is A, namely $[\mathrm{B} \vee \mathrm{C}]$, and the induction hypothesis yields,

$$\vdash \Gamma \rightarrow \mathrm{B}.$$

Continuing, using (7),

$$\frac{\Gamma \rightarrow \mathrm{B} \qquad \mathrm{B} \rightarrow [\mathrm{B} \vee \mathrm{C}]}{\Gamma \rightarrow [\mathrm{B} \vee \mathrm{C}]}\,\mathrm{Cut}$$

Case 4b. A is $[\mathrm{B} \vee \mathrm{C}]$, $\mathrm{C}^\circ(\gamma) = 1$. Handled symmetrically to case 4a, using (8).

Case 4c. A is $[\mathrm{B} \vee \mathrm{C}]$, $\mathrm{B}^\circ(\gamma) = 0$, $\mathrm{C}^\circ(\gamma) = 0$. Then, B^γ is $\sim \mathrm{B}$, C^γ is $\sim \mathrm{C}$, A^γ is $\sim \mathrm{A}$, namely $\sim [\mathrm{B} \vee \mathrm{C}]$, and the induction hypothesis yields:

$$\vdash \Gamma \rightarrow \sim \mathrm{B} \text{ and } \vdash \Gamma \rightarrow \sim \mathrm{C}.$$

Continuing, using (9),

$$\cfrac{\Gamma \rightarrow \sim \mathrm{C} \qquad \cfrac{\cfrac{\Gamma \rightarrow \sim \mathrm{B} \qquad \sim \mathrm{B}, \sim \mathrm{C} \rightarrow \sim [\mathrm{B} \vee \mathrm{C}]}{\Gamma, \sim \mathrm{C} \rightarrow \sim [\mathrm{B} \vee \mathrm{C}]}\,\mathrm{Cut}}{\sim \mathrm{C}, \Gamma \rightarrow \sim [\mathrm{B} \vee \mathrm{C}]}}{\cfrac{\Gamma, \Gamma \rightarrow \sim [\mathrm{B} \vee \mathrm{C}]}{\Gamma \rightarrow \sim [\mathrm{B} \vee \mathrm{C}].}}\,\mathrm{Cut}$$

(The steps from $\Gamma, \sim \mathrm{C} \rightarrow \sim [\mathrm{B} \vee \mathrm{C}]$ to $\sim \mathrm{C}, \Gamma \rightarrow \sim [\mathrm{B} \vee \mathrm{C}]$ and from $\Gamma, \Gamma \rightarrow \sim [\mathrm{B} \vee \mathrm{C}]$ to $\Gamma \rightarrow \sim [\mathrm{B} \vee \mathrm{C}]$ are marked with double lines.)

Case 5a. A is $[\mathrm{B} + \mathrm{C}]$, $\mathrm{B}^\circ(\gamma) = 1$, $\mathrm{C}^\circ(\gamma) = 1$. Then, B^γ is B, C^γ is C, A^γ is $\sim \mathrm{A}$, namely $\sim [\mathrm{B} + \mathrm{C}]$, and the induction hypothesis yields:

$$\vdash \Gamma \rightarrow \mathrm{B} \text{ and } \vdash \Gamma \rightarrow \mathrm{C}.$$

Continuing, using (10)

$$\cfrac{\Gamma \rightarrow \mathrm{C} \qquad \cfrac{\cfrac{\Gamma \rightarrow \mathrm{B} \qquad \mathrm{B}, \mathrm{C} \rightarrow \sim [\mathrm{B} + \mathrm{C}]}{\Gamma, \mathrm{C} \rightarrow \sim [\mathrm{B} + \mathrm{C}]}\,\mathrm{Cut}}{\mathrm{C}, \Gamma \rightarrow \sim [\mathrm{B} + \mathrm{C}]}}{\cfrac{\Gamma, \Gamma \rightarrow \sim [\mathrm{B} + \mathrm{C}]}{\Gamma \rightarrow \sim [\mathrm{B} + \mathrm{C}].}}\,\mathrm{Cut}$$

(The steps from $\Gamma, \mathrm{C} \rightarrow \sim [\mathrm{B} + \mathrm{C}]$ to $\mathrm{C}, \Gamma \rightarrow \sim [\mathrm{B} + \mathrm{C}]$ and from $\Gamma, \Gamma \rightarrow \sim [\mathrm{B} + \mathrm{C}]$ to $\Gamma \rightarrow \sim [\mathrm{B} + \mathrm{C}]$ are marked with double lines.)

Case 5b. A is $[\mathrm{B} + \mathrm{C}]$, $\mathrm{B}^\circ(\gamma) = 1$, $\mathrm{C}^\circ(\gamma) = 0$. Then B^γ is B, C^γ is C $\sim$ C, A^γ is A, namely $[\mathrm{B} + \mathrm{C}]$, and the induction hypothesis yields:

$$\vdash \Gamma \rightarrow \mathrm{B} \text{ and } \vdash \Gamma \rightarrow \sim \mathrm{C}.$$

Continuing, using (11),

$$\dfrac{\Gamma \to \sim C \qquad \dfrac{\dfrac{\Gamma \to B \qquad B, \sim C \to [B+C]}{\Gamma, \sim C \to [B+C]}\,\text{Cut}}{\sim C, \Gamma \to [B+C]}}{\dfrac{\Gamma, \Gamma \to [B+C]}{\Gamma \to [B+C].}}\,\text{Cut}$$

Case 5c. A is [B + C], $B°(\gamma) = 0$, $C°(\gamma) = 1$. Handled symmetrically to case 5b, using (12).

Case 5d. A is [B + C], $B°(\gamma) = 0$, $C°(\gamma) = 0$. Then, B^γ is $\sim$ B, C^γ is $\sim$ C, A^γ is A, namely $\sim$ [B + C], and the induction hypothesis yields:

$$\vdash \Gamma \to \sim B \text{ and } \vdash \Gamma \to \sim C.$$

Continuing, using (13),

$$\dfrac{\Gamma \to \sim C \qquad \dfrac{\dfrac{\Gamma \to \sim B \qquad \sim B, \sim C \to \sim [B+C]}{\Gamma, \sim C \to \sim [B+C]}\,\text{Cut}}{\sim C, \Gamma \to \sim [B+C]}}{\dfrac{\Gamma, \Gamma \to \sim [B+C]}{\Gamma \to \sim [B+C].}}\,\text{Cut}$$

Theorem 2.2. If $\vdash \Gamma, A \to \Delta$ and $\vdash \Gamma, \sim A \to \Delta$, then $\vdash \Gamma \to \Delta$.
Proof: We continue as follows:

$$\dfrac{\dfrac{\dfrac{\dfrac{\Gamma, A \to \Delta}{A, \Gamma \to \Delta}}{\Gamma \to \Delta, \sim A} \qquad \dfrac{\Gamma, \sim A \to \Delta}{\sim A, \Gamma \to \Delta}}{\Gamma, \Gamma \to \Delta, \Delta}\,\text{Cut}}{\Gamma \to \Delta}$$

Theorem 2.3. If A is a w.f.f. which is a tautology, then $\vdash$ A.
Proof: Let $b_1, b_2, \ldots, b_m$ be the variables which occur in A. Then, by Theorem 2.1,

$\vdash b_1^\gamma, b_2^\gamma, \ldots, b_m^\gamma, b_m^\gamma \to A$, for all γ. In particular, for any γ,
$\vdash b_1^\gamma, b_2^\gamma, \ldots, b_{m\text{-}1}^\gamma, b_m \to A$.
$\vdash b_1^\gamma, b_2^\gamma, \ldots, b_{m\text{-}1}^\gamma, \sim b_m \to A$.

Hence, by Theorem 2.2,

$$\vdash b_1^{\gamma}, b_2^{\gamma}, \cdots, b_{m\text{-}1}^{\gamma} \to A.$$

Continuing this process, we eventually obtain

$$\vdash \to A.$$

Theorem 2.4. If the sequent S is a tautology, then $\vdash$ S.
Proof: The proof divides into four cases:

Case 1. Neither antecedent nor succedent is vacuous. Let S be:

$$A_1, A_2, \cdots, A_n \quad \to \quad B_1, \cdots, B_m.$$

Since S is a tautology, the w.f.f.

$$[[A_1 \cdot [A_2 \cdot \ldots \cdot A_n] \ldots] \supset [B_1 \vee [B_2 \vee \cdots \vee B_m] \ldots]]$$

which we write [R $\supset$ S], is a tautology.
Hence, by Theorem 2.3, $\vdash \to$ [R $\supset$ S]. Then, we may continue as follows:

$$\dfrac{\to [R \supset S] \qquad \dfrac{\dfrac{R \to R}{R \to S, R} \qquad \dfrac{S \to S}{S, R \to S}}{[R \supset S], R \to S}}{R \to S}\ \text{Cut}$$

Finally, using $\cdot \to$ and $\to \vee$ the appropriate number of times, using cut as we go, we obtain $\vdash S$.

Case 2. Antecedent is vacuous; succedent is not.
Let S be

$$\to \quad B_1, \cdots, B_n$$

Then, as before

$$\vdash \to [B_1 \vee [B_2 \vee \cdots \vee B_m] \ldots],$$

and

$$\vdash \to [B_1, \cdots, B_m]$$

Case 3. Succedent is vacuous; antecedent is not. Similar to case 2.

Case 4. Sequent reduces to: $\to$.
This is not a tautology and so there is nothing to prove.

The converse of Theorem 2.4 is easily seen to hold; this shows that the present version of propositional calculus is complete.

3. *Elimination of Cut*

An examination of the proof given for the completeness of the present system makes it appear that "cut" must play a crucial role. And yet, as Gentzen first showed for a closely relate system, *the class of theorems is not diminished if* "cut" is entirely eliminated.

We begin by introducing a new rule of inference:

$$\frac{\Pi \rightarrow \Phi \qquad \Sigma \rightarrow \Omega}{\Pi, \Sigma_{\mathrm{M}} \rightarrow \Phi_{\mathrm{M}}\Omega}\ \text{Mix}$$

where the w.f.f. M is a formula of Φ and of Σ, and where Φ_{M} and Σ_{M} are the results of eliminating from Φ and Σ, respectively, *all* occurrences of M. M is then called the *mix formula*.

Theorem 3.1. If the rule "cut" is replaced by the rule "mix", the class of sequents which can be proved remains unchanged.

Proof: Since "mix" obviously preserves the property of being tautologous, it suffices to prove that each use of "cut" can be replaced by a use of "mix". But, this is easy since the "cut":

$$\frac{\Pi \rightarrow \Phi, \mathrm{M} \qquad \mathrm{M}, \Sigma \rightarrow \Omega}{\Pi, \Sigma \rightarrow \Phi, \Omega}\ \text{Cut}$$

can be accomplished as follows:

$$\frac{\dfrac{\Pi \rightarrow \Phi, \mathrm{M} \qquad \mathrm{M}, \Sigma \rightarrow \Omega}{\Pi, \Sigma_{\mathrm{M}} \rightarrow \Phi_{\mathrm{M}}, \Omega}\ \text{Mix}}{\Pi, \Sigma \rightarrow \Phi, \Omega}$$

This completes the proof.

Now we shall begin our proof that the class of provable formulas remain unchanged if the rule "mix" (and therefore, by what has just been shown, also "cut") is entirely eliminated. Let us suppose that we have a proof *which contains just one mix* and which terminates immediately after the mix. I.e. let the "bottom" of the proof be as follows:

$$\frac{\Pi \rightarrow \Phi \qquad \Sigma \rightarrow \Omega}{\Pi, \Sigma_{\mathrm{M}} \rightarrow \quad \Phi_{\mathrm{M}}, \Omega}\ \text{Mix}$$

We shall show how to replace this proof by another in which either no mixes occur, or in which the mixes which do occur are in a suitable sense simpler than the original mix.

In explaining the sense in which the new mixes will be simpler, we shall use the following terminology:

The *grade* of a mix is the number of propositional connectives in the mix formula M.

The mix is said to be of *left-rank p* in our proof if the proof contains the following structure above the mix:

$$\dfrac{\begin{array}{ccc}\Pi_{p\text{-}1} & \to & \Phi_{p\text{-}1} \\ \vdots & & \vdots \\ \Pi_2 & \to & \Phi_2 \\ \Pi_1 & \to & \Phi_1 \\ \Pi & \to & \Phi \end{array} \quad \Sigma \to \Omega}{\Pi,\ \Sigma_M \to \Phi_M,\ \Omega}\ \text{Mix}$$

where M is a formula of $\Phi_1, \Phi_2, \ldots, \Phi_{p\text{-}1}$ and is not a formula of the succedent in any sequent just above and used in the derivation of $\Pi_{p\text{-}1} \to \Phi_{p\text{-}1}$.

Right-rank is defined similarly but with reference to the antecedents of the sequents involved.

What we shall show is that the proof we are considering can be replaced by another in which, either:

(1) there are no mixes, or
(2) the mixes which occur have lower grade, and no greater left- or right-rank or
(3) the mixes which occur have lower left-rank, and no greater grade right-rank or
(4) the mixes which occur have lower right-rank and no greater grade or left-rank.

Now, once we succeed in showing this, our task will be completed. For, the process can be iterated. Since, the grade, left-rank and right-rank can each be decreased only a finite number of times, and since the process to be described can introduce only a finite number of new mixes, eventually the mix will be eliminated. In a proof with more than one mix, the top mix can be eliminated first, then the next, etc. Thus, we shall have proved:

Theorem 3.2: (Gentzen's Hauptsatz). The class of derivable sequents remains the same if the rules "cut" and "mix" are not used.

Proof: Our proof that (1), (2), (3), or (4) must occur will proceed by cases:

Case 1a. M *occurs in* Π. Then, the bottom of the proof can replaced as follows, thus eliminating the mix:

$$\dfrac{\Sigma \to \Omega}{\Pi,\ \Sigma_M \to \Phi_M,\ \Omega}$$

Case 1b. M *occurs in* Ω. Then, once again, the bottom of the proof can be made mixless:

$$\dfrac{\Pi \to \Phi}{\Pi,\ \Sigma_M \to \Phi_M,\ \Omega}$$

Case 2. M *occurs neither in* Π *nor in* Ω. *The mix is both of left-rank 1 and right-rank 1.*

Case 2a. Π → Φ *is derived by use of a structural rule.* Since M does not occur in the succedent in a sequent used in deriving Π → Φ, Φ must be of the form Θ, M, and the rule used must be a thinning as follows:

$$\dfrac{\dfrac{\Pi \to \Theta}{\Pi \to \Theta, \mathrm{M}}\,\mathrm{T} \qquad \Sigma \to \Omega}{\Pi, \Sigma_{\mathrm{M}} \to \Theta, \Omega}\,\mathrm{Mix}$$

But then, the bottom of the proof can be replaced by:

$$\dfrac{\Pi \to \Theta}{\Pi, \Sigma_{\mathrm{M}} \to \Theta, \Omega}$$

Case 2b. Σ → Ω *is derived by use of a structural rule.* As in Case 2a Σ must be of the form M, Δ and the bottom must be;

$$\dfrac{\Pi \to \Phi, \mathrm{M} \qquad \dfrac{\Delta \to \Omega}{\mathrm{M}, \Delta \to \Omega}}{\Pi, \Delta \to \Phi_{\mathrm{M}}, \Omega}\,\mathrm{Mix}$$

which can be replaced by:

$$\dfrac{\Delta \to \Omega}{\Pi, \Delta \to \Phi_{\mathrm{M}}, \Omega}$$

In the remaining subcases under Case 2, we assume that Cases 2a and 2b not arise.

Case 2c. Π → Φ *is derived by use of* →∼. Then, Σ → Ω is derived by ∼→, and the bottom of our proof must be as follows, where M is ∼ A and Φ is Θ, M and Δ is M, Σ:

$$\dfrac{\dfrac{\mathrm{A}, \Pi \to \Theta}{\Pi \to \Theta, \sim \mathrm{A}} \qquad \dfrac{\Delta \to \Omega, \mathrm{A}}{\sim \mathrm{A}, \Delta \to \Omega}}{\Pi, \Delta \to \Theta, \Omega}\,\mathrm{Mix}$$

But, we may then replace the bottom as follows, where the grade of the mix is smaller by 1:

$$\dfrac{\dfrac{\Delta \to \Omega, \mathrm{A} \qquad \mathrm{A}, \Pi \to \Theta}{\Delta, \Pi_{\mathrm{A}} \to \Omega_{\mathrm{A}}, \Theta}\,\mathrm{Mix}}{\Pi, \Delta, \to \Theta, \Omega\,.}$$

Case 2d. Π → Φ *is derived by use of* →⊃. Then, Σ → Ω is derived by ⊃→, and the bottom of our proof must be as follows, where M is [A ⊃ B]:

$$\dfrac{\dfrac{A, \Pi \to \Theta, B}{\Pi \to \Theta, [A \supset B]} \qquad \dfrac{\Delta \to \Omega, A \quad B, \Delta \to \Omega}{[A \supset B], \Delta \to \Omega}}{\Pi, \Delta \to \Theta, \Omega .}\,\text{Mix}$$

This bottom can be replaced as follows:

$$\begin{array}{c} \dfrac{\dfrac{\Delta \to \Omega, A \quad A, \Pi \to \Theta, B}{\Delta, \Pi_A \to \Omega_A, \Theta, B}\,\text{Mix} \qquad B, \Delta \to \Omega}{\Delta, \Pi_A, \Delta_B \to \Omega_A, \Theta_B, \Omega}\,\text{Mix} \\ \hline\hline \Pi, \Delta \to \Theta, \Omega \end{array}$$

where both mixes are of smaller grade.

Case 2e. $\Pi \to \Phi$ *is derived by use of* $\to \cdot$. Then $\Sigma \to \Omega$ is derived by $\cdot \to$, and the bottom of our proof must be as follows, where M is $[A \cdot B]$ and Q is either A or B.

$$\dfrac{\dfrac{\Delta \to \Theta, A \quad \Pi \to \Phi, B}{\Pi \to \Phi, [A \cdot B]} \qquad \dfrac{Q, \Delta \to \Omega}{[A \cdot B], \Delta \to \Omega}}{\Pi, \Delta \to \Theta, \Omega}\,\text{Mix}$$

The bottom can then be replaced by

$$\begin{array}{c} \dfrac{\Pi \to \Theta, Q \qquad Q, \Delta \to \Omega}{\Pi, \Delta_Q \to \Theta_Q, \Omega}\,\text{Mix} \\ \hline\hline \Pi, \Delta \to \Theta, \Omega \end{array}$$

where the mix is of lower grade.

Case 2f. Like 2e but with $\vee$ instead of $\cdot$.

Case 2g. $\Pi \to \Phi$ *is derived by use of* $\to +$. Then $\Sigma \to \Omega$ is derived by $+ \to$, and the bottom of our proof must be as follows where M is $[A + B]$

$$\dfrac{\Pi \to \Theta, [A + B] \qquad [A + B], \Delta \to \Omega}{\Pi, \Delta \to \Theta, \Omega}\,\text{Mix}$$

Examining the rules $\to +$ and $+ \to$ we see that the bottom of the proof can be replaced by one of the following:

$$\begin{array}{c} \dfrac{\Pi \to \Theta, A \qquad A, \Delta \to \Omega}{\Pi, \Delta_A \to \Theta_A, \Omega}\,\text{Mix} \\ \hline\hline \Pi, \Delta \to \Theta, \Omega \end{array}$$

or

$$\begin{array}{c} \dfrac{\Delta \to \Omega, B \qquad B, \Pi \to \Theta}{\Delta, \Pi_B \to \Omega_B, \Theta}\,\text{Mix} \\ \hline\hline \Pi, \Delta \to \Theta, \Omega \end{array}$$

or

$$\cfrac{\cfrac{\Pi \rightarrow \Theta, \mathrm{B} \qquad \mathrm{B}, \Delta \rightarrow \Omega}{\Pi, \Delta_{\mathrm{B}} \quad \rightarrow \quad \Theta_{\mathrm{B}}, \Omega}\ \mathrm{Mix}}{\Pi, \Delta \quad \rightarrow \quad \Theta, \Omega}$$

or

$$\cfrac{\cfrac{\Delta \rightarrow \Omega, \mathrm{A} \qquad \mathrm{A}, \Pi \rightarrow \Theta}{\Delta, \Pi_{\mathrm{A}} \quad \rightarrow \quad \Omega_{\mathrm{A}}, \Theta}\ \mathrm{Mix}}{\Pi, \Delta \quad \rightarrow \quad \Theta, \Omega}$$

Case 3. M *occurs neither in* Π *nor in* Ω. The mix is of left-rank >1.

Case 3a. $\Pi \rightarrow \phi$ *is derived by use of* T, C, *or* I *in the succedent and the formula* M *is the formula introduced, or the formula contracted, or one of the formulas interchanged*, respectively. Then, the bottom of the proof will be:

$$\cfrac{\cfrac{\Pi \rightarrow \Theta}{\Pi \rightarrow \Phi} \qquad \Sigma \rightarrow \Omega}{\Pi, \Sigma_{\mathrm{M}} \quad \rightarrow \quad \Phi_{\mathrm{M}}, \Omega}\ \mathrm{Mix}$$

Now, from the hypothesis it follows that Θ_{M} is identical with Φ_{M}. Hence, we may reduce the left-rank by 1 as follows:

$$\cfrac{\Pi \rightarrow \Theta \qquad \Sigma \rightarrow \Omega}{\Pi, \Sigma_{\mathrm{M}} \quad \rightarrow \quad \Phi_{\mathrm{M}}, \Omega}\ \mathrm{Mix}$$

Case 3b. $\Pi \rightarrow \Phi$ *is derived by use of* T, C, *or* I *but Case 3a does not hold.* Then, the bottom of the proof must be as follows where the letter S is T, or C, or I, whichever is appropriate:

$$\cfrac{\cfrac{\Lambda \rightarrow \Theta}{\Pi \rightarrow \Phi}\ \mathrm{S} \qquad \Sigma \rightarrow \Omega}{\Pi, \Sigma_{\mathrm{M}} \rightarrow \Phi_{\mathrm{M}}, \Omega}\ \mathrm{Mix}$$

Now, M is a formula of Θ (since otherwise the left-rank would = 1). Hence we may replace the bottom of the proof by the following where the left-rank has been decreased by 1:

$$\cfrac{\cfrac{\cfrac{\Lambda \rightarrow \Theta \qquad \Sigma \rightarrow \Omega}{\Lambda, \Sigma_{\mathrm{M}} \quad \rightarrow \quad \Theta_{\mathrm{M}}, \Omega}\ \mathrm{Mix}}{\Lambda, \Sigma_{\mathrm{M}} \rightarrow \Omega, \Theta_{\mathrm{M}}}}{\Pi, \Sigma_{\mathrm{M}} \rightarrow \Phi_{\mathrm{M}}, \Omega}$$

Case 3c. $\Pi \rightarrow \Phi$ *is derived by a one premise connective-introducing rule, introducing the formula* M *in the succedent.* Then the bottom of the proof is of the following form, where L is the rule used:

$$\dfrac{\dfrac{\Lambda_1, \Gamma \to \Theta, \Lambda_2}{\Gamma \to \Theta, \mathrm{M}}\,\mathrm{L} \qquad \Sigma \to \Omega}{\Gamma, \Sigma_{\mathrm{M}} \;\to\; \Theta_{\mathrm{M}}, \Omega}\,\mathrm{Mix}$$

By hypothesis, M is a formula of Θ, and is not a formula of Ω. Hence we may replace the above by:

$$\dfrac{\dfrac{\dfrac{\dfrac{\Lambda_1, \Gamma \to \Theta, \Lambda_2 \qquad \Sigma \to \Omega}{\Lambda_1, \Gamma, \Sigma_{\mathrm{M}} \to \Theta_{\mathrm{M}}, \Lambda_2, \Omega}\,\mathrm{Mix}}{\Lambda_1, \Gamma, \Sigma_{\mathrm{M}} \to \Theta_{\mathrm{M}}, \Omega, \Lambda_2}}{\Gamma, \Sigma_{\mathrm{M}} \to \Theta_M, \Omega, \mathrm{M}}\,\mathrm{L} \qquad \Sigma \to \Omega}{\dfrac{\Gamma, \Sigma_{\mathrm{M}}, \Sigma_{\mathrm{M}} \;\to\; \Theta_M, \Omega, \Omega}{\Gamma, \Sigma_{\mathrm{M}} \;\to\; \Theta_{\mathrm{M}}, \Omega}}\,\mathrm{Mix}$$

Case 3d. Like Case 3c, but M *is not brought into the succedent by the rule.* Then the bottom must be as follows:

$$\dfrac{\dfrac{\Lambda_1, \Gamma \to \Theta, \Lambda_2}{\Delta_1, \Lambda \to \Theta, \Delta_2}\,\mathrm{L} \qquad \Sigma \to \Omega}{\Delta_1, \Gamma, \Sigma_{\mathrm{M}} \;\to\; \Theta_{\mathrm{M}}, \Delta_2, \Omega}\,\mathrm{Mix}$$

Letting Λ_3 be either Λ_2 or empty according as Λ_2 is not or is M, we can replace the bottom by:

$$\dfrac{\dfrac{\dfrac{\Lambda_1, \Gamma \to \Theta, \Lambda_2 \qquad \Sigma \to \Omega}{\Lambda_1, \Gamma, \Sigma_{\mathrm{M}} \;\to\; \Theta_{\mathrm{M}}, \Lambda_3, \Omega}\,\mathrm{Mix}}{\dfrac{\Lambda_1, \Gamma, \Sigma_{\mathrm{M}} \;\to\; \Theta_{\mathrm{M}}, \Omega, \Lambda_2}{\Delta_1, \Gamma, \Sigma_{\mathrm{M}} \;\to\; \Theta_M, \Omega, \Delta_2}\,\mathrm{L}}}{\Delta_1, \Gamma, \Sigma_{\mathrm{M}} \;\to\; \Theta_M, \Omega_2, \Delta}$$

Case 3e. Like Case 3c, but with a two premise rule. Then, the bottom of the proof is of the following form where L is the rule used:

$$\dfrac{\dfrac{\Lambda_1, \Gamma \to \Theta, \Lambda_2 \qquad \Lambda_3, \Gamma \to \Theta, \Lambda_4}{\Gamma \to \Theta, \mathrm{M}}\,\mathrm{L} \qquad \Sigma \to \Omega}{\Gamma, \Sigma_{\mathrm{M}} \;\to\; \Theta_{\mathrm{M}}, \Omega}\,\mathrm{Mix}$$

Here M is a formula of Θ, and not of Ω. Hence, the above may be replaced by:

$$
\dfrac{\dfrac{\dfrac{\dfrac{\Lambda_1, \Gamma \rightarrow \Theta, \Lambda_2 \qquad \Sigma \rightarrow \Omega}{\Lambda_1, \Gamma, \Sigma_M \rightarrow \Theta_M, \Lambda_2, \Omega}\,\text{Mix}}{\Lambda_1, \Gamma, \Sigma_M \rightarrow \Theta_M, \Omega, \Lambda_2} \qquad \dfrac{\dfrac{\Lambda_3, \Gamma \rightarrow \Theta, \Lambda_4 \qquad \Sigma \rightarrow \Omega}{\Lambda_3, \Gamma, \Sigma_M \rightarrow \Theta_M, \Lambda_4, \Omega}\,\text{Mix}}{\Lambda_3, \Gamma, \Sigma_M \rightarrow \Theta_M, \Omega, \Lambda_4}}{\Gamma, \Sigma_M \rightarrow \Theta_M, \Omega, M}\,\text{L} \qquad \Sigma \rightarrow \Omega}{\dfrac{\Gamma, \Sigma_M, \Sigma_M \rightarrow \Theta_M, \Omega, \Omega}{\Gamma, \Sigma_M \rightarrow \Theta_M, \Omega}}\,\text{Mix}
$$

Case 3f. Like Case 3d, but with a two premise rule. Then the bottom of the proof is as follows where L is the rule used:

$$
\dfrac{\dfrac{\Lambda_1, \Gamma \rightarrow \Theta, \Lambda_2 \qquad \Lambda_3, \Gamma \rightarrow \Theta, \Lambda_4}{\Delta_1, \Gamma \rightarrow \Theta, \Delta_2}\,\text{L} \qquad \Sigma \rightarrow \Omega}{\Delta_1, \Gamma, \Sigma_M \rightarrow \Theta_M, \Delta_2, \Omega\,.}\,\text{Mix}
$$

Letting Λ_5 be Λ_2 or empty according as Λ_2 is not or is M, and Λ_6 likewise for Λ_4, we can replace the bottom by:

$$
\dfrac{\dfrac{\dfrac{\Lambda_1, \Gamma \rightarrow \Theta, \Lambda_2 \qquad \Sigma \rightarrow \Omega}{\Lambda_1, \Gamma, \Sigma_M \rightarrow \Theta, \Lambda_5\Omega}\,\text{Mix}}{\Lambda_1, \Gamma, \Sigma_M \rightarrow \Theta_M, \Omega, \Lambda_2} \qquad \dfrac{\dfrac{\Lambda_3, \Gamma \rightarrow \Theta, \Lambda_4 \qquad \Sigma \rightarrow \Omega}{\Lambda_3, \Gamma, \Sigma_M \rightarrow \Theta_M, \Lambda_6, \Omega}\,\text{Mix}}{\Lambda_3, \Gamma, \Sigma_M \rightarrow \Theta_M, \Omega, \Lambda_4}}{\dfrac{\Delta_1, \Gamma, \Sigma_M \rightarrow \Theta_M, \Omega, \Delta_2}{\Delta_1, \Gamma, \Sigma_M \rightarrow \Theta_M, \Delta, \Omega}}\,\text{L}
$$

Case 4. M *occurs neither in* Φ *nor in* Ω. *The mix is of right-rank* >1. Handled symmetrically to Case 3.

This completes the proof of the Gentzen Hauptsatz.

4. *Computational aspects of the Gentzen Hauptsatz*

Definition 4.1. A subformula of the w.f.f. A is a w.f.f. B which occurs as part of A.

Theorem 4.1. (Subformula Property). Each derivable sequent S may be proved by a proof such that all formulas occurring in sequents of the proof are subformulas of formulas of S.

Proof: Immediate from Gentzen's Hauptsatz and the effect of the rules of inference of our system other than cut.

Corollary. Any derivable sequent S may be proved by a proof in which no connectives occur that do not also occur in S.

In fact, as is easily seen, the Gentzen Haupstatz leads to a decision procedure for the propositional calculus. To see this, we construct a modified system as follows:

Axioms. C, Γ → Θ, C is an axiom.

Rules of inference.

	Introducing in succedent	in antecedent
$\sim$	$\frac{A,\Gamma\to\Theta,\sim A}{\Gamma\to\Theta,\sim A}$	$\frac{\sim A,\Gamma\to\Theta,A}{\sim A,\Gamma\to\Theta}$
$\supset$	$\frac{A,\Gamma\to\Theta,[A\supset B],B}{\Gamma\to\Theta,[A\supset B]}$	$\frac{[A\supset B],\Gamma\to\Theta,A \quad B,[A\supset B],\Gamma\to\Theta}{[A\supset B],\Gamma\to\Theta}$
	$\frac{\Gamma\to\Theta,[A\cdot B],A \quad \Gamma\to\Theta,[A\cdot B],B}{\Gamma\to\Theta,[A\cdot B]}$	$\frac{A,[A\cdot B],\Gamma\to\Theta}{[A\cdot B],\Gamma\to\Theta}$ $\frac{B,[A\cdot B],\Gamma\to\Theta}{[A\cdot B],\Gamma\to\Theta}$
$\vee$	$\frac{\Gamma\to\Theta,[A\vee B],A}{\Gamma\to\Theta,[A\vee B]}$ $\frac{\Gamma\to\Theta,[A\vee B],B}{\Gamma\to\Theta,[A\vee B]}$	$\frac{A,[A\vee B],\Gamma\to\Theta \quad B,[A\vee B],\Gamma\to\Theta}{[A\vee B],\Gamma\to\Theta}$
$+$	$\frac{\Gamma\to\Theta,[A+B],A \quad B,\Gamma\to\Theta,[A+B]}{\Gamma\to\Theta,[A+B]}$ $\frac{A,\Gamma\to\Theta,[A+B],\Gamma\to\Theta,[A+B,B]}{\Gamma\to\Theta,[A+B]}$	$\frac{A,[A+B],\Gamma\to\Theta \quad B,[A+B],\Gamma\to\Theta}{[A+B],\Gamma\to\Theta}$ $\frac{[A+B],\Gamma\to\Theta,A \quad [A+B],\Gamma\to\Theta,B}{[A+B],\Gamma\to\Theta}$

In this new formulation, there are no structural rules. *However, the antecedent and succedent are to be interpreted as finite sets rather than sequences of formulas.* Thus, no attention is to be paid to order or repetition. This convention gives the effect of I and C. It is left to the reader to verify that the change in introduction rules and in the axioms precisely gives the effect of T. In this new formulation, a decision procedure is immediate. It is only necessary to try all possible rules which apply, going backwards. This process (as is easily seen) must terminate. Then a sequent will be derivable if and only if it goes back to an axiom on at least one branch.

The decision procedure which this yields is, unfortunately, of little practical interest in its present form. This is because at each stage there will be branching corresponding to the different rules which could apply. And, this branching will produce exponentiation. It is possibly of interest that where truth-table methods lead to exponentiation on the number of variables, and normal form methods on the number of clauses, the present methods lead to exponentiation on the total number of connectives present (counting commas and $\to$).

The following is proposed as an interesting unsolved problem which would shed considerable light on the question of computation in the propositional calculus:

To construct a Gentzen-type of system which has the property that any sequent is immediately derivable from at most one other single sequent.

If such a system could be produced, branching would be eliminated, and a feasible computational method might well result.

Bibliography

CHURCH, ALONZO

[1] "Introduction to Mathematical Logic," Princeton University Press, Princeton, N.J., vol. 1, 1956.

HILBERT, DAVID, and WILHELM ACKERMANN

[1] "Grundzüge der theoretischen Logik", Springer-Verlag OHG, Berlin, 1928; 2nd ed., 1938; reprinted by Dover Publications, New York, 1946; 3rd ed., Springer, Berlin, 1949; English translation of 2nd ed. under the title "Principles of Mathematical Logic," Chelsea Publishing Company, New York, 1950.

KLEENE, STEPHEN C.

[1] "Introduction to Metamathematics," D. Van Nostrand Company, Inc. Princeton, N.J., 1952.

POST, EMIL L.

[1] "Introduction to a General Theory of Elementary Propositions," *American Journal of Mathematics*, vol. 43 (1921), pp. 163–185.

QUINE, W. V.

[1] "The Problem of Simplifying Truth Functions," *American Mathematical Monthly*, vol. 59 (1952), pp. 521–531.
SYMONDS, B. K. and R. M. CHISHOLM

[1] "Inference by Complementary Elimination," *Journal of Symbolic Logic*, vol. 22 (1957), pp. 233–236.

Appendix B
"Research on Hilbert's Tenth Problem", the Original Paper by M. Davis and H. Putnam

> "It was in the summer of 1959 that Hilary and I really hit the jackpot. We decided to see how far we could get with the approach we had used at the Logic Institute in Ithaca, if, following Julia Robinson's lead, we were willing to permit variable exponents in our Diophantine equations." (M. Davis, this volume p. 16)

The research report faithfully reproduced in this appendix was submitted for publication in 1959 by Martin Davis and Hilary Putnam, but then withdrawn, and gets published here for the first time. It proves that

> if *for every n there are n primes in arithmetic progression*, then every recursively enumerable predicate can be existentially defined in terms of polynomials and the function $y = 2^x$.

The premise of this proposition (here shown in italics), dubbed *P.A.P. hypothesis* at the time, will become a theorem in 2004 thanks to Ben Green and Terence Tao; thus the Davis-Putnam result can be seen, today, as a proof of the fact that every recursively enumerable predicate is existentially definable in terms of Diophantine 'exponential polynomials'. As recounted in Martin's autobiography in this volume, Julia Robinson accelerated the course of events by simplifying the proof found by Davis and Putnam and by making recourse to the P.A.P. hypothesis unnecessary; the three jointly published the celebrated Davis-Putnam-Robinson theorem in 1961.

To better highlight the bearing of this result on the study of Hilbert's 10th problem, let us indicate by $\mathfrak{D}$, $\mathfrak{E}$, and $\mathfrak{R}$ the collections of predicates which in the following paper are called: *Diophantine*, *existentially definable*, and *recursively enumerable*. The inclusions $\mathfrak{D} \subseteq \mathfrak{E} \subseteq \mathfrak{R}$ are readily seen; it is far from obvious, though, that they hold as equalities. Around 1950 Julia Robinson had proposed the hypothesis that a Diophantine predicate of exponential rate of growth exists (J.R.), and proved that it would imply $\mathfrak{D} = \mathfrak{E}$. At about the same time Martin Davis conjectured that $\mathfrak{D} = \mathfrak{R}$. He also found a normal form for recursively enumerable predicates that, at least superficially, seems close to the definition of $\mathfrak{D}$; namely where that definition stipulates a string of existential quantifiers, Davis's normal form interrupts that string with

E.G. Omodeo and A. Policriti (eds.), *Martin Davis on Computability, Computational Logic, and Mathematical Foundations*,
Outstanding Contributions to Logic 10, DOI 10.1007/978-3-319-41842-1

a single bounded universal quantifier. Davis and Putnam began with this normal form, and showed how the additional expressiveness made possible by the freedom to use variable exponents together with their assumption of P.A.P. made it possible to prove the Davis-Putnam-Robinson theorem, namely, $\mathfrak{E} = \mathfrak{R}$. Thus, Davis's conjecture—and, consequently, the algorithmic unsolvability of Hilbert's 10th problem—were reduced to the J.R. hypothesis, which Yuri Matiyasevich will prove in 1970.

Rensselaer Polytechnic Institute, Hartford Graduate Division.
Mathematical Sciences Directorate, Air Force Office of Scientific Research, Washington 25, D.C.

AFOSR TR59-124
A COMPUTATIONAL PROOF PROCEDURE;
AXIOMS FOR NUMBER THEORY;
RESEARCH ON HILBERT'S TENTH PROBLEM

Martin Davis
Associate Professor of Mathematics
Rensselaer Polytechnic Institute
Hartford Graduate Division
and
Hilary Putnam
Assistant Professor of Philosophy
Princeton University

Contract No.: AF 49(638)-527 October 1959

Abstract of Part III

On Hilbert's Tenth Problem

Hilbert's tenth problem is the problem of finding an algorithm for determining, given a diophantine equation, whether or not it has a solution. A closely related problem is that which arises if "diophantine equation" is taken in the following sense (which is wider than Hilbert's sense of the term): equation of the form $P = 0$, where P is a "polynomial" whose exponents may themselves be variables as well as constants, to be solved in integers. E.g., the Fermat equation $x^n + y^n = z^n$ is a "diophantine equation" in this wider sense, but not in Hilbert's sense, whereas $x^3 + y^3 = z^3$ is a diophantine equation in the Hilbert (and, so also in the wider) sense.

The present part establishes several theorems bearing on these problems. In particular, we show that *if* for every n there are n primes in arithmetic progression, then the problem of determining whether or not a diophantine equation in the "wider" sense possesses a solution is recursively unsolvable. In fact we show that if the above hypothesis about primes in arithmetic progression is true, then every recursively enumerable predicate can be existentially defined in terms of polynomials and the function $y = 2^x$. (By Julia Robinson's work, it follows that if for every n there are n primes in arithmetic progression, *and also* there exists a diophantine equation whose solutions are of "roughly exponential" rate of growth, then every recursively enumerable predicate is "diophantine".) In addition, we improve our results in [3],[12] by eliminating all but one of the "critical" predicates, there given.

Part III: On Hilbert's Tenth Problem

Hilbert's tenth problem is the problem of finding an algorithm for determining, given a diophantine equation, whether or not it has a solution. Here "diophantine equation" means equation of the form $P = 0$ (where P is a polynomial) to be solved in rational integers. A closely related problem is that which arises if "diophantine equation" is taken in the following sense (which is, of course, wider than Hilbert's sense of the term): equation of the form $P = 0$, where P is a "polynomial" whose exponents may themselves be variables as well as constants, to be solved in integers. E.g., the Fermat equation $x^n + y^n = z^n$ is a "diophantine equation" in this wider sense, but not in Hilbert's sense, whereas $x^3 + y^3 = z^3$ is a diophantine equation in the Hilbert (and, so also in the wider) sense.

Previous work (cf. bibliography at the end of this paper) has dealt with attempts to prove Hilbert's problem recursively unsolvable (i.e., to show the *non*-existence of the required algorithm), and, in effect, with the relation between these two problems. In particular, Julia Robinson, in a fundamental paper (cf. [8]) has shown that if there is any diophantine equation $P(x_1, \ldots, x_n, d) = 0$ with finitely many solutions (for a given value of d) whose solutions $x_1, \ldots, x_n$ are, in an appropriate sense, of "roughly exponential" rate of growth (in terms of d), then the decision problems for diophantine equation in the ordinary sense and that for diophantine equation in the "wider" sense are equivalent.

The present paper will be concerned with establishing several theorems bearing on these problems. In particular, we shall show that *if* for every n there are n primes in arithmetic progression, then the problem of determining whether or not a diophantine equation in the "wider" sense possesses a solution is recursively unsolvable. In fact we shall show that if the above hypothesis about primes in arithmetic progression is true, then every recursively enumerable predicate can be existentially defined in terms of polynomials and the function $y = 2^x$. (By Julia Robinson's work, it follows that if for every n there are n primes in arithmetic progression, *and also* there exists a diophantine equation whose solutions are of "roughly exponential" rate of growth,

---

[12]Cf. the bibliography at the end of Part III.

then every recursively enumerable predicate is "diophantine" in the sense defined immediately below.) In addition, we improve our results in [3], by eliminating all but one of the "critical" predicates, there given.

1. General Remarks. Roman letters will ordinarily stand for integers, and moreover, unless the context makes the contrary explicit, for *positive* integers. Greek letters will stand for positive real numbers. An upper case Roman letter with a superscript n, will abbreviate a corresponding sequence of lower case letters with subscripts; e.g. $X^{(n)}$ abbreviates $x_1, x_2, \ldots, x_n$. $[\alpha]$ is the greatest integer $\leqq \alpha$.

A *monomial* is an expression of the form

$$bx_1^{m_1} x_2^{m_2} \cdots x_n^{m_n}$$

where b is an integer positive, negative, or zero and $m_1, m_2, \ldots, m_n$ are fixed integers. An *exponential-monomial* is an expression of the same form in which the m's as well as the x's may be variables or constants. A *restricted exponential-monomial* is one in which for each factor $x_i^{m_i}$ either x_i or m_i is a constant. By a *polynomial* we mean a finite sum of monomials. Similarly an *exponential-polynomial* is a finite sum of exponential-monomials, and a *restricted exponential-polynomial* is a finite sum of restricted exponential-monomials.

A predicate $R(X^{(n)})$ of positive integers is *diophantine* if it can be written in the form

$$(\exists Y^{(m)}) \left\{P\left(X^{(n)}, Y^{(m)}\right) = 0\right\}$$

where P is a polynomial. It is *existentially definable*[13] if it can be written in the same form with P an exponential polynomial. It is *diophantine (existentially definable) in terms of predicates* $R_1\left(Y^{(m)}\right), \ldots, R_k\left(Y^{(m)}\right)$ if it can be written in the form

$$(\exists Y^{(m)} \left\{P\left(X^{(n)}, Y^{(m)}\right) = 0 \,\&\, R_1\left(Y^{(m)}\right) \;\&\; \cdots \;\&\, R_k\left(Y^{(m)}\right)\right\}$$

where P is a polynomial (exponential-polynomial). Thus, an existentially definable predicate is one which is diophantine in terms of predicates of the form $z = x^y$.

Julia Robinson [8] has proved that the predicate $y = x!$ is existentially definable. Moreover, in the same paper she has considered the question of predicates of *exponential rate of growth*. Such a predicate $R(u, v)$ is defined as one which satisfies the conditions:

(i) $R(u, v) \rightarrow v < u^u$

(ii) For each n, there are u, v such that:

$$R(u, v) \,\&\, v \geqq u^n\ .$$

[13]Julia Robinson [8] has employed the term *existentially definable* to mean what we are calling diophantine here (and in [1–3, 7]). However, in [8], the principal concern is with what we are here calling existentially definable predicates.

Robinson [8] has proved that an existentially definable predicate is *diophantine* in terms of any predicate R which is of exponential rate of growth. From this it follows readily, that in the definition of existential definability, P *can be taken to be a restricted-exponential polynomial.* (For e.g., the predicate $v = 2^u$ & $u > 2$ is of exponential rate of growth.) This suggests considering the following hypothesis, which we shall call J.R.:

There exists a diophantine predicate of exponential rate of growth.[14]

Then, J.R. implies that a predicate is *existentially definable if and only if it is diophantine.*

As is well-known,[15] the recursive unsolvability of Hilbert's tenth problem would follow at once if it could be shown that every recursively enumerable predicate is diophantine. In [3], we produced certain critical predicates with the property that if they are diophantine, so is every recursively enumerable predicate. Our first result is the following improvement of Theorem 3 of [3].[16]

Theorem 1. *If the predicate*

$$z = \sum_{k=1}^{y} \left[\frac{q}{1+ks}\right]$$

is diophantine, then so is every recursively enumerable predicate.

We also are able to prove:

Theorem 2. *If the predicate*[17]

$$(k)_y \ \{Rem(q, 1+ks) \leqq y\}$$

is existentially definable, then so is every recursively enumerable predicate.

[14]Robinson [8] has proved that an equivalent (though apparently weaker) hypothesis is obtained if u^u is replaced by u^{u^u} or $u^{u^{u^u}}$ etc.

[15]Cf. [1, 2].

[16]Equation (5) of [3] should be corrected as follows: The sequence of subscripts b, b, a should be, b, b, e. This (5) is essentially the critical predicate of our present Theorem 1.

[17]$(k)_y$ means: for all k between 1 and y. $Rem(q, 1+ks)$ is the least non-negative remainder on dividing q by $1+ks$. The predicate of this theorem is (7) of [3].

Theorem 2 has the immediate:

Corollary. If the predicate of Theorem 2 is diophantine and J.R. is true, then every recursively enumerable predicate is diophantine.

We have been able to completely eliminate the critical predicates only by assuming the following additional hypothesis which we shall call P.A.P.:

For each integer n, there are n primes in arithmetic progression.

Then, we have:

Theorem 3. If P.A.P. is true, then every recursively enumerable predicate is existentially definable.

From Theorem 3, remarks that have been made above, and the basic theorems of recursive function theory (cf. [2]), we infer:

Corollary. If P.A.P. is true, then there is a restricted exponential polynomial $P\left(x, Y^{(n)}\right)$ *such that, the predicate* $(\exists Y^{(n)})\ \left\{P\left(x, Y^{(n)}\right) = 0\right\}$ *is not recursive.*

Corollary. If P.A.P. is true, then the problem of determining for a given restricted exponential polynomial equation whether it has a solution in positive integers is recursively unsolvable.

Corollary. If P.A.P. is true, then there is a restricted exponential polynomial, $P(m, x, Y^{(n)})$ *such that for each recursively enumerable set S, there is m for which:*

$$S = \left\{x \mid (\exists Y^{(n)})\left(P\left(m, x, Y^{(n)}\right) = 0\right)\right\}$$

Theorem 3 also has the interesting consequence that if one could construct a recursively enumerable predicate which is not existentially definable, then it would follow that P.A.P. was false.

If we assume, in addition to P.A.P., that J.R. is true, Theorem 3 is strengthened as follows:

Theorem 4. If P.A.P. and J.R. are both true, then every recursively enumerable predicate is diophantine.

Similarly, the Corollaries to Theorem 3 can all be strengthened in the obvious way if J.R. is assumed true. In particular we have:

Corollary. If P.A.P. and J.R. are both true, then the problem of determining whether a given diophantine equation has a solution in positive integers is recursively unsolvable, and hence[18] *so is Hilbert's tenth problem.*

---

[18]Hilbert stated the problem for solutions in integers positive, negative, or zero. However, given an algorithm for Hilbert's tenth problem, one could determine whether or not $P(x_1, \ldots, x_n) = 0$ has a solution in positive integers by inquiring as to whether or not

$$P(p_1^2 + q_1^2 + r_1^2 + s_1^2 + 1, \ldots, p_n^2 + q_n^2 + r_n^2 + s_n^2 + 1) = 0$$

has a solution in integers, positive, negative, or zero.

Finally, using the results of [7], we have:

Corollary. If P.A.P. and J.R. are both true, then there is a polynomial $P\left(n, Y^{(m)}\right)$ *such that for each n the range of P includes* 0 *and all negative integers and such that for each non-empty recursively enumerable set S of positive integers, there is an* n_0, *for which S consists of all positive integers taken on by* $P\left(n_0, Y^{(m)}\right)$.

2. *Existential definability of some auxiliary predicates*. The methods used in Julia Robinson [8] in existentially defining $y = x!$ are extended below in order to existentially define certain predicates which are employed in deriving our present results.

For $a > 1,\ 0 < k \leqq n,\ \alpha = \dfrac{p}{q} > 0$, we have:

$$\begin{aligned} a^{\alpha k}(1+a^{-\alpha})^{\alpha} &= a^{\alpha k}\sum_{j=0}^{\infty}\binom{\alpha}{j}a^{-\alpha j} \\ &\leqq \sum_{j=0}^{k}\binom{\alpha}{j}a^{\alpha(k-j)} + \sum_{j=k+1}^{\infty}\binom{\alpha}{j}a^{-\alpha} \\ &\leqq \sum_{j=0}^{k}\binom{\alpha}{j}a^{\alpha(k-j)} + \left(\frac{2}{a}\right)^{\alpha}. \end{aligned}$$

I.e.

$$0 \leqq a^{\alpha k}(1+a^{-\alpha})^{\alpha} - \sum_{j=0}^{k}\binom{\alpha}{j}a^{\alpha(k-j)} \leqq \left(\frac{2}{a}\right)^{\alpha} \tag{B.1}$$

Letting, e.g.,

$$a = (3q^k k!)^q,\ \text{so that } a \text{ is a perfect } q\text{-th power and } \left(\frac{2}{a}\right)^{\alpha} < \frac{1}{q^k k!}, \tag{B.2}$$

we have:

$$\sum_{j=0}^{k}\binom{\alpha}{j}a^{\alpha(k-j)} = \frac{\left[a^{\alpha k}(1+a^{-\alpha})^{\alpha}q^k k!\right]}{q^k k!} \tag{B.3}$$

Replacing k by $k-1$ in (1) and using the same value of a,

$$0 \leqq a^{\alpha(k-1)}(1+a^{-\alpha})^{\alpha} - a^{-\alpha}\sum_{j=0}^{k-1}\binom{\alpha}{j}a^{\alpha(k-j)} < \frac{1}{q^k k!}. \tag{B.4}$$

Hence,

$$\sum_{j=0}^{k-1}\binom{\alpha}{j}a^{\alpha(k-j)} = \frac{a^{\alpha}\left[a^{\alpha(k-1)}(1+a^{-\alpha})^{\alpha}q^k k!\right]}{q^k k!} \tag{B.5}$$

From (3) and (5),

$$\binom{\alpha}{k} = \frac{\left[a^{\alpha k}(1+a^{-\alpha})^{\alpha}q^k k!\right] - a^{\alpha}\left[a^{\alpha(k-1)}(1+a^{-\alpha})^{\alpha}q^k k!\right]}{q^k k!} \tag{B.6}$$

Equation (6) enables us to prove:

Lemma 2.1. *The predicate*:

$$r/s = \binom{p/q}{k}$$

is existentially definable.

Proof. We first note that, for

$$\begin{aligned} & a = (3q^k k!)^q, \text{ we have} \\ & z = \left[a^{\frac{p}{q}k}(1+a^{-\frac{p}{q}})^{\frac{p}{q}}q^k k!\right] \\ \leftrightarrow \quad & z \leqq a^{\frac{p}{q}k}(1+a^{-\frac{p}{q}})^{\frac{p}{q}}q^k k! \ < z+1 \\ \leftrightarrow \quad & z^q(3q^k k!)^{p^2} \leqq (3q^k k!)^{pqk}((3q^k k!)^p+1)^p(q^k k!)^q \ < (z+1)^q(3q^k k!)^{p^2} \end{aligned}$$

Similarly,

$$\begin{aligned} & w = \left[a^{\frac{p}{q}(k-1)}(1+a^{-\frac{p}{q}})^{\frac{p}{q}}q^k k!\right] \\ \leftrightarrow \quad & w^q(3q^k k!)^{p^2} \leqq (3q^k k!)^{pq(k-1)}((3q^k k!)^p+1)^p(q^k k!)^q \ < (w+1)^q(3q^k k!)^{p^2} . \end{aligned}$$

But, by (6),

$$\begin{aligned} r/s = \binom{p/q}{k} \leftrightarrow (\exists a, z, w)\Big\{ & rq^k k! = sz - (3q^k k!)^p sw \\ & \& \ a = (3q^k k!)^a \ \& \ z = \left[a^{\frac{p}{q}k}(1+a^{-\frac{p}{q}})^{\frac{p}{q}}q^k k!\right] \\ & \& \ w = \left[a^{\frac{p}{q}(k-1)}(1+a^{-\frac{p}{q}})^{\frac{p}{q}}q^k k!\right]\Big\} . \end{aligned}$$

Lemma 2.2. *The predicate* $z = \prod_{k-1}^{y}(r+sk)$ *is existentially definable.*

Proof:

$$\prod_{k-1}^{y}(r+sk) = s^y \prod_{k-1}^{y}\left(\frac{r}{s}+k\right)$$
$$= s^y y!\binom{(r/s)+y}{y}.$$

Lemma 2.3. *The predicate*

$$\frac{p}{q} = \sum_{k=1}^{y} \frac{1}{r+ks} \;\&\; s \nmid r$$

is existentially definable.

The proof of Lemma 2.3 will be based on the following lemmas:

Lemma 2.4. $\sum_{k=1}^{y} \frac{1}{\alpha+k} = \frac{\Gamma'(\alpha+y+1)}{\Gamma(\alpha+y+1)} - \frac{\Gamma'(\alpha+1)}{\Gamma(\alpha+1)}.$

Proof: Differentiation of the equation $\log\,\Gamma(\alpha+k+1) - \log\,\Gamma(\alpha+k) = \log(\alpha+k)$ yields:

$$\frac{\Gamma'(\alpha+k+1)}{\Gamma(\alpha+k+1)} - \frac{\Gamma'(\alpha+k)}{\Gamma(\alpha+k)} = \frac{1}{\alpha+k}.$$

Now, sum from $k = 1$ to y.

Lemma 2.5. *For* $\alpha > 1,\ \Gamma''(a) < \left(\frac{1}{\alpha-1}+\alpha\right)\Gamma(\alpha)$.

Proof.

$$\begin{aligned}\Gamma''(a) &= \int_0^\infty (\log\,t)^2\, t^{\alpha-1}e^{-t}dt \\ &= \int_0^1 (\log\,t)^2\, t^{\alpha-1}e^{-t}dt + \int_1^\infty (\log\,t)^2\, t^{\alpha-1}e^{-t}dt \\ &< \int_0^1 t^{\alpha-2}e^{-t}dt + \int_1^\infty t^{\alpha}e^{-t}dt \\ &< \Gamma(\alpha-1) + \Gamma(\alpha+1) \\ &= \left(\frac{1}{\alpha-1}+\alpha\right)\Gamma(\alpha),\end{aligned}$$

where we have used the elementary inequality: $\log\,t < \sqrt{t}$.[19]

Lemma 2.6. *For* $\alpha > 1$, *and* $m > \frac{2}{\alpha-1}$,

[19]The function $\sqrt{t} - \log t$ clearly assumes its minimum value at $t = 4$, and is positive there.

$$0 < \frac{\Gamma'(\alpha)}{\Gamma(\alpha)} - m\left\{1 - \frac{\Gamma(\alpha - \frac{1}{m})}{\Gamma(\alpha)}\right\} < \frac{1}{2m}\left(\frac{2\alpha}{\alpha - 1} + \alpha^2\right).$$

Proof. By Lagrange's form of Taylor's theorem,

$$\Gamma(\alpha - \frac{1}{m}) = \Gamma(\alpha) - \frac{1}{m}\Gamma'(\alpha) + \frac{1}{2m^2}\Gamma''(\alpha - \frac{\theta}{m}),$$

where $0 < \theta < 1$. Hence,

$$\Gamma'(\alpha) = m\left\{\Gamma(\alpha) - \Gamma(\alpha - \frac{1}{m})\right\} + \frac{1}{2m}\Gamma''(\alpha - \frac{\theta}{m}).$$

Thus,

$$0 < \frac{\Gamma'(\alpha)}{\Gamma(\alpha)} - m\left\{1 - \frac{\Gamma(\alpha - \frac{1}{m})}{\Gamma(\alpha)}\right\} = \frac{1}{2m}\frac{\Gamma''(\alpha - \frac{\theta}{m})}{\Gamma(\alpha)}.$$

But, for $m > \dfrac{2}{\alpha - 1}$, we have $\alpha - \dfrac{\theta}{m} - 1 > \dfrac{1}{2}(\alpha - 1) > 0$. Hence, using Lemma 2.5, and the fact that Γ is an increasing function for arguments $\geqq 2$, we have for such m,

$$\begin{aligned}\frac{1}{2m}\frac{\Gamma''(\alpha - \frac{\theta}{m})}{\Gamma(\alpha)} &< \left(\frac{1}{\alpha - \frac{\theta}{m} - 1} + \alpha - \frac{\theta}{m}\right)\frac{\Gamma(\alpha - \frac{\theta}{m})}{2m\,\Gamma(\alpha)}\\ &< \frac{1}{2m}\left(\frac{2}{\alpha - 1} + \alpha\right)\frac{\Gamma(\alpha + 1 - \frac{\theta}{m})}{(\alpha - \frac{\theta}{m})\,\Gamma(\alpha)}\\ &< \frac{1}{2m}\left(\frac{2}{\alpha - 1} + \alpha\right)\frac{\Gamma(\alpha + 1)}{\Gamma(\alpha)}\\ &= \frac{1}{2m}\left(\frac{2\alpha}{\alpha - 1} + \alpha^2\right).\end{aligned}$$

Lemma 2.7 If α is not an integer, $m > \dfrac{2}{\alpha - [\alpha]}$, and if

$$\Delta = m\left\{\prod_{k=1}^{y}\frac{\alpha + k - \frac{1}{m}}{\alpha + k} - 1\right\}\prod_{j=1}^{[\alpha]+1}\frac{\alpha - [\alpha] - \frac{1}{m} + j - 1}{\alpha - [\alpha] + j - 1}, \tag{B.7}$$

then

$$\left|\sum_{k=1}^{y}\frac{1}{\alpha + k} - \Delta\frac{\Gamma(\alpha - [\alpha] - \frac{1}{m})}{\Gamma(\alpha - [\alpha])}\right| < \frac{1}{m}\left\{2 + \frac{2}{\alpha} + (\alpha + y + 1)^2\right\}.$$

Proof. By Lemmas 2.4 and 2.6,

$$\left|\sum_{k=1}^{y}\frac{1}{\alpha+k}-m\left\{\frac{\Gamma(\alpha+y+1-\frac{1}{m})}{\Gamma(\alpha+y+1)}-\frac{\Gamma(\alpha+1-\frac{1}{m})}{\Gamma(\alpha+1)}\right\}\right|$$
$$<\frac{1}{2m}\left\{\left(\frac{2(\alpha+y+1)}{\alpha+y}+(\alpha+y+1)^2\right)+\frac{2(\alpha+1)}{\alpha}+(\alpha+1)^2\right\}$$
$$<\frac{1}{m}\left\{2+\frac{2}{\alpha}+(\alpha+y+1)^2\right\}.$$

But, since $(\alpha-\frac{1}{m})-\left[\alpha-\frac{1}{m}\right]=\alpha-[\alpha]-\frac{1}{m}$,

$$m\left\{\frac{\Gamma(\alpha+y+1-\frac{1}{m})}{\Gamma(\alpha+y+1)}-\frac{\Gamma(\alpha+1-\frac{1}{m})}{\Gamma(\alpha+1)}\right\}$$
$$=m\left\{\prod_{k=1}^{y}\frac{\alpha+y+1-\frac{1}{m}-k}{\alpha+y+1-k}-1\right\}\frac{\Gamma(\alpha+1-\frac{1}{m})}{\Gamma(\alpha+1)}$$
$$=\Delta\frac{\Gamma(\alpha-[\alpha]-\frac{1}{m})}{\Gamma(\alpha-[\alpha])}.$$

Lemma 2.8. *For* $0<\alpha<1$,

$$\left|\frac{1}{\Gamma(\alpha)}-n^{1-\alpha}\binom{\alpha+n-1}{n}\right|<\frac{1}{\sqrt{n}}.$$

Proof. We begin with the inequality, valid for $0<\alpha<1$:

$$\left|\frac{\Gamma(\alpha+n)}{\Gamma(n)\,n^{\alpha}}-1\right|\leqq\frac{n^n}{e^n n!}<\frac{1}{\sqrt{2\pi n}} \tag{B.8}$$

(For the first part of this inequality, cf. [5], p. 350; the second follows from Stirling's formula.) Thus

$$\left|n^{1-\alpha}\binom{\alpha+n-1}{n}-\frac{1}{\Gamma(\alpha)}\right|<\frac{1}{\sqrt{2\pi n}\,\Gamma(\alpha)}<\frac{1}{\sqrt{n}},$$

since $\Gamma(\alpha)>\frac{1}{2}$.

Lemma 2.9. *If* $|L-1|<\varepsilon<\frac{1}{2}$, *then* $\left|\frac{1}{L}-1\right|<2\varepsilon$.

Proof. $\left|\frac{1}{L}-1\right|=\frac{|L-1|}{|L|}<2\varepsilon$.

Lemma 2.10. *For* $0<\alpha<1$,

$$\left|\Gamma(\alpha) - \frac{n^{\alpha-1}}{\binom{\alpha+n-1}{n}}\right| < \frac{1}{\alpha\sqrt{n}}\,.$$

Proof. By (8) and Lemma 2.9,

$$\left|\frac{\Gamma(n)\,n^{\alpha}}{\Gamma(\alpha+n)} - 1\right| < \frac{1}{\sqrt{n}}\,.$$

Hence,

$$\begin{aligned}\left|\frac{n^{\alpha-1}}{\binom{\alpha+n-1}{n}} - \Gamma(\alpha)\right| &< \frac{\Gamma(\alpha)}{\sqrt{n}}\\ &= \frac{\Gamma(\alpha+1)}{\alpha\sqrt{n}}\\ &< \frac{1}{\alpha\sqrt{n}}\,.\end{aligned}$$

Lemma 2.11. *If α is not an integer, $m > \frac{2}{\alpha-[\alpha]}$, if Δ is defined by* (7), *and if*

$$\Theta = n^{1-\alpha+[\alpha]}\binom{\alpha-[\alpha]+n-1}{n}\,, \tag{B.9}$$

$$\Omega = \frac{n^{\alpha-[\alpha]-\frac{1}{m}-1}}{\binom{\alpha-[\alpha]-\frac{1}{m}+n-1}{n}}\,, \tag{B.10}$$

then

$$\left|\sum_{k=1}^{y}\frac{1}{\alpha+k} - \Delta\Theta\Omega\right| < \frac{1}{m}\left\{2+\frac{2}{\alpha}+(\alpha+y+1)^2\right\} + \frac{3\Delta}{(\alpha-[\alpha]-\frac{1}{m})\sqrt{n}}\,.$$

Proof. By Lemma 2.7, $\left|\sum_{k=1}^{y}\frac{1}{\alpha+k} - \Delta\Theta\Omega\right|$

$$< \frac{1}{m}\left\{2+\frac{2}{\alpha}+(\alpha+y+1)^2\right\} + \Delta\left|\frac{\Gamma(\alpha-[\alpha]-\frac{1}{m})}{\Gamma(\alpha-[\alpha])} - \Theta\Omega\right|\,.$$

But, using Lemmas 2.8 and 2.10,

$$\left|\frac{\Gamma(\alpha-[\alpha]-\frac{1}{m})}{\Gamma(\alpha-[\alpha])}-\Theta\Omega\right| \leqq \left|\Gamma(\alpha-[\alpha]-\frac{1}{m})-\Omega\right| + \Omega\left|\frac{1}{\Gamma(\alpha-[\alpha])}-\Theta\right|$$
$$< \frac{1}{(\alpha-[\alpha]-\frac{1}{m})\sqrt{n}} + \frac{\Omega}{\sqrt{n}}$$
$$< \frac{1}{\sqrt{n}}\left\{\frac{1}{(\alpha-[\alpha]-\frac{1}{m})} + \Gamma(\alpha-[\alpha]-\frac{1}{m}) + \frac{1}{\sqrt{n}(\alpha-[\alpha]-\frac{1}{m})}\right\}$$
$$= \frac{1}{(\alpha-[\alpha]-\frac{1}{m})\sqrt{n}}\left(1+\Gamma(\alpha-[\alpha]-\frac{1}{m}+1)+\frac{1}{\sqrt{n}}\right)$$
$$\leqq \frac{3}{(\alpha-[\alpha]-\frac{1}{m})\sqrt{n}}\,.$$

Lemma 2.12. $\sum_{k=1}^{y}\frac{1}{r+ks} = \frac{1}{s}\frac{\left[\Delta\Theta\Omega P+\frac{1}{2}\right]}{P}$, *where* Δ, Θ, Ω *are defined in* (7), (9), (10), $P = \prod_{k=1}^{y}(r+ks)$, *and* $\alpha = \frac{r}{s}$ *is not an integer, if*

$$m > \frac{2}{\alpha-[\alpha]}\,,$$
$$m > 4P\left(2+\frac{2}{\alpha}+(\alpha+y+1)^2\right),\ \textit{and}$$
$$n > 144\,P^2\Delta^2 m^2\,.$$

Proof. By Lemma 2.11, the stated conditions on m, n, imply that,

$$\left|\sum_{k=1}^{y}\frac{1}{\frac{r}{s}+k}-\Delta\Theta\Omega\right| < \frac{1}{2P}\,.$$

Hence,

$$P\sum_{k=1}^{y}\frac{1}{\frac{r}{s}+k} < P\Delta\Theta\Omega+\frac{1}{2} < P\sum_{k=1}^{y}\frac{1}{\frac{r}{s}+k}+1\,,$$

i.e.

$$P\sum_{k=1}^{y}\frac{1}{\frac{r}{s}+k} = \left[P\Delta\Theta\Omega+\frac{1}{2}\right]\,.$$

Proof of Lemma 2.3. By Lemma 2.12, we have:

$$\frac{p}{q} = \sum_{k=1}^{y}\frac{1}{r+ks}\ \&\ s \nmid r$$

$$\leftrightarrow \quad (\exists m, n, a, b, c, d, e, f, P, Q, R, S, T)\left\{a = \left[\frac{r}{s}\right] \;\&\; r \neq as\right.$$

$$\& \; P = \prod_{k=1}^{y}(r + ks) \;\&\; \frac{b}{c} = \binom{\frac{r}{s} - a + n - 1}{n}$$

$$\& \; \frac{d}{e} = \binom{\frac{r}{s} - a - \frac{1}{m} + n - 1}{n} \;\&\; Q = \prod_{k=1}^{y} (mr + mks - s)$$

$$\& \; R = \prod_{k=1}^{y}(mr + mks) \;\&\; S = \prod_{j=1}^{a+1}(mr - ams - s + msj - ms)$$

$$\& \; T = \prod_{j=1}^{a+1}(mr - ams - s + msj) \;\&\; m \;>\; \frac{2}{\frac{r}{s} - a}$$

$$\& \; m \;>\; 4P\left(2 + \frac{2s}{r} + \left(\frac{r}{s} + y + 1\right)^2\right) \;\&\; n > \; 144\, P^2 \left(\frac{Q}{R} - 1\right)^2 \frac{S^2}{T^2} m^4$$

$$\left.\& \; psP = qf \;\&\; f = \left[m\left(\frac{Q}{R} - 1\right)\frac{SbeP}{Tcd}\, n^{-\frac{1}{m}} + \frac{1}{2}\right]\right\} .$$

In this expression, the clauses involving binomial coefficients and $\prod$ are existentially definable by Lemmas 2.1 and 2.2. The clauses involving inequalities between rational numbers are equivalent, in an obvious way, to inequalities involving integers, which can be handled in the usual manner. Finally,

$$a = \left[\frac{r}{s}\right] \;\leftrightarrow\; as \leqq r \;<\; (a+1)\, s, \;\text{ and}$$

$$f = \left[m\left\{\frac{Q}{R} - 1\right\} \frac{SbeP}{Tcd} n^{-\frac{1}{m}} + \frac{1}{2}\right]$$

$$\leftrightarrow \quad (2f - 1)^m\, n \;\leqq\; \left\{m\left(\frac{Q}{R} - 1\right)\frac{SbeP}{Tcd}\right\}^m \;<\; (2f + 1)^m\, n \,.$$

3. Reduction to two critical predicates. We begin with two easy number-theoretic lemmas.

Lemma 3.1. *Let* $r_k < m_k$, $1 \leqq k \leqq y$, *where the* m_k*'s are relatively prime in pairs. Then,* $\sum_{k=1}^{y} \frac{r_k}{m_k}$ *is an integer if and only if* $r_k = 0$ *for* $k = 1, 2, \ldots, y$.

Proof. Let $\sum_{k=1}^{y} \frac{r_k}{m_k} = N$. Then,

$$r_1 m_2 \cdots m_y + m_1 r_2 \cdots m_y + \cdots + m_1 m_2 \cdots r_y = N m_1 m_2 \cdots m_y \,.$$

Hence, e.g., $m_1 \mid r_1 m_2 \cdots m_y$, i.e. $m_1 \mid r_1$ which implies $r_1 = 0$.

Similarly, $r_2 = r_3 = \cdots = r_y = 0$.

Lemma 3.2. $\sum_{k=1}^{y} Rem(A, r+ks) \equiv Ay - r\sum_{k=1}^{y}\left[\frac{A}{r+ks}\right] \pmod{s}$.

Proof. We have

$$A = \left[\frac{A}{r+ks}\right](r+ks) + Rem(A, r+ks).$$

Hence,

$$Ay = r\sum_{k=1}^{y}\left[\frac{A}{r+ks}\right] + s\sum_{k=1}^{y} k\left[\frac{A}{r+ks}\right] + \sum_{k=1}^{y} Rem(A, r+ks),$$

which yields the desired result.

As in [3], we recall that every recursively enumerable predicate $H(X^{(n)})$ may be represented in the form

$$(\exists y)(k)_y(\exists Z^{(m)})\ \left\{P(y, k, X^{(n)}, Z^{(m)}) = 0\right\},$$

where

$$P(y, k, X^{(n)}, Z^{(m)}) = 0 \rightarrow (i)_m(z_i \leqq y).\ ^{8} \tag{B.11}$$

Let[20]

$$P^2(y, k, X^{(n)}, Z^{(m)}) = \sum_{v_0, v_1, \ldots, v_m = 0}^{N} Q_{v_0,\ldots,v_m}(y, X^{(n)})\, k^{v_0} z_1^{v_1} \cdots z_m^{v_m}.$$

We let $\sum$ stand for $\sum_{v_0, v_1, \ldots, v_m = 0}^{N}$ and $Q(y, X_n)$ for $Q_{v_0,\ldots,v_m}(y, X_n)$.

Let $RP(r, s, y)$ be a predicate with the properties:

(a) $RP(r, s, y) \rightarrow \begin{cases} r+ks \textit{ is relatively prime} \\ \textit{to } r+js \textit{ for } 1 \leqq j < k \leqq y. \end{cases}$

(b) *For every* y, *there are arbitrarily large values of* $r+s$ *for which* r, s satisfy $RP(r, s, y)$.

(c) $RP(r, s, y) \rightarrow\ s \nmid r$.

In particular, we may take for $RP(r, s, y)$ the existentially definable predicate:

$$r = 1\ \&\ y! \mid s\ \&\ s \neq 1.\ ^{9} \tag{B.12}$$

[20]That H can be represented as asserted may be seen e.g. fro the proof of Theorem 3.8, pp. 113–114, [2].

Then, using the Chinese remainder theorem as in [3], and setting $K = 2N(m+1)$, we have [21]

$$H(X^{(n)}) \leftrightarrow (\exists y)(\exists Z_1^{(m)})(\exists Z_2^{(m)})\cdots(\exists Z_y^{(m)})\Big\{\sum\sum_{k=1}^{y} Q(y, X^{(n)})k^{v_0}z_1^{v_1}\cdots z_m^{v_m} = 0\Big\}$$

$$\leftrightarrow (\exists y)(\exists r)(\exists s)(\exists a_0)\cdots(\exists a_m)\Big\{\sum Q(y, X^{(n)})\sum_{k=1}^{y}\prod_{i=0}^{m} Rem^{v_i}(a_i, r+sk) = 0$$

$$\& RP(r, s, y) \& r+s > y^K \& (k)_y\big(Rem(a_0, r+sk) = k\big)\Big\}$$

$$\leftrightarrow (\exists y)(\exists r)(\exists s)(\exists a_0)\cdots(\exists a_m)\Big\{$$

$$\sum Q(y, X^{(n)})\sum_{k=1}^{y} Rem(a_0^{v_0}a_1^{v_1}\cdots a_m^{v_m}, r+sk) = 0$$

$$\& RP(r, s, y) \& r+s > y^K \& (k)_y(Rem(a_0, r+sk) = k)$$

$$\& (i)_m(Rem(a_i, r+sk) \leqq y)\Big\}.$$

To see that the last equivalence is correct, we note that (using (11)) the matrix of each side of the equivalence implies $r+s > y^K$ and $(i)_m(Rem(a_i, r+sk) \leqq y)$. But, then $\prod_{i=0}^{m} Rem^{v_i}(a_i, r+sk) \leqq y^{N(m+1)} < y^K < r+s \leqq r+sk$, so that, since

$$\prod_{i=0}^{m} Rem^{v_i}(a_i, r+sk) \equiv Rem(a_0^{v_0}a_1^{v_1}\cdots a_m^{v_m}, r+sk) \pmod{r+sk},$$

we have

$$\prod_{i=0}^{m} Rem^{v_i}(a_i, r+sk) = Rem(a_0^{v_0}a_1^{v_1}\cdots a_m^{v_m}, r+sk).$$

We now introduce the variables $t_{v_0,v_1,\ldots,v_m}$ where each v_i has the range $0 \leqq v_i \leqq N$, with the understanding that, e.g. $(\exists t)$ abbreviates $(\exists t_{0,0,\ldots,0})(\exists t_{1,0,\ldots,0})\cdots(\exists t_{N,N,\ldots,N})$, and that $t_{v_0,\ldots,v_m} = Q(v_0,\ldots,v_m)$ abbreviates $t_{0,0,\ldots,0} = Q(0,0,\ldots,0) \& \cdots \& t_{N,N,\ldots,N} = Q(N,N,\ldots,N)$.

Then,

[21] Cf. [4], or [2], p. 45, Lemma.

$$
\begin{aligned}
H(X^{(n)}) \;&\leftrightarrow\; (\exists y)(\exists r)(\exists s)(\exists a_0)\cdots(\exists a_m)(\exists t) \\
&\Big\{ \sum Q(y, X^{(n)})\, t = 0 \;\&\; t_{v_0,\ldots,v_m} = \sum_{k=1}^{y} Rem(a_0^{v_0}\cdots a_m^{v_m}, r + ks) \\
&\& \; RP(r, s, y) \;\&\; r + s > y^K \;\&\; (k)_y\big(Rem(a_0, r + sk) = k\big) \\
&\& \; (i)_m\big(Rem(a_i, r + sk) \leqq y\big)\Big\}.
\end{aligned}
\tag{B.13}
$$

But, the conditions on $Rem(a_i, r+sk)$ and s in (13) imply $t_{v_0,\ldots,v_m} < s$. Hence, using Lemma 3.2,

$$
\begin{aligned}
H(X^{(n)}) \;&\leftrightarrow\; (\exists y)(\exists r)(\exists s)(\exists a_0)\cdots(\exists a_m)(\exists t) \\
&\Big\{ \sum Q(y, X^{(n)})\, t = 0 \;\&\; t_{v_0,\ldots,v_m} \equiv \sum_{k=1}^{y} \left[\frac{a_0^{v_0}\cdots a_m^{v_m}}{r + sk}\right] \quad (mod\; r + sk) \\
&\& \; t_{v_0,\ldots,v_m} < s \;\&\; RP(r, s, y) \;\&\; r + s > y^K \;\&\; (k)_y\big(Rem(a_0, r + sk) = k\big) \\
&\& \; (i)_m\big(Rem(a_i, r + sk) \leqq y\big)\Big\}.
\end{aligned}
\tag{B.14}
$$

Now, using Lemma 3.1, we note that $RP(r, s, y)$ implies:

$$
\begin{aligned}
&(k)_y\big(Rem(a_0, r + sk) = k\big) \\
&\leftrightarrow (k)_y\big(Rem(a_0 - k, r + sk) = 0\big) \\
&\leftrightarrow \sum_{k=1}^{y} \frac{Rem(a_0 - k, r + sk)}{r + sk} \text{ is an integer} \\
&\leftrightarrow \sum_{k=1}^{y} \frac{a_0 - k}{r + sk} \text{ is an integer} \\
&\leftrightarrow \left(a_0 + \frac{r}{s}\right) \sum_{k=1}^{y} \frac{1}{r + sk} - \frac{y}{s} \text{ is an integer} \\
&\leftrightarrow (\exists p)(\exists q) \left\{ \frac{p}{q} = \sum_{k=1}^{y} \frac{1}{r + sk} \;\&\; qs \mid (pa_0 s + pr - yq) \right\}.
\end{aligned}
$$

where, by Lemma 2.3, this last predicate is existentially definable. Thus (14) enables us to obtain an existential definition of $H(X^{(n)})$ in terms of the two critical predicates:

$$
z = \sum_{k=1}^{y} \left[\frac{A}{r + ks}\right] \tag{B.15}
$$

$$
(k)_y\big(Rem(A, r + ks) \leqq y\big). \tag{B.16}
$$

In fact, if we take (12) for $RP(r, s, y)$, we may set $r = 1$ in (15) and (16) obtaining the two critical predicates:

$$z = \sum_{k=1}^{y} \left[\frac{A}{1+ks}\right] \tag{B.17}$$

$$(k)_y(Rem(A, 1+ks) \leqq y), \tag{B.18}$$

in terms of which all recursively enumerable predicates can be existentially defined.

4. Proof of Theorem 1. According to [6], page 117, problem 4, if $y+1 < 1+ks$, then

$$\left[\frac{A}{1+ks}\right] - \left[\frac{A-y-1}{1+ks}\right] = 0 \text{ or } 1,$$

according as

$$\frac{Rem(A, 1+ks)}{1+ks} \geqq \frac{y+1}{1+ks} \text{ or } \frac{y+1}{1+ks}.$$

Hence,

$$(k)_y\big(Rem(A, 1+ks) \leqq y\big) \leftrightarrow (k)_y\big(Rem(A, 1+ks) < y+1\big)$$
$$\leftrightarrow y = \sum_{k=1}^{y}\left[\frac{A}{1+ks}\right] - \sum_{k=1}^{y}\left[\frac{A-y-1}{1+ks}\right],$$

so that all instances of (18), can be replaced in (14) by predicates which are existentially definable in terms of (17).

Theorem 1 now follows at once from the following:

Lemma 4.1. *If* $z = \sum_{k=1}^{y}\left[\frac{A}{1+ks}\right]$ *is diophantine, then J.R. is true.*

Proof. Let

$$P(z, y) \leftrightarrow yz = \sum_{k=1}^{y}\left[\frac{y}{1+k}\right] \;\&\; z > 6.$$

Then, the hypothesis clearly implies that $P(z, y)$ is diophantine.

Now, $P(z, y) \rightarrow \sum_{k=1}^{y} \frac{y}{k+1} - y < zy \leqq \sum_{k=1}^{y} \frac{y}{k+1}$, i.e., estimating the sums by integrals,

$$y \log\left(\frac{y+2}{2}\right) - y < zy < y\log(y+1),$$
$$e^z - 1 < y < 2e^{z+1} - 2.$$

Hence, if for some n, and all y, z,

$$P(z, y) \rightarrow y < z^n,$$

we should have for a sequence of z's approaching ∞, $e^z - 1 < z^n$ which is impossible. Moreover, since $P(z, y) \rightarrow z > 6$ we have:

$$P(z, y) \rightarrow y < 2e^{z+1} - 2 < 2e^{z+1} < z^z.$$

Thus, $P(z, y)$ satisfies J.R.

5. Proof of Theorem 2. Returning to (14), we note that each occurrence in it of a sum of the form $\sum_{k=1}^{y} \left[\frac{A}{r+ks}\right]$ is in a context which implies that:

$$(k)_y \left(Rem(A, r+ks) \leqq y^{\frac{K}{2}}\right).$$

But since the matrix of (14) implies $r + ks > y^K$, we have:

$$\sum_{k=1}^{y} \frac{Rem(A, r+ks)}{r+ks} < \sum_{k=1}^{y} \frac{y^{\frac{K}{2}}}{y^K} = \frac{1}{y^{\frac{K}{2}-1}} < 1.$$

Hence,

$$\sum_{k=1}^{y} \left[\frac{A}{r+sk}\right] = \left[\sum_{k=1}^{y} \frac{A}{r+sk}\right].$$

But, by Lemma 2.3,

$$z = \left[\sum_{k=1}^{y} \frac{A}{r+sk}\right] \leftrightarrow z \leqq \sum_{k=1}^{y} \frac{A}{r+sk} < z+1$$

is existentially definable.

Hence, we have:

Lemma 5.1. *Every recursively enumerable predicate is existentially definable in terms of* (16), *or even, using* (12) *for* $RP(r, s, y)$, *in terms of* (18).

Lemma 5.1 immediately implies Theorem 2.

6. Proof of Theorem 3. In this section, we shall employ the *hypothesis P.A.P.* Then, using the notation (a, b) for the greatest common divisor of a and b, we may take for $RP(r, s, y)$ the predicate:

$$\left(\prod_{k=1}^{y}(r+ks), (r+s-1)!\right) = 1 \ \&\ y < r \ \&\ r > 1. \qquad \text{(B.19)}$$

For, this predicate implies that $\sqrt{r+ks} < \sqrt{r+rs} < \sqrt{r^2+2rs+s^2} = r+s$, so that none of the $r+ks$ can have a divisor less than its square root, and so each $r+ks$ must be a prime. Moreover, since $r+rs$ is composite for $r>1$, P.A.P. implies that for each y there are r, s with $r>y$ such that (19) is satisfied. Finally, that r (and hence[22] $r+s$) may be made arbitrarily large for given y is clear from the fact that, taking (19) for $RP(r, s, y)$:

$$RP(r, s, y_2) \;\&\; y_1 < y_2 \rightarrow RP(r, s, y_1).$$

Now, (19) may be written:

$$(\exists u)(\exists v)\left\{\left(u\prod_{k=1}^{y}(r+ks) - v(r+s-1)!\right)^2 = 1 \;\&\; y < r \;\&\; r > 1\right\},$$

and so, is (cf. Lemma 2.2) existentially definable. Finally, P.A.P. and (19) implies:

$$\begin{aligned}(k)_y\big(Rem(A, r+ks) \leqq y\big) &\leftrightarrow (k)_y(\exists i)_{y+1}(r+ks) \mid (A-i+1)\\ &\leftrightarrow (k)_y(r+ks) \mid \prod_{i=1}^{y+1}(A-i+1)\\ &\leftrightarrow \prod_{k=1}^{y}(r+ks) \mid \prod_{j=1}^{y+1}(A-y-1+j)\,.^{11}\end{aligned}$$

Lemma 2.2 and 5.1 now suffice for the proof of Theorem 3.[23]

References

1. Davis, M. (1953). Arithmetical problems and recursively enumerable predicates. *Journal of Symbolic Logic, 18*, 33–41.
2. Davis, M. (1958). *Computability and unsolvability*. McGraw-Hill.
3. Davis, M., & Putnam, H. (1958). Reductions of Hilbert's tenth problem. *Journal of Symbolic Logic, 23*, 183–187.
4. Gödel, K. (1931). Über formal unentscheidbare Sätze der Principia Mathematica und verwandter Systeme. *I, Monatshefte für Mathematik und Physik, 38*, 173–198.
5. Landau, E. (1951). *Differential and integral calculus*. Chelsea Publishing Company.
6. Polya, G., & Szegö, G. (1945). *Aufgaben und Lehrsätze aus der Analysis*, vol. 2. Dover Publications.
7. Putnam, H. *An unsolvable problem in number theory*. submitted to the Journal of Symbolic Logic.

[22] It is not difficult to see that s may also be taken arbitrarily large. But this fact is not needed here.

[23] This simple expression for $(k)_y(Rem(A, r+ks) \leqq y)$ was suggested by a remark of H.S. Shapiro. Our original proof of Theorem 3 used a considerably more complicated expression.

8. Robinson, J. (1952). Existential definability in arithmetic. *Transactions of the American Mathematical Society*, *72*, 437–449.
9. Robinson, R. M. (1956). Arithmetical representation of the recursively enumerable sets. *Journal of Symbolic Logic*, *21*, 162–186.

Subject Index

E.G. Omodeo and A. Policriti (eds.), *Martin Davis on Computability, Computational Logic, and Mathematical Foundations*,
Outstanding Contributions to Logic 10, DOI 10.1007/978-3-319-41842-1

Author Index

E.G. Omodeo and A. Policriti (eds.), *Martin Davis on Computability, Computational Logic, and Mathematical Foundations*,
Outstanding Contributions to Logic 10, DOI 10.1007/978-3-319-41842-1

FSC
www.fsc.org
MIX
Papier | Fördert
gute Waldnutzung
FSC® C083411

Kolibri 360 GmbH
Ferdinand-Jühlke-Straße 7
99095 Erfurt, Deutschland
produktsicherheit@kolibri360.de